BIODIVERSITY OF PAKISTAN

Edited by

SHAHZAD A. MUFTI
CHARLES A. WOODS
SYED AZHAR HASAN

Pakistan Museum of Natural History
Islamabad

Florida Museum of Natural History
Gainesville

1997

BIODIVERSITY OF PAKISTAN

Published December, 1997

ISBN: 0-9660913-0-2

Copies of BIODIVERSITY OF PAKISTAN may be ordered from:
Pakistan Museum of Natural History, Garden Ave., Shakarparian,
Islamabad, 44000, Pakistan
E-mail Nathist @ pmnh.sdnpk.undp.org
Fax: (051) 9221864
Price: US $ 75/-

FOREWORD

This book is the result of an International Symposium on "Biodiversity of Pakistan" held in Islamabad during November 1995. During this 2-day Symposium a large number of experts who have been working on the systematics, biology, biogeography and conservation of the fauna and flora of Pakistan presented the results of their research in the form of detailed oral presentations. These presentations were followed by extensive discussions and all manuscripts were then revised in the light of the proceeding of the Symposium. A total 41 manuscripts were then refereed and selected for inclusion in this book.

It is evident that this book should serve as an excellent resource for the latest information about the biodiversity of Pakistan. However, this book can not be considered as exhaustive by any means. It is hoped that endeavours such as the present Symposium and the resultant book, will continue on a regular basis so that the condition of biodiversity and its conservation in Pakistan be regularly documented and updated.

I am sure this book will prove to be extremely useful for all scientists, researchers and conservationists all over the world for better planning and execution of work in this important field of science.

Prof. Dr. Khalid Mahmood Khan
Chairman
Pakistan Science Foundation
Islamabad

PREFACE

Pakistan is located at the junction of the ancient continent of Gondwanaland and Eurasia. The flora and fauna of the country come together to form one of the most diverse and richest biological assemblages in the world. Few areas have experienced as much geological change, or have as many endemic species. The biodiversity of this country is a national and international treasure.

We believe that it is time to bring the story of biodiversity of Pakistan together where it can serve as a foundation for efforts to protect and conserve the natural resources of the country and point the way towards future studies. Much still needs to be learned about the biodiversity of this area, but a remarkable amount of research has been done already. Our mission has been to bring together specialists from around the country, as well as from international programmes, and to cover as broad a spectrum of Pakistan biodiversity as possible.

We began this overall project in 1992. The culmination was a two- day Symposium on the Biodiversity of Pakistan held in Islamabad in November, 1995, hosted by the Pakistan Science Foundation, while Pakistan Museum of Natural History took the lead in organizing this event in association with the Florida Museum of Natural History, with which it has collaborated on a number of natural history projects. Contributors have come from many universities, programmes and research centres.

We hope that this volume expends the knowledge of Pakistan's biodiversity, and stimulates future studies on the flora and fauna of the country. We especially hope that the Book will promote interactive projects on biodiversity, and stimulate efforts to preserve and protect the natural patrimony of Pakistan. It is also hoped that this Book will help make it possible to focus on what is most important about the flora and fauna of the country, and how to go about identifying and protecting the rich and unique biodiversity of Pakistan.

We would like to thank the Pakistan Science Foundation (PSF) and the office of International Affairs of the U.S. Fish and Wildlife Services (Washington) for their support and encouragement. The staff of the Pakistan Museum of Natural History was instrumental in making the Symposium possible, and the PSF provided their facilities and sponsorship. The Florida Museum of Natural History coordinated activities associated with the Symposium outside Pakistan. We would like to thank in particular David Ferguson, who has been a longtime supporter of activities associated with biodiversity and conservation in Pakistan. Our special thanks are extended to both the present and past Chairmen of Pakistan Science Foundation, Prof. Dr. Khalid Mahmood Khan and Dr. Bashir A. Sheikh, whose support made all this possible.

Shahzad A. Mufti
Charles A. Woods
S. Azhar Hasan

CONTENTS

SECTION I REGIONAL BIODIVERSITY

SECTION I

REGIONAL BIODIVERSITY

BIODIVERSITY

BASHIR A. SHEIKH AND MOHAMMAD AFZAL*

Pakistan Science Foundation, Islamabad
**Pakistan Scientific and Technological Information Centre, Islamabad*

INTRODUCTION

An astounding degree of unity of structure and function has been discerned in all known living organisms on earth. The structure of flagella is essentially the same in all cells having nuclei. The molecules involved in growth and metabolism are remarkably similar, and often they are constructed of identical subunits. Furthermore, enzymes, the catalysts of biological chemistry, are now known to act similarly in all organisms. Phenomena such as cell division and the transmission of the genetic code also appear to be universal. However, an equally astounding diversity can be seen in the organisms that have been discovered.

The fossil record vividly demonstrates that life on earth has chaged greatly during geologic time. Paleoecologists ask, however, whether any regular patterns in the distribution of organisms are maintained even though individual species evolve and become extinct. One such pattern is suggested by the observation that different kinds of environments support different numbers of species. Many more species, for example, exist in the tropics than in polar regions. Similar diversity gradients exist among benthic marine species of both ancient and modern seas. Generally, very few species (but large numbers of individuals) live in environments--like beaches, bays, and lagoons--that are subjected to stressful conditions, such as high wave activity or large variations in temperature or salinity. In offshore environments, where physical conditions are more stable, species diversities are higher, but numbers of individuals per species are fewer.

The most satisfactory explanation of this pattern appears to be as follows. For a species to avoid extinction in stressful environments, individuals of that species must be adapted to reproduce rapidly and to disperse over wide geographic areas. The species thus make up for the vulnerability of local populations to unfavorable conditions. Where environmental conditions are nearly constant, species need not be adapted for broad dispersal and rapid reproduction. The resulting smaller population sizes of species permit a given habitat to contain individuals of a greater number of different species.

Generally speaking biodiversity is the sum total of genotypes of living organisms. Wilson (1992) has given a somewhat comprehensive definition of the term:

"the variety of organisms considered at all levels, from genetic variants belonging to the same species through arrays of genera, families, and still higher taxonomic levels; includes the variety of ecosystems, which comprise both the communities of organisms within particular habitats and the physical conditions under which they live". He has implicitly incorporated the abiotic components of the environment into the concept of biodiversity. This approach actually satisfies ecologists on large extent who would not merely depend on taxonomic composition in their ecosystems.

Various authors (Wilson, 1992; Ehrlich and Ehrlich, 1992) have perceived the subject of biodiversity from their own point of view including its economics and utilies but the fact the of matter is that it is biota within ecosystems which purifies water and air, thus even species which have no conceivable value in the provision of raw materials or amenities may still be of value to society.

The World Commission on Environment and Development during its meeting in (1987) discussed in detail the preservation of species and ecosystems. Strong anthropocentrical arguments, concerning economical as well as scientific, cultural, esthetic and ethical values were forwaded and where considered to be important. Even a species version of anthropocentrism concerning the value of preserving wild species and ecosystems and "transformation value", which is the power of human contact with wild species, was also taken into account.

Recently some authors including Hagvar (1994) have proposed a system for dividing the arguments regarding biodiversity into three main blocks as follows:

1) Ecological: describe the need of biodiversity to support basic functions of nature. They could be summed up as follows:

Species richness is needed to maintain food chains and food webs.

Pollination by insects and other animals is necessary for the existence of a great variety of the plant material.

Biodiversity is vital to the fertility and formation of soil through decomposition of dead material.

A species-rich pool of parasites and predators stabilizes population levels of many species.

A high genetic diversity represents a great potential if conditions change. For instance, changed climatic conditions may favour certain species or subspecies which are rare today. In an evolutionary perspective, species must have the inherent capacity of continuous adaptation.

2) Egoistical, or Utilitarian: where one looks to the nature as means of fulfilling human needs and wishes including recreational and esthetic values. They are exemplified as follows:

A high biodiversity ensures many possible food sources for man.

Biodiversity represents important basis and potential for human medicines.

Many valuable products are achieved from special species (timber, fibers, dyes, oils, rubber, various industrial chemicals, etc., from plants, and silk, beeswax, pearls, sponges, etc., from animals).

Although certain insects are agricultural pests, many species support agriculture through pollination and biological control.

Many species are "indicator organisms", serving as sensitive "red lamps" in nature. For instance, lichens are well-known indicators of air quality.

Both in soil and water, a high biodiversity helps to keep the environment clean. Certain species have the ability to detoxify human waste products.

Research and education: Nature's biodiversity represents an enormous source of knowledge in many different disciplines (medicine, genetics, chemistry, physics, ecology, evolution, etc.)

The psychological value of biodiversity: Contact with other life forms, often increases the recreational value of nature. Many species have a high esthetic value for man. There is also a value in simply knowing that certain species or life forms exist.

3) Ethical: Encompass the inherent value of all life forms and our responsibility to future generations and future evolution. They are given below:

Intrinsic value: All life forms can be considered to have a specific, or intrinsic value. This value is independent of whether or not the species has a direct value for man.

Man is new on earth: Man has existed for only a few million years, while many other life forms are old "inventions" of nature (e.g.

dragonflies, turtles or crocodiles) and have long traditions on the living planet".

Ethics of future evolution: It is wrong to terminate evolutionary lines, and to reduce nature's potential for further evolution.

Ethics of future generations: It is wrong to reduce the potential of future generations to use biodiversity as a basis for their life quality.

DIVERSITY IN ECOSYSTEMS

An ecosystem is characterized by two distinctive features. The first one is the number of species in a given area compared to the number of individuals. In other words, it is a way of measuring variety. Sometimes this is called the species richness of an ecosystem. The second feature is dominance. This term measures relative numbers of individuals present in each species. Two ecosystems may have the same number of species present, but in one there may be approximately equal numbers of individuals of each species (high eveness or equatibility) while in the second most of the individuals may be of the same species (high dominance). The first ecosystem is then considered to have greater diversity. Diversity is an important indicator of human interference with a natural ecosystem which often leads to reduced diversity. Fewer kinds of species are found in a polluted ecosystem than in a similar ecosystem that is relatively cleaner. Measurements of diversity in a given area over a period of time can be a fair measurement of the effects of pollution.

The diverse environmental conditions over longer periods of time lead to the distribution and existence of different niches. The longer sustained environmental diversification usually leads to evolutionary changes culminating into variety of species and families. It should be emphasized, though, that there are basic biological processes common to all organisms. These processes work best at certain optimum temperatures, light intensities, and availability of resources for food, etc. The conditions which lead to provision of lesser biological optimal environment are called farther conditions. During this situation, only fewer species are found. To put it another way, fewer species are able to use the available niches in harsh climates. Thus, fewer species are found in arctic regions, on icy mountain tops, and in deserts than in more moderate environments such as temperate, wet or tropical forests. Three important factors, viz., variation in the environment, its harshness and longevity in a system are important influences in an ecosystem which determine its floral or faunal diversification.

Human interference may also move conditions farther from biological optima. For instance, the addition of sewage to a stream reduces the amount of oxygen available in the water. Only a few organisms are adapted to low oxygen concentrations. As a result, the diversity in the polluted stream is much lower than in a normal, well-oxygenated stream. It should be emphasized that the actual number of organisms in the polluted and nonpolluted habitat may be similar, but the organisms in the polluted habitat will mostly be members of one or a few species. It is the number of different species that is reduced in that environment.

The ever-increasing loss of plant and animal species due to human interference and environmental degradation represents a major global conservation concern. Habitat loss, especially in tropical forest areas, is the greatest threat. Some species, such as whales and the rhinoceros, are threatened by overexploitation. The Convention on Trade in Endangered Species of Flora and Fauna has worked well to control trade in most threatened species. A more fundamental solution, however, must be the establishment of a global network of areas that protect and maintain representative samples of the world's ecosystems. Substantial progress has been made toward this goal, and there are protected areas--on paper at least--in virtually all nations. Maintaining species of plants and animals in botanical gardens, zoos, or gene banks is a secondary solution. It may work well for some species (although, kept in storage, even seeds of domesticated plants and their wild relatives tend to lose genetic vigor), but relative to the magnitude of the problems, such efforts are very limited.

GENETIC DIVERSITY

Gene bank is a popular term used to describe repositories for genes of living organisms. It is commonly used in the context of plant breeding, but it also applies to the freezing and storage of animal sperm and embryos for use in animal breeding techniques.

Genetic variation is the raw material for the breeder, who must often select from primitive and wild plants and animals including related species, in the search for new genes. The appearance of new diseases, pests, or new virulent forms of existing disease-causing organisms makes it imperative that biota variation be preserved, because it offers a potential for the presence of disease-resistant genes. Also, there are demands for new characters-for example, high protein, improved nutritional factors, and fertility restoration. As a result, breeders require a large and diverse gene pool to meet ever-changing demands.

An understanding of crop and animal origins and variations is necessary in assembling genetic diversity in them. In certain geographical areas there has existed a wide variability in flora and fauna, but the encroachment of civilization has reduced the natural variability inherent in primitive forms and related species of the flora and fauna.

Agricultural progress, on the other hand, as a result of new breeding techniques, has reduced rather than increased crop variability as improved cultivars, or varieties, are planted in wider and wider areas and old cultivars, which may contain valuable genes, are lost. Crop failures, which result in a smaller gene pool, have also led to an increased awareness of the need to preserve genetic diversity in plants.

Efforts are under way to increase collections of plant materials in various forms. Usually these are preserved as seeds, but living plants, pollen, and cell cultures may also be used. In most gene banks, seeds are usually preserved under conditions of low temperature and humidity. These collections must be periodically renewed by growing the plants and producing new seeds. Increasing emphasis is also being placed on preserving living collections of asexually propagated crops, such as fruit and nut species.

At no time in their history have zoos experienced such rapid and profound change as that seen in recent years. Advances are occurring in exhibit design, animal management, keeper training, visitor education, and, most importantly, in international conservation programs (in situ conservation). As wild populations continue to shrink in their dwindling habitats, zoos must now fill the roles of breeding grounds and reservoirs of genetic diversity for the many species in danger of becoming extinct. As is all too often the case nowadays, there are larger populations of some endangered species (Siberian tiger, Arabian oryx, California condors) in zoos than in the wild. While some of these species have been successfully returned to the wild, others must wait until the wild habitat can be reclaimed and protected. The populations of many species in the wild are already being carefully managed in ways similar to the collective management strategies employed by zoo professionals.

HABITAT DIVERSITY

Habitat is the place where a particular animal or plant species dwells. Habitats of similar climate and vegetation form land complexes called biomes. The natural habitats on the earth and the great variety of species are a product of the changes that have occurred over long geological time periods; however, not all habitats are natural. Humans can alter nature and thereby promote the welfare of certain species that would otherwise not occur in the same numbers. Some ecologists consider an organism's total physical and chemical surroundings (the environment) synonymous with habitat. No description of habitat is complete without including some environmental parameters such as temperature and dissolved oxygen. In all habitats, large or small, the biota has adapted to the physical and chemical features of the environment. Wind, moisture, slinity, concentration gradients of chemicals, and even other organisms affect the habitat and influence the success of a species trying to survive in nature.

OCEANS

Oceans are the largest major habitat and cover 70 percent of the Earth's surface. Along marine coastal areas is a littoral zone. Tidal action, upwelling currents, and the influx of fresh water from rivers significantly influence the biota (flora and fauna) that can exist. Tides circulate loose organic particles, or detritus, to detritus feeders such as marine crabs. Detritus is consumed by oyster, zooplanktons, and barnacles. Phytoplanktons thrive where upwelling currents circulate phosphorus into the upper layers of a body of water (the euphotic zone) in which plant growth is possible. Blue crabs have adapted to the coastal river estuary habitat with the associated problems of regulating osmotic pressure. Below the euphotic zone, from parts of the continental shelf bottom down to the abyssal plain of perhaps 5,000 m (16,400 ft), is an aphotic zone, or light-lacking zone, that is inhabited by glass sponges, lamp shells, brittle stars, and microorganisms. Plant life, as humans know it, does not function here. The aphotic zone nekton, or free-swimming aquatic organisms such as fishes, prawns, and squids, have biolumenescent organs or symbiotic bacteria that supply illumination, or both. The angler fish has huge jaws and teeth and uses a luminescent lure that it dangles in front of its jaws.

FRESHWATER LAKES

The freshwater lake, or lentic, habitat has a littoral zone that is euphotic and hence may be heavily populated by aquatic plants. Very deep lakes have low phytoplankton density and few rooted aquatic plants because nutrients are scarce and the bottom is dimly lighted. The benthos (bottom) of many lakes consists of oxygen-demanding organic detritus and is thus anaerobic. Benthic life such as bacteria and midge larvae, however, thrive and utilize the organic matter as food. With time, lakes change physically and chemically. The classic concept of aging in lakes involves a progression of changes. Deep, nutrient-poor oligotrophic lakes with trout, whitefish, and cisco evolve into shallow, richly organic eutrophic lakes with blue gills, black bass, and pike.

STREAM HABITATS

A current of water is the main physical feature of stream, or lotic habitats. As a stream flows towards sea or lake level, suspended particles and chemical nutrients accumulate and dissolved oxygen decreases. Algae and mosses attach to rocks and contribute to community organic production, but basically the lotic habitat depends on the importation of organic foods such as tree leaves. The imported organics form the base of an important detritus food chain populated by snails, midge larvae, and mayflies. With a longitdinal change in physical and chemical features, the biota also changes. Trout and various species of darters give way to bass, pike, and catfish as the stream approaches sea or lake level.

TERRESTIAL HABITATS

A temperature and moisture complex dictates the major terrestrial habitat types that range from the verdant tropical rain forest to the snow-covered polar regions. Tropical rain forests are tall and highly stratified, and each stratum offers shelter to specialized animal life forms. Temperate climates support deciduous forests of such trees as beech, oak, and maple. The deciduous forest once covered eastern North America and Europe. Coniferous forests of pines, spruces, and fir most typically occur in cool climates. Further poleward in the Northern Hemisphere is the tundra, a marshy-plain habitat that is frozen for much of the year. The vegetation consists of lichens, grasses, and dwarf woody plants with species of animals such as the lemming, snowy owl, and Arctic hare.

CULTURAL DIVERSITY

Human cultures vary widely over the Earth, from the so-called primitive (nonliterate) societies and cultures of the Amazon rain forest or the valleys of highland New Guinea to the literate and civilized societies and cultures. In some societies people practice monogamy (one spouse at a time); in others the practice of polyandry (multiple husbands) or polygamy (multiple wives) is common. The languages of some cultures have numbers only from one to five (with anything larger referred to as "many"); those of others count into the millions and billions and more by means of a decimal system.

Inspite of the great diversity of human cultures around the world, in all societies certain universals of culture reflect basic responses to the needs of human beings as social organism. These universals include a primary means of subsistence--for example, hunting and gathering, agriculture, industrialized labor; some form of the primary family; a system of kinship; a set of rules of social conduct; religion; material culture (tools, weapons, clothing); forms of art; and many other institutions that indicate the common adaptations of all human societies to varied natural environments.

HUMAN ENVIRONMENT AND BIODIVERSITY

During the millions of years that preceded the appearance of human life, extinction of organisms was linked to large-scale geologic and climatic changes, the effects of which were translated into major alteration of the environment. Environmental change is still the primary cause of the extinction of animals, but now the changes are greatly accelerated by human activity. Clearing land for farms and towns, lumbering, mining, building dams, and draining wetlands all alter the environment so extensively that ecosystems may be completely destroyed. With a burgeoning human population requiring food, shelter, and clothing and constantly demanding more energy-using devices, the temptation to exploit land for human use without regard for consequences is great.

Although it has been known ever since the beginning of mankind that we depend upon available natural resources for existence, the contribution of biodiversity to our sustainable development was formally acknowledged recently. In spite of the recognition of the fact that biological resources are essential for human subsistence and welfare, they are under constant threat. Biodiversity is threatened both at local levels, in the form of poaching, encroachment of protected areas and habitat destruction and pollution, as well as at global level in the form of ozone depletion, deforestation and global warming. Despite great efforts by the biologists, ecologists, wildlife managers and park managers, the depletion of biodiversity continues unabated and is under siege throughout the world. More and more species are threatened with extinction, with hundreds of species disappearing from the face of the earth each year. The genetic resources of earth's natural heritage are barely understood, let alone utilized. One such glaring example is that of Marine flora & fauna. Of 4000 species of plants and animals in the seas and oceans, hardly 50% have been identified & documented. It is also feared that the rate at which environmental degradation, deforestation and habitat destruction is taking place all around the world, much of this treasure is liable to be lost even before it is discovered.

The air, water and soil all around us are being irreversibly destroyed and degraded due to our daily activities. With the ever growing population the need for living space and food has grown to such an extent that many ecologically well balanced habitats have been eliminated with the concurrent elimination of biodiversity. Thus one has to tackle the issue of biodiversity on two fronts. We not only have to find, identify and study our biodiversity but we also have to take measures not to allow the destruction of the biodiversity that we have. This is the challenge which faces all of us all around the world.

Pakistan is quite fortunate in being one of the few countries of the world which have extremely varied landscape comprising many ecological zones rich in biodiversity. For example, so far about 6000 species of plants have been described from Pakistan out of which almost 400 are endemic to this country. Among animals, no less that 25,000 species of Invertebrates, including insects, have been recorded. Similarly over 750 species of fish, amphibian & reptiles, and 850 species of birds mammals have been described.

Several forms of environmental changes are responsible for the disappearance of species in our country. For example, as forests are cut down, wild animals have progressively smaller feeding and living spaces. They also become more accessible to hunters, who kill or trap them for eating or for sale as pests, research animals, and zoo specimens. Some plant and animal species may move into human communities as soon as their immediate economic importance is discovered. Extermination of marauding monkeys, roaming leopards, or foraging deer is easy to justify by people whose livelihood is threatened.

Pollution is another form of environmental change. Scores of species of birds lay thin-shelled or shell-less eggs as a result of ingesting degradation products of DDT and some of the other chlorinated hydrocarbon insecticides that make their way into the food chain. Species of amphibians are dying out because the ponds in which they breed and the moist soil in which they must live are watered by "acid rain"--water that combines with pollutants in the air to form nitrous acid, sulfuric acid, and other corrosive compounds. Industrial wastes dumped in our sea have so depleted the oxygen supply that some species of bacteria that decompose sewage have been wiped out and the nutrient cycles disturbed.

It has become increasingly clear that resources and resource uses are intimately interrelated. A forest contains lumber, which is a valuable economic commodity; it also, however, serves as a watershed, keeps soil from eroding, provides habitat for wildlife, provides recreation, and ameliorates local climate. Indiscriminate cutting of trees may destroy the forest, but it will also have corollary, and potentially far more serious, effects. A paramount principle of conservation is that the use of any resource requires consideration of the impacts of that use on associated resources, and on the environment as a whole. Therefore the goals of resource conservation must be based upon principles of the maintenance of essential ecological processes (which range from the global cycles of nitrogen, carbon dioxide, and water to the localized regeneration of soil, recycling of nutrients, and cleansing of waters and air) and life-support systems, such as agricultural systems, coastal and freshwater systems, and forests; the preservation of biodiversity; and the assurance that utilization of species and ecosystems such as forests and grazing lands is sustainable. The consumption of nonrenewable resources should insure that scarce minerals are used conservatively and recycled where possible, and that their mining and use have the least possible adverse impact on other resources, and on environmental quality.

REFERENCES

EHRLICH, P.R. AND EHRLICH, A.H., 1992. *The value of biodiversity.* Ambio.

WILSON, E.O., 1992. *The Diversity of Life.* Allen Lane, London.

BIODIVERSITY: AWARENESS AND EDUCATION THROUGH NATURAL HISTORY MUSEUM

SHAHZAD A. MUFTI

Pakistan Museum of Natural History, Garden Avenue, Shakarparian, Islamabad

Abstract: Environmental pollution through rapid industrialization and urbanization has rendered environment quite unfit for human health and prosperity. In many parts of the world, including some big cities of Pakistan, the environmental conditions have deteriorated to such an extent that a major effort is required to rehabilitate these. In order to bring about this rehabilitation and to stop further environmental degradation, public awareness and active participation is imperative. Many governmental and private institutions and organizations have already started work in this direction. However, a natural history museum, by virtue of its direct activity in environmental research and education, can play an important role in the creation of mass awareness about our environmental and biodiversity problems. It is therefore proposed to set up such natural history musea all over the country so that public at large can be educated about various environmental and conservation issues.

INTRODUCTION

The modern human history is not very old. It has been estimated that the first so called "modern man" appeared on this planet only about 1.5 million years ago (Colbert, 1966). Throughout its existence man had to deal with whatever was present around him-his environment as we now recognize it. The human race soon realized that all its basic needs for survival have to be attained from whatever resources are present in its surroundings. This was especially true in terms of food and space. The early humans found abundant food, in the form of grain, fruit and water, all around them. There was plenty of space too, where they could hide and sleep, escaping the cold, the rain, wild animals and many other harsh realities of nature. As their number grew the natural laws of "competition" and "survival" started operating and thus tools, elementary agricultural practices and other mechanisms for maximal "utilization" and "exploitation" of the resources present around them came into operation. There was apparently neither an understanding nor a need for "conservation" of natural resources. When all natural resources were exhausted at any one place the groups of people just moved on to "greener pastures", so to speak.

Within a few thousand more years, however, with a substantial increase in their numbers, humans found themselves faced with a situation where natural resources around them had to be "regenerated" or "sustained" for them to survive. In the meantime agricultural practices had also advanced to the extent where crop raising had been established to a considerable extent. This allowed them to have permanent settlements now and villages, towns and cities took root. One of the earliest such settlements can be traced to have existed about 5000 - 6000 years ago and we do have records of these times. It appears that till that time natural resources were still aplenty and various environmental components were able to sustain themselves through their natural "resilience" properties (Miller, 1985). However, there were indications that a basic understanding of environmental components as well as natural resources had been evolved and elementary "conservation" practices were adopted. Agriculture was now established as a routine practice and so did the storage of grains.

The advent of modern times was identified with a rapid increase in world population, rapid urbanization, intensive agriculture and most of all, expanding industrialization. As we entered the 15th & 16th century, the deleterious effects of human population on the environment started becoming apparent and thus a cause of concern. It is however only during the last one hundred years or so when the health of the earth's environment has become so bad that there has been felt an urgent need to take all necessary measures to look after our environment properly.

As one of the first steps in this direction, a thorough understanding of our environment has been developed. It has now been established that our environment consists of an abiotic and a biotic component. Abiotic components comprise light and heat (from sun), air, soil and water while biotic components of environment are the plants, animals, the viruses and bacteria and above all, fellow human beings. We now know that all these components exist in an intricate relationship with each other and comprise what is known as an "ecological system or ecosystem". We have now known that due to our agricultural and industrial practices over the last few decades, we the humans have already caused grave environmental degradation. Ozone layer depletion, greenhouse effect, deforestation, desertification and gross water, soil and air pollution are now all too familiar concerns for us. We have reached a situation where we have destroyed our environment to an extent that our own existence on this planet has become perilous. No wonder all nations of the world are now ready to do whatever it takes to rehabilitate our environment, so that mankind could survive.

Rehabilitation of our environment, however, is a gigantic task. It obviously is not possible to mend in an instant the damage which we caused over thousands of years. But at least a beginning has been made. It is in a way quite ironic that all tools for the rehabilitation of the environment are provided by nature itself all around us. Biodiversity is one such tool, which if properly used, can bring about betterment of our environment and of ourselves. Most simply put, biodiversity means the existence of all animals and plants in any given area of the planet earth. It has been estimated that there are 5-30 million species of plants and animals in the world (Kapoor-Vijay, 1992). Inspite of hundreds of years of efforts, it is estimated that out of these about 1.4 million species have so far been known (UNEP-Profile 1990). This means that there are still literally millions of species of plants and animals which are yet to be discovered. With the enormous amount of benefits obtained from the known species of plants and animals, it is easy to surmise that knowledge about more species will be of tremendous utilization for mankind. Presently out of all the known plant species, only about 5000 species have been utilized for their medicinal properties (Kapoor-Vijay, 1992). One can only imagine the benefits which additional discoveries of biodiversity of plants and animals can generate. Another extremely important aspect of biodiversity is that the wild (as yet unknown) species of plants and animals are a great genetic resource. It is well known that persistent inbreeding of plants and animals leads to many problems such as low yield, decrease in pest resistance and even longevity. In nature there is a constant cross breeding between various varieties and subspecies with their wild relatives. This produces what is known as "hybrid vigor". Basically this introduction of wild genes renders the offspring more productive, more arable and more resistant to disease. It is thus obvious that wild plants and animals provide an extremely important source of genetic variability. In addition, knowledge of biodiversity of any given area is essential for understanding its health and dynamics and thus the health of the total ecosystem operating there. A rich biodiversity of an area is a clear indication of a sound and well balanced ecosystem and vice versa.

A complete knowledge of biodiversity is thus not only essential for its own sake, as a great natural resource, but is also important in the betterment of our crops and animal farming and for proper management of our environment. Pakistan is one of the most fortunate countries in the world in terms of its geological and biological history. Within the few hundred miles of its territory,

Pakistan represents parts of at least three distinct zoogeographical zones, namely Palaearctic, Ethiopian and Oriental (Roberts, 1977). There are practically all types of habitats represented, including dry cold, tropical, subtropical, plains, deserts, estuaries and marine etc. Correspondingly a variety of plants and animals are present in these areas. As can be expected from world wide statistics, hardly 10-15% of the biodiversity has yet been known in this country. There are relatively very few organizations which are involved in the collation, collection, preservation, identification and carrying out research on the biodiversity of this country. Major among these being, the Zoological Survey Department, The National Herbarium and the Pakistan Museum of Natural History. In addition to these, almost all zoology, botany and geology Department of the Universities of Pakistan also carry out small amount of research in this direction. There is, thus, an obvious need for many more high powered research organizations to conduct research on the biodiversity of Pakistan.

Another aspect of a lack of information about biodiversity all over the world is that there is relatively very little public awareness and input in this direction. Although animal and plant systematics constitutes a part of curriculum at University level, public at larger remains quite ignorant about the biodiversity present around them and completely fails to understand its importance for their well-being. As a result they not only do nothing to help conserve biodiversity, but on the other hand, contribute towards its annihilation. Overpopulation, with the consequent urbanization and deforestation has played havoc with the biodiversity all over the world. There has, thus been felt an acute need to educate public in general and young students in particular about the biodiversity around them and emphasize its importance in terms of economics and environment. This kind of public support and interaction is necessary if any society intends o make any programme successful.

There are a number of mechanisms through which public awareness and education can be brought about, both formal and nonformal. The formal education can be imparted at school, college and/or university level. Indeed almost all schools and colleges are now introducing environmental studies as part of their curriculum. The nonformal education can and is being imparted through mass media, both print and audiovisual. Many newspapers bring out regular “Environment” pages, which discuss all aspects of environmental problems, including pollution and habitat degradation and its impacts on biodiversity. There are also regular radio and TV programmes about similar topics. There have also been established many Non-Governmental Organizations (NGO’s) dealing specifically with environmental issues. For example Margalla Hills Society, Pakistan Wildlife Conservation Foundation and Sustainable Development Policy Institute etc. have all been quite active in this direction. There are of course, many multinational NGO's such as IUCN, WWF and EPA etc. which have been quite active in this field over the last several years.

One very important organization, which has played a major role in imparting education and creating mass awareness about environment and biodiversity in the developed world, is a Natural History Museum. This Organization is unique, in the sense that it is both a highly professional research organization in the field of biodiversity, as well as the centre of a well organized public education programme. There are a number of highly qualified and trained scientists in such a Museum, whose job is to carry out research exclusively in the filed of biodiversity. The discoveries they make and the new information they generate is then handled by professional exhibit designers and educators who then translate that information in the form of an easy-to-understand and attractive medium so that the information not only reaches general public but is also easily understood and digested by them. Usually such mass communication medium is in the form of 3-dimensional diorama, which are visually quite attractive and informative. Other communicational vehicles, such as films, models and charts etc. are also routinely used. More recently computerized interactive programmes are being used more and more extensively to communicate information. In short the natural history musea all over the world have become the primary sources of mass education and awareness in terms of biodiversity.

Unfortunately, Pakistan did not have any natural history museum for a long time. There were a few zoological mesea, such as the one associated with the Zoological Survey Department, Karachi, or the one located in the Government College, Lahore. There were also some plant collections, given the name of herbaria such as the one in Gorden College, Rawalpindi and at the PARC in Islamabad. The first and the only "real" Natural History Museum was established as recently as 1976 in Islamabad. It is still in formative stages, but has already become a national centre of a moderately large biodiversity collection and a bonafide place of public education and awareness. This Museum, officially known as Pakistan Museum of Natural History (PMNH), has already collected more then 200,000 specimens of plants, animals, rocks, minerals and fossils, from various parts of Pakistan. More than half of this collection has been identified, curated and described. The Museum also has a separate Public Services Division which is responsible exclusively for mass education and awareness about environmental and biodiversity issues of Pakistan.

As many as 10,000 - 15,000 persons, mostly school going children, visit the two display centres, on a monthly basis. Thus, PMNH is playing its due role in the creation of awareness among general public, about biodiversity of Pakistan.

But, is it enough? When we look at the statistics about such natural history museums around the world, the answer is a simple "no". As can be seen from Table I, Pakistan lags far behind other countries in terms of having Musea as such, what to speak to natural history musea. It is clear from the Table that Pakistan with its population reaching almost 130 million needs to set up many many more musea, if it were to compete with the rest of the world.

Table I: Museums and population

Country	Population	Museums
Australia	14.92 m.	999
Canada	24.60 m.	1514
Denmark	5.12 m.	331
France	54.09 m.	1921
Germany	16.74 m.	748
Great Britain	55.93 m.	2127
Italy	56.00 m.	1848
Japan	117.8 m.	807
Korea	38.72 m.	81
Pakistan	85.00 m.	54
U.S.A.	226.5 m.	7892

(Data source:The Directory of Museums & Living Displays, Kenneth Hudson and Ann Nicholls, Third edition 1985)

More specifically, the situation is even worse when the number of Natural History Musea is compared with other countries (Table II).

Table II: GNP AND and Natural History Museum Establishments

Country	GNP US$	Natural History Museums	Total Museums
Bangladesh	120	3	58
China	290	5	429
India	240	7	459
Indonesia	420	3	89
Pakistan	300	1	54

(Data source:TheDirectory of Museums & Living Displays, Kenneth Hudson and Ann Nicholls, Third edition 1985)

Even the so-called third world countries, including the ones of the subcontinent have higher number of natural history musea than in Pakistan. This situation needs urgent attention. The number of musea, especially the natural history musea, around the world clearly indicates a public as well as private commitment on the part of various countries for science in general and environment in particular. Pakistan by all means, needs that kind of public and private support, if it is to be successful in its endeavors towards the betterment of its environment and its biodiversity. This means that not only the Natural History Museum of the country needs strengthening, but many more natural history musea need to be established all over the country. If we fail to do so, then Pakistan is likely to continue to suffer in terms of its environmental health and its biodiversity. It is feared that if the present trends in governmental and public apathy are allowed to continue, Pakistan will lose much of its biodiversity, even before it is discovered, and that will be a tremendous national loss, both scientific and economic.

REFERENCES

COLBERT, E.H., 1966. *Evolution of the Vertebrates.* 1-479, John Wiley & Sons, Inc., New York.

UNEP-PROFILE, Biological Diversity, 1990. 1-18.

KAPOOR-VIJAY, P., 1992. *Biological Diversity and Genetic Resources.* The Commonwealth Science Council, London.

MILLER, G.T.JR., 1985. *Living in the Environment.* Wadsworth, California.

ROBERTS, T. J., 1977. *The Mammals of Pakistan.* Ernest Benn Ltd., London & Tonbridge.

CONVENTION ON BIOLOGICAL DIVERSITY - PAKISTAN'S PERSPECTIVE

KALIMULLAH SHIRAZI

Ministry of Environment, Urban Affairs, Forestry & Wildlife, Islamabad

Since the early 1970's several members of the World Conservation Union's (IUCN) Commission on Environmental Law (CEL) had agreed that the natural environment as a whole was not being adequately conserved through the existing legal instruments of the time. IUCN therefore took up the matter further in its General Assembly and Commission meetings. The IUCN General Assembly, held in Madrid in 1984, requested the IUCN Secretariat to develop a number of principles to serve as a basis for a preliminary draft of a global instrument on the conservation of the World's genetic resources. After intensive consultations within IUCN, a final draft was completed in June, 1989.

The IUCN proposals centered on three main points:

- A general obligation for all States to conserve biological diversity;
- The principle of freedom of access to wild genetic resources;
- The principle that the cost of conservation should be shared equitably by all nations.

In parallel, the World Commission on Environment and Development recommended in its report, "Our Common Future" that Governments should "investigate the prospect of agreeing to a 'Species Convention' similar in spirit and scope to the Law of the Sea Treaty and other international conventions reflecting the principles of 'universal resources'.

The Governing Council of the United Nations Environment Programme (UNEP) adopted a Resolution in 1989 under which an Inter-governmental Negotiating Committee (INC) was set up to negotiate and develop an international legal instrument on the conservation of biological diversity. The INC held five negotiating sessions and agreed to adopt the final draft of the Convention on Biological Diversity in 1992.

The Convention on Biological Diversity was placed for signatures by the Heads of governments in Rio in June 1992 wherein it was signed by 166 States, including Pakistan. So far 118 States have deposited their instruments of ratification. This Convention entered into force on 29 December, 1993. Pakistan became Party to this Convention on 26 July, 1994 with Ministry of Environment, Urban Affairs, Forestry and Wildlife as focal point.

OBJECTIVES

The objective of the Convention is the conservation of biological diversity, the sustainable use of its components and fair and equitable sharing of the benefits arising out of utilization of genetic resources, including appropriate access to genetic resources and appropriate transfer of relevant technologies, taking into account all rights over those resources and to technologies.

COMMITMENTS OF DEVELOPED COUNTRIES

The Member States have been divided into "developing country parties" and "developed country parties" with definite commitments and obligations to achieve the objective of the Convention. The commitments of the developed countries, inter alia, are to:-

- Provide new and additional financial resources to developing country parties to meet the agreed full incremental costs of fulfilling obligations under the Convention through bilateral, regional and other multilateral channels;
- Make voluntary contributions
- Transfer technology, including bio-technology, under fair and most favourable terms to developing country parties.

OBLIGATIONS OF THE PARTIES

All parties to the Convention, i.e. both the 'developing' and the 'developed countries' are obliged to:

- Develop national strategies, plans or programmes for the conservation and sustainable use of biological diversity.
- Identify components of biological diversity important for conservation and sustainable use.
- Take measures for the establishment of a system of Protected Areas as well as their management and protection;

- Share fairly and equitably benefits arising out of the utilization of genetic resources between the suppliers and users;
- Provide appropriate access to genetic resources on mutually agreed terms, subject to prior informed consent of the Contracting Parties providing such sources; and
- Develop or maintain necessary legislation for the protection of threatened species.

PROVISION OF FINANCIAL MECHANISM

A permanent mechanism for the provision of financial resources on grant or concessional basis, including transfer of technology, has been envisaged in the Convention through the setting up of an international entity for the purpose. In the Interim period this function is being discharged by the Global Environment Facility (GEF). The developed country parties are required to provide financial resources related to the implementation of the Convention, through bilateral, regional and multilateral channels.

BENEFITS TO THE PARTIES

- New and additional funds from international sources for projects relating to conservation and sustainable use of its biodiversity;
- Technology, including bio-technology, and its transfer on preferential terms from the developed countries;
- Global opportunities for the development and strengthening of national capability for the conservation of bio-diversity and its sustainable use through human resource development and institution building; and
- The initial deliberations of the Conference of the Parties and thereby influence its future role, procedures and determination of eligibility criteria for funding of projects by the GEF or the permanent future mechanism for provision of financial resources.

CATALYST FOR BIODIVERSITY CONSERVATION IN SOUTH ASIA: THE ROLE OF THE U.S. FISH AND WILDLIFE SERVICE

DAVID A. FERGUSON

Office of International Affairs, U.S. Fish and Wildlife Services, Arlington, Virginia, U.S.A.

The state of our natural environment is a matter of international concern. The United States Fish and Wildlife Service's mission is to promote fish and wildlife conservation and species enhancement, both inside and outside of U.S. boundaries. In 1973, a special directive was given to the Service with the passage of the Endangered Species Act. The Act allows U.S. assistance to be offered to foreign nations for the development and management of programs necessary or useful for threatened or endangered species conservation.

Starting in the late 1970s, the U.S. Fish and Wildlife Service has been assisting with wildlife conservation efforts in South Asia focusing on the Indian sub-continent and the countries of India and Pakistan. This assistance has been broad-based, but can generally be grouped into the three categories of resource management, research, and education and training. Together with various governmental and non-governmental organizations, the Fish and Wildlife Service works to identify projects of interest to local governments where assistance can best be utilized. Project support can range from technical consultation, to training, to direct financial sponsorship. Such support is provided with the intent to strengthen institutions and increase capacity within the host country to cope with wildlife management and conservation issues. In all cases, projects are initiated to ensure that they will make some permanent contribution to the knowledge base in the country or be self-sustaining.

General Background

The Indian subcontinent is a landmass diverse in its physical landscape, climatic conditions and flora and fauna. Conditions range from the moist environs of tropical rain forests, to the sun-baked deserts of Sind and Balochistan, to the cold, dry reaches of the

Himalayan peaks. The diverse physical features mountains, desert, plains, lowlands and coastal areas have spawned the rich biological diversity of the region which are reflected in over 20,000 species of plants, 500 species of mammals ' 1300 species of birds and at least 20,000 species of insects.

The subcontinent's favorable biodiversity probably had a lot to do with the establishment of one of the earth's early human civilizations which flourished in the Indus Valley over 5,000 years ago. The human population has continued to flourish ever since to the point where the sheer numbers of people and their demands upon their environment have slowly, but effectively, led to the degradation of natural habitats and the demise of species. Today, approximately 100 of the wildlife species found in the subcontinent are considered to be endangered or threatened.

U.S. Fish And Wildlife Service Assistance

Early in 1977, the U.S. Fish and Wildlife Service planned separate study missions to Pakistan and India to investigate the interests of the host countries in cooperating in joint programs of wildlife conservation. Contacts were made with a variety of existing governmental and non-governmental organizations in the respective countries and successfully identified a number of activities of mutual interest. Implementation of the programs,, particularly in Pakistan, however, proved more difficult due to political fluctuations, vicissitudes of governmental priorities, and perhaps a general apprehension for engaging in any new relationship.

Despite the difficulties, we persisted and opportunistically encouraged any activity from any local organization that showed some benefit for wildlife conservation. Initially modest, the proposals for support of activities such as environmental education materials and workshops, attendance at international symposia, short field surveys soon expanded to long-term conservation research and management exercises, institutional support and training, and strong bilateral relationships.

A few of the projects undertaken, especially in Pakistan, are highlighted here:

Crane Research and Education

Of the fifteen crane species in the world, the Siberian crane (*Grus leucogeranus*) is one of the most critically endangered. The present population is split into two groups. Both breed in Siberian Russia with one group (the larger - about 3,000 birds) migrating to China to over-winter along the middle Yangtze River, primarily at Poyang Lake. The other and much smaller group, known as the western flock, seems to be further split with 9-10 birds (7 in 1993/94) that winter in the southeast corner of the Caspian Sea region at Ferreydoon Kenar, Iran. The other is an ever decreasing group, that up to the 1993/94 season, passed through Pakistan to over-winter in India at the Keoladeo National Park, Bharatpur.

In attempting to put together the pieces of where the exact migratory route of the India-bound Siberians was located and what dangers the birds faced to try to determine the causes for their steady decline, the Service initiated a two-pronged approach in India and Pakistan starting in 1980. In cooperation with the Bombay Natural History Society, an Indian non-governmental organization, a long-term research study was begun at Keoladeo National Park, the only known Siberian wintering grounds at that time. The study was designed to look at how the park ecosystem functioned and to develop recommendations to address management implications.

At the same time, the Service enlisted the help of Dr. Steven Landfried, an environmental educator, then associated with the International Crane Foundation, to visit the region to seek information on public awareness of the Siberians' plight and to design programs to address this need. One of the early major findings of this work was the discovery that crane hunting (mainly of Common crane *Grus grus* and Demoiselle crane *Anthropoides vimo*) was a widespread and basically un-managed practice in Pakistan. There was no evidence that Siberian cranes were being brought down by the hunters, but because of their low numbers, the risk was always there as the birds crossed to India. A plan was developed and implemented in Pakistan's Northwest Frontier Province to try to determine the numbers of crane hunters engaged in the sport, locations of their hunting grounds, and quantifying the number and kinds of birds being taken.

The Service joined with other international groups working on crane conservation to complement efforts. In 1981, the Russians discovered a breeding population of Siberian cranes in the Kunovat Basin of western Siberia and in subsequent years the Service joined them and the International Crane Foundation to catch and fit representatives of this population with color markers and satellite transmitters.

Starting in 1983, the Service has supported a series of international workshops on cranes in India, China, Estonia, Pakistan and Russia which serve to bring together researchers and managers from all the crane range states to exchange information and strengthen combined efforts.

To raise awareness, the Service has produced a 55 minute color film in Hindi and English about the ecological cycle at Keoladeo National Park, featuring a section on the Siberian crane as well as a conservation message which is shown every night at the park and has been on Indian television; a series of audio-visual programs on cranes, crane hunting, and wetland conservation translated into three Pakistani language dialects and distributed to crane hunter groups; a video on crane hunting in Pakistan shown on U.S. public television networks and translated into local languages and distributed; hunter education and crane management practices with the Northwest Frontier Province including a recently completed two-day workshop in October 1995. Expanding from a small research project initiated to collect data, Service efforts have grown into full-fledged support for research, training, management and public education with respect to crane and wetland conservation not only in Pakistan, but on a worldwide scale.

Wildlife Clubs Offer An Innovative Approach To Education

At the request of the Government of Pakistan and in cooperation with the World Wide Fund for Nature in Pakistan and several provinces, the U.S. Fish and Wildlife Service has offered assistance in developing a conservation education program aimed specifically at the nation's youth. Youth education was deemed critical for sustainable natural resource usage. The education program, as envisioned by its Pakistani collaborators, would create a network of wildlife clubs for school-aged youth. Club members would become engaged in many aspects of conservation from nature study, to wildlife protection, to environmental education outreach programs.

Since wildlife clubs were a new concept in Pakistan, leader workshops were visualized as a means for preparing interested individuals to work as club organizers.

The first workshop was held in Peshawar, Northwest Frontier Province in 1986 and included 30 participants and a dozen observers from public secondary schools and colleges, provincial forest and wildlife departments, and others from the agricultural and natural resource sectors. This workshop served as a model to test the basic organization and content using environmental concepts and issues relevant to Pakistan, engaging participants in activities designed to demonstrate how these concepts can be taught as part of interesting exciting projects and activities for youth. A session dealt with organization and administration of youth clubs and a field trip served to provide examples of many of the ecological concepts and environmental issues discussed in the classroom as well as providing opportunities for the participants to apply in a field setting some of the information gained.

A second workshop in Quetta, Balochistan in 1987 drew 30 participants, and a third in Lahore, Punjab in 1990 with nearly 50 participants. The focus of this workshop was on activities for clubs, club organization, regional coordination, and national communication and coordination.

This project has been continued, at least in the Northwest Frontier Province, by the provincial wildlife department which has created wildlife clubs in a number of local schools.

The Small Mammals Of Pakistan

An early cooperator in Pakistan has been the Zoological Survey Department. In the late 1980s, the Director of the Zoological Survey asked us for some help in assisting that institution in strengthening its abilities to carry out its role in conducting zoological surveys. The result has been a multi-year collaboration between the Zoological Survey Department, the Pakistan Museum of Natural History, the Florida Museum of Natural History, the University of Vermont and the U.S. Fish and Wildlife Service in two fields of endeavor, herpetology and mammalogy.

In the previous session, Dr. Charles Woods described some of this work in his papers on "The Status of the Giant Woolly Flying Squirrel in Northern Pakistan" and "Biodiversity of Small Mammals in the High Himalaya".

The objectives of the two U.S.-Pakistan projects supported by the Service are similar:

- To study the distribution of reptiles and amphibians and small mammals in the country through extensive field work;
- Use the projects as training vehicles for providing colleagues at the local institutions with the techniques they need to become experienced herpetologists and mammalogists;
- Improve the status of the herp and mammal collections through improved curatorial techniques;
- Develop publications and field guides for the groups of animals for use in Pakistan and internationally;

- Use the data and emphasis on the projects to promote interest in the biogeography of Pakistan.

The Herpetology project field work is completed and a comprehensive book of the results is nearly completed. Two individuals, Dr. Khalid Baig from the Pakistan Museum and Dr. Hafizur Rahman from the Zoological Survey Department have received extensive on-the-job training as well as additional training in the United States and Germany and both have completed their doctorates as a direct result of this project. Dr. Walter Auffenberg from the Florida Museum of Natural History is the U.S. Technical counterpart.

The small mammal project is winding up its first phase with focus on the northern part of the country. One individual from the Pakistan museum of Natural History, Mr. Mohammad Rafique and two from the Zoological Survey Department, Mr. M Farooq Ahmad and Mr. Syed Mohammad Shamim Fakhri have been the main recipients in the training in the U.S. and Pakistan. Attention in Phase II of the project will shift to the south. This two-day international symposia was stimulated by the project work. Dr. Charles Woods from the Florida Museum of Natural History and Dr. William Kilpatrick of the University of Vermont are the two U.S. technical counterparts.

Other Activities Supporting Biodiversity Conservation

Other activities the U.S. Fish and Wildlife Service has supported in Pakistan to enhance biodiversity conservation have included:

- International workshops on bustards, snow leopard, environmental education, wetland conservation, marine parks and endangered marine mammals, CITES implementation;
- In-country training workshops on wildlife conservation and management, snow leopard survey techniques, crane conservation, raptor identification;
- training in wildlife field research and conservation;
- In-country institutional support to federal and provincial organizations through provisions of book, publication and environmental education materials;
- In-country research/survey and inventory of houbara bustard and its habitat, snow leopard, marine mammals, marine turtles, ecological field study of endangered monitor lizard, ecological field study of Kirthar National Park focusing on ibex and urial status;

- Support of the first phase of the Himalayan Jungle Project in the Palas Valley.
- In addition we have sent technical experts to work with Pakistani authorities on game bird captive breeding, to help develop a national zoological park and botanical garden in Islamabad, and we have assisted Pakistani wildlife personnel obtain advanced degrees in wildlife and natural resource management from U.S. universities.

CONCLUSION

I can assume that most of you in the audience have not heard much about the Service's biodiversity assistance role in Pakistan that I have just described and you may wonder why. I will go back to the first sentences in my presentation. 'The state of our environment is a matter of international concern. The U.S. Fish and Wildlife Service's mission is to promote fish and wildlife conservation and species enhancement, both inside and outside of U.S. boundaries'.

We believe that the best way we can promote this conservation and biodiversity enhancement is through the strengthening of the existing institutions and organizations in Pakistan and through the empowerment of local expertise. If our efforts have helped, then it is the environment and natural resources that benefit; and this is to the advantage of all of us.

BIODIVERSITY AND CONSERVATION OF THE DEOSAI PLATEAU, NORTHERN AREAS, PAKISTAN

CHARLES A. WOODS[1], C. W. KILPATRICK[2], M. RAFIQUE[3], MUQARRAB SHAH[3] AND WASEEM KHAN[4]

1. Florida Museum of Natural History, Gainesville, FL 32611, USA
2. University of Vermont, Burlington, VT 05401, USA;
3. Pakistan Museum of Natural History, Islamabad
4. Government College, Islamabad

Abstract: The Deosai Plateau is an area of rolling country at elevations mostly above 4,000 meters. It has a cold climate and is covered with snow for nine months of the year. It is located at the junction of four major mountain ranges, the Karakoram, Ladakh, Zanskar and Himalaya. The known flora includes 342 species in 36 families and 142 genera. The documented fauna includes 1 frog, 3 fish, 1 or 2 reptiles, 108 birds and 18 mammals. The area was declared a national park in 1995, and a management plan for the area is currently being prepared. There is an active program underway to study brown bears on the plateau, and to protect them. The reasons for the high level of biodiversity on the Deosai Plateau are multiple, and include topography, location at the junction of 4 major mountain ranges and the adaptations of its plants and animals. Some species, such as the burrowing vole (*Hyperacrius*) serve as keystone species, allowing numerous other species to make use of underground burrow systems for shelter and foraging activities, thus maximizing species numbers and diversity in an otherwise austere region.

INTRODUCTION

The Deosai Plateau is an uninhabited mountain wilderness surrounded by deep valleys and snowy peaks in the far north of Pakistan. In this rugged terrain the plateau stands apart as an expansive open area of grass, flowers, streams and lakes. LaPersonne (1928) vividly described the area when he wrote:

"It was a relief being able to see such an expanse of open country after months of confined valleys and dreadful gorges. To see the sun set without intervening mountains or the gold orb of the moon rise over the rim of the world."

This remote plateau possesses a unique flora and fauna, and is situated at the confluence of three distinct mountain ranges of the Himalaya, the Ladakh, the Zanskar and the main Himalaya crests. These three ranges merge together to form a complex (= knot) of ranges that channel species onto the Deosai from Ladakh, central Kashmir, and the high Himalaya. Unlike the more famous Pamir Knot to the northwest, however, this merging complex of mountains join together to form a high, vast alpine plateau that is characterized by abundant precipitation and deep soils. As a consequence, the Deosai Plateau is a center of unique biodiversity in northern Pakistan.

The Deosai Plateau is situated between two of the major mountain ranges of the world, the Karakorams and Himalayas. The rich biodiversity includes plants and animals from both regions, as well as from the Indus Valley. These regions have been subjected to extensive exploitation by humans during the past 50 - 100 years, with population growth, pastoral grazing, hunting, tourism and development having led to a great reduction of biodiversity. However, on the Deosai, most of these activities have had very little impact, and as a result, the flora and fauna of the Deosai are largely intact.

The plateau is located about 25 km south of Skardu in the Northern Areas of Pakistan in Baltistan. It is bounded on the south by the Deosai Mountains, on the west by the Astor Valley, and on the north by a series of high ridges south of the Indus River. A jeep road passes across the plateau from Ali Malik Mar Pass (4,084 m) to Chakor Pass (4,266 m) just above Sheosar Lake, and down to the village of Chilam (3,350 m). The Deosai Plateau occupies an area approximately 35 km from east to west and 20 km from north to south, and ranges in elevation from 3,400 m to 4,300 m, with occasional peaks over 5,000 m.

The Deosai Plateau is an area of high plains and river valleys sloping southeast. There are three main rivers draining the Deosai. The Astor River originates on the western slopes of Burzil Pass. It also drains Chilam Nullah and the Deosai Mountains surrounding the Parashing Valley north of Astor, and roughly forms the western boundary of the plateau.

The Shigar River is the major river of the interior Deosai Plateau. It is also known as the Bara Deosai. It originates in the Shigar mountains, and drains most of the area of the northern, northeastern and northwestern Deosai. In the upper reaches it is divided into many smaller branches. The Shatung River is the main branch, and drains the eastern uplands and flows westward to join the Shigar upstream of the wooden suspension bridge across the Shigar River which is a major landmark on the Deosai. Another major tributary of the Shigar

River is the Kala Pani River, which drains the northern and western uplands, including the Sheosar Lake area, and joins the main river below the wooden suspension bridge. These rivers drain a region of open rolling hills, spring-fed nullahs, and mountain peaks and lakes that form the upland regions of the Deosai. This overall region is often called the Bara Deosai, which means "big Deosai".

The Gultari River is on the southern and southeastern side of the Deosai. The river originates on the northern slopes of Burzil Pass, and flows eastward. Below the village of Gultari the river joins the main Shigar River. It flows on to the north and east, joining the Suru River from Zanskar, and finally joins the Indus upstream of Kargil. This lower area of the Deosai in the region of the Gultari River is called the Shota Deosai ("small Deosai").

The Shota Deosai, as defined by Khan and Zakaria (1995), is restricted to an area south and west of the main Deosai Plateau. It begins near the village of Chilam, and runs eastward as a narrow valley to Gultari. This area of valleys and lower plains ranges in elevation from 3,300 - 4,000 m, and is more populated and ecologically degraded than the upper reaches of the Deosai Plateau proper.

The area receives abundant snow fall during the winter, and rain during the brief summer season. The moisture percolates through the course soil, and emerges in springs along nullahs and in open grassy valleys. In areas where springs emerge, deep rich grasslands and numerous flowers abound. The high elevation and strong winds combine to prevent trees from growing on the higher areas of the plateau (Bara Deosai).

OBSERVATIONS AND RESULTS

Biogeographical Affinities

The Deosai Plateau is at the southern edge of the Palaearctic Realm. Two Biogeographical Provinces characterize the zone: 1). The Himalayan; which occupies the southern and western regions of the plateau in the area of Burzil Pass, the upper Astor Valley, the upper Parashing Valley and Deosai Mountains. 2). The Pamir-Karakorum Highlands; occupy the northern ridges of the plateau including Katich La, Burji La and Ali Malik Mar passes, and high alpine habitats of the Bara Deosai.

The small mammal fauna is mainly Western Palaearctic in affinity, although the Chinese birch mouse (*Sicista concolor*) and Tibetan red-toothed shrew (*Sorex thibetanus*) have affinities with the Eastern Palaearctic.

The fish fauna is completely High Asian (Banarescu, 1990), but it has affinities with Southeast Asia. These fishes evolved with the change in drainage system due to the uplift of the Himalaya (Menon, 1954).

The alpine and subalpine plants of the Himalaya belong to Central Asian region (Engler, 1879). This conclusion was followed by Kitamura (1960), Breckle (1974) and Dhar and Kachroo (1984). These alpine plants also have links with Sino-Tibetan flora (Katamura, 1960) and with the Mediterranean region (Meusal, 1971). It appears that the alpine flora of Deosai leans heavily towards Central Asia, Afghanistan, South China and Europe. However, some other adjacent areas such as Siberia, the Caucasus region and the eastern Himalaya also have close links with the flora of northern Pakistan.

Political frontiers are not generally regarded as absolute criteria for endemic species since the particular species may spread to the adjoining regions while remaining restricted to a particular geographical area (Hedge and Wendelbo, 1978). However here endemic species means, endemic to Himalaya. For example, among the 342 represented plant species occurring on the Deosai Plateau 91 (about one fourth) are endemic to Himalaya. About 73 plant species are also distributed in the adjoining countries whereas the rest have worldwide distribution.

FAUNA OF DEOSAI PLATEAU

Fishes

Triplophysa stoliczkae (Steindachner, 1866) High Altitude Loach

Distribution: This high altitude fish was originally described from Lake Tsho Mararai at 4,728 m elevation in western Tibet. Its distribution extends from Central Asia in the Oxus and Tarim river systems (Day, 1876), both of which occur at high elevations. In the Indus River system it has only been recorded from headwaters areas at elevations from 4,179 m to 4,905 m (Horas, 1936). Reports of this species as occurring at Skardu and in Hazara (Mirza and Khan, 1974; Mirza ,1975, 1980) are probably incorrect since most areas of Hazara are too low in elevation, and specimens of *Triplophysa* collected at Skardu are *T. gracilis*. So, the upper Gultari River is the only documentable site for the occurrence of this species in Pakistan. Nineteen specimens of this species were collected in the Gultari

River at an altitude of 3,680 m from a site located 10 km upstream from where Sufaid Nullah joins the Gultari.

Natural History: Even though this species was originally described from a lake in Tibet, it has never been collected again in a lake environment. In 1932 the Yale North India Expedition searched for the fish at the type locality at Lake Tso Mararai without success, and Hutchinson (in Hora, 1936) suggested that it is likely that the fish were originally collected in a stream flowing into the lake rather than the lake itself. As the species lacks an air bladder, it appears that it is specialized for life in torrential streams and rivers (Hora, 1936).

On the Deosai we observed this species in a small, clear side stream with a stony and gravely bottom. The fish were in schools of 40-50 individuals, and frequented shallow pools. When disturbed they used stones and rocks for shelter.

Fries and fingerlings were observed in a side channel of the Gultari River, suggesting that the species breeds in the area, and is a permanent resident of the Deosai Plateau. During the cold winter months when water freezes, these fish probably find refuge under rocks in deep water.

Even though the species lacks a well developed air bladder (Hora, 1936), it does not appear to be totally specialized for living in torrential streams. The body is rounded and flexible, which are specializations for moving through the stony river bottom, and its pectoral fins are not strong. Moreover, it has a sub-terminal mouth specialized for feeding on insect larvae and algae on the bottom. However, it does have an asymmetrical caudal fin, the lower lobe of which is better developed than the upper, enabling the fish to move forward in fast flowing water, and to rotate its anterior end upward when moving from stone to stone in shallow, fast flowing water.

This is an omnivorous fish and its specialized bottom feeding habits on insect larvae and algae enable it to survive throughout the year under very harsh conditions, even at times of low water levels. It does not appear to have major competition from other fish species, or to have any predators.

Body Size and Proportions: The six specimens collected by the Yale North India Expedition in 1932 ranged in length from 84.5 to 121 mm. The specimens we collected on the Deosai Plateau ranged from 105-190 mm in total length. However, we observed many individuals in schools in the Gultari River that were much small than the ones we collected. Our specimens differed from the distinguishing specific characters given by Hora (1936) in the following ways:

Hora, 1936	Deosai Series
1). Ventral fin extends considerably beyond anal opening.	1). Ventrals do not cross anal opening;
2). Least height of caudal peduncle 3-4 times length.	2). Least height 2-2.5 times length

Diptychus maculatus Stiendachner, 1866 Slate-coloured Snow Trout

Distribution: This is one of the common fish of the upper Indus River system, as well as of the Tarim and Yarkhand Rivers at higher elevations. It is also found in Nepal and western Tibet. In Pakistan the species has been collected in the upper Hushe River in Ghanche, and at several locations in upper Hunza, as well as in the Astor River in Diamer and on the Deosai Plateau.

Natural History: Although this species is relatively uncommon in most areas of its range, it is extremely common on the Deosai Plateau. Small fast flowing streams on the Deosai Plateau are full of this species, and many individuals are of large size. It is the dominant fish of the plateau, and inhabits all aquatic habitats from narrow spring fed water channels to swift streams, rivers and lakes. It is most common in pools along torrential streams.

This species breeds on the Deosai Plateau. Fries of this fish were observed in small side channels and still water during the first week of August, indicating that the species breeds in July on the Deosai. This is further confirmed by our observation that most females collected in August were spent, while males were still secreting milt. The Shatung, Bara Pani, Kala Pani and Gultari watersheds are natural breeding grounds for the species.

Fries and fingerlings were most commonly observed in shallow quiet areas of streams, while medium sized fish were most commonly observed in shallow fast flowing areas with many stones. Larger individuals were most common in the main rivers, the deeper pools in small streams and in lakes.

As noted above, this species on the Deosai Plateau reaches very large body size. The fish was originally described on the basis of four specimens from Leh ranging in length from 76 - 114 mm. The largest specimen collected by the Yale North India Expedition in 1932 was 270 mm, whereas on the Deosai Plateau many specimens up to 400 mm in length were collected. The specimens collected on the Deosai represent the largest examples known of this species.

This fish prefers fast flowing, cold, and clear mountain streams. The species has a high oxygen demand. Most areas within its range are too turbid to offer ideal conditions. However, on the Deosai Plateau extensive areas of clear, cold and fast flowing oxygen rich water abound presenting a perfect habitat for the fish.

It feeds on a variety of sources. Immature fish feed on nymphs and larvae of insects, while adults feed predominantly on slime, filamentous algae and other vegetable material (Mukerjei, 1936).

Taxonomic Status: We recognize *Diptychus* as a monotypic genus, with only *D. maculatus*. Mirza and Awan (1979) described a second species from the Northern Areas as *D. pakistanicus*. Jayaram (1981) treated this form as *D. maculatus*. We have collected this species from a number of locations in the Northern Areas, and find that it is not considerably different from the original description. Specimens are nearly uniform in size and colour throughout its distribution in the Northern Areas except for minor variations associated with sexual dimorphism and age.

Ptychobarbus conirostris Steindachner, 1866

Fleshy-mouthed Snow Trout

Distribution: This species is widely distributed in northern Pakistan, and occurs in the Yasin, Gilgit, Hunza, Shigar, Shyok and Indus Rivers. The presence of this species on the Deosai Plateau represents the highest known range of this species in Pakistan. On the Deosai it is more common on the Shota Deosai near Gultari than on the Bara Deosai in the Shigar River watershed.

Natural History : This large fish prefers deep and fast flowing rivers, and almost never inhabits shallow streams or spring fed brooks. It is one of the largest fish of the upper Indus, with one fish collected on the Deosai being 500 mm in length. It is omnivorous, feeding on slime, algae and insect larvae. We observed many individuals with black patches on their sides and back, as well as stunted and deformed barbles. We interpret these deformities as weather related, probably the result of tissue damage from cold water during the harsh winter of the Deosai.

Reptiles & Amphibians

The number of reptiles and amphibians from the Deosai plateau is very limited because of the high elevation. On the plateau itself only three herps are know, one frog and two lizard.

Frog

Scutiger occidentalis Kashmiri Mountain Toad

This frog is classified in the family Pelobatidae (subfamily Megophryinae) by some authorities, while others consider it to have characters associating it with the family Bufonidae. It is specialized for high altitude habitats. Its main distribution is Ladakh and Kashmir. The species is the most western representative of the subfamily Megophryinae, which is most diverse in China, Indochina and the eastern Himalayas. As a result of recent intensive surveys in Pakistan, it appears to occur east and south of the Indus River at very high elevations. We collected a large number of specimens from moist areas of the Deosai Plateau near Sheosar Lake (4,150 m), and from just over Babusar Pass in the first meadow on the Chilas side. At Deosai the specimens were collected near the edge of Sheosar Lake, and in tall grass around small pools of water. Near Babusar Pass it occurred in the same kinds of wet habitats under rocks at the edge of stream channels. Therefore, we predict that this frog may be widely distributed on the Deosai Plateau in wet, grassy habitats up to highest elevations at which such habitats are found (± 5,000 m). According to Auffenberg (pers. comm.) this frog can occur as high as 5,643 m in some areas where there are pockets of open ground in the permanent snowline.

Little is known about its natural history. A specimen from Babusar had a wasp in its mouth when collected. The tadpoles may take two years to develop into adults (Auffenberg, pers. comm.), and can be observed in deeper still pools of water in grassy areas where springs emerge. Tadpoles are found in both still and flowing water. Adult frogs are very slow moving, and are not usually observed moving about until after the sun has warmed the surface of the ground after about 10 in the morning.

It appears to be restricted to alpine habitats. Dubois (1978) found it in high mountain torrents between 2,680-3,900 m in India. In addition to finding refuge from cold in water (streams and pools) and under rocks, this frog was observed emerging from the burrows of *Hyperacrius* (burrowing voles) and from rock crevices. The burrows of *Hyperacrius* also are known to shelter *Sicista* (birch mice) and *Alticola* (rock voles) and numerous insects, and may play a major role in the survival and distribution of these frogs on the Deosai.

Lizards

A number of small skinks were collected in grassy areas near Sheosar Lake and the Kala Pani River. According to Auffenberg et al. (in press) the subspecies designations of all Pakistani populations of the genus *Skincella* are in need of further study, and so it is not possible to identify the skinks with certainty. There are two possible subspecies according to Auffenberg, and it is likely that one or both occur on the Deosai Plateau.

Family Scincidae

Skincella (Lygosoma) ladacensis himalayanus

Himalayan Ground Skink

This skink is common in upland areas of the Himalayas. It occurs in the Pamirs, eastern Afghanistan east to western Nepal. In northern Pakistan it is most common south of the main Himalaya ranges. We found it under stones and in patches of grasses on the Deosai. According to Auffenberg et al. (in press) it is common in Pakistan between 2,600-4,500 m in damp grassy meadows and open pine-covered hillsides. It was found in the burrows of *Hyperacrius* (burrowing voles) on the Deosai, and reported to hide in mouse burrows by Gruber (1981).

Skincella (Lygosoma) ladacensis ladacensis Glacier Skink

This form is likely a Tibetan species. It has been collected north of the main Himalaya, and is more common in Karakoram habitats. It is an alpine species that occurs only above timberline (3,200 - 6,000 m depending on exposure and geographic location). According to Auffenberg et al. (in press) it is sometimes collected near snow fields, and apparently lives at the highest elevation of any known reptile. The range of this species is listed by Das (1996) as India and Nepal, but Hafizur Rahman (personal communication) reports its presence on the Deosai.

Birds

The actual number of birds occurring on the Deosai Plateau is confusing because many migratory species pass through the area, and there is a significant amount of altitudinal migration. Khan and Zakaria (1995) report that over 100 bird species have been observed on the Deosai Plateau. Matthews (1941) studied birds in the area between 15 July and 20 August, 1941. From his overall list of 58 birds, he documented 23 as occurring on the Deosai Plateau proper. Khan et al. (1996) report that they have observed 108 species of birds over a three year period, 81 of which were confirmed breeding, 23 were passing through the area as migrants, and 4 were vagrants. In our list of the 41 birds that are likely to be observed in the overall area of the Deosai Plateau we have followed published records, and the range maps and discussion in Roberts (1992). Our list is not a complete record. The lack of reliable published data on the birds of the Deosai region is an indication of the need for a long-term ornithological program to determine the distribution and abundance of species. Without the data generated by such a long-term study it is not possible to discuss the conservation status, seasonal presence, or systematics of the birds of the Deosai, or the possible significance of the plateau as a refugium for montane species that are suffering from habitat destruction in other mountainous regions of Pakistan. Data on the distribution of other vertebrate groups suggest that the Deosai is an important biogeographical region, and part of an invasion route into the Western Himalaya. However, the information on birds of the region is too limited to allow such an analysis of the avifauna of the Deosai, and is reflected in the lack of any indication of such an invasion route in Roberts (Figure 5, 1992).

Additional field work and data on the birds of the Deosai Plateau, such as is being generated by Dr. Aleem Khan and his colleagues will answer many important questions.

Some of the more easily observed or ecologically significant species recorded on the Deosai Plateau are listed below. We have followed the nomenclature in Roberts (1992).

Order Accipitriformes
Family Accipitridae
Haliaeetus leucoryphus — Palla`s or Ring-tailed Fish Eagle
Gypaetus barbatus — Lammergeier or Bearded Vulture
Gyps himalayensis — Himalayan Griffon Vulture
Accipiter nisus melaschistos — Eurasian Sparrow Hawk
Aquila chrysaetos — Golden Eagle
Order Falconiformes
Family Falconidae
Falco tinnunculus — Eurasian Kestrel
Order Galliformes
Family Phasianidae
Tetraogallus himalayensis — Grey Himalayan Snowcock
Order Charadriiformes
Family Charadriidae
Charadrius mongolus — Lesser Sand Plover or Mongolian Plover

Family Scolopacidae
Actitis hypoleucos — Common Sandpiper
Tringa totanus — Common Redshank
Family Laridae
Larus brunnicephalus — Brown-headed or Tibetan Gull
Family Sternidae
Sterna hirundo tibetana — Tibetan Common Tern

Order Columbiformes	
Family Columbidae	
Columba rupestris	Turkestan Hill Pigeon
Columba leuconota	Snow Pigeon
Order Cuculiformes	
Family Cuculidae	
Cucules canorus	Eurasian Cuckoo
Order Caprimulgiformes	
Family Caprimulgidae	
**Caprimulgus europaeus unwini*	Hume's European Nightjar
Order Apodiformes	
Family Apodidae	
Apus apus pekinensis	Eastern Swift
Order Coraciiformes	
Family Upupidae	
Upupa epops	Hoopoe
Order Passeriformes	
Family Alaudidae	
**Alauda gulgula*	Lesser Skylark
Eremophila alpestris longirostris	Long-billed Horned lark
Family Hirundinidae	
Delichon dasypus · cashmeriensis	Kashmir House Martin
Family Motacillidae	
**Anthus roseatus*	Vinous-breasted Pipit
**Anthus trivialis*	Tree Pipit
**Motacilla citreola*	Yellow-headed Wagtail
Family Cinclidae	
Cinclus cinclus	Common Dipper
Family	Prunellidae
**Prunella rubeculoides*	Robin Accentor
Family Turdidae	
**Luscinia svecica*	Blue-throat
Phoenicurus ochruros phoenicuroides	Kashmir Redstart
Phoenicurus erythrogaster	Guldenstadt's Redstart
Family Sylviidae	
**Phylloscopus affinis*	Chinese Leaf Warbler
**Phylloscopus sindianus*	Mountain Chiffchaff
Family Tichodromadidae	
Tichodroma muraria	Wall Creeper
Family Corvidae	
Pyrrhocorax pyrrhocorax	Red-billed Chough
Pyrrhocorax graculus	Yellow-billed Chough
Corvus corax tibetanus	Tibet Raven

Family Passeridae
(Subfamily Carduelinae)

Carduelis cardeulis caniceps	Grey-headed Goldfinch
Carduelis flavirostris	Twite
Serinus pusillus	Red-fronted Serin
**Leucosticte nemoricola*	Plain Mountain Finch
Carpodacus rubicilla severtzovi	Spotted-crowned Rosefinch
(Subfamily Emberizinae)	
Emberiza leucocephalos	Pine Bunting

Mammals

Information on the mammals of the Deosai Plateau is very limited, with only Roberts (1977) available as a source of reliable information. However, since few investigators have collected or studied mammals on the Deosai, even Roberts has little information on the mammals of the plateau. We have been able to document 18 mammals living on the Deosai Plateau, ranging in size from the tiny Tibet red-toothed shrew (*Sorex thibetanus*) weighing less than 7 grams to the Himalayan brown bear (*Ursus arctos)* weighing approximately 150 kilograms. Both species are relatively rare, however, and it is uncommon to encounter either. On the other hand, the long-tailed marmot (*Marmota caudata)* is extremely common on the Deosai, and at times a dozen or more individuals can be observed across the rolling terrain. The marmot is the mammalian symbol of the area, sitting at the entrance to its extensive burrow system, vocalizing loudly. Few places in Pakistan have such large concentrations of marmots. However, in terms of total biomass, the most dominant mammalian species on the Deosai Plateau is the small burrowing vole (*Hyperacrius fertilis*). Even though most individuals of this species weigh less than 20 grams, they live in huge colonies that cover large areas where the soils are deep and not too compacted. It is possible to find large areas of excavated soil and small burrow openings throughout the area of both the Bara and Shota Deosai, and when the total number of individuals is multiplied by the average mass of 20 grams, the total biomass of this species on the Deosai is extraordinary.

The activities of marmots and burrowing mice are important to the ecology of the Deosai in many ways. They excavate the soil, keeping it porous and loamy. They disperse plant materials. They serve as the major source of food for carnivores such as brown bears, foxes, predatory birds, and even the snow leopard. In addition, their deep burrows serve as shelters and nesting areas for a number of insects, reptiles, amphibians and even other mammals.

A complete list of the mammals of the Deosai Plateau follows:

Order Insectivora	
Family Soricidae	
Sorex thibetanus	Tibetan Red-toothed Shrew
Crocidura pergrisea	Baltistan Shrew
Crocidura suaveolens	Kashmir White-toothed Shrew
Order Chiroptera	
Family Vespertilionidae	
Pipistrellus sp	Pipistrelle Bat
Order Carnivora	
Family Canidae	
Canis lupus pallipes	Tibetan Wolf
Vulpes vulpes montana	Tibetan Red Fox
Family Ursidae	
Ursus arctos	Himalayan Brown Bear
Family Mustelidae	
Mustela erminea	Short-tailed Weasel
Family Felidae	
Uncia uncia	Snow Leopard
Order Rodentia	
Family Sciuridae	
Marmota caudata	Long-tailed Marmot
Eoglaucomys fimbriatus	Kashmir Flying Squirrel
Family Muridae	
Apodemus wardi	Wood Mouse
Rattus turkestanicus	Turkestan Rat
Hyperacrius fertilis	Burrowing Vole
Alticola argentatus	Alpine Vole
Family Zapodidae	
Sicista concolor	Chinese Birch Mouse
Order Lagomorpha	
Family Ochotonidae	
Ochotona roylei .	Pika
Order Artiodactyla	
Family Bovidae	
Capra ibex sibirica	Himalayan Ibex

Small Mammals

A list of the small mammals known to occur on the Deosai is given in Table I, along with their numbers, localities and respective elevation of capture site.

The distribution of the small mammals more or less coincides with the boundaries of the Deosai Plateau. The alpine vole (*Alticola argentatus*) is found only among rocky outcrops and talus slopes. It descends as low as 3,333 meters where habitats are suitable, but it is generally restricted to higher elevations. The burrowing vole (*Hyperacrius fertilis*) is also clearly associated with the Deosai Plateau. This species requires deep, rich soils with grass. It is most frequently found where the soil is not too compacted, and near moisture. These habitats are widespread on the Deosai from 3,333 m to well above 4,200 meters. Several species are sympatric with burrowing voles, and utilize their extensive burrow systems for shelter. For example, the Kashmir mountain toad (*Scutiger occidentalis*), Chinese birch mouse (*Sicista concolor*) and Tibetan red-toothed shrew (*Sorex thibetanus*) all frequently utilize *Hyperacrius* burrows on the Deosai.

The absence of the wood mouse (*Apodemus wardi*), white-toothed shrews (*Crocidura suaveolens* and *C. pergrisea*) and the Kashmir flying squirrel (*Eoglaucomys fimbriatus*) mark the actual boundary of the upper and lower regions of the plateau. The above species are predominantly forest dwelling, and all occur in forested areas of the Shota Deosai (Gutumsar in the Parashing Valley and Chilam). Turkestan rats (*Rattus turkestanicus*) occur in the transitional zone between the Shota and Bara Deosai, but are not found on the true reaches of the upper Deosai.

The reason for the absence of *Apodemus wardi* on the upper treeless expanses of the Bara Deosai is probably more than just because of the lack of forest cover, but also may be ecological or

physiological. Kalabuchov (1937) showed that *Apodemus sylvaticus* (a species closely related to *A. wardi*) is limited in its altitudinal range in the Caucasus Mountains by its ability to increase the number and volume of red blood cells in response to high elevations (lower barometric pressure). It is possible that the species of *Apodemus* found at Chilam and the Astor Valley is not capable of producing enough additional RBC to survive on the Bara Deosai. There could also be a dynamic equilibrium between *Apodeumus* and *Sicista*. Our surveys indicate that the two forms rarely occur together. *Sicista* are abundant on the Bara Deosai, and may be excluding *Apodemus*.

The highest we have found *Apodemus* living in Pakistan is at Dalsangpa (4,050 m) in the upper Hushe Valley on the slopes of Masherbrum mountain. This is very high for this species. We did not collect *Sicista* at Dalsangpa, and the distribution of many species in this high, arid region of northwestern Pakistan (Baltistan) are 600 to 700 meters higher than farther south (Astor, Nanga Parbat, Kohistan). These observations seem to refute the hypothesis that *Apodemus* is limited in altitudinal distribution only by a physiological

Table I : Relative Numbers of Small Mammals on Deosai Plateau

	Bara Deosai				Shota Deosai				
	Shatung La (4,050m)	Shatung Jct. (3,880m)	Kala Pani (3,970m)	Sheosar Lake (4,150m)	Burzil Pass (4,200m)	Sufaid Nullah (3,650m)	. Parashing (3,333m)	Chilam (3,350m)	Total #s
Sorex thibetanus	0	2	2	2	0	2	0	1	9
Crocidura pergrisea	0	0	0	0	0	0	0	1	1
C. suaveolens	0	0	0	0	0	0	1	0	1
Mustela erminea	1	1	0	0	0	0	0	0	2
Ochotona roylei	1	0	0	0	0	1	2	0	4
Hyperacrius fertilis	21	15	28	16	9	4	9	8	110
Alticola argentatus	26	0	7	1	0	2	15	0	51
Sicista concolor	0	0	5	4	0	1	0	0	10
Apodemus wardi	0	0	0	0	0	0	18	4	22
Rattus turkestanicus	0	0	0	0	0	0	0	2	2
Marmota caudata	8	8	10	10	0	0	0	0	36
E. fimbriatus	0	0	0	0	0	0	1	0	1
Pipistrellus sp	0	0	0	1	0	0	0	0	1
Total #s	57	26	52	34	9	10	46	16	250

ability to produce RBCs at higher elevations. However, the *Apodemus* found at Dalsangpa are pale in coloration, and are possibly a distinct species. So the answer to the conundrum as to why *Apodemus* does not occur on the upper (=Bara) Deosai remains unresolved.

The open grassy habitats of the Deosai would appear to be excellent habitat for gray hamsters (*Cricetulus migratorius*), but none were captured. Kalabuchov (1937) has demonstrated that this species also has a very poor ability to increase red blood cell numbers in reaction to low barometric pressures, and that it is not found in high mountain areas in the Caucasus mountains. The factors limiting the distribution and abundance of small mammals in the area of the Deosai Plateau are complex, and we recommend additional studies on the small mammals of the area, such as transect surveys along altitudinal gradients between Chilam and Sheosar Lake, and Satpara Lake and Ali Malik Mar Pass.

Species in addition to the Wood Mouse and Gray Hamster that we would expect to occur on the Deosai, but which we did not observe or collect include:

Stone marten	*Martes foina*
Himalayan lynx	*Lynx lynx isabellina*
Pallas's cat	*Otocolobus manul*

The Indian porcupine (*Hystrix indica*) is found in the Skardu area but does not occur on the Deosai Plateau.

Large Mammals

There are few large mammals on the Deosai, and most species, such as the wolf and snow leopard, are uncommon. There are reports of wolves from some areas of the Deosai, but we never saw any signs. There are no recent reports of snow leopards. Recent surveys of brown bears indicate that 21 individuals survive, and that reproduction is still taking place, with 2 or 3 young a year being recruited into the population (Khan and Zakaria, 1995). A program to study and promote the conservation of brown bears has been established (Khan and Zakaria, 1995), and Pakistani and international bear biologists are becoming actively involved (Dave Ferguson, personal communication). According to Khan and Zakaria (1995) 14 Himalayan ibex were sighted in the core area of the Deosai in 1994, and scat was observed near "Wolf Peak" in 1993.. The most frequently observed larger mammal on the Deosai is the Tibetan red fox, which has been photographed as well as seen. A list of the larger mammals that are most likely to occur on the Deosai include:

Ursus arctos	Himalayan Brown Bear
Canis lupus pallipes	Tibetan Wolf
Uncia uncia	Snow Leopard
Vulpes vulpes montana	Tibetan Red fox
Capra ibex sibirica	Himalayan Ibex

FLORA OF THE DEOSAI PLATEAU

The flora of Deosai can be divided in to three categories: weeds, desert type native plants and high alpine plants. Weeds are mostly found near the cultivated fields. Desert type native plants grow on cliffs, sandy soils below springs and along streams. The climax vegetation on the Deosai Plateau is represented mostly by herbs and small shrubs. The dominant genera being *Rheum, Polygonum, Oxyria, Pedicularis, Stellaria, Cerastium, Arenaria, Aquilegia, Delphinium, Saxifraga, Sibbaldia, Potentilla, Draba, Arabis, Oxytropis, Geranium, Primula, Nepeta, Carex, Astragalus, Gentianodes, Artemisia, Aconitum, Sedum, Lonicera, Lectuca Taraxacum* and *Saussurea.*

Most high alpine plants are found near the melting snow and glaciers along the moraines, and are herbs and small shrubs. The only trees are birches and junipers, and they occur in valleys that form the lower limits of the Deosai region, such as the Satpara, Chilam and Parashing valleys. These trees are being rapidly cut down for firewood and building materials, and are becoming very rare.

Most of the alpine plants are perennial, having very brief growing periods. The plants are generally dwarf and caespitose and these conditions are caused by severe wind and frost. Nevertheless they survive in harsh conditions due to their very thick rootstock. Most of the species are densely hairy, an adaptation probably for protection from high intensity ultraviolet rays. A variety of habitats are met within the plateau like open sunny sites, rock slopes, steppes and marshy places. The species differ in their ecological preferences even when occurring within the same general area.

Goats and yaks are the main grazing animals of the Deosai Plateau. Grazing pressure, becoming more intense each year, and the habitat is being modified as a result in many regions of the Deosai. Non-palatable plant species have tended to spread and flourish at the expense of the palatable ones. In addition, the cutting of shrubs by shepherds, and the digging of valuable medicinal herbs are increasingly altering the composition and distribution of plants on the plateau and its surrounding valleys.

This checklist of Deosai plants (Table II) has been prepared based on previous reports and publications (Stewart 1972, 1982, Shah 1993,

Shinwari and Chaudhri 1994, Nasir and Rafiq 1995, Khan & Rajput 1995) plus comprehensive field work in 1987, 1992 and 1993 in the Deosai Plateau area. In the present checklist a total of 342 species belonging to 36 families and 142 genera are listed. The families are arranged in the evolutionary sequence whereas the genera and species in each family are arranged alphabetically.

Table II: Checklist of plants of Deosai

Gymonosperms
Family Cupressaceae
Juniperus communis L. var. *saxatilis* Pallas**
Ephedra gerardiana Wall. ex Stapf.**

Angiosperms
(Monocotyledons)

Family: Juncaceae
Juncus articulatus L.
J. membranaceus Royle ex Don
J. Sphacelatus Decne
Luzula spicata (Linn.) D.C.

Family: Alliaceae
Allium carolinianum DC.**
A. fedtschenkoanum Regel**
A. tuberosum Rottl. ex Spreng.

Family: Orchidaceae
Dactylorhiza hatagirea (D.Don.) Soo
Hermenium monorchis (L.) R.Br.

Family: Cyperaceae
Carex borii Nelmes.*
C. cruenta Nees.
C. curta Gooden.*
C. diluta M.Bieb.
C. haematostoma Nees.
C. infuscata Nees.
C. karoi Freyn.
C. microglochin Wahl.
C. moorcroftii Falc.**
C. nivalis Boot.
C. obscura Nees.
C. obscuriceps Kuk.
C. oligocarpa Clarke*
C. orbicularis Boot.
C. parva Nees
C. pseudobicolor Boeck.*
C. pseudofoetida Kuk.
C. setosa Boot.
C. songorica Kar. & Kir.**
C. tristis M. Bieb.**
C. vulpinaris Nees*
Kobresia capillifolia (Dcne) Clarke.**
K. nitens Clarke*
K. pamiroalaica Ivanova*

Family: Poaceae/Graminae
Agrostis vinealis Schreh.
Alopecurus arundinaceus Poir
Briza media Linn.
Calamagrostis arundinacea (Linn). Roth
C. decora Hook.f.*
C. epigejos (Linn) Roth.
Dactylis glomerata Linn.
Danthonia cachemeryiana Jaub. & Spach.*
Deschampsia caespitosa (Linn.) P. Beauv.
D. koelerioides Regel.**
Digitaria sanguinalis (Linn.) Scop.
Elymus nutans Griseb.**
E. dentatus (Hook.f.) T.A. Cope.*
E. longi-aristatus (Boiss.) Tzvelev.**
Eremopoa altaica (Trin.) Rozhev.
Festuca alaica Drobov.**
F. hartmannii (Markgr - Dannenb.) Alexecv.
F. kashmiriana Stapf.**
F. nitidula Stapf.**
F. olgae (Regel.) Krivot.**
Helictotrichon pratense (Linn.) Pilger.
Hyalopoa nutans (Stapf.) Alex*
Leymus secalinus (Georgi) Tzv elev.**
Milium effusum L.
Panicum milliaceum L.
Pennisetum lanatum Kl.
Paracolpodium altaicum (Trin.)Tzv.**
Piptatherium gracile Mez.**
Poa alpina L.
P. attenuata Trin.**
P. glauca Vahl.**
P. nemoralis L.
P. stepfiana L.*
P. versicolor M. Bieb.
Saccharum filifolium Nees ex steud.**
Stipa capillata Linn.
Trisetum aeneum (Hook. f.) R.R. Stewart.**
T. clarkei (Hook. f.) R.R. Stewart.**

Dicotyledons

Family: Salicaceae
Salix denticulata N.J. Anderss.
S. flabellaris N. J. Anderss.**

Family: Betulaceae
Betula utilis D.Don.

Family: Polygonaceae
Oxyria digyna (L.) Hill.
Polygonum affine D.Don.**
P. alpinum All.
P. amphibium L.
P. convolvulus L.
P. decatulum Meissn.
P. filicaule Wall. ex Meissn.
P. glaciale (Meissn.) Hook.f.
P. lapathifolium L.
P. nepalensis Meissn.
P. paranycoides C.A.Mey.**
P. persicaria L.
P. rumicifolium Royle ex Bab.

P. viviparum L.
Rheum webbianum Royle.
R. spiciforme Royle.

Family: Caryophyllaceae
Arenaria griffithii Boiss.**
Cerastium cerastioides (L.) Britton.
Minuartia biflora (L.) Schinz & Thell.
M. kashmirica (Edgew.) Mattf*
Silene gonosperma (Rupr.) Bacquet.
S. tenuis Willd.
Stellaria decumbens Edgen.
S. perbica Boiss.
S. subumbellata Edgew.*

Family: Ranunculaceae
Aconitum chasmanthum Stapf. ex Holmes*
A. fragrans Benth*
A. heterophyllum Wall. ex Royle.
A. laeve Royle.
A. violaceum Jacq.
Actaea spicata L.
Aquilegia moorcroftiana Wall. ex Royle.
A. nivalis Falce*
Anemone rupicola Camb.*
Caltha alba Camb.
Delphinium cashmerianum Royle*
Paraaquilegia anemonioides (Willd.) Ulbr.**
Pulsatilla wallichiana (Royle) Ulbr.*
Ranunculus hirtellus Royle.
R. karakoramicola Tamura*
R. stewartii H. Riedl.*
Thalictrum alpinum L.
T. foetidum L.

Family: Papaveraceae
Papaver nudicaule L.

Family: Fumariaceae
Corydalis clarkei Prain.
C. erithmifolia Royle.
C. gortschakovii Sch.
C. longipes D.C.
C. thyrsiflora Prain.
C. tibetica Hook.f.

Family: Crassulaceae
Rhodiola heterodenta (Hook.f. & Thoms.) Boiss.
R. himalensis (D.Don.) S.H.Fu.
Sedum crassipes Wall. ex H. & T.
S. ewersii Ledeb.
S. oreades (Dcne) Hamet*
S. quadrifidum Pall.

Family: Grossulariaceae
Ribes himalense Decne
R. nigrum Linn.

Family: Parnassiaceae
Parnesia palustris Linn.

Family: Saxifragaceae
Bergenia stracheyi (Hook.f. & Thoms.) Engl.**
Saxifraga androsacea L.
S. flagellaris Willd. ex Sternb.
S. hirculus L. var. *indica* Clarke
S. jacquemontiana Decne*

Family: Rosaceae
Cotoneaster humilis Dunn.
C. integerrima Medick
Geum elatum G. Don.
Potentilla agrimonioides M. Bieb.**
P. cathaclines Lehm.**
P. desertorum Bunge.**
P. doubjouneana Camb.**
P. drydanthoides (Juz.) Vir.**
P. evestita Wolf.**
P. gelida C.A. Mey.**
P. grisea Juz.**
P. multifida L.
P. pamirica Juz.**
P. salesoviana Steph.**
P. stewartiana Shah & Wilcock*
Rosa webbiana Wall. ex Royle.
Rubus antennifer Hk.f.*
R. irritans Focke
R. saxatilis L.
Sibbaldia cuneata Kunze
S. tetandra Bunge
Spiraea affinis Parker

Family: Papilionaceae
Astragalus frigidus (L.) A. Gray.
A. himalayenus Kl.*
A. maddenianus Bth. ex Baker.**
A. melanostachys Bth. ex Bunge.**
A. oxydon Baker*
A. rhizanthus Royle ex Benth.**
A. tribulifolius Bth. ex Bunge*
Caragana conferta Bth. ex Baker*
C. versicolor Bth.
Indigofera heterantha Wall. ex Brand.**
Medicago sativa L.
Ononis antiquerum L.
Oxytropis cachemeriensis Camb.*
O. glabra DC.
O. lapponica Willd.
O. mollis Royle ex Benth.*
O. tartarica Camb.**

Family: Geraniaceae
Geranium himalayense Kl.*
G. pratense L. ssp. *stewartianum* Y. Nasir*

Family: Euphorbiaceae
Euphorbia cornigera Boiss.
E. kanaorica Boiss.
E. thomsonianum Boiss.
E. tibetica Boiss.*

Family: Umbelliferae
Angelica archangelica Linn.*
Anthriscus nemorosa (M.Bieb.) Spreng.
Bupleurum longicaule Wall. ex DC.
B. himaleyense (Kl) C.B. Clarke
B. thomsonii C.B. Clarke*
Carum carvi L.
Chaerophyllum acuminatum Lindl.
Heracleum candicans Wall. ex DC.
H. pinnatum C.B. Clarke.
Platytaenia lasiocarpa (Boiss.) Rech. f.
Pleurospermum stylosum C.B. Clarke*
P. hookeri C.B. Clarke*
Schultzia dissecta (Clarke) Norman*
Selinum vaginatum (Edgew) C.B. Clarke*
Trachydium roylei Lindl.*

Family: Ericaceae
Gaultherea trichophylla Royle.
Rhododendron penanthum Balf. f.

Family: Primulaceae
Androsace aizoon Duby*
A. baltistanica Y. Nasir*
A. macronifolia Watt.
A. sempervivoides Jacq.
A. septentrionalis L.
Primula duthieana Balf.f. *
P. elliptica Royle*
P. inayatii Duthie*
P. macrophylla D. Don.**
P. rosea Royle**
P. schlagintweitiana Pax**

Family: Plumbaginaceae
Acantholimon lycopodioides (Girard.) Boiss.
Limonum macrorhabdon (Boiss.) O. Kuntz.

Family: Gentianaceae
Comastoma borealis Bunge
C. pseudopulmonarium Omer*
Gentianodes alii Omer & Qaiser*
G. eumarginata Omer.**
G. intermedia (Clarke) Omer & Ali & Qaiser*
G. pedicellata (D.Don.) Omer, Ali & Qaiser
G. tianschanica (Rupr. ex Kusn.) Omer, Ali & Qaiser
Gentianopsis paludosa (Munro ex Hook.f.)**
Jaeschkea oligosperma (Griseb.) Knob.**
Swertia petiolata D.Don.*
Qaisera carinata (D.Don.) Omer.**

Family: Polemoniaceae
Polemonium caeruleum L.

Family: Boraginaceae
Arnebia enchroma (Royle ex Benth.) I.M. Johnston
A. benthamii (Wall. ex G.Don.) I.M. Johnston*
Lappula heterantha (Ledeb.) Gurke
L. nanum (L.) Schrad.
Lindelofia anchusoides (Lindl.) Lehm.

Family: Labiatae
Clinopodium umbrosum (M.Bieb.) C. Koch.
Dracocephalum stamineum Kar. & Kir.**
D. nutans L.
Mentha longifolia (L.) L.
Nepeta adenophyta Hedge*
N. astorensis Shinwari & Chaudhri
N. connata Royle ex Benth*
N. discolor Royle ex Benth.
N. floccosa Benth.**
N. gilgitica Shinwari & Chaudhri*
N. glutinosa Benth.
N. govaniana (Wall. ex Benth) Benth.*
N. kokanica Regel.
N. linearis Royle ex Benth.**
Perovoskia abrotanoides Karel.
Salvia hians Royle ex Benth.*
Thymus linearis Benth.

Family: Solanaceae
Hyoscymus nigar L.

Family: Scrophulariaceae
Euphrasia aristulata Penn.*
E. paucifolia Wett.*
E. schlagintweitii Wettst.
Lagotis cashmeriana (Royle) Rupr.*
L. globosa (Kurz.) Hook.f.*
L. kunawurensis (Royle) Rupr.*
Pedicularis albida Penn.
P. bicornuta Kl.**
P. brevifolia D.Don.
P. brevirostris Penn.
P. kashmiriana Penn.**
P. oederi Vahl.**
P. pectinata Wall. ex Bth.**
P. punctata Dcne.**
P. pyramidata Royle.**
P. rhinanthoides Schrenk ex Fisch. & Mey.
P. roylei Maxim
P. tenuirostris Bth
Picrorhiza kurrooa Royle ex Bth.*
Scrophularia koelzii Penn.**
Verbascum thapsus L.
Veronica lanosa Royle ex Bth.
V. lasiocarpa Penn.*

Family: Orobanchaceae
Orobanche amonena C.A.Mey.
O. kotschya Rent.

Family: Caprifoliaceae
Lonicera asperifolia (Decne) Hook.f.
L. heterophylla Decne*
L. microphylla Willd. ex Roem.
L. semenovii Regel.
L. vaccinioides Rehder*

Family: Valerianaceae
Valeriana himalayana Grub.*

Family: Campanulaceae
Campanula aristata Wall.*
Codonopsis clematidea (Sch.) C.B.Clarke

Family: Compositae
Achillea millefolium L.
Anaphalis nepalensis (Spreng.) Hand. Mazz.
A. virgata T.T. ex Clarke
Arctium lappa L.**
Artemisia absinthium L.**
A. brevifolia Wall. ex DC**
A. dracunculus L.
A. maritima L.
A. roxburghiana Wall. ex Besser.**
A. stricta Edgew.**
Aster falconeri (Clarke) Hutch*
Brachyactis roylei (DC) Wend.**
Carduus edelbergii Rech.f.*
Chrysanthemum pyrethroides (Kar. & Kir.) B. Fed.**
C. wallichii DC.
Cremanthodium arnicoides (Wall. ex DC) R. Good*
C. decaisnei Clarke*
C. plantagineum Max.**
C. flexuosa (DC.) Bth.*
C. kashmirica Bob.*
Erigeron multicaule. Wall. ex DC
E. roylei DC*
Gnaphalium luteo-album L.
G. stewartii Clarke**
Heteropappus altaicus (Willd.) Nov.
H. vulgatum Fries
Inula grandiflora Willd.**
Jurinea ceretocarpa (Dcne) Bth.**
J. himalaica R.R.S.*
Lectuca decipiens (H & T) Clarke.**
L. lessertiana (Wall. ex DC) Clarke*
L. rapunculoides (DC) Clarke**
L. sikkimensis (Hk.f.) Stebb.*
Psychrogeton andryaloides (DC.) Nov.
Saussurea albesecens (DC) Sch.
S. atkinsonii Clarke*
S. bracteata Dcne*
S. falconeri Hook.f.*
S. gnaphalodes (Royle) Sch.
S. jacea (Klot.) Clarke
S. obvallata (DC) Sch.*
S. roylei (DC) Sch.*
S. virgata DC.
Senecio jacquemontianus (Dcne) Bth.
S. kraschenninikovii Schis.*
S. tibeticus Hk.f.**
Solidago virgaurea L.
Soroseris deasyi (S. Moore) Stebb.
Tanacetum artemisoides Sch.*
T. falconeri Hook.f.*
T. longifolium Wall. ex DC.*
T. tomentosum DC.*
Taraxacum baltistanicum Soest.*
T. dissectum Ledeb.
T. eriopodum (D.Don.) DC.
T. kashmirense Soest.*
T. nasiri V.S..*
T. stenolepium Hand.Mazz.*
Tussilago farfara L.
Waldheimia glabrata (Dcne) Regel.**

* Endemic to Himalaya
** Also occur in adjoining countries.

DISCUSSION AND RECOMMENDATIONS

Environmental conditions on the Deosai Plateau are very severe. It is high in elevation, windy, snow covered for most of the year, and has a very short period of intense biological activity. Since it is above the tree-line, and ground cover is sparse in many areas, it has the appearance of a high, cold desert in many regions. However, it is rich in biodiversity, and serves as a refugium for many species of plants and animals. It is noted for its rich diversity of medicinal plants, and the air is sweet with the fragrance of flowing plants during July. It has the highest abundance of marmots in Pakistan, and the last remaining viable breeding population of Himalayan brown bear. Among small mammals it has the highest concentration of Chinese birch mice (*Sicista concolor*) in Pakistan, as well as one of only two known sites of Abbott's high mountain shrew, *Crocidura pergrisea.* Burrowing voles (*Hyperacrius*) reach enormous numbers. Among reptiles and amphibians, two species are common on the Deosai in spite of the severe cold and long winter. *Scutiger occidentalis,* the Kashmiri mountain toad, lives at extraordinarily high elevations for an amphibian, and *Skincella (Lygosoma) ladacensis ladacensis,* the glacier skink, lives at the highest elevation of any known reptile. The high altitude hill stream loach (*Triplophysa stoliczkae*) in Pakistan is found only on the Deosai Plateau and the snow trout (*Diptychus maculatus*) reaches its maximum size and abundance on the plateau.

In addition to the large number of permanent residents, the Deosai Plateau also is a refuge for large numbers of migratory birds. The explosion of insect species during the summer months, as well as the rich flora support a large number of locally migratory birds from surrounding areas which nest on the plateau during the brief summer season. Therefore, in spite of the rigorous climate, the Deosai Plateau is rich in biodiversity, and is a high altitude refugium of great importance.

The great diversity of plants and animals of the Deosai can be attributed to several factors. Some species originate in Tibet or Central Asia and have special adaptations for living at high elevations that enable them to survive the extreme environmental conditions found on the Deosai Plateau. They are not able to survive in the dryer valleys at lower elevations that surround the Deosai, nor on the austere mountain peaks that form the rest of the upland areas of the Himalayas and Karakorams. However, they are specialized for the extreme conditions on the Deosai. brown bears, marmots and Chinese birch mice, for example, hibernate. High altitude fishes secrete antifreeze hormones into their blood that prevents their tissues from freezing.

Another factor that allows so many vertebrates to survive the extreme conditions of the Deosai is the presence of "keystone species" such as the burrowing vole (*Hyperacrius fertilis*). This species constructs extensive underground burrows that protect them from the harsh environment above ground. These burrows also provide shelter for other species that would otherwise be vulnerable to the stressful conditions. We have observed birch mice, alpine voles, shrews, insects and even toads in the shelter of *Hyperacrius* burrows on the Deosai Plateau

The Deosai Plateau is also a relatively simple ecosystem, and the numbers of predators and competing species is reduced. This reduced level of competition and predation may increase the chances of survival for the species occurring on the Deosai. The high elevation and extreme cold, wind and snowfall combine to make the Deosai unsuitable for most human activities. As a result, for most of the year, there is little or no hunting, habitat destruction or overgrazing. As a consequence, species that are endangered in other more populated areas of Pakistan abound on the Deosai, such as the Himalayan brown bear.

The unique biological and geological features of the Deosai Plateau, therefore, represent something very special. No other place in Pakistan has such an extensive undisturbed high mountain ecosystem. Few other regions of the country are as pristine and important to the overall biodiversity of Pakistan.The Deosai Plateau is an island of rich biodiversity surrounded by the perennial snows of the mountain peaks and the aridity of the surrounding valleys and gorges. It is a "sky island". As with other islands, it has been protected by its isolation and inaccessibility. However, as with islands around the world, it is also very vulnerable to rapid degradation, and the potential for local extinctions is very high (Quammen, 1996). Now is the time for action and special attention.

Recommendations for Conservation

Alpine fauna and flora of Deosai Plateau deserve special attention because of its unique and relictual characteristics and diverse floristic composition. The area is currently being used as summer pasture for yaks, goats and sheep during the brief summer months, and the jeep track between Chilam and Skardu that crosses the center of the plateau is becoming a major route. In 1995 parts of the plateau were set aside as a conservation zone for Himalayan brown bear by the Northern Areas, and a management plan has been written for Deosai National Park (Wilderness Park) by Aleem Khan and Vaqar Zakaria (1995). Their recommendations for management are mainly restricted to activities that promote the status of the brown bears, although the overall goals of the management plan are far ranging and go a long way towards protecting the area. In addition to the recommendations presented by Khan and Zakaria, we recommend the following activities as ways of promoting the biodiversity of the Deosai Plateau and its surrounding valleys and mountain ranges.

1). Survey of Biodiversity- An in depth survey of the flora and fauna of the Deosai should be organized, and coordinated by a single organization (such as the Pakistan Museum of Natural History or the Zoological Survey Department) so that there is as much uniformity of methodology as possible. These surveys should seek to identify ecosystems and species that are important for conservation and sustainable use of the area. The collections should be deposited at the PMNH in Islamabad so that voucher specimens are available for documentation and future reference. These surveys should be done over a five year period in order to document any cycles, and to point out if some species are declining in numbers.

2). Geological Survey- A geological survey of the area should be planned to document the geological relationships of the plateau. It is unclear how the Ladakh, Zanskar and Himalaya ranges combine to form the elements of the Deosai Plateau. It is also unclear how and when the soils and main topography of gentle slopes and deep soils of the plateau developed. A geological survey and analysis would reveal the nature of the Deosai Plateau, and be useful in developing a plan for the developmental activities that can be undertaken without leading to increased soil erosion, stream degradation or habitat loss.

3). Land Use Survey- A land use survey should be undertaken to document the current use of the plateau, and its potential for grazing, ecotourism and transportation. The plateau does not appear to be over-exploited at the present time, but this could change rapidly. If areas of over-exploitation are revealed, the survey should include a habitat rehabilitation plan for the vulnerable and degraded area.

4). Advisory Committee- An honorary advisory committee including qualified Pakistani and international experts should be established. This advisory committee should be charged with providing recommendations for long term planning and management. The group could develop lists of all actual and potential threats to the biodiversity of the Deosai Plateau, and provide suggestions to combat these threats as they might develop.

5). Biosphere Reserve- The entire Deosai Plateau should be declared a national park or biosphere reserve.

6). Eco-tourism- A non-evasive eco-tourism plan should be developed. This plan should take into account the five year biodiversity surveys, and the recommendations of the advisory committee. It is important that no eco-tourism plans be initiated before the biodiversity studies are completed, and have been carefully evaluated by the advisory committee.

7). Brown Bear Programme- A practical programme should be implemented to enhance the population of brown bears. The goal of this programme should be the creation of a species recovery plan for brown bears on the Deosai Plateau and surrounding areas of suitable habitat where bears might currently or potentially occur. A programme in environmental education featuring the brown bear should be designed. This program can be presented to villagers and city dwellers in the surrounding area, as well as in local schools. By educating people about the importance of saving the brown bear, other species that depend on the same habitat can be protected. Brown bears are an excellent species to focus on because they are large, attractive, and true symbols of the wild that people in the area can easily relate to.

9). Trout Fishing Programme- The entire Deosai Plateau, from the Astor River to Sheosar Lake are excellent waters for trout. Sheosar Lake and all of the major rivers of the Deosai should be stocked with trout. A major trout fishing programme could be developed on the Deosai with very little overall cost, and with little environmental impact. Limited fishing licenses could be sold as part of the management program, and the proceeds from these fees could be used to further enhance the fishing programme. Brown trout and rainbow trout are strictly carnivores, and will not be in direct competition with the local fish fauna. The local snow trout of the Deosai, which reach enormous size, and are also very good to eat, also could be developed as a game fish on the Deosai.

10). Wild Yak Reserve- The high altitude areas of the Karakorams and Ladakh Ranges of Pakistan were once important in the range of the wild yak. The species is now almost extinct in the wild, having been replaced by the domestic yak, and yak-cow hybrids. A thorough genetic survey should be conduced to locate any remaining populations of wild yak. The Deosai Plateau is an ideal place to undertake an active breeding program to increase numbers of wild yak. Properly managed and funded, such a program could become a world-class component of the Deosai Biosphere Reserve and bring important notoriety to the programme.

11). Medicinal Plants- The Deosai Plateau is a vast rolling plain full of wildflowers. Many of these plants are medicinally important. With a team of experts documenting the potential use of these plants, the flora of the Deosai could be managed for better yield, and to protect species from over-exploitation. These valuable medicinal plants could be regularly sold by local peoples associated with the Deosai Biosphere Reserve, providing local community enhancement, and funds for improving the region (people, plants and animals).

The wild plants should be classified in general on the basis of their distribution and risk of extinction using a typical IUCN categories of Red data book viz Abundant, Extinct, Endangered, Vulnerable and Rare. Both *in situ* and *ex situ* conservation methods should be adopted for a long term plant conservation programme. *In situ* conservation is protecting plant populations in their natural habitats. E*x situ* programmes grow plants under controlled conditions and then reintroduce these plants in the wild. These techniques (especially *in situ* conservation), require an understanding of the life cycles of the species targeted for protection, the relationship between the species and the habitat and the pollination and dispersal mechanisms.

REFERENCES

AUFFENBERG, W., K. AND RAHMAN, H. In Press. The Herpetology of Pakistan.

BANARESCU, P. 1990. *Zoogeography of fresh waters* (Vol. I). General distribution and dispersal of freshwater animals. AULA-verlag, Wiesbaden

BRECKLE, S.W. 1974. *Notes on alpine and nival flora of Hindukush, East Afghanistan. Botaniska Notiser* **127**:278-284.

DAS, INDRANEIL. 1996. *Biogeography of the Reptiles of south Asia.* Krieger, Malabar, FL. 87 pp.

DAY, F. 1876. *On the fishes of Yarkand.* Proceedings of Zoological Society of London, 1876:781-807

DHAR, U. AND KACHROO, P. 1984. *Alpine flora of Kashmir* - its phytogeographic assessment. Jodhpur.

DUBOIS, A. 1978. *Une espece nouvelle de Scutiger Theobald 1868 de 'Himalaye occidentale (Anura: Petobatidae). Senckenberg Biol.,* **59**:163-171.

ENGLER, A.1879. *Versuche einer entwicklungsgeschite der pflanzenwelt insbesondere der Florengebiete seit Tertiar Period* 1.Versuch einer entwicklungsgeschichte der extropischen der nordliichen Hemesphere. Leibzig.

GRUBER, U. 1981 *Notes on the herpetofauna of Kashmir and Ladakh. British Journal of Herpetology* **6**:145-150.

HEDGE, I.C. AND WENDELBO, P. 1978. *Patterns of distribution and ndemicism in Iran. Notes from Royal Botanic Garden, Edinburgh* **36**:441-464.

HORA, S.L. 1936. *Yale North India Expedition. Article xvii, Report on fishes, Part 1. Cobitidae. Memoirs of Connecticut Academy of Science*, **10**:299-321.

JAYARAM, K. C. 1981. *Freshwater fishes of India, Pakistan, Bangladesh, Burma and Sri Lanka.* Handbook of Zoological Survey of India, No. 2. Calcutta. xii + 475 pp.

KALABUCHOV, N. J. 1937. *Some physiological adaptations of the mountain and plain forms of the wood-mouse (Apodemus sylvaticus) and other species of mouse-like rodents. Journal of Animal Ecology*, **6**:254-272.

KHAN, A. A. AND RAJPUT, R.A. 1995. *Biodiversity of Deosai Plateau, Baltistan- Northern Areas, Pakistan. Proceedings of the International Symposium on the Himalaya, Karakorum and Hindukush* - Dynamics of Change, Islamabad, Pakistan.

KHAN, A. A. AND ZAKARIA, V. 1995. *Management plan for Deosai National Park.* Himalayan Wildlife Project, Islamabad. 52 pp. plus appendices.

KHAN, A. A., RAJPUT, R. A. AND KHALID, U. *1996. Birdlife at Deosai: A source of sustenance and sustainability for local people.* Abstract, Birdlife Asia Conference. 9-16 November 1996. Coimbatore, India.

KITAMURA, S. 1960. *Flora of Afghanistan.* Results of Kyoto University Scientific Expedition to Karakorum and Hindukush. 1955, Vol. II. Kyoto University.

LAPERSONNE, V. S. 1928. *A collecting trip to Ladak, Part II. Journal Bombay Natural History Society*, **32**:650-659.

MENON, A. G. K. 1954. *Fish geography of the Himalaya.* Proceedings of National Institute Science, India, 21(4):467-493.

MATTHEWS, W. H. 1941. Bird notes from Baltistan. *Journal Bombay Natural History Society.* **42**:658-663.

MEUSAL, H. 1971. *Mediterranean elements in the flora and vegetation of W. Himalaya.* In Davis, P.H., P.C. Harper and I. C. Hedge. Plant Life of South West Asia. pp. 53-72.

MIRZA, M. R. AND AWAN, A. A. 1979. *Fishes of the genus Diptychus Steindachner, 1866 from Pakistan with the description of a new species. Biologia* (Pakistan), **25**:135-140.

MIRZA, M. R. AND KHAN, S. A. 1974. *Fish and fishes of Northern Areas, Pakistan.* pp. 1-14 (Report submitted to the Pakistan Science Foundation, Islamabad, Pakistan).

MIRZA, M. R. 1975. *Freshwater fishes and zoogeography of Pakistan. Bijdr. Dierk.*, **45**:143-180.

MIRZA, M. R. 1980. *The systematics and zoogeography of the freshwater fishes of Pakistan and Azad Kashmir.* Proceedings of Pakistan Congress of Zoology.,**1**: 1-41.

MUKERJI, D. D. 1936. *Yale North India Expedition, xviii. Report on Fishes. Part II. Sisoridae and Cyprinidae. Memoirs of Connecticut Academy of Arts and Science,* **10**:323-359.

NASIR, Y. J. AND RAFIQ, R. A. 1995. *Wild Flowers of Pakistan.* T.J. Roberts (ed.), Oxford University Press. New York. 298 pp.

QUAMMEN, D. 1996. *The Song of the Dodo: Island Biogeography in an Age of Extinctions.* Scribner, New York. 702 pp.

ROBERTS, T. J. 1992. *The Birds of Pakistan.* Oxford University Press, New York. 2 Volumes. Vol. I, 598 pp.; Vol. II, 617 pp.

ROBERTS, T. J. 1977. *The Mammals of Pakistan.* Ernest Benn Limited, London, 361 pp.

SHAH, M. 1993. *A preliminary checklist of Potentilla L. (Roseacea) occurring in Pakistan and Kashmir.Natural History Bulletin* **1**: 5-7.

SHINWARI, Z. K. AND CHAUDHRI, M. N. 1994. *New species of genus Nepeta (Labiatae) from Pakistan. Pakistan Journal of Science.* **37**: 477-478.

STEINDACHNER, F. 1866. *Ichthyologische Mitter-lungen (ix) vi Zur Fisch-f auna Kaschmirs und der benachbarten Landerstriche. Verh. Zool Bot. Ges Wien,* **16:**789-796.

STEWART, R. R. 1972. *An annotated catalogue of vascular plants of West Pakistan and Kashmir.* Fakhri Printing Press.

STEWART, R. R. 1982. *Flora of Pakistan - history and exploration of plants of Pakistan and adjoining area.* University of Michigan.

BIODIVERSITY IN A TYPICAL SUB-MOUNTAINOUS PROTECTED AREA - CHHUMBI-SURLA WILDLIFE SANCTUARY, PUNJAB, PAKISTAN

ABDUL ALEEM CHAUDHRY, IZAZ IBRAHIM AGHA, ANWAR HUSSAIN, RIAZ AHMAD AND MANSOOR HAMEED

Punjab Wildlife Research Institute, Faisalabad

Abstract: Chhumbi-Surla Wildlife Sanctuary, 55987 ha, declared in 1978, was established with a view to conserving the population of Punjab urial (*Ovis orientalis punjabiensis*). Located in the Salt Range of the Punjab, the area supports Tropical thorn and Semi-arid sub tropical scrub vegetation. Core area 4931 ha is the Reserved Forest, located almost in the middle of the sanctuary area mainly on the hill slopes ranging in elevation from 460 m to 1050 m above main sea level, the rest of the area surrounding the Core Zone is mainly community forests or Shamilats, extended on flat lands put under agricultural crops, rain fed or irrigated through installation of lift pumps at the perennial nullahs and through construction of small dams. Core area has a dense vegetation typical of the forest types, and offers excellent habitat conditions not only for the urial but also to a number of vertebrate species. The community forests are poorly preserved and have been subjected to pressures including clearing of land for agriculture, livestock grazing and brushwood cutting for fuel and fodder. The denudation has also helped increase water erosion, further deteriorating the vegetation structure. A survey was undertaken in August 1995 to record the Biodiversity of the area. During the survey, 114 plant species including 28 shrubs/ trees, 34 herbs, 41 grasses and single fern were recorded from the area. *Acacia modesta* was the only dominant tree species, *Dodonaea viscosa, Justicia adhatoda, Lantana indica, Lespedeza floribunda* and *Opuntia monacantha* were dominant species among shrubs, *Dicliptera bupleuroides* and *Pupalia lappacea* were the dominant herbs, *Aristida mutabilis, Cenchrus pennisetiformis, Chrysopogon serrulatus, Cymbopogon jwarancusa, Cynodon dactylon, Dactyloctenium scindicum, Desmostachya bipinnata, Dichanthium foveolatum, Digitaria sanguinalis, Heteropogon contortus, Hordeum murinum, Imperata cylindrica, Saccharum bengalense, Saccharum spontaneum* and *Sporobolus* ioclados were the dominant grasses and *Cyperus* neavus was the only dominant sedge species. Comparison of the data revealed a strong effect of grazing and wood cutting. Plant species preferred for grazing for economical point of view such as *Aristida mutabilis, Cenchrus pennisetiformis, Desmostachya bipinnata, Diclyptera bupleuroides, Dodonaea viscosa, Justicia adhatoda, Lantana indica, Lespedeza floribunda, Ochthochloa compressa, Olea ferruginea, Opuntia monacantha* and *Sporobolus* ioclados tend to decrease in the peripheral area, whereas non-palatable/ unimportant species like *Cyperus naevua, Imperata cylindrica* and *Saccharum spontaneum* tend to increase in the peripheral area. Some of the species like, *Zizyphus nummularia* and *Cynodon dactylon* also showed an abundance in the peripheral area probably due to the effect of cultivation. Urial population was the single largest in one block (estimated as 150) in the Punjab. Besides urial twelve species of mammals belonging to nine families and five orders were observed. Apart form some important game species like urial, grey and black francolins, seesee partridge, quail and rock pigeon, 76 bird species representing 57 genera and 34 families belonging to 14 orders were recorded. The ecology of the sanctuary is threatened by the construction of a metalled road bisecting the core zone. The impact of visitors on the population of Punjab urial and francolins may be negative. Mitigatory measures will have to be taken to offset this impact.

INTRODUCTION

Located in the heart of Salt Range, Punjab, Chhumbi-Surla Forests and Shamilats (community grazing lands), 55,987 ha were declared a Wildlife Sanctuary in 1978, with the objective of conserving one of the largest intact populations of Punjab urial (*Ovis orientalis punjabiensis*) in the area.

A survey was carried out in August, 1995, to record biodiversity occuring in the sanctuary area with a view to devise management strategies for the area.

Study Area

Chhumbi-Surla is located 15 km north-east of Chakwal town in close proximity to Kallar Kahar Lake, 32° 47’ north latitude and 67° 42’ east longitude. The core zone (5342 ha), Surla and Bakshiwala Reserved Forests is surrounded by 49912 ha of community forest, ‘Shamilats’ or other lands (privately owned), comprising of hill slopes ranging from 460 to 1050 m in altitude above mean sea level. The mean maximum and mean minimum year round temperatures range between 18.9 to 41.1 and 10.6 to 30.0 ^{o} C with June the hottest and January the coldest month of the year. The mean relative humidity varies between 15 to 29 % during the driest (May) and 56 to 76 % during the most humid months of the year. Average annual rain fall is 4994 mm. Distribution of rain in time is twin-peak type, 64 % received during the monsoon (June to September) and 25 % during the winter rainy season, January to April .

The rocks forming the hills belong to tertiary formations, 40 million years in age, consisting of sandstone and limestone, tilted throughout the area at a very high angle, sometimes reaching 90°. At places the rock is not pure and is found mixed with shale, clay or sand. The sedimentary configuration of rock strata is highly susceptible to water erosion especially on vegetation free hilly slopes. Erosion is further accelerated due to unchecked over gazing by the residents of the area on their own lands, shamilats as well as on the state owned forest areas. Thus the grazing incidence, being higher than the area can support is deteriorating the situation day after day. The area comprises of highly undulating surfaces, and cultivation is possible only on flat or moderately sloping lands, as such less than 25% area is under cultivation.

A broad valley Kahoon exists in the southern part of the sanctuary area and is extensively cultivated. The area around Dhok Ban Amir Khatoon, Tharpal, Bhadlla, Kariala, Khokharzir and Shamsabad is rather flat and cultivated. The areas around Dhok Sahla and Chhumbi village are also cultivable. The total population of adjacent villages and towns exceeds over one hundred thousand.

The agriculture mainly depends upon rains. Irrigation facilities have also been provided by the construction of small dams and lifting of water from perennial nullahs. The main crops are wheat, sorghum, millet and mustard.

The state forest area comprises of four blocks; Surla-I, Surla-II and Bakhshiwala in the centre of the sanctuary area (4932 ha) and Choa Saiden Shah in the south east (410 ha). The core forest land is surrounded by the community lands of different villages and the area under cultivation is situated on the periphery of the wildlife sanctuary. The settlements are also situated on the periphery of the wildlife sanctuary and no village is situated in or near the core forest area. No road or jeepable tract existed in that area which could pass through the forest. In 1994 work was started for the construction of 16 km long hill road from Khokharzir to Khokharbala bisecting the forest in almost two equal halves. The funds for the construction of road were provided by the Agency for Barani Area Development under Project, Barani-II on the behest of local politicians. The central core zone is the prime habitat for urial. With the construction of this road, approach of hunters has now been facilitated which was extremely difficult previously and was the main reason of the still existing urial population.

The forests, mainly comprise of phulai *Acacia modesta* but on higher altitudes wild olive *Olea ferruginea* and sanatha *Dodonaea viscosa* also exist. The forest is generally termed as 'Phulai Forest'. Vegetation of the area is Dry Sub-tropical Semi-evergreen Scrub Forest type, with an open canopy of wild olive and phulai trees/ shrubs. Other important shrub species are bhekar (*Justicia adhatoda*) sanatha (*Dodonaea* viscosa) and pataki (*Maytenus royleanus*). Important grass species include *Chrysopogon serrulatus, Heteropogon contortus, Digitaria sanguinalis, Dichanthium foveolatum* and *Dactyloctenium scindicum.*

The fauna of the area dominate in Oriental and Palaeartic affinities, though the Ethiopian element is also fairly well represented.

MATERIALS AND METHODS

Chhumbi Surla Wildlife Sanctuary, perhaps having the largest intact Punjab urial population, is situated in Salt Range hills of district Chakwal. The scrub forest covering the hills is dry sub-tropical semi-evergreen in nature. Vegetation was studied at five relatively homogenous sites within the Surla forest and at two sites in the peripheral zone

Vegetation study sites were selected on the basis of differences in ecological parameter such as soil type and texture, slope, aspect, etc. in order to make the sample representative of the area. Criteria for selection of various sites have been summarized in Table 1. Samples of all the plant species found in the area were collected for identification and herbarium record.

Flora of Pakistan by Nasir and Ali was consulted and followed for the identification and nomenclature of the plant species. Ten quadrates, each having 5x5 m area, were laid out at each study site along a transect line, 20 m apart from each other. Data for density, frequency and coverage were recorded and importance value of each species calculated in accordance with Hussain (1983).

Importance value = Relative Density + Relative frequency + Relative Cover

Plant communities were established on the basis of dominant species having a larger importance value in the given ecological community.

Table I: Summary of ecological parameters at different vegetation sampling sites at Chhumbi-Surla Wildlife Sanctuary, Chakwal

Site	Topography	Slope	Aspect	Soil texture
1	Top of the hill where surface was more or less flattened	15%	Western	Clayey sand with sand stones
2	Plain areas between the hills	5%	Northern	Sandy clay
3	Hills with less steep slopes	45%	Southwestern	Sand stone
4	Hills with steep slopes	60%	Northern	Sand stone
5	Slightly inclined plain at the foot hills	15%	Southeastern	Red sandy clay with sand stones
6	Slightly inclined surface in the periphery of the Surla forest	20%	Western	Sandy clay
7	Plain surface in the periphery of the Surla forest	5%	Southern	Sandy clay

To study and record the fauna of the wildlife sanctuary, the area was surveyed as per accessibility and travel paths with a view to making the sample representative of all micro habitats found in the area qualitatively. The area was surveyed for four days by 12 workers, approximately for eight hours daily on the average by each worker to record various bird species occurring in the area. Each bird species was assigned a status on the basis of relative qualitative occurrence in the area during the survey period.

Mammals are rather shy creatures and therefore difficult to observe in the field. Therefore for mammals of the area and their status, along with few visual observations, information was collected from the local people and compiled to have some idea of the mammalian fauna of the area. Punjab urial being key species of the region was given special consideration.

As it is often not possible to record all the biodiversity of an area in single visit over a short period of time frame, therefore literature was consulted to build as comprehensive a check-list of fauna inhabiting the area as it was possible to have a fair idea of the biodiversity.

RESULTS

Vegetation

During the vegetation survey of the core zone (Surla forest) and the peripheral areas, 113 plant species were recorded belonging to 35 families (Table II). The largest family was Poaceae containing 41 grass species, while the other major families were Cyperaceae comprising of nine species and Papilionaceae eight species. The area was predominantly occupied by nine grass species viz. *Chrysopogon serrulatus, Dactyloctenium scindicum, Cymbopogon jwarancusa, Sporobolus ioclados, Digitaria sanguinalis* and *Dichanthium foveolatum* with some tree or shrub species like *Acacia modesta, Dodonaea viscosa, Justicia adhatoda* and *Lespedeza floribunda* and only a single herb, *Dicliptera bupleuroides.* In the peripheral area of the Surla forest *Cynodon dactylon, Imperata cylindrica, Saccharum spontaneum* and *Cyperus niveus* also shared the major part of the habitat. These species were occasionally recorded from the forest area.

Table II: Floral list of Chhumbi Surla Wildlife Sanctuary, Chakwal.

Family	Plant species
Acanthaceae	*Dicliptera bupleuroides var. ciliata, Justicia adhatoda*
Aizoaceae	*Trianthema portulacastrum*
Amaranthaceae	*Aerva javanica var. bovei, Aerva javanica var. javanica, Digera muricata, Pupalia lappacea*
Apocynaceae	*Nerium oleander*
Araliaceae	*Hedera nepalensis*
Asclepiadaceae	*Calotropis procera ssp. hamiltonii, Cynanchum auriculatum, Periploca aphylla, Periploca hydaspidis*
Boraginaceae	*Cynoglossum lanceolatum, Heliotropium rariflorum, Trichodesma indicum*
Cactaceae	*Opuntia monacantha*
Capparidaceae	*Capparis decidua, Capparis spinosa, Cleome scaposa*
Celastraceae	*Maytenus royleanus*
Chenopodiaceae	*Chenopodium botrys*
Commelinaceae	*Commelina albescens*
Compositae	*Bidens pilosa, Cnicus arvensis, Vernonia cinerascens*
Convolvulaceae	*Evolvulus alsinoides, Ipomoea carnea, Ipomoea eriocarpa*
Cucurbitaceae	*Corallocarpus epigaeus*
Cyperaceae	*Cyperus compressus, Cyperus iria, Cyperus naevus, Cyperus rotundus, Kyllinga triceps, Scirpus littoralis, Scirpus maritimus, Scirpus michelianus, Scirpus mucronatus, Scirpus roylei*
Euphorbiaceae	*Euphorbia clarkeana*
Labiatae	*Leucas nutans*
Liliaceae	*Asparagus adscendens*
Malvaceae	*Abutilon fruticosum, Malvastrum coromendelianum*
Mimosaceae	*Acacia hydaspica, Aaccia modesta*

Nyctaginaceae	*Boerhavia procumbens*
Oleaceae	*Olea ferruginea*
Oxalidaceae	*Oxalis corniculata*
Papilionaceae	*Argyrolobium stenophyllum, Butea monosperma, Caragana ambigua, Caragana gerardiana, Lespedeza floribunda, Lespedeza juncea, Lotus corniculatus var. corniculatus, Rhynchosia capitata*
Plantaginaceae	*Plantago major*
Poaceae	*Acrachne racemosa, Aristida adscensionis, Aristida mutabilis, Brachiaria deflexa, Brachiaria ramosa, Cenchrus pennisetiformis, Cenchrus setigerus, Chrysopogon serrulatus, Cymbopogon jwarancusa ssp. jwarancusa, Cynodon dactylon, Dactyloctenium aegyptium, Dactyloctenium scindicum, Desmostachya bipinnata, Dichanthium annulatum, Dichanthium foveolatum, Digitaria ciliaris, Digitaria sanguinalis, Echinochloa colona, Echinochloa crus-galli, Enneapogon persicus, Eragrostis cilianensis, Eragrostis minor, Eragrostis pilosa, Hemarthria compressa, Heteropogon contortus, Hordeum murinum ssp. glaucum, Imperata cylindrica, Ochthochloa compressa, Panicum atrosanguineum, Panicum miliaceum, Paspalum paspaloides, Phragmites karka, Rhynchelytrum repens,.Saccharum bengalense, Saccharum spontaneum, Setaria viridis, Sorghum halepense, Sporobolus coromandelianus, Sporobolus ioclados, Stipagrostis hirtigluma, Tragus roxburghii*
Polygalaceae	*Polygala arvensis, Polygala erioptera*
Polypodiaceae	*Adiantum capillus-veneris*
Rhamnaceae	*Rhamnus pentapomica, Ziziphus nummularia,*
Sapindaceae	*Dodonaea viscosa*
Solanaceae	*Solanum incanum, Solanum surattense*
Verbenaceae	*Lantana camara, Lantana indica,*
Typhaceae	*Typha domingensis*
Zygophyllaceae	*Fagonia bruguieri, Fagonia indica var. indica, Tribulus terrestris*

Different type of plant communities were recorded at each study site. Plant communities were identified on the basis of their importance value. *Chrysopogon serrulatus* was a unique case which showed complete dominance in all the habitats of Surla forest, while shared the habitat with some other species in the periphery (Table III).

Table III: Importance value of some dominant plant species at various sites of study in Chhumbi Surla Wildlife Sanctuary

(Importance value = Relative frequency + Relative density + Relative cover)

	Study Sites						
Plant Species	**I**	**II**	**III**	**IV**	**V**	**VI**	**VII**
Acacia modesta	19.4	16.3	32.2	28.6	20.9	16.8	8.4
Aristida mutabilis	1.2	4.0	0.0	2.1	11.7	6.1	1.3
Cenchrus pennisetiformis	0.0	5.7	15.2	10.0	13.8	0.0	7.0
Chrysopogon serrulatus	63.2	60.3	58.3	50.1	32.2	60.1	33.0
Cymbopogon jwarancusa	12.4	1.4	9.8	40.1	16.1	6.5	18.6
Cynodon dactylon	2.2	12.2	0.0	0.0	0.0	7.9	9.5
Cyperus naevus	2.0	1.7	0.0	4.8	0.0	7.8	13.7
Dactyloctenium scindicum	7.4	19.9	13.6	14.3	8.9	12.8	18.1
esmostachya bipinnata	5.9	23.9	0.0	2.1	0.0	0.0	0.0
Dichanthium foveolatum	4.8	13.1	11.8	3.6	12.3	7.8	6.5
Dicliptera bupleuroides	21.4	0.0	0.0	0.0	0.0	3.1	8.4
Digitaria sanguinalis	24.0	3.3	19.9	26.8	8.3	12.8	2.4
Dodonaea viscosa	0.0	33.9	0.0	25.3	4.5	0.0	0.0
Heteropogon contortus	27.6	11.8	7.2	20.8	17.0	17.0	22.7
Hordeum murinum	11.3	3.4	4.9	1.8	5.0	2.8	21.2
Imperata cylindrica	0.0	0.0	0.0	0.0	0.0	0.0	38.5
Justicia adhatoda	0.0	0.0	0.0	0.0	27.9	0.0	0.0
Lantana indica	2.3	3.0	0.0	12.9	0.0	0.0	0.0
Lespedeza floribunda	0.0	11.9	28.6	5.1	19.8	0.0	0.0
Ochthochloa compressa	0.0	1.8	7.3	0.0	22.0	0.0	0.0
Opuntia monacantha	13.6	13.3	0.0	0.0	0.0	0.0	0.0
Pupalia lappacea	19.1	0.0	0.0	0.0	1.7	0.0	0.0
Saccharum bengalense	0.0	3.5	0.0	9.8	17.4	0.0	5.3
Saccharum spontaneum	0.0	7.6	4.9	10.6	0.0	0.0	18.7
Sporobolus ioclados	10.3	23.5	40.9	12.2	23.8	9.4	5.8

Site 1. Chrysopogon serrulatus community

Top of hills on the western side of the wildlife sanctuary were more or less plain. Maximum vegetation was recorded here. Thirty eight plant species were noted including 16 grass species, one sedge, 7 herbs, 12 shrubs and 2 tree species. *Chrysopogon serrulatus* dominated the area covering more than 60% of the total vegetation. Other important species were *Cymbopogon jwarancusa, Digitaria sanguinalis, Heteropogon contortus, Hordeum murinum* and *Sporobolus ioclados* among grasses, *Acacia modesta* among trees, *Opuntia monacantha* among shrubs and *Dicliptera bupleuroides* and *Pupalia lappacea* among herbs. Vegetation cover was highly overlapping and hardly any bare land could be seen.

Site 2. Chrysopogon serrulatus and Dodonaea viscosa community

This site was studied at more or less flat plains between the hills on the northern side of the Surla forest. *Chrysopogon serrulatus* and *Dodonaea viscosa* dominated all the other species. Thirty two plant species were recorded including 20 grass species, one sedge, 5 herb, 10 shrub and 2 tree species. Other major grass species were *Desmostachya bipinnata, Sporobolus ioclados, Dactyloctenium scindicum, Cynodon dactylon, Heteropogon contortus* and *Dichanthium foveolatum,* shrub species were *Dodonaea viscosa* and *Opuntia monacantha* and tree species was *Acacia modesta.* Vegetation cover was much thinner than that of site 1.

Site 3. Chrysopogon serrulatus-Sporobolus ioclados community

On the southwestern side slopes were about 45 %. Here 29 plant species were collected. Dominant grass species were *Chrysopogon serrulatus, Sporobolus ioclados, Dichanthium foveolatum, Digitaria sanguinalis, Dactyloctenium scindicum* and *Cenchrus penniset-iformis.* Only single dominant tree species was *Acacia modesta* and shrub pecies was *Lespedeza floribunda.* Vegetation comprised of few dominant species and observed in patches only.

Site 4. Chrysopogon serrulatus-Cymbopogon jwarancusa community

Slopes in this habitat were the steepest (60 %) on the northern aspect. Twenty nine plant species were observed; dominant grass species were *Chrysopogon serrulatus, Cymbopogon jwarancusa, Sporobolus ioclados, Cenchrus pennisetiformis, Dactyloctenium scindicum, Digitaria sanguinalis, Heteropogon contortus* and *Saccharum spontaneum.* Dominant tree/ shrub species were *Acacia modesta, Dodonaea viscosa* and *Lantana indica* while herb species were only recorded in a very small number.

Site 5. Chrysopogon serrulatus-Justicia adhatoda community

This site was selected in the valleys on the southeastern side of the wildlife sanctuary where soil was typically reddish sandy clay. Twenty eight species were recorded at this site. Vegetation predominantly comprised of grasses like *Chrysopogon serrulatus, Heteropogon contortus, Dichanthium foveolatum, Cymbopogon jwarancusa, Sporobolus ioclados, Cenchrus pennisetiformis, Ochthochloa compressa, Saccharum benghalensis* and *Aristida mutabilis.*

Dominant trees/ shrubs were *Justicia adhatoda, Lespedeza floribunda* and *Acacia modesta.*

Site 6. Cynodon dactylon-Chrysopogon serrulatus community

Western periphery of the wildlife sanctuary supported 30 species. Dominant grass species were *Cynodon dactylon, Chrysopogon serrulatus, Heteropogon contortus, Dactyloctenium scindicum* and *Digitaria sanguinalis* with only singular dominant tree species *Acacia modesta* which attained a height up to 1-2 m only. The vegetation cover was the thinnest in the area and comprised of only a few dominant species. Other species were occasionally reported.

Site 7. Imperata cylindrica-Chrysopogon serrulatus community

The site was located on the southern side of the sanctuary where maximum diversity of the plant species were recorded. Forty two species were noted at this site. Dominant grass/ sedge species were *Imperata cylindrica, Chrysopogon serrulatus, Cymbopogon jwarancusa, Dactyloctenium scindicum, Hordeum murinum, Heteropogon contortus, Saccharum spontaneum* and *Cyperus naevus* while not a single dicot species could be strictly regarded as the dominant or co-dominant.

Avifauna

During the survey period 71 species belonging to 57 genera from 34 families representing 14 orders were recorded from the sanctuary area (Table IV).

Table IV. Bird species recorded from Chhumbi-Surla Wildlife Sanctuary, during August 1995

(A= ++++, >100 ; C= +++, >50 <100; F= ++ >10 <50 and R= +, <10 sightings)

Taxonomic summary : 13, orders; 32, families; 57, genra; 70, species

Status summary : C = Common 24 (34 %), **F** = Fair 19 (27 %), **A** = Abundant 16 (23 %), **R** = Rare 11 (16 %); **Rt** = Residents 49 (70.0 %), **W** = Wintering 6 (8.9 %), **OM** = Ordinary migrants 5 (7.0 %), **SB** = Summer breeding 5 (7.0 %), **OW** = Occasional wintering 2 (2.9 %), **PM** = Passage migrants 1 (1.4 %), **V** = Vagrant 1

Name of bird species	English Name	Sightings	Status	Residence
Family Ardeidae				
Bubulcus ibis	Cattle egre	3	R	Rt
Ardeola greyii	Indian pond heron	7	R	Rt

Egretta garzetta	Little egret	18	F	Rt
Ardea cinerea	Grey heron	4	R	W
Family Accipitridae				
Pernis apivorus	Honey buzzard	2	R	I
Elanus caeruleus	Black-winged kite	3	R	Rt
Milvus migrans	Black kite	68	C	Rt
Gyps bengalensis	Oriental white-backed vulture	12	F	Rt
Family Falconidae				
Falco tinnunculus	Common kestrel	2	R	W
Family Phasianidae				
Alectoris chukar	Chukar	4	R	Rt
Ammoperdix griseogularis	See-see partridge	10	F	Rt
Francolinus francolinus	Black partridge	91	C	Rt
Francolinus pondicerianus	Grey partridge	71	C	Rt
Coturnix coturnix	Common quail	5	R	OM
Pavo cristatus	Peafowl	10	F	Rt
Family Charadriidae				
Hoplopterus indicus	Red-wattled lapwing	19	F	Rt
Family Columbidae				
Columba livia	Blue rock pigeon	33	F	Rt
Columba palumbus	Common wood pigeon	41	F	Rt
Streptopelia decaocto	Indian ring dove	145	A	Rt
Streptopelia tranquebarica	Red turtle dove	104	A	SB
Family Psittasidae				
Psittacula krameri	Rose-ringed parakeet	46	F	Rt
Family Cuculiformes				
Clamator jacobinus	Pied-crested cuckoo	19	F	SB
Hierococcyx varius	Common hawk cuckoo	4	R	SB
Eudynamys scolopacea	Common koel	3	R	SB
Taccocua ieschenaultii	Sirkeer cuckoo	5	R	Rt
Centopus sinensis	Common crow pheasant	4	R	Rt
Family Tytonidae				
Athene brama	Spotted owlet	2	R	Rt
Family Caprimulgidae				
Caprimulgus europaeus	European nightjar	11	F	OW
Family Apodidae				
Apus affinis	House swift	155	A	Rt
Family Alcidinidae				
Halcyon smyrnensis	White-breasted kingfisher	10	F	Rt
Alcedo atthis	Common kingfisher	2	R	Rt
Family Meropidae				
Merops orientalis	Little green bee-eater	113	A	Rt
Merops superciliosus	Blue-cheeked bee-eater	20	F	SB

Family Coraciidae				
Coracias benghalensis	Blue jay	31	F	Rt
Family Upupidae				
Upupa epops	Hoopoe	45	F	RT
Family Picidae				
Dinopium benghalense	Golden-backed woodpecker	1	R	Rt
Family Alaudidae				
Galerida cristata	Crested lark	14	F	RT
Alauda gulgula	Lesser skylark	13	F	Rt
Family Motacilidae				
Motacilla flava	Yellow wagtail	2	R	PM
Motacilla cinerea	Grey wagtail	8	R	W
Motacilla alba	Pied wagtail	2	R	W
Family Campephagidae				
Pericrocotus cinnamomeus	Small minivet	22	F	Rt
Family Pycnonotidae				
Picnonotus cafer	Red-vented bulbul	24	F	Rt
Picnonotus leucogenys	White-cheeked bulbul	96	C	Rt
Family Turdidae				
Saxicola caprata	Pied bush-chat	46	F	RT
Saxicoloides fulicate	Indian robin	70	C	Rt
Myiophoneus caeruleus	Blue whistling thrush	2	R	W
Family Sylvidae				
Prinia buchanani	Rufous-fronted wren warbler	35	F	Rt
Prinia gracilis	Streaked long-tailed warbler	9	R	Rt
Prinia inornata	Tawny prinia	11	F	Rt
Family Rhipiduridae				
Rhipidura aureola	White-browed fantail flycatcher	18	F	Rt
Family Monarchidae				
Terpsiphone paradisi	Asian paradise flycatcher	1	R	OM
Family Timaliidae				
Turdoides caudatus	Common babbler	60	C	Rt
Turdoides striatus	Jungle babbler	68	C	Rt
Family Nectariniidae				
Nectarinia asiatica	Purple sunbird	10	F	SB
Family Oriolidae				
Oriolus oriolus	Golden oriole	130	A	OM
Family Laniidae				
Lanius vittatus	Bay-backed shrike	73	C	Rt
Lanius schach	Rufous-backed shrike	52	C	Rt
Lanius excubitor	Great grey shrike	3	R	Rt
Family Dicruridae				
Dicrurus macrocercus	Black drongo, King crow	118	A	Rt

Family Corvidae				
Dendrocitta vagabunda	Indian tree pie	127	A	Rt
Corvus macrorhyncos	Jungle crow	4	R	Rt
Family Sturnidae				
Sturnus pagodarumi	Brahminy myna	2	R	V
Acridotheres tristis	Common myna	560	A	Rt
Acridotheres ginginianus	Bank myna	35	F	Rt
Family Passeridae				
Passer domesticus	House sparrow	64	C	Rt
Family Ploceidae				
Ploceus philippinus	Baya weaver	159	A	Rt
Ploceus manyar	Streaked weaver	1500	A	Rt
Euodice malabarica	White-throated munia	8	R	Rt
Lonchura punctulata	Spotted munia	14	F	I
Family Emberizidae				
Emberiza bruniceps	Red-headed bunting	149	A	OM

Of all the families Phasianidae was recorded to be the most diversified at the time of survey, represented by six species. It was followed by Cuculidae, having five species. Four families, Ardeidae, Accipitridae, Columbidae and Ploceidae were represented by four species each whereas five families, Motacilidae, Turnidae, Sylvidae, Laniidae and Sturnidae had three species each. Six families, Alcidinidae, Meropidae, Alaudidae, Pycnonotidae, Timaliidae and Corvidae, each had two species, 17 families had a representation each by single species.

Passeriformes was the most dominant bird order contributing half the number of families i.e. 18 as well as species (35 species) followed by Coraciiformes, represented by six species from four different families. Galliformes had six representative species all belonging to a single family; followed by Cuculiformes with five species of family Cuculidae. Except for Passeriformes and Coraciiformes all the bird orders were represented, each by a single family.

Depending upon the number of sightings during the survey period each species was assigned a population status in the area (Table IV): Abundant 16 (23 %), Common 24 (34 %), Fair 19 (27 %), Rare 11 (16%).

Based on the reports of Roberts, 1991, 1992 permanent resident species numbered 51 (72 %); summer breeders 6 (8.4 %), double passage migrants 4 (5.6 %), irregular year round visitors 2 (2.8 %), and only one species could be the occasional winter visitor, passage migrant and vagrant.

URIAL POPULATION

Urial, the lone sheep species found in the Punjab is the most important big game animal of the area. The Chhumbi- Surla area was declared a Wildlife Sanctuary to provide maximum protection and multiplication opportunities to the urial population. Surveys in the last few years revealed its population as given in Table V. With this sighting data in different parts of the sanctuary and reports by the local people and the wildlife staff of the area, it can be safely presumed that some more than 150 urials may be thriving within the wildlife sanctuary. However whole of this population is presently concentrated in the core zone, in the state owned forests.

Table V: Records of urial sightings in Chhumbi - Surla Wildlife Sanctuary, Chakwal during recent years

S. No.	Survey Time	Urial sightings	Remarks
1.	--/--/1979	50	Reported by Game Inspector at the time of declaration of Wildlife Sanctuary
2.	---/10/1987	--	Main emphasis was to record the birds
3.	12/11/1989	17	
4.	29/04/94	20	Survey along four different 4-6 km long tracks in south east corner of the sanctuary
5.	01/05/94	1	A very short survey track near southern boundary
6.	06/05/94	59	Survey along four different tracks covering 25-30 % of core area of the sanctuary
7.	10/09/94	23	Survey along 4-5 km long single track in the core area of the sanctuary
8.	30/09/94	12	Survey along an approximately 18-20 km long single track
9.	26-30/08/1995	85	Survey effort as given in methodology section and in survey rout map

OTHER MAMMALS

Beside urial 12 other mammal species were recorded from the area either through direct observation or indirect evidence like presence of foot prints, scats or faecal pellets or gathering information from Wildlife and Forest Staff or from residents of the area (Table VI). Four of them have a fair population in the area. Four more occur commonly while three species are a rare sight. Leopard was reported to pay rare wanderings while chinkara has become totally exterminated from the area for over hunting almost for last ten years or so.

Check list of terrestrial vertebrates developed after consulting the literature about the region is given as Appendix to have some idea about the total diversity in the fauna of the region. It includes four

Amphibians, 16 Reptiles, 247 Birds and 33 Mammals constituting 1.3, 5.3, 82.3 and 11 % respectively to the total terrestrial vertebrate fauna for the region respectively. Efforts have also been made to assign a status to each species, and also to record their habits and breeding status in the area.

Table VI: Mammal species recorded during survey in 1995

Generic name	Sightings	Status *	Remarks
Ovis orientalis	85	Fare	Key species
Sus scrofa	10	Common	Pest on crops in the area
Ghazella bennetti	---	Extinct	For last 10 years
Canis aureus	6	Common	Can damage bird eggs
Vulpes vulpes	2	Fare	Can damage bird eggs
Panthera pardus	---	Vagrant	Rare wanderings
Felis chaus	2	Fare	Nocturnal
Martes flavigula	---	Fare	Foot prints seen most often
Herpestes edwardsi	8	Common	Damages bird eggs
Herpestes auropunctatus	2	Rare	Damages bird eggs
Lepus nigricollis	7	Common	Hunted for meat
Hystrix indica	1	Rare	A vermin
Manis crassicaudata	---	Rare	Nocturnal

* Status as reported by local people and Wildlife and Forest Staff of the area

DISCUSSION

The variety of life forms in the area is indicative of richness of Biodiversity, thus being entitled to a conservation status of the highest order i.e., Wildlife Sanctuary.

Vegetation studies at Chhumbi-Surla Wildlife Sanctuary suggested that it is a good habitat for Punjab urial having ample number of palatable grass species and consumable herbs. *Chrysopogon serrulatus*, a dominant grass at all study sites, is palatable and may be the major food item for the urial. Other important grass species were *Cymbopogon jwarancusa, Dactyloctenium scindicum, Dichanthium foveolatum, Digitaria sanguinalis, Heteropogon contortus, Hordeum murinum* and *Sporobolus ioclados*. Most of these species are also palatable and may contribute to the urial forage. Therefore it can safely be assumed that urial may not be suffering any serious food shortage in the area throughout the year and the habitat in its present form could support the present population. An excercise in determining the carrying capacity would however be beneficial.getation deep inside the Surla forest was very dense and at some places multistoried due to overlapping vegetation. Particularly on the tops of the hills, there was hardly any bare ground. Urial prefers an open canopy normally but such areas are preferred by the females at the time of parturition.

Places in between the hills and along the route to water points had patchy vegetation, often very dense in nature,which alternating with open areas thus making it an ideal place for urial survival. High grazing pressure in the periphery of the forests has resulted in the loss of desired vegetation cover. Here the vegetation was thin and trees like *Acacia modesta,* only up to 1-2 m high. In the peripheral zone the vegetation was also scanty, only a few specific species tolerant of the marginal conditions for survival.

In spite of such an excellent habitat urial population in the area faces some serious risks, like illegal hunting and poaching, habitat deterioration from live stock grazing and fire wood cutting, and habitat fragmentation because of road construction. Illegal grazing is one of the major causes of habitat destruction, affecting the urial population directly as well as indirectly. One of the major risks for urial is the prolonged drought seasons when the water availability is restricted to a few areas only. Such a situation provides a fair opportunity to the hunters to shoot down thirsty animals around these water points. Another potential damage to the urial population is the capturing of new born lambs. Local people, particularly the herdsmen of the area are involved in this mal-practice, and sell urial lambs to the influential persons who keep them as pets, as status symbol, or they barter these lambs for some other benefits.

Even though the population of the species has greatly improved in the area since its declaration as a wildlife sanctuary yet the situation could have been still better if a control could have been excercised on illegal hunting and lamb picking. The two factors are also responsible for sharp decline in urial population in other parts of the Salt Range.

The wild hare (*Lepus nigricollis*) is also an important game animal. Its hunting is legally allowed but only outside the sanctuary-area and the buffer zone. It provides good sport when chased with hounds. Presently it occurs in abundance however indiscriminate hunting with coursing dogs and shooting with shot guns, after sun set, with the help of search lights may affect its present status negatively.

Occurrance of six major game bird species, *viz.,* chakor, seesee, black and grey partridges, common quail and black-breasted rain quail makes the area all important from biodiversity conservation point of view. All these species are common except the rain quail which occurs frequently. Peafowl, occurring in the area, though not naturally (Roberts, 1991), also deserve conservation status. It is also worth mentioning that it was introduced at a holy shrine at Kalar Kahar where it multiplied and spread over the adjacent areas. Grey and seesee partridge populations are prone to illegal hunting and capturing round the year.

The construction of a road bisecting the core zone of the sanctuary area will also have far reaching effects on the ecology and biodiversity of the area. Increased disturbance may render the habitat unavailable to the shy species such as the urial and other wild birds and mammals. Further it will facilitate poachers to raid the core zone easily and will result in the sharp decline of its numbers. Concerned quarters and the nature conservation NGOs are required to take the matter seriously, come forward and play their role in educating the public for nature conservation in order to minimize the hazards.

REFERENCES

HUSSAIN, F., 1983. *Manual of plant ecology,* Univ. Grants Commission, Sector H-8, Islamabad, Pakistan.

NASIR, E. AND S. I., ALI, *Flora of Pakistan,* National Herbarium, Agricultural Research Council, Islamabad.

ROBERTS, T. J., 1977. *The mammals of Pakistan,* Ernest Benn Ltd., London

ROBERTS, T. J., 1991. *The birds of Pakistan,* Vol. **1**, Oxford Univ. Press, Karachi.

ROBERTS, T. J., 1992. *The birds of Pakistan,* Vol. **2,** Oxford Univ. Press, Karachi.

SAID, M., 1951. *Working plan for the Forests of the Jhelum, Mianwali and Shahpur Forest Divisions.*

KHAN, M. S., 1991. *Reptiles, Pakistan Ki Jangli Hayat* (in urdu), Urdu Science Board, 299-Upper Mal, Lahore

KHAN, M. S., 1991. *Snakes of Pakistan, Pakistan Ki Jangli Hayat* (in urdu), Urdu Science Board, 299-Upper Mal, Lahore

KHAN, M. S., 1991. *Tortoises, Pakistan Ki Jangli Hayat* (in urdu), Urdu Science Board, 299-Upper Mal, Lahore

KHAN, M. S., 1991. *Amphibia, Pakistan Ki Jangli Hayat* (in urdu), Urdu Science Board, 299-Upper Mal, Lahore

BIODIVERSITY IN CHOLISTAN DESERT, PUNJAB, PAKISTAN

ABDUL ALEEM CHAUDHRY, ANWAR HUSSAIN, MANSOOR HAMEED AND RIAZ AHMAD

Punjab Wildlife Research Institute, Faisalabad, Pakistan.

Abstract: Cholistan desert located in the southeast of the Punjab, Pakistan, covering 26,000 km^2 is a part of greater Thar desert. The soil types characteristically include: sand dunes, non-saline non-sodic sandy soils with small patches of non-saline, non-sodic loamy soils and sodic clayey soils. Vegetation cover varies from place to place depending upon the texture and structure of the soil. Vegetation structure and density is greatly influenced by the rainfall. During low rainfall years even drinking water gets scarce and both the plant and animal communities are adversely affected. Flora of Cholistan is characterized by special xerophytic adaptations due to which it can withstand very severe drought conditions. This provides food and shelter to wildlife and livestock in the area. Cholistan is an important grazing ground for the livestock. The area is also favoured by sport hunters, especially the falconers from the Middle East. Their hunting safaris, however, effect the biodiversity of the area. Biodiversity surveys were undertaken in October 1992 to September 1995. Vegetation was studied at 26 different sites along jeepable tracks and 23 distinct plant communities were recognized on the basis of their importance value. Dominant species among grasses were *Aristida adscensionis, Ochthochloa compressa, Lasiurus scindicus, Cymbopogon jwarancusa, Cenchrus biflorus* and *Aeluropus lagopoides*. Among herbs *Suaeda fruticosa, Salsola baryosma, Dipterygium glaucum* and *Crotalaria burhia* and among shrubs *Haloxylon recurvum, Haloxylon salicornicum* and *Calligonum polygonoides* were the important species. In the latest surveys 58 bird species were recorded representing 42 genera, 26 families and 12 orders, including, *Chlamydotis undulata, Pterocles orientalis, Elanus caeruleus, Accipiter badius, Cursorius coromandilicus, Eremopterix grisea, Calandrella brachydactyla, Alauda arvensis, Alaemon alaudips,* and *Lanius excubitor.* Presence of vultures and birds of prey is a special feature. Eleven species of reptiles belonging to 10 ġenra representative of nine families have been reported in the area. Apart from these 30 species of mammals belonging to 24 genra and 14 families have also been reported from the area which include *Gazella bennetti, Herpestes edwardsi, Herpestes auropunctatus, Felis chaus, Felis caracal, Vulpes bengalensis, Hemiechinus auritus* and *Lepus* *nigricoll.* Biodiversity in the area is being affected by livestock grazing, cutting of shrubs for fuel wood, hunting, both legal and illegal, falconry and related operations by the Middle Eastern dignitaries and by the climatic vagaries. This is a fragile ecosystem and any change in land use practice could drastically reduce the viability of the system.

INTRODUCTION

Cholistan Desert stretching along the southern border of the Punjab, Pakistan, 26,000 km^2, is a part of the world's seventh largest desert, The Thar or Great Indian Desert which covers an area of 368,000 km^2. Summer monsoons pass nearby the Thar towards the eastern side without giving due share to the area. Cholistan merging with Indus Valley that has been the home of one of the world's oldest civilizations, Mohenjo Daro and Harappa, 4000 to 5000 years ago, on its western side ,possibly at that time was sharing monsoon downpours on the Indus Valley and was not so arid as today. A gradual drift in the monsoon winds declined the area into a desert (Leopold, 1963).

The huge land mass comprises of hard textured clayey flat grounds 'dahars' interspersed among semi stable and unstable sand dunes that may attain a height of 100 m. The area supports a characteristic xeric vegetation. Ḍahar soils vary in texture, structure, salinity and sodicity. Marked differences in physico-chemical characteristics of soil and its water holding capacity form basis for development of different floral communities and their associated fauna within short distances. The area is classified into Greater Cholistan (18,130 km^2) characterized with comparatively frequent and high sand dunes and Lesser Cholistan (7,770 km^2) with comparatively less high sparse sand dunes (Khan 1992). This division is quite arbitrary and lacks any clear line differentiating the two zones. Rainfall is erratic, varying from 100 mm in the north to 200 mm in the south annually and may reach 250 mm near the Pakistan-India border. Average winter temperature ranges between 14 and 16 ^oC, with December and January being the coldest months when the temperature frequently drops below zero. Mean summer temperatures range between 34 to 37 ^oC with maximum temperature in May / June when it may rise up to 50 ^oC or even more. The relative humidity varies from 50 % to 65 % (Khan, 1992). During the summer months, strong winds are a characteristic feature that frequently result in sand-storms and shifting of sand dunes from one place to the other.

Human population, as projected for the year 1991, based on 1981 census was 97,000 with a population density of 3.73 individuals per km^2 , (FAO, 1993). Deep interior of the desert is much thinly populated than the peripheral zone. Semi permanent and nomad inhabitants roam about measuring length and breadth of the desert continuously looking for forage for their livestock and more strongly for drinkable water. According to the figures provided by Divisional Forest Officer, Cholistan Range Management Division, during May, 1994, total livestock in Cholistan was 262,430 (63,095 cattle; 114,421 sheep; 72,726 goats and 12,188 camels) with generally low annual growth, 2.7% for cattle, 3.4% for goats and a slight decline in the number of sheep and camels. Last twenty years (1964-94) combined average growth remained less than 1%. Prolonged droughts in combination with high temperatures, evaporating the last drops of water from the 'tobas' force the inhabitants to leave the region and move towards irrigated areas until the next showers. Erratic rainfall influences grass cover, which in spite of its great sprouting potentials, does not match masses of livestock dependent on it and therefore, mostly is present in the form of over grazed stubbles unable to stabilize the sand dunes properly.

The wilderness of Cholistan has its own charms and beauty and has the fame of a rewarding hunting ground. Chinkara (*Gazella bennetti*), blackbuck (*Antilope cervicapra*), nilgai (*Boselaphus tragocamelus*), houbara bustard (*Chlamydotis undulata*), great Indian bustard (*Ardeotis nigriceps*), grey partridge (*Francolinus pondicerianus*), imperial sandgrouse (*Pterocles orientalis*) and common quail (*Coturnix coturnix*) are some of the important game species and attract the attention of local hunters as well as foreign dignitaries, specially from the Middle East, who visit Cholistan to practice falconry on houbara.

Over exploitation of vegetation cover by livestock grazing, brush wood cutting and soda ash extraction; wind erosion, poaching by local hunters and falconry by Middle Eastern dignitaries are the main threats to the socio-economic and ecological system of the area. There is a great need to asses the resources of the area thoroughly on scientific grounds and design a safe system to utilize the resources at a sustainable level.

This study was undertaken in September 1995, with a view to assessing the biodiversity existing in Cholistan desert, to form the basis for the management of the area on a sustainable yield basis.

The Study Area

The Cholistan, a protected forest, is being managed as a protected area for wildlife under the provisions of Punjab Wildlife (Protection, Preservation, Conservation and Management) Act-1974 since 1983. The initial notification was for five years period which has been extended for an other five year periods regularly. A part of Cholistan, 6533 km^2 in Bahawalnagar and Bahawalpur districts has been declared a wildlife sanctuary. The remaining 20,184 km^2 area being managed as a wildlife reserve. According to the legislation, hunting is permitted in a wildlife reserve under a special permit, where as in a wildlife sanctuary all types of population exploitation like hunting, poaching, killing, capturing or netting is strictly prohibited as is any other land use changing the ecology of the area.

The habitat in general is semi-desert to desertic in nature. Micro habitats support a variety of wildlife species, specially the birds. Some of the species found in Cholistan have enormous game value and regularly exploited by the hunters and poachers local as well as from abroad. Most important of these species are chinkara antelope, nilgai, blackbuck, houbara bustard, great Indian bustard and the imperial sandgrouse.

Houbara bustard is a globally vulnerable species and presently a matter of debate among the conservationists. This widely occurring migratory bustard is a quarry to the falconers all over the world wherever it is found. Royal hunting safaris from Gulf states visit Cholistan every winter to hunt houbara. Therefore a multi-national conservation strategy is required to save the future of the species.

Since the establishment of Punjab wildlife Research & Training Institute, Faisalabad, in the year 1986, the area is being surveyed once or twice a year to assess its biodiversity and the biotic potential of the habitat and its occupants, specially the commercially important game species to develop a long term management policy for the area. Such surveys, with the passage of time will provide sufficient base information to develop and adopt a comprehensive management plan. Ultimately the biodiversity of the area will be conserved and its sustainable harvest will become possible.

MATERIALS AND METHODS

Vegetation Sampling

Vegetation of the area was sampled by quadrats laid along a transect line in three distinct habitat types viz. sand dunes (height of dunes at some areas is up to 100 m but in lesser Cholistan it is only up to 25 m), interdunal flats or 'Dahars' (flat clayey pieces of land measuring several km^2) and saline patches (thick saline surfaces occurring in small patches of about 100 m^2 or more). Each transect was separated from the previous one by 20 km, traveled by 4x4 jeep (see map). Along each transect line 10 quadrats, 5x5 m each were taken perpendicular to the transact line, with a 5 m distance in between two consecutive quadrats. Frequency, density and per cent cover of all the species were recorded and their importance value was calculated in accordance with Hussain (1983).

Importance value = Relative frequency + Relative density + Relative cover

Plant species were collected for identification and herbarium record. Plant communities were studied in all habitat types keeping in view their socio-botanical aspect.

Sampling of Fauna

King's Strip Method was used to record the bird species and their population. The bird species flushed while traveling in the desert were recorded for their number and distance from the vehicle. The length of the strip was 10 km. After the end of a strip, new strip was started without any inter-strip distance. Binoculars (10 x 50 mm) were used to spot, observe and identify the bird species following Roberts (1991,1992) and Ali and Ripley (1985).

The data were standardized for distance and pooled to calculate average flushing distance, Y, and number of birds flushed per transact, Z. Keeping X, the length of the transact 10 km, the density, D was calculated using the following formula:

$$D = Z / 2YX$$

Literature was consulted to enlist all the vertebrate fauna recorded from the region, to develop a checklist for Cholistan for comparison and review purposes.

RESULTS

Vegetation of the Area

Vegetation in Cholistan is characterized by xerophytic adaptations, and depends largely on the erratic rain fall.. During the study 64 plant species belonging to 24 families were recorded (Table I).

Table I: Plant species recorded from Cholistan desert

Family	Plant species
Aizoaceae	*Aizoon canariense, Gisekia pharnaceoides, Limeum indicum, Sesuvium sesuvioides, Trianthema triquetra, Zaleya pentandra*
Amaranthaceae	*Achyranthes aspera* var. *aspera, Aerva javanica* var. *bovei*

Asclepiadaceae	*Calotropis procera* ssp. *hamiltonii, Leptadenia pyrotechnica*
Boraginaceae	*Heliotropium crispum, Heliotropium strigosum*
Brassicaceae	*Farsetia hamiltonii*
Capparidaceae	*Capparis decidua, Cleome brachycarpa, Cleome scaposa, Dipterygium glaucum*
Caryophyllaceae	*Cerastium fontanum*
Chenopodiaceae	*Haloxylon recurvum, Haloxylon salicornicum, Salsola baryosma, Suaeda fruticosa*
Convolvulaceae	*Cressa cretica*
Cucurbitaceae	*Citrullus colocynthis, Mukia maderaspatana*
Cyperaceae	*Cyperus conglomeratus*
Euphorbiaceae	*Euphorbia prostrata*
Malvaceae	*Abutilon muticum*
Mimosaceae	*Prosopis cineraria, Prosopis glandulosa*
Molluginaceae	*Mollugo cerviana*
Nyctaginaceae	*Boerhavia procumbens*
Papilionaceae	*Crotalaria burhia, Indigofera argentea*
Poaceae	*Aeluropus lagopoides, Aristida adscensionis, Aristida hystricula, Aristida mutabilis, Cenchrus biflorus, Cenchrus ciliaris, Cenchrus pennisetiformis, Cenchrus prieurii, Cymbopogon jwarancusa* ssp. *olivieri, Echinochloa colona, Eragrostis barrelieri, Eragrostis ciliaris, Lasiurus scindicus, Ochthochloa compressa, Panicum antidotale, Panicum turgidum, Schoenefeldia gracilis, Sporobolus ioclados, Stipagrostis plumosa*
Polygalaceae	*Polygala erioptera*
Portulacaceae	*Trianthema triquetra*
Polygonaceae	*Calligonum polygonoides*
Rhamnaceae	*Ziziphus mauritiana* var. *spontanea*
Scrophulariaceae	*Anticharis linearis*

Tamaricaceae	*Tamarix androssowii, Tamarix aphylla*
Tiliaceae	*Corchorus depressus*
Zygophyllaceae	. *Fagonia indica* var. *indica, Fagonia indica* var. *schweinfurthii, Tribulus longipetalus.* ssp. *longipetalus, Zygophyllum simplex*

The largest family was Poaceae where 19 grass species were recorded. Dominant plant species were *Aristida adscensionis, Cymbopogon jwarancusa, Ochthochloa compressa, Lasiurus scindicus, Sporobolus ioclados, Cenchrus biflorus, Aeluropus lagopoides, Calligonum polygonoides, Crotalaria burhia, Suaeda fruticosa, Salsola baryosma, Leptadenia pyrotechnica, Haloxylon recurvum, Haloxylon salicornicum, Dipterygium glaucum, Zaleya pentandra and Trianthema triquetra.*

Three habitat types were identified in the Cholistan desert *viz.*, sand dunes, interdunal flats, and saline patches. Plant species recorded from each of the three habitats are shown in Table II.

Table II: Some dominant plant species in different habitat types

Family	**Species**	**Habitat Types**		
		Sand dune	Interdunal flats	Saline Patches
Aizoaceae	*Trianthema triquetra*	--	+	+
Amaranthaceae	*Aerva javanica*	+	+	--
Asclepiadaceae	*Calotropis procera*	--	+	--
Capparidaceae	*Capparis decidua*	+	+	--
	Dipterygium glaucum	+	+	--
Chenopodiaceae	*Haloxylon recurvum*	--	--	+
	Haloxylon salicornicum	--	+	--
	Salsola baryosma	--	+	+
	Suaeda fruticosa	--	+	+
Cucurbitaceae	*Citrullus colocynthis*	+	--	--
	Mukia maderaspatana	+	--	--
Cyperaceae	*Cyperus conglomeratus*	+	--	--
Euphorbiaceae	*Euphorbia prostrata*	--	+	--
Mimosaceae	*Prosopis cineraria*	+	+	--
Molluginaceae	*Mollugo cerviana*	+	+	--
Papilionaceae	*Crotalaria burhia*	+	+	--
Poaceae	*Aeluropus lagopoides*	--	--	+
	Aristida adscensionis	+	+	--
	Cenchrus biflorus	+	+	--
	Cenchrus pennisetiformis	+	+	--
	Cymbopogon jwarancusa	--	+	--
	Lasiurus scindicus	+	+	--
	Ochthochloa compressa	+	+	+
	Panicum antidotale	+	+	--
	Panicum turgidum	+	+	--
	Sporobolus ioclados	--	+	--
Polygonaceae	*Calligonum polygonoides*	+	--	--
Zygophyllaceae	*Fagonia indica*	+	+	--
	Tribulus longipetalus	+	+	--

+ = present, -- = absent

Common plant species growing on sand dunes were *Leptadenia pyrotechnica* and *Calligonum polygonoides,* herb species were *Dipterygium glaucum, Mollugo cerviana, Polygala erioptera* and *Tribulus longipetalus* and grasses/ sedges were *Cyperus conglomeratus, Stipagrostis plumosa* and *Panicum turgidum.*

Vegetation of interdunal area consisted of *Aerva javanica, Calotropis procera, Farsetia hamiltonii* and *Capparis decidua* among shrubs, *Prosopis cineraria* a tree species, *Fagonia indica* a herb species and *Aristida adscensionis, Aristida mutabilis, Cenchrus biflorus, Cenchrus ciliaris, Panicum antidotale* and *Sporobolus ioclados* among grasses. Vegetation of saline area was specific where common species of shrubs were *Salsola baryosma* and, *Suaeda fruticosa,* herb species was *Trianthema triquetra* and grasses were *Aeluropus lagopoides* and *Ochthochloa compressa.*

The data were analyzed on the basis of importance value, 23 plant communities were identified (TableIII-IV).

Aristida adscensionis and *Ochthochloa compressa* were the most important grasses dominating most of the communities. *Aristida adscensionis* showed the importance value over 80 in seven communities while it was absent in *Suaeda fruticosa* and *Cymbopogon jwarancusa-Suaeda fruticosa* communities. *Ochthochloa compressa* depicted its dominance in four communities and was absent in six communities. Another important species among grasses was *Cymbopogon jwarancusa* showing its representation in most of the communities. *Dipterygium glaucum* and *Haloxylon recurvum* were the most frequent species among dicots. Other important species were *Crotalaria burhia* and *Calligonum polygonoides.*

Table III: Importance value of some dominant herb/ shrub species in various communities in Cholistan desert

Plant community	Name of plant species					
	Calligonum polygonoides	*Crotalaria burhia*	*Suaeda fruticosa*	*Salsola baryosma*	*Haloxylon recurvum*	*Dipterygium glaucum*
Aristida	13.7	9.9	1.8	8.5	18.7	13.4
Aristida-Crotalaria	24.8	67.5	0.0	0.0	6.3	17.5
Aristida-Salsola	21.9	0.0	0.0	0.0	71.3	31.7
Cenchrus-Lasiurus-Aristida	6.0	12.7	0.0	0.0	17.2	36.9
Cymbopogon	0.0	2.7	0.0	0.0	68.5	18.9
Cymbopogon-Aristida	0.0	10.4	0.0	0.0	24.8	8.8
Cymbopogon-Dipterygium-Aristida	6.4	5.3	0.0	3.1	15.8	40.7
Cymbopogon-Suaeda	4.0	0.0	109.0	21.2	19.9	0.0
Dipterygium	31.4	2.6	0.0	0.0	43.9	64.6
Dipterygium-Aristida	9.0	9.9	1.0	0.0	27.5	57.0
Dipterygium-Calligonum	106.9	0.0	0.0	0.0	16.2	102.3
Dipterygium-Lasiurus-Aristida	0.0	15.7	0.0	2.1	25.5	39.8
Haloxylon-Lasiurus	4.5	24.6	0.0	0.0	48.9	25.3
Lasiurus-Cymbopogon	0.0	2.3	0.0	0.0	36.0	34.0
Ochthochloa	1.7	0.0	0.0	0.0	20.6	0.9
Ochthochloa-Aeluropus	16.1	3.3	15.5	27.0	0.0	13.5
Ochthochloa-Aristida	15.6	26.9	0.0	0.0	4.7	11.2
Ochthochloa-Suaeda-Aeluropus	16.9	3.8	45.0	5.0	4.3	23.6
Ochthochloa-Zaleya-Aristida	27.9	2.1	0.0	0.0	2.1	20.4
Sporobolus-Ochthochloa	0.0	0.0	31.3	46.6	7.0	0.0
Suaeda	0.0	0.0	80.1	69.7	25.8	3.8
Suaeda-Lasiurus-Ochthochloa	10.7	20.7	27.1	0.0	19.1	17.9
Suaeda-Trianthema	0.0	6.9	63.2	9.6	8.1	0.0

Table IV: Importance value of some grass species in various communities in Cholistan desert

Plant community	Name of plant species					
	Aeluropus lagopoides	*Aristida adscensionis*	*Cymbopogon jwarancusa*	*Lasiurus scindicus*	*Ochthochloa compressa*	*Sporobolus ioclados*
Aristida	0.0	103.9	24.7	12.9	6.5	5.6
Aristida-Crotalaria	0.0	53.7	3.5	39.2	14.5	0.0
Aristida-Salsola	0.0	51.4	0.0	0.0	7.1	0.0
Cenchrus-Lasiurus-Aristida	0.0	70.2	1.8	40.0	24.3	0.0
Cymbopogon	0.0	10.0	80.1	60.9	0.0	0.0
Cymbopogon-Aristida	0.0	119.2	75.8	13.7	0.0	0.0
Cymbopogon-Dipterygium-Aristida	0.0	82.5	37.4	20.3	14.4	0.0
Cymbopogon-Suaeda	0.0	0.0	105.9	7.8	0.0	0.0
Dipterygium	0.0	43.2	5.7	15.9	0.0	0.0
Dipterygium-Aristida	0.0	88.9	12.0	11.4	13.5	0.0
Dipterygium-Calligonum	0.0	28.2	0.0	2.3	0.0	0.0
Dipterygium-Lasiurus-Aristida	0.0	81.4	15.4	43.6	16.3	0.0
Haloxylon-Lasiurus	0.0	34.9	9.0	57.4	4.4	0.0
Lasiurus-Cymbopogon	0.0	25.9	58.0	69.2	31.8	0.0
Ochthochloa	0.0	17.9	2.6	9.9	176.9	4.3
Ochthochloa-Aeluropus	92.3	3.1	0.0	0.0	89.7	0.0
Ochthochloa-Aristida	0.0	105.6	21.0	28.4	64.8	0.0
Ochthochloa-Suaeda-Aeluropus	61.4	24.6	0.0	11.4	33.0	0.0
Ochthochloa-Zaleya-Aristida	0.0	89.5	3.5	8.7	42.9	0.0
Sporobolus-Ochthochloa	17.2	22.9	4.7	2.5	86.2	56.1
Suaeda	0.0	0.0	38.7	12.5	50.8	11.4
Suaeda-Lasiurus-Ochthochloa	0.0	16.3	24.0	30.0	80.2	0.0
Suaeda-Trianthema	3.6	17.6	9.6	0.0	0.0	0.0

Fauna of the area

During the study period 58 bird species from 42 genera, 26 families and 12 orders were recorded (Table V).

Table V: Estimated density of different bird species in Cholistan Desert during February and September, 1995

Scientific Name	Density / Km^2	
	Feb. 95	Sep. 95
CICONIIFORMES		
Ardeidae		
Egretta garzetta	0.10	1.80
ACCIPITRIFORMES		
Accitpitridae		
Pernis apivorus	0.25	-
Elanus caeruleus	0.18	0.99
Milvus migrans migrans	0.17	9.58
Neophron percnopterrus	0.14	4.09
Gyps bengalensis	0.18	-
Gyps indicus	-	0.01
Circaetus gallicus	0.22	6.70
Circus macrourus	0.04	1.98
Circus pygargus	0.01	2.83
Accipiter badius cenchroides	0.17	10.00
Butastur teesa	-	0.04
Aquila chrysaetos	-	0.002
Aquila heliaca	-	0.01
Aquila clanga	-	0.02
Hieraaetus pennatus	0.008	4.49
FALCONIFORMES		
Falconidae		
Falco tinnunculus	0.16	0.33
Falco peregrinus	-	0.02
GALIFORMES		
Phasianidae		
Francolinus pondicerianus	-	22.24
GRUIFORMES		
Otididae		
Clamydotis undulata	0.12	0.16
CHARADRIIFORMES		
Recurvirostridae		
Himantopus himantopus	0.11	6.65
Glareolidae		
Cursorius coromadilicus	0.73	-
Charadriidae		
Hoplopterus indicus	0.57	19.41
PTEROCLIDIFORMES		
Pteroclidiae		
Pterocles orientalis	5.95	-
Pterocles exustus	1.13	-

COLUMBIFORMES		
Columbidae		
Columba livia	-	0.61
Streptopelia decaocto	2.62	10.66
Streptopelia tranquebarica	-	0.08
STRIGIFORMES		
Strigidae		
Athene brama	-	0.05
APODIFORMES		
Apodidae		
Apus affinis	1.41	7.49
CORACIIFORMES		
Meropidae		
Merops orientalis	-	1.60
Merops superciliosus	-	1.42
Coraciidae		
Coracias benghalensis	1.09	5.94
Upupidae		
Upupa epops	0.16	13.40
PASSERIFORMES		
Alaudidae		
Eremopterix grisea	4.11	30.27
Alaemon alaudipes	2.41	52.43
Callendrella brachydactyla	-	78.56
Callendrella rufescens	-	28.68
Galerida cristada	1.04	17.45
Alauda gulgula	1.67	2.70
Hirundinidae		
Hirundo fluvicola	-	2.04
Motacillidae		
Anthus campestris	0.60	12.05
Pycnonotidae		
Pycnonotus leucogenys leucotis	0.22	13.20
Pycnonotus cafer	-	0.11
Turdidae		
Oenanthe isabellina	2.64	13.23
Oenanthe deserti	0.71	-
Oenanthe picata	1.02	9.90
Timaliidae		
Turdoides caudatus	2.48	87.53
Laniidae		
Lanius exculeitor	0.63	5.48
Lanius schach	0.11	6.70
Dicruridae		
Dicrurus macrocereus	0.02	17.09
Corvidae		
Corvus splendens	0.32	6.48
Corvus corax	0.003	1.98
Sturnidae		
Sturnus roseus	1.25	74.63
Acridotheres tristis	0.38	22.73
Acridotheres ginginianus	-	1.14
Passeridae		
Passer domesticus	58.6	143.00

Passeriformes is the most dominant order represented by 24 species belonging to 11 families, followed by Accipitriformes represented by 15 species all belonging to a single family, Accipitridae. Other important avian orders found in Cholistan are Coraciiformes, three families represented by four species and Charadriiformes, three families with a representation of one species each. All other orders are represented by single family having one species only.

A comprehensive check list of vertebrate fauna (Reptiles, Birds and Mammals) has been prepared from the literature (Table VI).

Table VI. Checklist of terrestrial vertebrates of Cholistan desert

Habit: R = resident, W = wintering, I = irregular year round visitor, OM = ordinary migrant, V = vagrant, SM = summer migrant, OW = occasional wintering, PM = passage migrant, SpM = spring migrant; Breeding: + = breeds in the area, - = does not breed in the area; Status: A^+ = very abundant, A = abundant, F = frequent, C = common, U = uncertain, S = scarce, S^+ = very scarce, SR = scarce becoming rare, R = rare

Family	Name of species	Habits	Breeding	Status
	AMPHIBIAN*			
Bufonidae	*Bufo stomaticus*	R	+	U

* Status as given in Khan, M. S. 1992. Reptiles and amphibians of Pakistan. In 'Pakistan ki Jangli Hayat', Urdu Science Board, Lahore. (Urdu Language):

Family	Name of species	Habits	Breeding	Status
	REPTILES			
Varanidae	*Varanus bengalensis*	R	+	U
Agamidae	*Uromastrix hardwicki*	R	+	U
Chamoeleonidae	*Chamaeleo zeylanicus*	R	+	U
Gekkonidae	*Hemilacytylus brooki*	R	+	U
Elapidae	*Naja naja*	R	+	U
	Bungarus caereleus	R	+	U
Viperidae	*Echis carinatus*	R	+	U
Typhlopidae	*Typhlops porractus*	R	+	U
	Typhlops braminus	R	+	U
Boidae	*Eryx conicus*	R	+	U
Calubridae	*Natrix piscator*	R	+	U
	BIRDS**			
Accitpitridae	*Elanus caeruleus*	R	+	C
	Neophron percnopterrus	R	+	C
	Gyps bengalensis	R	+	A
	Gyps fulvus	W	-	F
	Aegypius monachus	W	-	S
	Circaetus gallicus	R, I	+	F
	Circus macrourus	W	-	C
	Butastur teesa	R	+	A
	Buteo buteo	W	-	F

	Buteo rufinus	W	-	C
	Aquila rapax vindhiana	R	+	C
	Aquila rapax nipalensis	R	+	C
	Aquila heliaca	W	-	S
	Hieraaetus fasciatus	W	-	S
Falconidae	*Falco tinnunculus*	W	-	C
	Falco chicquera	R	+	U
	Falco biarmicus jugger	R	+	SR
	Falco cherrug	W	-	R
	Falco cherrug milvipes	W	-	R
	Falco peregrinus	W	-	S
	Falco peregrinoides	W	-	S
Phasianidae	*Francolinus pondicerianus*	R	+	C
	Coturnix coturnix	OM	-	C
Gruidae	*Anthropoides virgo*	OM	-	C
Otididae	*Clamydotis undulata*	W	-	SR
	Ardeotis nigriceps	V	-	R
Recurvirostridae	*Himantopus himantopus*	R	-	A
Glareolidae	*Cursorius coromadilicus*	W	-	S
Charadriidae	*Hoplopterus indicus*	R	+	A
Scolopacidae	*Tringa ochropus*	W	-	C
Pteroclidae	*Pterocles senegallus*	W	-	F-C
	Pterocles exustus	R	+	C-A
	Pterocles orientalis	W	-	C
Columbidae	*Columba livia*	R	+	A
	Streptopelia decaocto	R	+	A
	Streptopelia tranquebarica	SM	+	A
	Streptopelia senegalensis	R	+	A
Psittacidae	*Psittacula krameri*	R	+	A
Cuculidae	*Clamator jacobinus*	SM	+	C
Tytonidae	*Athene brama*	R	+	C
	Asio flammeus	W	-	S
Caprimulgidae	*Caprimulgus maharattensis*	R	+	C
	Caprimulgus europaeus unwini	OW	-	C
Apodidae	*Apus affinis*	R	+	A
Alcedinidae	*Halcyon smyrnensis*	R	+	C
Meropidae	*Merops orientalis*	R	+	A
	Merops superciliosus	SM	+	A
Coraciidae	*Coracias benghalensis*	R	+	C
Upupidae	*Upupa epops*	W	-	C
Picidae	*Dendrocopos assimilis*	R	+	F
	Dendrocopos maharattensis	R	+	F
Alaudidae	*Eremopterix nigriceps*	I	-	C-F
	Eremopterix grisea	R	+	C
	Alaemon alaudipes	R	+	S
	Callendrella brachydactyla	W	-	A
	Callendrella rufescens	W	-	F
	Callendrella raytal	R	+	C
	Galerida cristata	R	+	A
	Alauda gulgula,	R	+	A
Hirundinidae	*Riparia paludicola*	R	+	A
	Hirundo rustica	W	-	A
Motacillidae	*Anthus campestris*	W	-	C
	Anthus trivialis	OM	-	F
	Motacilla flava beema	PM	-	U
	Motacilla cinerea	W	-	F-C
	Motacilla alba dukhunensis	W	-	
Campephagidae	*Tephrodornis pondicerianus*	R	+	C
Pycnonotidae	*Pycnonotus leucogenys leucotis*	R	+	A
	Pycnonotus cafer	R	+	A
Turdidae	*Luscinia svecica*	W	-	C
	Phoenicurus ochruros	W	-	C
	Saxicola torquata	OM	-	C
	Saxicola caprata	R	+	C-A

	Oenanthe isabellina	W	-	C
	Oenanthe deserti	W	-	C
	Oenanthe xanthoprymna	W	-	U
	Oenanthe picata	W	-	C
	Saxicoloides fulicata	R	+	C
	Turdus ruficollis	W	-	A
Sylvidae	*Prinia gracilis*	R	+	C
	Prinia buchanani	R	+	C-A
	Prinia inornata	R	+	C
	Prinia burnesii	R	+	F
	Orthotomus sutorius	R	+	C
	Hippolais caligata	PM	-	C
	Sylvia nana	W	-	C
	Sylvia curruca	W	-	A
	Phylloscopus neglectus	W	-	U
	Phylloscopus collybita	W	-	A
Muscicapidae	*Ficedula superciliaris*	W	-	U
Timaliidae	*Chrysomma sinense*	R	-	F-C
	Turdoides caudatus	R	+	A
	Turdoides earlei	R	+	C-A
	Turdoides striatus	R	+	A
Nectariniidae	*Nectarinia asiatica*	R	+	C
Oriolidae	*Oriolus oriolus*	OM	-	C
Lanidae	*Lanius isabellinus*	W	-	F
	Lanius excubitor	SM	+	C
Dicruridae	*Dicrurus macrocereus*	R	+	A
Corvidae	*Dendrocitta vagabunda*	R	+	C
	Corvus splendens	R	+	A^+
	Corvus corax	R	+	F
Sturnidae	*Sturnus roseus*	OM	-	A^+
	Acridotheres tristis	R	+	A^+
	Acridotheres ginginianus	R	+	A
Passeridae	*Passer domesticus*	R	+	A^+
	Passer hispaniolensis	W	-	C
	Petronia xanthocollis	SpM	+	C
Ploceidae	*Ploceus philippinus*	R	+	A
Estrildidae	*Eodice malabarica*	R	+	C
	Carpodacus erythrinus	OM	-	A
	Emberiza bruniceps	OM	-	A

** For all the bird species, status as given in Roberts, T. J. 1991 & 1992. The birds of Pakistan, vol. 1 & vol. 2, Oxford Univ. Press, Karachi.

MAMMALS***

Erinaceidae	*Hemiechinus auritus*	R	+	U
Soricidae	*Suncus murinus*	R	+	U
Rhinopomatidae	*Rhinopoma microphyllum*	R	+	U
	Rhinopoma hardwickei	R	+	U
Vespertilionidae	*Pipistrellus mimus*	R	+	U
	Pipistrellus kuhli	R	+	A
	Scotophilus heathi	R	+	U
Canidae	*Canis lupus*	R	+	S
	Canis aureus	R	+	S
	Vulpes vulpes	R	+	A
Mustelidae	*Mellivora capensis*	R	+	S^+
Viverridae	*Viverricula indica*	R	+	U
	Herpestes edwardsi	R	+	C
Felidae	*Felis libyca*	R	+	S
	Felis chaus	R	+	C
	Felis caracal	R	+	S
Suidae	*Sus scrofa cristatus*	R	+	C
Bovidae	*Boselaphus tragocamelus*	R	+	S
	Antilope cervicapra	R	+	S^+
	Gazella bennetti	R	+	C

Leporidae	*Lepus nigricollis*	R	+	F
Sciuridae	*Funambulus pennanti*	R	+	U
Hystricida	*Hystrix indica*	R	+	S
Muridae	*Rattus meltada*	R	+	S
	Mus booduga	R	+	U
	Nesokia indica	R		U
	Gerbillus cheesmani	R	+	U
	Gerbillus nanus	R	+	U
	Tatera indica	R	+	C
	Meriones hurrianae	R	+	U

*** For all the mammal species, status as given in Roberts, T. J. 1977. The mammals of Pakistan, Ernest Benn Ltd., London.

Birds are the dominating group in the area represented by 112 species from 73 genera belonging to 39 families and 15 orders, followed by mammals represented by 31 species of 24 genera, 14 families and 6 orders. The reptile fauna comprises of 11 species all from different genera representative of 9 families of only one order Squamata, whereas the Amphibians are represented by a single species.

The relationship with habitat and status of all the species enlisted in Table VI have been summarized in Table VII. Among birds 55 species (49 %) are resident of the area, 38 species (34 %) use to winter in the area, 8 species (7 %) are ordinary or double passage migrants, 2 species (2 %) are passage migrants, 1 species (1 %) is each vagrant.

Table VII: Summary of status categories of Amphibians, Reptiles, Birds, and Mammals in Cholistan desert

Status Category	**Number of Species (% of the total)**			
	Amphibians	Reptiles	Birds	Mammals
HABITS				
Resident	1 (100)	11 (100)	55 (49)	31 (100)
Wintering	-	-	38 (34)	-
Irregular year round visitors	-	-	2 (02)	-
Ordinary migrants	-	-	8 (07)	-
Vagrants	-	-	1 (01)	-
Summer migrants	-	-	4 (03)	-
Occasional wintering	-	-	1 (01)	-
Passage migrants	-	-	2 (02)	-
Spring migrants	-	-	1 (01)	-
BREEDING				
Breeding in the area	1 (100)	11 (100)	58 (52)	31 (100)
Not breeding in the area	-	-	54 (48)	-
POPULATION STATUS				
Very abundant	-	-	4 (03)	-
Abundant	-	-	30 (27)	2 (6.5)
Frequent	-	-	10 (09)	1 (03)
Common	-	-	42 (38)	5 (16)
Scarce	-	-	8 (07)	7 (23)

Very scarce	-	-	-	2 (6.5)
Scarce becoming rare	-	-	2 (02)	-
Rare	-	-	3 (03)	-
Variable	-	-	8 (07)	-
Uncertain	1 (100)	11 (100)	5 (04)	14 (45)

Occasional wintering, spring migrant whereas 2 species (2 %) are irregular year round visitors. All mammals, reptiles and amphibians are resident of the area. Fifty eight species of birds (52 %) breed in the area. All mammals, reptiles and amphibians also breed locally.

All species have varied population status in the area. Among birds 4 species (3 %) are very abundant, 30 species (27 %) are abundant, 10 species (9 %) are met with frequently, 42 species (38 %) are common, 8 species (7 %) are scarce, 2 species (2 %) are scarce becoming rare and 3 species (3 %) are rare. Eight species of bird (7 %) showed a variable status in the area where the status of 5 species (4 %) is uncertain (Roberts, 1991 & 1992).

Out of 31 species of mammals, 2 species (6 %) are abundant, 1 species (3 %) is found frequently, 5 species (16 %) are common, 7 species (23 %) are scarce and 2 species (6 %) are very scarce, where the status of 14 species (45 %) is uncertain (Roberts, 1977).

No studies have been carried out on the status of reptiles and amphibians in the area, therefore, their status may be considered as uncertain on the basis of incomplete data available.

DISCUSSION

Vegetation of Cholistan was specific at different soil types. Only *Ochthochloa compressa* was found in all the three types. Vegetation at saline/ sodic soils was very specific and very few species were recorded there. Jalal-ud Din and Farooq (1975) reported *Salsola baryosma, Suaeda fruticosa, Tamarix aphylla* and *Prosopis glandulosa* at Lalsohanra but in the present investigation in addition to *Salsola* and *Suaeda, Aeluropus lagopoides, Ochthochloa compressa* and *Trianthema triquetra* were also recorded quite frequently.

Vegetation at interdunal flats varied greatly from the other habitats and also maximum number of species were recorded there. Rao *et al.* (1989) reported two dominant grasses, *Cymbopogon jwarancusa* and *Lasiurus scindicus,* whereas in our case *Aristida adscensionis* dominated all the species. Other dominant species were *Cenchrus biflorus* and *Sporobolus ioclados, Tamarix aphylla, Acacia nilotica* and *Prosopis cineraria* were also occasionally reported which were according to Rao *et al.* (1989) dominant species.

Unlike general conceptions about desert areas the Cholistan is quite adaptable as a living place for a variety of animals. Birds having greater mobility to avoid unfavourable circumstances and find and exploit means of livelihood, are the most dominating vertebrates of the area, dominating with respect to parameters of diversity namely greater number of species, genera, families and orders, i.e., 112 species, 73 genera, 39 families and 15 orders. Closest competitors of birds are mammals represented by 31 species, 24 genera, 14 families and 16 orders.

According to the present compilation of records, 11 species of reptiles all representing different genera from nine families of order Squamata (snakes and lizards) occur in the area. The figure seems quite unimpressive while considering the adaptability of the group to the drier conditions. It is probably due to the fact that very few studies have been carried out to enlist the biodiversity of the region, especially reptiles. The knowledge about amphibious fauna is still more discouraging. Only one species of toad, *Bufo stomaticus* is reported to occur in the area (Khan, 1991). The presence of 'Tobas' is an important feature in the biodiversity of Cholistan. 'Tobas' are man-maintained ponds where rain water is collected and stored. After rains the water in these tobas persists for months. It is used as drinking water by the desert dwellers and their livestock. The presence of such ponds in the area may have been supporting greater variety of amphibians but serious systematic studies are required to discover the facts.

Mass migrations have not been reported among mammals and reptiles of the region which may be considered as permanent residents of the area. Among the birds 49 % of the species are reported to be resident and 34 % regularly winter in the area. Many of these species are important game species such as chinkara antelope, nilgai, hare, houbara bustard, great Indian bustard, imperial sandgrouse, grey partridge and blue rock pigeon. These demand prime consideration while implementing biodiversity conservation programmes in the area.

Houbara bustard and falcons are internationally important species. Every winter royal families from the Kingdom of Saudi Arabia and Gulf States visit the area to practice falconry on houbara. During their hunting safaris they hardly ever observe the hunting regulations and the poor quarry is hunted indiscriminately. Houbara is also netted by local people to be sold to Arab Sheikhs who use these to train their falcons for falconry. Due to such practices the population of houbara has greatly declined during the past few decades. Similar is the case with falcons which are netted to be sold to falconers at very high prices. Reptiles, monitor lizard and the spiny-tailed lizard are also taken for skin/ hide trade and food by local non-Muslim tribes. Spiny-tailed lizard is exploited also for the medicinal value attached to its fat.

Over exploitation of vegetation resources by grazing livestock, fire wood, soda ash extraction (from *Haloxylon recurvum*) and disturbance caused by activities of man such as mechanized hunting, oil explorations and other related activities are affecting the biodiversity of the region persistently. Fewer species are found abundantly, commonly or fairly. Status of all reptiles and 45 % mammals of the region is still uncertain. There is a need to conduct detailed investigations to assess the resources of the region on scientific grounds and adopt comprehensive measures to conserve, manage and exploit the biodiversity of the region.

There is a growing concern in the world to involve private sector in the management of natural resources. In most of the cases, people of the area are in direct interaction with the habitat, benefiting from it since generations. Management operations interfere with the needs of the people and they view such management as contrary to their rights. Therefore for successful management of large areas such as Cholistan, carefully designed Participatory Rural Appraisal (PRA) should be carried out with a view to involving local population in implementing management options.

The wildlife resources of Cholistan, specially houbara bustard and the interest of 'Arab Shaikhs' in hunting houbara collectively can generate great economic gains for the area. Agreements should be worked out amongst the visiting falconers, peoples of the area and the Punjab Wildlife and Parks Department on the following lines:

a) The falconers to invest capital for the welfare of the people, and the management of wildlife and their habitats.

b) Local population in collaboration with the Punjab Wildlife & Parks Department to protect the houbara population.

c) Scientific studies be undertaken to determine the population trends and harvestable surpluses for houbara and other game species.

d) Management decisions be taken in collaboration with local communities.

e) Falconry be allowed under the prescribed limits of exploitation.

The proposed scheme have workability and potential for saving future of wildlife of the area in special and whole biodiversity complex in general but its success can only be expected after all the concerned parties work in close collaboration, with honesty and fulfill their respective obligations on time.

REFERENCES

ALI, S. I. 1983. *Flora of Pakistan*, No. 150, Asclepiadaceae. Department of Botany, University of Karachi, Karachi.

BAIG, M. S., M. AKRAM, M. A. HASSAN. 1980. Possibilities for range development in Cholistan desert as reflected by its physiography and soils. *Pak. J. Forestry*, **30**: 61-71.

FOOD AND AGRICULTURE ORGANIZATION OF THE UN. 1993. Cholistan Area Development Project, Socio-economic and production systems, Diagnostic study, Investment Centre, FAO/Asian Development Bank Cooperative Programme.Food and Agriculture Organization of the UN. 1993.

HUSSAIN, F. 1983. *Field and Laboratory manual of plant ecology*, National academy of Higher Education, UGC, H-9, Islamabad, Pakistan.

KHAN, M. S. 1992. *Reptiles and amphibians of Pakistan. In 'Pakistan ki Jangli Hayat'*, Urdu Science Board, Lahore. (Urdu Language):

KHAN, S. R. A. 1992. *Agricultural development potential of Cholistan Desert,* 213-N, LCCHS, Lahore Cantt., Pakistan.

JALAL-UD DIN, M. FAROOQ. 1975. Soil variations in relation to forest management in Lalsohanra irrigated plantation. *Pak. J. Forestry*, **25**: 5-13.

LEOPOLD, A. S. 1963. The desert. *Time-life Int.*, Netherlands.

QAISER, M. 1982. *Flora of Pakistan, No. 141, Tamaricaceae.* Department of Botany, University of Karachi, Karachi.

RAO, A. R., M. ARSHAD. 1991. Perennial grasses of Cholistan Desert and their distribution. People's participation and management of resources in arid lands, *CHIDS, Islamia Univ., Bahawalpur.*

RAO, A. R., M. ARSHAD, M. SHAFIQ. 1989. Perennial grass germplasm of Cholistan Desert and its phytosociology, *CHIDS, Islamia Univ., Bahawalpur.*

ROBERTS, T. J. 1977. *The mammals of Pakistan*, Ernest Benn Ltd., London.

ROBERTS, T. J. 1991. *The birds of Pakistan*, vol. **1**, Oxford Univ. Press, Karachi.

ROBERTS, T. J. 1992. *The birds of Pakistan*, vol. **2**, Oxford Univ. Press, Karachi.

SCHEMNITZ, S. D. 1980. *Wildlife management techniques manual,* 4th Ed., The Wildlife Soc., Inc., USA.

THE STATE OF MAJOR ECOSYSTEMS OF PAKISTAN

SHAHZAD A. MUFTI AND S. AZHAR HASAN

Pakistan Museum of Natural History, Garden Avenue, Skakarparian, Islamabad

Pakistan covers a land area of 310, 400 square miles, or 79.1 million hectares. In the north, are present the outer mountain chains and foothills which form the watershed of the great Indus river. This river covers a distance of nearly 2,900 kilometers (1,800 miles) inside Pakistan territory. Almost the entire country, thus can be considered as the watershed of this mighty river. The heartland of the country is the flat alluvial flood-plain which forms the drainage basin of the Indus, including its five major tributaries fanning out in a north-easterly direction across the Punjab. The Indus basin, for it can hardly be called a valley, covers an area of about 360,000 square miles (GOP Report, 1978). It is in this region that the bulk of the human population is settled, and well over 80 per cent of the nation's agricultural wealth and food production is derived. Roughly two-thirds of Pakistan's surface area is mountainous. The Indus drainage basin is bounded not only by range upon range of high mountains to the north, but also by jagged rocky mountain ranges extending southwards and down its western flanks to the Arabian seacoast, starting from Chitral in the north, and down through the Safed Koh and other ranges in the North West Frontier and Balochistan provinces.

Rainfall, tends to be highly seasonal over most of the country and with few exceptions is not well distributed. It ranges from as low as 20-40 mm annually in the Chaghai and Sibi deserts, as well as the main valleys of Ishkoman, Yasin and Hunza in Gilgit and the Indus and Shyok valleys in Baltistan, to as high as 1,350 mm. (53 inches) in the Murree hill range. This hill range is unique in having a fairly evenly distributed annual rainfall with up to 6.5 meters (22 feet) of winter snow falling in the north-western part of the range, which is known as the Galis. Over the Indus plains, rain falls mainly during the monsoon season, whilst in the North West Frontier Province there is distinctly Mediterranean character to the climate (and some Mediterranean plant affinities also) with up to 60 per cent of the rainfall occurring in the cold winter months. Further north, in Chitral and Dir, up to 90 per cent of the annual precipitation (snow and rain) falls during the winter. In the foothill regions of the Punjab and Hazara districts, rainfall is more evenly distributed with a pronounced monsoon influence, but at least 25 per cent of rain does fall in the winter and there is also some rainfall in almost every month of the year.

ECOSYSTEMS

Climatic variations significantly affect the vegetation and its associated fauna. Pakistan therefore comprises many different ecosystems including marine, coastal, mangrove, deltaic, riverine wetland, dry desert, tropical thorn, mountain and cold desert. An ecosystem, comprising both biotic and abiotic factors of the environment, inseparably inter-related and interactive in their nature, is a very precious thing, providing man with all the basic ingredients of everyday living. Pakistan , a developing nation with limited natural resources is faced with many challenges and hopes. In Pakistan, natural systems such as rivers, mountains, land, forests, water, and soils are holistic and interconnected. They provide food, fiber, and other products for about 130 million people in an otherwise arid and semi-arid ecosystem. This population will double itself in next 15 years or so and the vulnerable environmental resources will be under tremendous pressure. If these resources are depleted or destroyed rapidly or are not sustained, the ecological support system for mankind may erode and eventually collapse. For many centuries primitive mankind existed in this region in a state of relative ecological equilibrium. It was not till the last four or five decades of this century that Pakistan experienced profound and accelerated ecological changes resulting form the greatly increased human population, industrialization, massive deforestation, salinity and water-logging combined with progressive desertification, decline of living resources and wildlife. In fact our environmental problems rank among the most serious in this part of the world. Various aspects of development and environmental management have been dealt with on an ad hoc basis in the past. Consequently, development plans have fallen short of our expectations in agriculture, housing, water, energy and industrial sectors. In many cases, developments have also brought about unexpected environmental, and public health related problems. The extent and assessment of the pervasive environmental stress is not only widespread, but also complex to compute in terms of resource depletion and lost human and land productivity.

TERRESTRIAL ECOSYSTEM

More than 60% of Pakistan's terrestrial ecosystem consists of rugged mountains or desert. The land-use pattern. (Table I) shows that an area of 26 million hectares is either under water or covered by cities, towns, villages roads and other use patterns. The land available for cultivation is 33 million hectares though only 21 million of this is actually cultivated, accounting for 24% of the total land area and 64%

of cultivable land. There are about 4-5 million hectares of forest land lying outside cultivated area.

Table I: Land-use pattern

Land-use category	Area(million ha)	% share
Cultivated	20.7	23.6
Cultivable waste	11.8	13.4
Not available for cultivation	25.9	29.5
Forests	04.5	05.0
Not classified/reported	25.0	28.0
Total	**87.9**	**100.0**

Source: Govt. Of Pakistan, M/O Food, Agriculture and Cooperation (1984): Forestry Range and Wildlife Management in Pakistan.

AQUATIC ECOSYSTEM

The aquatic ecosystem in Pakistan can be divided into inland waters, estuarine waters, and marine waters. The inland water bodies include the rivers, streams (mainly those of Indus system), lakes and other wetlands as well as underground waters. The estuarine water occupies lower delta of the river Indus where it splits into numerous distributaries and is often inundated by tidal floods. The marine waters spread over an area of 105,400 sq. nautical miles in the Exclusive Economic Zone of Pakistan extending for a distance of 200 nautical miles parallel to the coast from Kutch border in the east to Iranian border in the west.

The process of development has brought some major impact on the aquatic ecosystem of Pakistan. More than 95% of inland water is being used for irrigation. Although irrigation plays a significant role in increasing agricultural production, several ecological problems such as water-logging, salination, alkalization, lost of forest cover and associated fauna have been associated with improper irrigation practices. The water-table has risen due to seepage from reservoir and irrigation channels at an average rate of 15 to 35 cm per year. Furthermore, in an area where the underground water has a salinity of 1000 part per million, evaporation at the rate of about half meter per year will increase the salt content to about one percent in 20 years (Spooner, 1982).

Main factors playing a major role in the degradation of our land and aquatic ecosystems can thus be summarized as follows:

Deforestation and forest degradation

With recent increase in the prices of oil, wood, charcoal and alternate fuel, people have been indiscriminately cutting and selling trees on a large scale, thus causing extensive deforestation. Pakistan is already in an unhappy position of being amongst the group of countries with the lowest areas under forest in the world, the forest area being less than 0.05 hectares per capita (IUCN, 1989; IUCN/ GOP, 1991). Further depletion of forest resources will be simply disastrous. Extensive deforestation particularly in the water-sheds is a major ecological threat to the nation. Due to the loss of tree cover on slopes the silt load in the river is rising and choking up reservoirs behind major dams, causing increasing severity of floods in the country.

Afforestation and reforestation of hill sides and of land unsuitable for field crop production are essential for the effective protection of the water sheds of rivers and reservoirs, for protecting the fragile ecology and especially the water regime in many areas, for combating the increasingly critical fuel scarcity, for stabilizing hill slopes threatened with severe erosion and mass movements etc. The quality of afforestation work has improved significantly, and many recently reforested hill sides are now healing and have come to acquire a contiguous tree cover (IBRD Report, 1985).

Soil erosion

Soil erosion is another major scourge of the terrestrial environment in Pakistan. The eroded land in the country is about 1.2 million ha. (GOP Report, 1988). About 40% of the land is affected to varied intensity by wind and 36% by water.

Desertification

Desertification is a serious problem globally but it is more acute in case of Pakistan where almost three fourth of the land is either already affected or likely to be affected by it. Desertification is a process or a set of processes which cause diminution of the biological potential of land resources. The resources include land-use and land cover, crop production, forest cover, livestock, soils etc.

Water-logging and salinity of land

Water-logging and salinity, the twin menace, affect productivity of about 9.5 million hectares of irrigated land in the Indus Basin. Losses of about Rs. 36 billion are estimated for decline in crop productivity due to salinity and water-logging annually in the Indus Plains. In 60 per cent of the irrigated lands in Sindh, the water table is within 0-3 meters of the surface, in 15 per cent of land in Punjab, in 10 percent of land in NWFP. The lower Indus Plains of Sindh will become a saline and swamp land similar to the Rann of Kutch as manifested by increasing salinity and water-logging, restriction of downstream flow of the Indus River due to upstream impounding in the mega reservoirs, degradation of quality of river waters, and marine intrusion in the Indus Delta (Gazdar, 1987).

River pollution, erosion, silting and sedimentation

Water development and planning has been perceived considering water as a technical element. Water is an element of landscape and its use and control requires conservation, preservation of its quality and beneficial usage. River pollution has reached alarming proportions in major and minor streams in the Indus River System from the point and non-point sources. Dumping of farm effluents, pesticides, toxic industrial effluents, wastes and metals, and of raw sewage and garbage into streams and rivers is a common practice. The pollution of surface and groundwater is increasing and about 40 per cent of reported sickness cases are attributed to bad quality of drinking water in the country.

Floods, draughts and severe water phenomena

Without the natural resources of Himalayan-Karakoram-Hindu Kush Mountains and the Indus River System and its plains, the landscape of Pakistan would be a barren desert instead of productive plains of the Indus Basin. The fragile balance of nature, the erratic behavior of monsoons and the aggregation of climatic events such as floods and draughts is very much highlighted by climate. The climatic change has unavoidable consequences for the development process. The draught of 1984-88 in the Thar and Coastal Makran-Las Bela afflicted thousands of people with starvation and disease with loss of 4,000 lives, and almost all of the crops and live-stock were lost. Flooding of Indus Plains is a severe recurring problem accentuated by the deforestation and mismanagement of the vast irrigation system of canals and link canals. Two major floods in 1988 monsoon season affecting 4Mha in the Indus Plains have caused loss of human lives and misery in both the rural and urban areas. Crops and livestock damages were estimated over Rs. 5 billion in Punjab alone.

It is quite clear from the above description that both land and aquatic ecosystems are threatened in Pakistan and are undergoing deterioration at a fairly rapid rate. Consequently the biodiversity which is an integral part of a healthy ecosystem is also undergoing severe stress. There is thus an urgent need that various habitats and ecosystems present in Pakistan be saved from further deterioration and rehabilitated. Only after these endeavours, the biodiversity of Pakistan can be maintained in a sustainable way.

REFERENCES

GAZDAR, M. N. 1987. Natural Resources Development and Environmental Management in Pakistan. The Open Press., Kaula Lumpur, Malaysia.

GOP PLANNING COMMISSION. 1978. The report of the Indus Basin. Research Assessment Group, Islamabad

GOP REPORT.1988. Environmental Profile of Pakistan. Environment and Urban Affairs Division, Islamabad.

IBRD 1985. Pakistan Environmental Rehabilitation, Protection and Management: Reconnaissance Mission Report.

IUCN 1989. Pakistan Fact Sheets: Deserts, Mountains, Lakes, Forests, and Water. December 1989. Journalists' Resource Centre, IUCN, Pakistan

IUCN/ GOP 1991. National Conservation Strategy. Where we are, where we should be and how to get there. World Conservation Union IUCN)/ Government of Pakistan.

SPOONER, B. 1982. Environmental Problems and The Planning of Development in Arid South West Asia in *Environment and Development in Asia and The Pacific: Experience and Prospects.* UNDP Reports and Proceedings Series 6.

GENETIC DIVERSITY AND ITS ROLE IN AFFORESTATION IN PAKISTAN

SHAMS R. KHAN

Pakistan Forest Institute, Peshawar

Abstract: Forest trees are integral part of human society which provide timber, fuel, fodder, food, fiber and medicines among many other things. Several species in Pakistan exhibit variation in many traits in their native habitats as these grow under varied edaphic and geoclimatic conditions. As a result, different vegetation types originating from coastal, sub-tropical, tropical and temperate areas including those of arid and semi-arid zones are found in the country. Thus there is a need that the extent, nature and pattern of variation in important tree species may be tapped and used in afforestation in order to achieve the aforementioned goods and services as forests are rich reservoirs of biological diversity.

A Himalayan white pine *Pinus wallichiana* A.B. Jacks is an important high hill forest tree species of Pakistan which covers a wide range of distribution starting from Gilgit / Chitral in the north and extending upto Zhob in the south with an altitudinal range from 1500-3100 m. In the east-west direction, it touches the borders of India and Afghanistan. The growth data of 12 provenances in their natural habitat revealed that the trees in high rainfall areas (> 725 mm/ annum) had higher MAI as compared with those of low rainfall areas (< 725 mm/annum) indicating that the trees of former habitat are growing faster than the latter. Height growth data on two-year-old seedlings of provenances originating from mesic and xeric habitats had also shown similar trend under controlled conditions as for MAI, verifying the genetic adaptation of two varieties in Pakistan. Marked differences between two habitats in several morphological and anatomical characters in the natural stands of the species had already been reported confirming the ecotypic differentiation in this pine.

These results therefore, suggest that the seed from low rainfall areas should not be transferred to high rainfall areas and vice versa for afforestation as currently practiced. Such restrictions on transfer of germplasm are equally applicable to another important coniferous species viz. spruce *(Picea smithiana)* which possesses more or less similar ecological and silvicultural requirements in Pakistan. It is expected that by following these recommendations in two valuable timber species, the survival and growth rate in the plantations could be improved.

INTRODUCTION

Natural forests in general and those in Himalayas in particular are great reservoirs of diversity because many species in these lofty mountainous areas are growing under varying geo-climatic conditions and are spread over several countries. Some of the tree species exhibiting such gross variation in several traits and occupying diversified ecogeographical areas are: *Abies pindrow, Acacia nilotica, Betula utilis, Cedrus deodara, Picea smithiana, Pinus roxburghii, P. wallichiana, Prosopis cineraria* and P. *juliflora* etc. Since the quantum of variation is the basis of any tree improvement programme, it is imperative that the extent, nature and pattern of diversity existing in any species may be properly explored.

In view of the importance of pines in the national economy of Pakistan, one of the extensively distributed and valuable timber species, P. *wallichiana A.B.* Jackson, has been chosen for this purpose. This pine was found to grow under diverse climatic and edaphic conditions and exhibits differentiation at the ecotypic level in its native habitat (Shams 1986a, 1986b, 1991, 1992, 1994 and 1995). Some other pines like *Pinus caribaea* (Barret and Golfari 1962); P. *elliottii* (Squillace 1966); *P. contorta* (Critchfield 1957); and *P. merkusii* (Cooling 1968) have also been known to show ecotypic variation. In most cases these authors have endorsed their findings on the basis of variability found in the natural stands which evidently need further investigations to confirm whether the sub-specific/varietal level of differentiation is due to genetical or environmental factors. To confirm this, mean annual increment (MAI) of 12 natural stands of the species (6 originating from low rainfall and 6 from high rainfall areas) was compared with two years' growth trend of 4 provenances (2 xeric and 2 mesic) grown under controlled conditions.

MATERIALS AND METHODS

In order to determine MAI for three age classes viz. 30,60 and 90 years in 6 mesic and 6 xeric provenances of *P. wallichiana,* one tree in each of the specified age was selected, sampled and marked. The geo-climatic data of 12 selected sites in Pakistan is given in the following Table. Height (ft) and age (years) were recorded to determine MAI for each locality.

$$\text{MAI} = \frac{\text{Height}}{\text{age}}$$

Table I: Geographic origin of the provenances of *P. Wallichiana*

Localities/ Provenances	Latitude./ Longitude	Alt (m)	Annual rainfall (mm)	Smr. rainfall (mm)	Ann. temp. (C°)	Smr. temp. (C°)	Met. Station	Habitat
Baghi Laila	33.55 / 70.01	2165	739	254	15.0	23.4	Parachinar	Xeric
Bamburet	35.44 / 71.40	2346	462	68	16.7	27.3	Darosh	Xeric
Beha	34.52 / 72.10	2170	881	381	12.5	19.7	Kalam	Mesic
Bhurban	33.58 / 73.22	2015	1506	904	11.9	19.2	Murree	Mesic
Kalkot	35.22 / 72.10	2316	462	68	16.7	27.3	Darosh	Xeric
Lilonai	34.58 / 72.42	2170	881	508	12.5	19.7	Kalam	Mesic
Madain	35.10 / 72.30	1705	881	508	12.5	19.7	Kalam	Mesic
Mermai Tangi	34.01 / 69.58	2377	739	254	15.0	23.4	Parachinar	Xeric
Miandam	35.04 / 72.38	2077	881	381	12.5	19.7	Kalam	Xeric
Naran	34.58 / 73.50	2499	647	254	8.6	17.8	Naran	Xeric
Siway Pungai	31.38 / 69.49	2635	271	124	19.2	29.0	Zhob	Xeric
Yakhtangi	34.54 / 72.39	2170	881	508	12.5	19.7	Kalam	Mesic

In order to confirm whether differences in MAI of provenances of two distinct habitats of *P. wallichiana* follow the same trend under controlled conditions, seedlings of 4 provenances (2 mesic and 2 xeric) viz. Kuza Gali (Mesic), Sathan Gali (Mesic), Baghilaila (Xeric) and Rama (Xeric) were grown in a replicated trial in green house for two years. The needle length (cm) was also recorded as this was the most prominent character in the natural stands for ecotypic differentiation.

RESULTS AND DISCUSSION

A study on 32 provenances of *P. wallichiana* had shown significant differences for 18 morphological and anatomical characters of seeds, apophyses and needles (Shams, 1994). The variation within and between provenances could be attributed to several environmental and genetic factors. There was thus a need to grow these provenances under controlled conditions to find out the extent of genetic variation. In the natural stands, seed size and needle length were found as very important characters in bringing about maximum variation in the species which led to the differentiation at the ecotypic level. The distinguishing features of these two varieties in *P. wallichiana* are described below:

Key to varieties of *Pinus wallichiana* A. B. Jackson

- Needles 13:6-21.5 cm. long; seed length 0.75 to 0.98 cm; breadth 0.46 to 0.57 cm; and thickness 0.31 to 0.40 cm; 100 - seed weight 4.4 to 9.2 g ---------------------*Pinus wallichiana* var. *wallichiana*

Needles 7.5 to 13.8 cm long; seed length 0.59 to 0.81 cm; breadth 0.31 to 0.50 cm; and thickness 0.26 to 0.35 cm; 100 - seed weight 2.1 to 4.7 g.----------------------*Pinus wallichiana* var. *karakorama*

Pinus wallichiana* var. *wallichiana *A.B. Jackson, Kew Bull. 1938: 85, 1938. var. wallichiana*

Urdu name: Biar

Flowers: Male catkins appear in the first or second week of April at lower altitudes and fertilization completes by second or third week of June.

Habitat: Found at lower altitude varying from 1475-2500 m with *Abies pindrow, Acer caesium* and *Aesculus indica* as dominant species.

Distribution: Exhibits almost continuous distribution in areas where the annual rainfall is 729-1500 mm mostly falling in monsoon season.

Pinus wallichiana A.B. Jackson ***var. karakorama*** Shams R. Khan Proc. IUFRO Conf. N. Car. Uni.Tree Imp. Prog: 325-333, 1986.

Urdu name: Kail

Flowers: Male catkins appear in the last week of April at higher altitudes, fertilization completed by Ist week of July.

Habitat: Found at higher altitude between 2500-3300 m; *Pinus gerardiana* and *Picea smithiana* as dominant species.

Distribution: Occurs in isolated patches with discontinuous distribution where the rainfall is less than 729 mm per annum. The areas are rainshadow and are outside the range of monsoon rains.

Coniferous species generally need long time in producing seed and therefore,seed size and needle length differences exhibited by the parents would need at least 2-3 decades to confirm juvenile-mature correlations. In order to determine whether the growth differences are genetically controlled and are not influenced by the environmental factors, the MAI in the natural stands/provenances were compared with those grown under controlled conditions. The MAI of six provenances in each habitat are given in Table II.

Table II: Variation in MAI(ft) among 3 age classes in two distinct habitats of *P. wallichiana*

Xeric (An. rainfall from 200-729 ± 25 mm)				Mesic (An. rainfall from 729 - 1500 ± 50 mm)			
Localities	30 yr	60 yr	90 yr	Localities	30 yr	60 yr	90 yr
Baghi Laila	0.90	0.97	0.83	Beha	0.98	1.23	1.60
Bamburet	0.87	1.28	1.33	Bhurban	1.17	1.23	1.22
Kal Kot	0.93	1.23	1.28	Lilonai	0.98	1.27	1.07
Mermai Tangi	0.80	1.25	1.17	Madain	1.83	1.92	1.72
Naran	1.20	1.12	1.02	Miandam	1.20	1.85	1.64
Siway Pungai	1.06	1.17	1.01	Yakhtangi	1.23	1.17	1.04
Range	0.80 - 1.20	0.97 - 1.28	0.83 - 1.33	Range	0.98 - 1.83	1.17 - 1.83	1.04 - 1.72
Mean	0.96	1.17	1.11	Mean	1.23	1.45	1.38

Table II indicates that MAI during early ages for the mesic provenances was 28 % higher than those of xeric sources. In the later stages of development, these differences are gradually reduced but still at age 90, the provenances of low rainfall areas had shown less MAI than those of high rainfall areas.

The growth trend of 4 provenances of the species during their 2 year's period in the green house has shown that provenances of mesic habitat were growing faster than xeric (Fig. 1). These results therefore, suggest that aforementioned 2 varieties of *P. wallichiana (P. wallichiana* var. *wallichiana* and *P. wallichiana* var. *karakorama)* have shown genetic adaptation with little or no environmental influence when grown under controlled conditions. In the natural stands too, the first named variety had higher MAI than the latter.

The needle length was rated as one of the suitable characters to differentiate populations from different habitats but it was ineffective with 2-year-old seedlings under colltrolled environmental conditions. Probably the age. of seedlings and the environmental factors like excessive water supply and constant temperature were the major factors for such a lack of difference. Critchfield (1957) also observed similar reversal conditions in young seedlings of *P. contorta* when he compared the results with needles studied in the natural stands and reported that young age could be one of the causes to bring about such a reversal of conditions in the species.

Based upon the aforementioned discussion it is therefore, suggested that seed from high rainfall areas should not be transferred to low rainfall areas and vice versa. Seed stands, seed production areas and seed orchards should be separately established in two habitats with strict avoidance of transfer of germplasm from one habitat to another. Following these recommendations, a higher genetic gain with better survival of seedlings in the plantations could be expected. Another conifer, spruce *(Picea smithiana),* occurs mixed with *P. wallichiana* under almost similar ecological and silvicultural requirements. It is therefore, suggested that this species might also exhibit the same trend of variation as observed in *P. wallichiana.* Therefore, the following recommendations are also applicable to spruce. However, the results in spruce need further confirmation.

REFERENCES:

AITCHISON J.ET., 1880. On the flora of Kurram valley and Afghanistan. *Linnean Society's Journal Botany,* **18**: 1- 200.

BARRETT, W. AND GOLFARI, L., 1962. Descripcion de dos nuevas variedales del pino del caribe *(Pinus caribaea* Morelet). *Caribbean Forester* **23:** 59-69.

BRANDIS, D., 1906. *Indian trees.* 767pp. London Archibald Constable.

COOLING, E., 1968. *Pinus merkusii.* Oxford, Commonwealth *Forestry Inst. Tech. Paper* **4:**1- 169.

CRITCHFIELD, W.B., 1957. Geographic variation in *Pinus contorta.* Maria Moors Cabot Foundation Publ., **3**: 1- 118

HOOKER, J.D., 1890. *Flora of British India.* **5**: 1- 657. London, Reeve.

HOOKER. J.D. AND THOMPSON, T., 1855. *Flora Indica.* **1:** 1-285 London, Pamplin.

SHAMS, R. K., 1986a. Ecotypic differentiation in *Pinus wallichiana* A.B. Jackson. *Proceedings IUFRO Conference Williamsburg, Va: North Carolina State University.* Industry Cooperative Tree Improvement Program, 325 - 333.

_____. 1986b. Provenance trials of forest species: A new approach. In *New Genetical Approaches to Crop Improvement.* (eds.)

K.A. Siddiqui and A.M. Faruqui PIDC Printing Press Karachi. 721-725.

____. 1991. The establishment of seed zones for blue pine *(Pinus wallichiana* A.B. Jackson) in: Pakistan, A first attempt. *Proc.lOth World Forestry Congress,Paris.* **5:** 72-76.

____. 1992. The correspondence between geoclimatic and edaphic variables to enhance infraspecific variation in P. *wallichiana. Proc. IUFRO Conference, Cali, Columbia*

____. 1994. Key structural determinants in enhancing infraspecific variability in *P. wallichiana. Proc. Symp. On genetic improvement of forest tree species. Beijing, China,* In press.

____. 1995. Ecotypic differentiation by phenological differences in P. *wallichiana.*Paper presented in IUFRO World Congress, Tampere, Finland; (In press).

SQUILLACE, A.E., 1966. Geographic variation in slash pine. *For.Sci. Monograph* **10:** 1- 56.

STEWART, R.R., 1972 *An Ann. Cat. Vasc. Pl.; W. Pak. & Kashmir.* 1028 pp. Fakhri Printing Press.

TAUNSA WILDLIFE SANCTUARY: A CANDIDATE RAMSAR SITE ON DECLINE

ALEEM AHMED KHAN AND UMEED KHALID*

Pakistan Science Foundation, Islamabad
** National Council for Conservation of Wildlife, Islamabad*

Abstract:Taunsa Wildlife Sanctuary (TWS), a candidate Ramsar site is continuously facing biotic and abiotic pressure which may lead to deleting of another important wetland of Pakistan. During the last two decades Pakistan has lost a number of important wetlands including three Ramsar sites due to pollution, eutrophication, over-exploitation of wetland resources and weak management. TWS is renowned for a number of migratory and breeding bird species, Smooth-coated Otter, Hog Deer and many other animals. Among the waterfowl; Marbled Teal, Cotton Teal, Ferruginous Duck, Whistling Teal, Spoonbill, Greylag and Bar-headed Geese and a variety of waders and raptors utilize the TWS as a transit, wintering or breeding refuge.

In the province of Punjab a total of 41,793 hectare area is designated as protected area of wildlife sanctuaries in which TWS shares a total of 6,567 ha. The highest waterfowl count at TWS was 28,255 birds, while the average population of waterfowl from 1972 to 1989 was 20,110 birds. Due to the enhanced threats of eutrophication, encroachment, physical disturbance and poaching, there has been a considerable decline in the number of breeding birds like Lesser Whistling Teal, herons, egrets, cormorants and Cotton Teal in particular and other migratory waterfowl in general. It has also been observed that the populations of endangered Smooth-coated Otter and Hog Deer are also struggling to survive in the area. Also during the last three years, egrets, cormorants, herons and Lesser Whistling Teal breeding populations have left the sanctuary area and started breeding in spots outside the periphery of the sanctuary. This clearly indicates that Taunsa as wildlife sanctuary in its present state has failed to provide safeguards for breeding wild fauna. There is an urgent need to safeguard the very fragile ecosystem and the overall biodiversity of TWS.

INTRODUCTION

Pakistan has some of the world's hottest low lying areas with some of the highest and the coldest. The country thus possesses many of the world's major climatic and vegetational zones or biomes within a relatively small area. After independence in 1947 extensive water management programmes were undertaken to ensure regular supply of water. The present irrigation system is mainly dependent upon river Indus and its tributaries. In this regard three water storage reservoirs at Tarbela, Mangla and Chashma, 16 barrages, 12 interlink canals, 2 siphons and 43 main channels were built to prosper the agro-based economy of the country (IUCN, 1989).

This man made irrigation system has provided a great number of important wetland sites in close vicinity of the river Indus. Taunsa Barrage is situated in the upper Indus basin as a large water storage reservoir behind a barrage on the Indus river near the town of Taunsa. TWS is located 30° 42′ N, 70° 50′ E, 20 kilometers northwest of Kot Addu, district Muzaffargarh, Dera Ghazi Khan division, Punjab, Pakistan. Many embankments project out into the reservoir and shallow lagoons as the water level in the main river channel falls. Land exposed at low water level is leased to local farmers for cultivation. The depth of water in the main channel varies between 5-11.5 meters, depending on flood levels, while in the seepage lagoons, the depth varies between 0.2-5 meters. The pH is 6.5-7. The climate of TWS is dry sub-tropical with an annual rainfall of 200-450 mm, and relative humidity ranging between 25-88%. The average minimum temperature is 4.5-5.5^{o}C, and the average maximum in June is 42-45°C. Taunsa was declared as wildlife sanctuary in 1972 for the protection of Hog Deer, Indus Dolphin and Marsh Crocodile. In the province of Punjab a total of 41,793 hectare area is designated as protected area of wildlife sanctuaries in which TWS shares a total of 6,567 hectares at an altitude of 139 meter above sea level. Most of the sanctuary area is state owned with some patches of privately owned adjacent land, most of TWS area is controlled by the irrigation department (Scott, 1989).

The area is very important source for irrigation and supports fishery worth more than two million rupees per annum. Also a very important wintering area for waterfowl notably Anatidae; a breeding area for several species like *Dendrocygna javanica* (Lesser Whistling Teal) and a staging refuge for cranes including waders. TWS is most important wintering area for the Bar-headed Geese and Ruddy Shelduck; Marbled Teal is a regular passage migrant or transit migrant (Green, 1993). Indus Dolphin, and Smooth-coated Otter, Hog

Deer, Wild Boar and other mammals are also the resident of TWS (Reeves, 1991). The area has immense potential for scientific research and conservation education.

METHODOLOGY

Some reconnaissance surveys of the area were done during 1993-95. The survey methods include, on site observation and informal discussions and dialogues with local people. The historical data and information were gathered from a number of concerned departments and by consulting the appropriate literature.

RESULTS AND DISCUSSIONS

TWS has been ignored in the recent past of its effective management and is showing decline in overall biodiversity due to the following reasons.

Eutrophication or habitat deterioration:

In the last ten years it has been observed that the required amount of water is not being allowed to TWS. The lowering of water-table in the lagoons is due to building of seasonal dams near inlet of river. This has resulted in tremendous increase of wild bushes and plants like typha, reeds, lotus and *tamarix* spp. invading the wetlands. Also seasonal burning of the vegetation contributes towards increased level of inorganic eutrophication in shallow waters. Such eutrophication reflects poor light penetration and more turbidity and subsequently results in overall decrease in aquatic life which is the main food source of macro consumers.

Poaching and disturbance caused by nomads:

Almost ten years back some nomad fishermen started coming upstream in River Indus from Sindh, in search of, contract jobs for commercial fishery services. Since last five years that is 1990, the trend of these Sindhi nomads has increased tremendously towards Upper Indus Basin (UIB), particularly, at Taunsa and Chashma Barrages due to the enhanced potential of commercial freshwater fishery in the area. Initially these nomads used to come only during the winter season which is the peak fish harvest time in Pakistan but later on during early 90s they have started settling down with their nomadic life style in UIB. When these nomads arrived in UIB, they brought new and versatile techniques of harvesting fish, wildlife and other natural aquatic resources. As for example, they introduced the use of explosives for poaching of fish, poisoning and netting of waterfowl, killing of jackals, Red Fox and otters as well as Hog Deer. These people have an edge for marketing and harvesting their catch due to the commercial fishery facility available to them. At TWS, in summer time these nomads have developed another source of income by spreading the lotus in state (inundated) waters. In this connection they live with their families in temporary huts at the closest vicinity of the lotus, in spite of harsh living conditions. This continued disturbance has forced the breeding waterfowl to emigrate from the TWS.

Over-exploitation of available resources:

The natural resources of TWS are being over-exploited by the local and nomad Sindhis. Locals were used to practice agriculture in non flooding time i.e., winter season but in last few years they have started encroaching the TWS and adjacent lands in peak flooding season by building dams at some water channels. Additionally, pumping out the inundated water for irrigation from nearby creeks is resulting in mass killing of fish seed every year. Also permanent presence of nomads in the area has resulted enhanced pressure on wood cutting, and wildlife poaching.

Lack of management infrastructure:

All these above mentioned activities are not being monitored by the law enforcing authorities from the last five years or so. The government installations are symbolizing as ruins due to the lack of interest in sanctuary management and non-availability of staff in the area.

Though TWS was originally designated as wildlife sanctuary for the protection of mammalian wildlife species like marsh crocodile, Hog Deer and Indus Dolphin but it got more importance due to the rich ornithological fauna particularly waterfowl (Table I).

The average population of waterfowl from 1972 to 1989 was 20,110 with a highest ever count of 28,255 (Khurshid, 1991). Hence, according to one of the criteria for designating a site as RAMSAR site, the wetland should cater an average of 20,000 waterfowl each year (Scott, 1989). In July 1976 when Pakistan joined the Ramsar Convention, a total of 9 wetland sites were listed as Ramsar sites of

Table I: The Overall Situation of Waterfowl at TWS from 1987-1995 mid winter count)

No.	Scientific name	Common name	1987	1988	1989	1990	1991	1992	1993	1994	1995
01.	*Podiceps cristatus*	Great Crested Grebe			-				2		
02.	*Tachybaptus ruficollis*	Little Grebe		10	30	9	52	30		200	19
03.	*Podiceps nigricollis*	Black-necked Grebe			-		4				
04.	*Phalacrocorax carbo*	Great Cormorant	621		-		8	40			
05.	*Phalacrocorax fuscicollis*	Indian Shag	-		-						11
06.	*Anhinga melanogaster*	Oriental Darter	-	22	5						
07.	*Phalacrocorax niger*	Little Cormorant	-	450	400	105	35	425	652	90	119
08.	*Egretta garzetta*	Little Egret	78	197	150	265	70	65	177	160	126
09.	*Egretta intermedia*	Intermediate Egret	23	25	40	20	10	22	13	10	62
10.	*Egretta alba*	Great White Egret	-	-	-	-		10	18		43
11.	*Bubulcus ibis*	Cattle Egret	-	-	-	-		40			
12.	*Ardea cineria*	Grey Heron	23	14	10	1	2	14	3	15	12

No.	Scientific name	Common name	1987	1988	1989	1990	1991	1992	1993	1994	1995
13.	*Ardeola grayii*	Indian Pond Heron	36	5	8	13		13	48	8	150
14.	*Ardea purpurea*	Purple Heron	-	-	-			6	5		
15.	*Ciconia nigra*	Black Stork	-	5	-						13
16.	*Anser indicus*	Bar-headed Goose	-	-	2						100
17.	*Tadorna tadorna*	Common Shelduck	-	8	-						
18.	*Anas penelope*	Eurasian Wigeon	2776	150	2100	840	466	250	873	80	10
19.	*Anas strepera*	Gadwalls	771	5	150	37	-118	65	768	340	4
20.	*Anas crecca*	Common Teal	4884	3100	4400	2340	14		9790	1100	2320
21.	*Anas platyrhynchos*	Mallard	274	1000	3000	2	12	1350	263		1296
22.	*Anas acuta*	Pintail	1659	40	120	45	8	600	600	60	1
23.	*Anas clypeata*	Shoveler	388	15	60	25	100		790	40	83
24.	*Netta rufina*	Red-Crested Pochard	31	100	20	6	250			250	
25.	*Aythya ferina*	Pochard	4685	3100	250	40	216	1000		10	1
26.	*Aythya nyroca*	Ferroginous Duck	3	2	8	2					

No.	Scientific name	Common name	1987	1988	1989	1990	1991	1992	1993	1994	1995
27.	*Aythya fuligula*	Tufted Duck		-	-					10	20
28.	*Fulica atra*	Coot	7511	7600	4000	782	680	330	308	650	693
29.	*Gallinula chloropus*	Indian Moorehen		-	-			22	27		18
30.	*Porphyrio Porphyrio*	Purpule Moorehen	150	-	15			10	25	10	2
31.	*Himantopus himantopus*	Black-winged Stilt	47	50	35		60	30	87	26	120
32.	*Limosa limosa*	Black-tailed Godwit	40	-	3		2	2			
33.	*Vanellus vanellus*	Lapwing	39	1	5		12		5		3
34.	*Vanellus malabaricus*	Yellow Wattled Lapwing	-	-	-		5				-
35.	*Vanellus indicus*	Red Wattled Lapwing	-	-	-		4	20		5	6
36.	*Vanellus leucurus*	White-tailed Plover	-	-	-						23
37.	*Charadrius dubius*	Little-ringed Plover		-	-			17			-
38.	*Chalidris ferruginea*	Curlew Sandpiper	6	-	-					200	-
39.	*Chalidris alba*	Little Stint	-	-	-						3
40.	*Tringa nebularia*	Green shank	-	-	-			4	35	6	183

No.	Scientific name	Common name	1987	1988	1989	1990	1991	1992	1993	1994	1995
41.	*Tringa totanus*	Red shank	64	90	30	5	20	18	12	36	296
42.	*Gallinago gallinago*	Common Snipe	-	-	-		1				21
43.	*Actitis hypoleucus*	Common Sandpiper	3	5	20	3		8	34	40	41
44.	*Larus ridibundus*	Black-headed Gull	-	-	-			11	2	14	8
45.	*Larus argentatus*	Herring Gull	-	-	-		30				
46.	*Sterna aurantia*	Indian River Tern	-	-	-		8	8	14	4	52
47	*Ceryle rudis*	Pied Kingfisher			-				2		
48.	*Hydrophasianus chirurgus*	Pheasant-tailed Jacana	-	-				8			
			24112	15994	14861	4540	2187	4418	14553	3364	5859

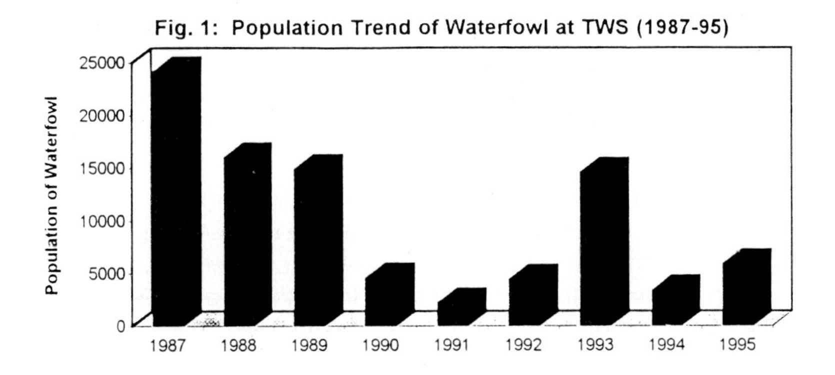

Fig. 1: Population Trend of Waterfowl at TWS (1987-95)

Population of Waterfowl
25000
20000
15000
10000
5000
0
1987
1988
1989
1990
1991
1992
1993
1994
1995

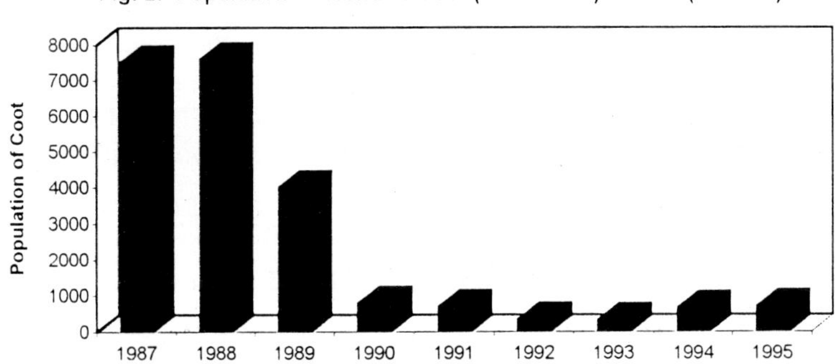

Fig. 2: Population Trend of Coot (*Fulica atra*) at TWS (1987-95)

Population of Coot
8000
7000
6000
5000
4000
3000
2000
1000
0
1987
1988
1989
1990
1991
1992
1993
1994
1995

Fig. 3: Population Trend of Eurasian Wigeon at TWS (1987-95)

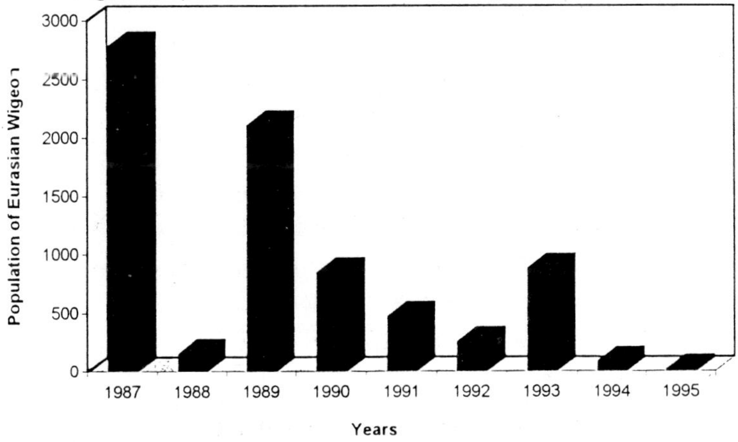

Population of Eurasian Wigeon
3000
2500
2000
1500
1000
500
0
1987
1988
1989
1990
1991
1992
1993
1994
1995
Years

Pakistan (Rao, 1989). Monitoring mission of Ramsar Bureau conducted a detailed survey of wetlands of Pakistan in May 1990 and enlisted Taunsa barrage a candidate site as future Ramsar site (Scott *et. al*, 1990). The government of Punjab has given the consent to protect this wetland as Ramsar site.

In this regard the formal notification by the federal government and UNESCO is awaited in near future (Government of the Punjab, 1995). According to the mid winter counts, the TWS is on continuous decline from 1987 onward with drastic decline from the year 1990. Since 1990 till recent the number of waterfowl has decreased 2/3rd. of the average previous two decades population (Fig. 1). This drastic decline has not only affected the birds diversity but has also decreased the overall wildlife diversity like Hog Deer, Wild Boar, Jackal, wild hare, Red Fox and Smooth-coated Otter. The overall swim in species diversity has out maneuvered some of the internationally important species like Marbled Teal, Ferruginous Duck, Cotton Teal, Crane, Coot and Eurasian Wigeon (Figs 2 and 3), and some breeding birds like Lesser Whistling Teal, Purple Moorhen, egrets, cormorants and jacana have been forced to evacuate their parental sites.

ACKNOWLEDGMENT

Authors are grateful to Punjab Wildlife Research Center, Gatwala - Faisalabad for borrowing the waterfowl census data related to TWS.

REFERENCES

GOVERNMENT OF THE PUNJAB, Forestry, Wildlife and Tourism Department. 1995. Nomination for the Ramsar List. SOFT (EXT) XII - 5/94, dated 9.4.95.

GREEN, A.J., 1993. The Status and Conservation of the Marbled Teal (*Marmaronetta angustirostris*). IWRB, special Publication No. 23. Slimbridge, Gloucester, GL2 7BX, UK

IUCN, 1989. Pakistan Fact Sheet; Water. Journalist Resource Center for the Environment. IUCN-Pakistan

KHURSHID, N., 1991. A Step Towards Wetland Conservation. WWF-Pakistan.

RAO, A.L., 1989. A Directory of Asian Wetland (Pakistan). IUCN occasional paper No.18, IUCN, Gland, Switzerland

REEVES, R.R., 1991. Conservation of the Bhulan (Blind River Dolphin) in the Punjab. *Natura*, Jan. 1991, WWF-Pakistan

SCOTT. D.A., 1989. A Directory of Asian Wetlands. IUCN.

SCOTT. D.TA., RAO, A.L, AND BEG, A.R., 1990. The Wetland of Pakistan and the Ramsar Convention. Ramsar Bureau, Slim bridge, Oucester, GL2 7BX, UK.

CONSERVATION: A MUST TO SAFEGUARD THE BIODIVERSITY OF RANGLA WETLAND COMPLEX

KASHIF M. SHEIKH, JAVED A. MAHMOOD* AND M. SAJID NADEEM**

Biological Sciences, Quaid-e-Azam University, Islamabad
**Department of Fisheries-Punjab,y, Rawal Town, Islamabad*
***Zoology Department, University of the Punjab, Lahore,*

Abstract: Rangla wetland complex is located at a distance of 15 to 20 km. in the north-west of Muzaffargarh District in the province of Punjab-Pakistan, which provides refuge to a diversity of fauna and flora. It is an excellent habitat for some internationally threatened duck species such as Marbeled Teal and Ferruginous duck. Populations of these and other important waterfowl were recorded during 1993-94. Fish diversity was also found out. People rely on these wetlands for fuelwood extraction, livestock grazing, fishing, earth-cutting, waterfowl hunting and agricultural use etc. By knowing the perceptions of the local peoples, it was concluded that the conservation of existing natural resources, by employing the up-to-date tools of nature resource conservation and development, is urgently needed.

INTRODUCTION

Simply, biological diversity means the wealth of life forms found on earth; millions of different plants, animals and micro-organisms, the genes they contain and the intricate ecosystems they form. Biological diversity occurs even in the middle of deserts (Markham et.al., 1993).

Rangla wetland complex is an example of rich biodiversity in the desert. This complex is present in Muzaffargarh District in the province of Punjab, Pakistan. It is situated at 30° 40' N, 70° 11' E (Khan and Shah, 1993). The wetlands included in the complex are Rangla Dhand, Drowle, Bari wala Dhand, Kutae Wala Dhand, Bhando, Dori Head, Bhudena, Mubarik Wala, Drawae Wala, Bhangan and Nai Wala and many others. Most of these wetlands were originated after the construction of Muzaffargarh canal. Before this, these were shallow agricultural lands providing the local needs. All these wetlands are present in an area of 25 to 30 kms. At most of the sites, it is not easy to approach the open water and so the waterfowl populations find good refuge for them. The wetlands are under heavy grazing pressure and livestock trampling. During monsoon season, each year, the flood water from Indus river reaches the wetlands through Muzaffargarh canal and thus adds to its biodiversity particularly fishes. The present study deals with the biodiversity of birds, fish, mammals and macrophytes found in the Rangla wetland complex. This wetland is still a neglected area and keeping in view the ecological importance of the wetlands it is strongly recommended that a comprehensive research plan should be launched for its long-term conservation and sustainable use.

MATERIALS AND METHODS

The study was carried out during April to September,1994 to find out the biodiversity with special reference to birds, fishes and macrophytes in and around the wetlands. All the potential approachable sites of the Rangla wetland complex were surveyed by foot and boat surveys. Macrophytes along the banks and on the islands were collected and identified with the consultation of "Plant Taxonomy Section" of the Botany Department, University of Punjab, Lahore. Fishes were sampled by using cast and drag nets and preserved in 8% formalin solution. Injections of 20% formalin were also given in the gut region to the large sized fishes. Systematic keys were used to identify the fishes. Notes and onsight observations were also recorded such as fuelwood extraction, livestock grazing, earth-cutting, fishing, waterfowl hunting and land use for agricultural purposes. The locals were also consulted about the natural resources of the wetlands and their management.

RESULTS AND DISCUSSION

During the study the following species diversity of birds, fishes, mammals and plants (macrophytes) was found:

Birds

Family Anatidae

Anas penelope	Wigeon
Anas strepera	Gadwall
Anas creca	Common Teal
Anas platyrhynchos	Mallard
Anas acuta	Pintail

Anas clypeata	Shoveler
Marmarronetta anguistirostris	Marbled Teal
Netta rufina	Red-crested Pochard
Aythya ferina	Common Pochard
Aythya nyroca	Ferruginous Duck
Family Rallidae	
Fulica atra	Black or Eurasian Coot
Gallinula chloropus	Moorhen or Waterhen
Family Gruidae	
Anthropoides virgo	Demaiselle Crane
Family Recurvirostridae	
Himantopus himantopus	Black-winged Stilt
Recurvirostra avosetta	Pied Avocet
Larus ridibundus	Black-headed Gul
Charadrius alexandrinus	Kentish Plover
Haplopterus indicus	Red-wattled Lapwing
Family Scolopacidae	
Calidrius minuta	Little Stint
Family Tringidae	
Tringa totanus	Red Shank
Tringa stagnatilis	Marsh Sandpiper
Tringa nebularia	Green Shank
Actilis hypoleucos	Common Sandpiper
Family Alcedinidae	
Halcyon smyrenensis	White-breasted Kingfisher
Ceryle rudis	Pied Kingfisher
Family Meropidae	
Merops orientalis	Little Green Bee-eater
Famliy Coraciidae	
Coracias benghalensis	Indian Roller
Family Upupidae	
Upupa epops	Hoopoe
Family Phalacrocoracidae	
Phalacrocorax niger	Little Cormorant
Family Ardeidae	
Ardeola grayii	Pond Heron
Ixobrychus sinensis	Yellow Bittern
Ixobrychus cinnamomeus	Chestnut Bittern
Egretta garzetta	Little Egret
Egretta intermedia	Intermediate Egret
Ardea cinerea	Grey Heron
Family Caprimulgidae	
Caprimulgus mahrattensis	Sindh Nightjar
Family Apodidae	
Apus affinis	Little Swift

Family Columbidae	
Streptopelia senegalensis	Little brown Dove
Treron phoenicaptera	Common green Pigeon
Family Phasinidae	
Francolinus francolinus	Black Patridge
Francolinus pondicerianus	Grey Patridge
Coturnix coturnix	Common Quail

Marmarronetta anguistirostris or Marbled Teal were observed in flocks of 8 to 15 birds. But a maximum of 35 Marbled Teal were observed during the study. Males were larger than females and were identified by their prominent nuchal crest (Roberts, 1991). The fact that Marbled Teal were present on these wetlands but no eggs or ducklings were seen, indicates that it may breed in other wetlands in the nearby areas (Sheikh et.al., 1995). Ferruginous Duck is another internationally threatened duck species found in the complex (IWRB, 1995). A handsome population of this Duck was seen flying from one wetland to the other in the whole wetland complex. The breeding birds at Rangla complex include *Tachybaptus ruficollis*, *Himantopus himantopus*, *Vanellus indicus*, *Fulica atra*, and *Gallinule chloropus* (Sheikh et.al., 1995).

Fish

Family Notopteridae
Notopterus notopterus (Hamilton)
Notopterus chitala (Hamilton)
Family Cyprinidae
Labeo rohita (Hamilton)
Labeo calbasu (Hamilton)
Cirrhinus mrigala (Hamilton)
Puntius sophore (Hamilton)
Cyprinus carpio (Linnaeus)
Family Bagridae
Aorichthys aor (Hamilton)
Mystus cavasius (Hamilton)
Mystus vittatus (Bloch)
Family Siluridae
Wallago attu (Bloch & Schneider)
Family Schilbeidae
Pseudeutropius atherinoides (Bloch)
Family Channidae
Channa marulius (Hamilton)
Channa punctata (Bloch)

Family Chandidae
Chanda baculis (Hamilton)
Family Mastacembelidae
Mastacembelus armatus (Lacepede)

All the above mentioned fishes have also been recorded from river Chenab near District Multan (Khan, et.al., 1991). *Notopterus notopterus*, *Cyprinus carpio*, *Mystus vittatus* and *Channa marulius* have not been reported from river Jhelum in Sargodha District (Mirza et.al., 1987).

Mammals

Jackal, Fox, Wild Boar, Rabbit, Jungle Cat, Hedgehog, Porcupine, Rat

Plants

Tamarix galica, Typha angustata, Phragmites karka, Sueda fruticosa, Capparis aphylla, Aeura jawanica, Aeluropus lagopoides, Salvadora oleoides, Prosopus cineraria, Acacia nilotica

All the wetlands in the complex share some common features among the vegetation, for example, *Phragmites karka*, *Typha angustata*, *Tamarix galica*, and *Aeluropus lagopoides*. Vegetation was dominated by *Typha angustata*, *Phragmites karka*, and *Tamarix galica* all over the wetlands providing a thick cover to the waterfowl population. Vegetation is often exploited by the locals for roof-making, household, for example mats, Moora (chair), fence around the houses etc.

ACKNOWLEDGMENTS

The authors are grateful to the staff of Taunsa Wildlife Sanctuary, Sanawan and Kot Adu. Also special thanks to WWF - Pakistan for providing financial help for the project.

REFERENCES

KHAN, A.A. AND SHAH, M.A., 1993. Marbled Teal Breeding in Punjab Pakistan. IWRB, *Threatened Waterfowl research Group Newsletter,* **4**. UK.

KHAN, M.I., IRSHAD, R. AND SAGA, F.H., 1991. Fishes of the River Chenab in Multan District. *Biologia,* **37**: 23-25.

MARKHAM, A. DUDLEY, N. AND STOLTON, S., 1993. Some Like it Hot. WWF. Int.CH-1196 Gland, Switzerland.

MIRZA, M.R., 1990. *Pakistan Main Taza Pani Ki Muchlian* (in Urdu). Urdu Science Board. Lahore.

MIRZA, M.R. AND AHMAD, I., 1987. Fishes of the River Jhelum in Sargodha District. *Biologia*, **33**: 253-263.

NELSON, J.S., 1984. *Fishes of the World.* 2nd.ed. Jhon Willey. New York.

ROBERTS, T.J., 1991. *Birds of Pakistan:* Nonpesseriformes.Vol. **I**. Elite Publishers, Karachi-Pakistan.

SHEIKH, K.M., NADEEM, M.S. AND GASHKORI, M.H., 1995. Marbled Teal study at the Rangla Complex, Pakistan. *TWRG Newsletter for IWRB.,* **7**. UK.

PHYTOREMEDIATION - A STRATEGY TO DECONTAMINATE HEAVY METAL POLLUTED SOILS AND TO CONSERVE THE BIODIVERSITY OF PAKISTAN SOILS

A.G. KHAN, A. BARI*, T.M. CHAUDHRY** AND A.A. QAZILBASH***

* *Biological Sciences, Quaid-i-Azam University, Islamabad*

** *Biological Sciences, University of Western Sydney, Australia.*

Abstract: This paper reviews the literature on pollution of soils due to various mining, smelting and manufacturing activities during the last 50 years or so in Pakistan. Various remedial strategies, with particular reference to "phytoremediation", i.e. removal of heavy metals using hyperaccumulator mycorrhizal plants to decontaminate soil environment has been discussed in an attempt to preserve the biodiversity of Pakistan soils. Emphasis is placed on the urgent need to prioritise the existing and potential problems associated with soil, plants, animals and human safety and health in Pakistan.

INTRODUCTION

The population of Pakistan is increasing at 3% yr^{-1} and was estimated to be 124 million in 1994 and its major economic activity is agriculture. The ever increasing population together with increased food demand has led to greater use of fertilizers and pesticides, which in turn have increased soil pollutant levels.

Most of the soils in Pakistan are characteristically deficient in N, P, and organic matter and, as a result of intense cropping, are becoming deficient in micro-nutrients as well. As with most of the developed countries, there are a wide range of inorganic and organic metal soil contaminants arising from both agricultural and industrial practices in Pakistan. Soils are creating serious environmental problems. The extensive use of pesticides in Pakistan since 1954 has given rise to environmental concern such as residues in food and in soil, pest resistance and health hazards to animals and man. Mubarik and

Jabbar (1992) found that soils in cotton growing areas of Punjab are generally contaminated with pesticides by varying amounts.

In the recent past, there is a shift from agriculture to industry in Pakistan. A number of textile, tannery, edible oil, cement, fertilizer and other chemical factories have been established in various parts of the country. Environment Protection Agency studies (EPA, 1990) have shown that industrial effluents from electroplating and tannery industries containing toxic metals such as Cr, Ni, As, Hg, Cu, Pb, etc., have contaminated soil and biota of such soils in Pakistan. Khan *et al.* (1992) reported that Faisalabad city effluents, comprising domestic and industrial wastes, contained toxic materials such as Pb, Cu, Cd and other heavy metals. The authors found that there was a considerable build up of such metals in soils subjected to effluent irrigation and that vast areas of useful land has been contaminated by heavy metals as a result of these manufacturing and urban activities.

Recently, Hussain *et al.* (1996) have reviewed the literature on occurrence of toxic substances and pollutants in some industries in Karachi, Multan, Faisalabad, Kala Shah Kaku, Peshawar and Nowshera. This poses potential threat to the biodiversity of soil and plants, as well as to the animal and human health. Their studies have highlighted the point that heavy metal contaminated land is becoming increasingly important environmental, health, economic and planning issue in Pakistan.

There is an urgent need to not only take curative, but also preventive measures to remediate such soils. Due to various mining, smelting and manufacturing activities during the last 50 years or so in Pakistan for urban or agricultural development there is need to develop a safe and efficient decontamination strategy. This also includes a need to study, identify and prioritize the existing and potential problems associated with plant, animal and human safety and health in Pakistan and preserve the country's biological diversity.

This paper discusses various remedial strategies, with particular reference to "green remediation" *i.e.,* removal of heavy metals using hyperaccumulator mycorrhizal green plants to decontaminate soil environments in an attempt to preserve biodiversity of Pakistan soils.

REMEDIAL STRATEGIES

Metal contaminants of soil are notoriously persistent in the soil and various physio-chemical and bioremedial strategies are employed. Some of the commonly used techniques include:

Excavation and Land Filling

A solid waste is dumped and allowed to decompose. In this process solid waste containing both organic and inorganic material is deposited and covered with soil, in low lying and hence low value land (U.S. EPA, 1989). Exposed waste causes various aesthetic and public health problems, attracts insects and rodents and poses a fire hazard. The "sanitary landfill" is an improvement over this in which each day's waste deposit is covered with a layer of soil (U.S. Department of Health, Education and Welfare, 1970). After completion of the landfill, the site becomes useable for recreation and eventually for construction.

Physical Separation of contaminants

Physical separation techniques have been used commonly in the mining industry for many years. These techniques involve the physical separation of particles from each other based on certain particle characteristics (Perry and Chilton, 1984; Willis, 1985). It depends upon the particle characteristics, which are particle size (screening), particle density (sedimentation thickening), surface properties of particles (froth flotation) and magnetic properties (magnetic separation).

Physical separation has long been used in mineral beneficiation to extract the desired metal from a mineral ore. It usually involves a series of steps that lead to successive products containing increasing concentrations of the desired metal. Each separation technique thus results in the feed being divided into at least two streams, concentrate and tailings.

Recently, there has been a growing interest in applying physical separation techniques to soil remediation. Physical separation is applicable to remediation primarily in two situations. Firstly, if the metal contamination is in the form of discrete particles in the soil, a technique could be applied to physically separate the metal from the soil. Secondly, if the metal contamination is molecular (adsorbed onto soil particles) and it is limited to a specific particle-size range, physical separation based on particle size often is used as a pretreatment to reduce the total amount of contaminated soil that would have to undergo final (chemical) treatment (Smith *et al.*, 1995).

High Temperature Thermal Treatments

(a) Vitrification Technologies:

These technologies apply high temperature treatment primarily to reduce the mobility of metals by their incorporation in a vitreous material such as stable oxide solid (U.S. EPA, 1992).

(b) *Polymer Microencapsulation:*

This involves application of asphalt and similar organic binders to treat HM contaminated substrates. Solidification / stabilization by polymer microencapsulation can include application of thermoplastic or thermosetting resins. This process involves heating and mixing the waste material and the resin at elevated temperature, typically 130°C to 230°C, in an extrusion machine. Any water or volatile organics in the waste boil off during extrusion and are collected for treatment or disposal. Because the final product is a stiff, yet plastic resin, the treated material typically is discharged from the extruder into a drum or other container. Asphalt-treated soils or abrasives contaminated with metals have been reused as paving material (Means *et al.*, 1993).

(c) *Pyrometallurgical Separation:*

The term *pyrometallurgy* encompasses techniques for processing metals at elevated temperature to treat a metal contaminated solid for recovery of metals as metal, metal oxide, ceramic product, or other useful form. It is the oldest type of metal processing, dating back to the origins of extracting useful metals from ore. Pyrometallurgy offers a well-developed method for recovery of metals from waste materials (Nilmani *et al.*, 1994).

Microbiological Treatments

(a) *Bioremediation:*

The ability of heavy metal biodegrading microbes to reclaim soils and waters polluted by substances hazardous to human health and/or the environment, can be exploited for remedial purposes. Several micro-organisms, including *Pseudomonas putida* (Horn *et al.,* 1992) and *Thiobacillus ferrooxidans* (Hansen and Stevens, 1992), have been tested for application to chemical reduction and recovery of mercury from wastewater. Biological mechanisms are available to accomplish the reduction of chromium by direct microbial production of sulfide (Mattison, 1993).

(b) *Composting:*

It is the decomposition of organic matter in a heap by microorganisms, and a method of solid waste disposal. Composting is a microbial process that converts putrefyable organic waste materials into a stable, sanitary, humus-like product that is reduced in bulk and can be used for soil improvement. Composting is accomplished in static piles (windrows), aerated piles, or continuous feed reactors (Atlas and Bartha, 1993).

The composting process is initiated by mesophilic heterotrophs. As the temperature rises, these are replaced by thermophilic forms. Thermophilic bacteria prominent in the composting process are *Bacillus stearothermophilus, Thermomonospora, Thermoactinomyce, and Clostridium thermocellum.* Important fungi in the thermophilic composting process are *Geotrichum candidum, Aspergillus fumigatus, Mucor pusillus, Chaetomium thermophile, Thermoascus auranticus, and Torula thermophila* (Finstein and Morris, 1975).

(c) *Slurry Reactors:*

Slurry-phase treatment of contaminated soils is a recent innovation in bioremediation. In this form of remediation, the excavated soil is treated as a water-based slurry in a bioreactor. The potential for optimization and the possibility for continuous-mode operation can minimize treatment times and make the bioreactor an attractive remediation alternative (Cacciatore and McNeil, 1995).

Chemical and Electro-chemical Processes

Chemical treatment of contaminated solids entails the reaction of contaminants with reagents to form products that are less toxic or less mobile (Smith *et al.*, 1995).

(a) *Hydrolysis:*

One of the chemical ways to remove HM compounds that are reactive with water, e.g. metal carbides, hydrides, amides, alkoxides, and halides; and nonmetal oxyhalides and sulfides, is to allow them to react with water under controlled conditions.

***(b) Chemical Extraction and Leaching*:**

Heavy metal ions in soil contaminated by hazardous wastes may be present in a co-precipitated form with insoluble iron and manganese oxides. Those oxides can be dissolved from soil by reducing agents, such as solutions of sodium dithionate/citrate or hydroxylamine. This result in the production of soluble Fe^{2+} and Mn^{2+} and the release of heavy metal ions, such as Cd^{2+} or Ni^{2+}, which are removed with the water (Manahan, 1994).

(c) Electrolysis:

Electrolytic removal of contaminants from solution can be by direct electro-deposition, particularly of reduced metals, and as the result of secondary reactions of electrolytically generated precipitating agents. An example of such strategy is the electrolytic removal of both cadmium and nickel from wastewater contaminated by nickel/cadmium battery manufacture using fibrous carbon electrodes (Abda and Oren, 1993).

Phytoremediation - An alternative strategy for decontaminating HM polluted soils of Pakistan

In situ phytoremediation (a green plant based strategy) of heavy metal contaminated soils is an alternative method to decontaminate. It is not a new concept; constructed wetlands, reed beds and floating-plant systems have been common for treatment of heavy metal contaminated wastelands for many years. Current research efforts now focus on expanding phytoremediation to address contaminated soils and atmospheric pollutants. As far as soils are concerned, phytoremediation includes 'phytodecontamination' and 'phytostabilization' techniques as a long term strategy.

Phytodecontamination strategies involve:

- *phytoextraction,* where plants accumulate the contaminants and are harvested for processing. Post-harvest processing of contaminants includes thermal, microbial and chemical treatments (Cunningham *et al.*, 1995).
- *phytodegradation,* where plants, or plant-associated microflora, convert pollutants into non-toxic materials.
- *phytostabilization* is where pollutants precipitate from solution or are absorbed or entrapped in either plant tissue or the soil matrix. Sequestration can be enhanced either by amendments to the soil or through the action of the plant and its microflora.

Cunningham *et al.* (1995) have redefined plants as 'solar-driven pumping and filtering systems' and roots as 'exploratory, liquid-phase extractors'. This has given birth to a 'new' technology termed 'phytoextraction', 'phytoaccumulation', or 'phytoremediation' of heavy metal contaminated soils.

Contaminated areas often support characteristic plant species growing in these heavy metal enriched environments. Some of these species can accumulate usually high concentrations of toxic metals (Baker and Brooks, 1989). Use of these rare heavy metal tolerant accumulator plant species in decontamination strategies provides an alternative, economically cheap and environmentally friendly remediation technology which maintains biological properties and physical structure of remedied soils and offers the possibility of biorecovery of metals. Small scale field trials with wild hyperaccumulators collected from naturally contaminated soils have demonstrated the feasibility of the phytoremediation approach (Baker *et al.*, 1994).

Although many reviews have appeared in the books of Markert (1994), Adriano (1992), Shaw (1990) and Wolf (1989), the potential for application of hyperaccumulator plants in phytoremediation of heavy metal polluted soils is limited by lack of knowledge about their distribution, agronomic characteristics, metabolic capabilities and growth habits. Therefore, as pointed out by Kumar *et al.* (1995), we need to identify and know more about these plant species that can accumulate heavy metals while producing high biomass. Metal uptake and tolerance by plants depend on both plant and soil factors and generalization about metal uptake and phytotoxicity requires information on many aspects of metal-soil-plant interactions.

Recently, the concept of using hyperaccumulator plants to decontaminate industrially HM-contaminated soils was tested over a period of 4 years by McGrath *et al.* (1996) who found that accumulation of Zn in the plants did not decline with time. The potential exploitation of metal uptake into plant biomass as a means of decontamination of soil can be enhanced by inoculating plants with appropriate VA mycorrhizal propagules (Dueck *et al.*, 1986). Some plants, growing on contaminated sites, may possess the hyperaccumulation property, but they are highly tolerant to high concentrations of heavy metals. These tolerant plants may also be employed for revegetation of HM-contaminated mine wastes and tailing. This would prevent further pollution of nearby soils by wind and water erosion. However, there is again no data on the mycorrhizal status of plants and their rhizospheres in heavy metal polluted sites in Pakistan. Furthermore, the mechanism of metal tolerance versus metal hyperaccumulation also requires further studies due to the conflicting reports in the literature. The mechanisms of plant tolerance are attributed to internal inactivation and cellular/sub-cellular compartmentation (Vogal-Lange and Wagner, 1989), or partial exclusion and reduced translocation from root to shoot (Paliouris and Hutchinson, 199l).

In view of the potential importance of symbiotic VA-mycorrhizal associations in the amelioration of heavy metals (Galli *et al.* 1994; Brown and Wilkins, 1985) and in increasing plant biomass at higher levels of heavy metals (Shetty *et al.*, 1995), a detailed survey of the mycorrhizal characterization of roots and rhizospheres of the plants growing on the heavy metals contaminated waste sites in various parts of Pakistan is also needed before this strategy can be applied at commercial level.

VA ecotype from heavy metal contaminated soils seem to be more tolerant to heavy metals than reference strains from uncontaminated soils (Leyval *et al.*, 1991; Weissenhorn and Leyval, 1995; Griffioen *et al.*, 1994; Galli *et al.*, 1994). The potential exploitation of HM uptake into plants can be enhanced by inoculating plants with appropriate VA-mycorrhizal propagules (Dueck *et al.*, 1986).

CONCLUSION

So far the search for the hyperaccumulator plants have been carried out by few people in the USA (RL Chaney), UK (AJM Baker), New Zealand (RD Reeves) and the Netherlands (E Ernest). A recent survey of a zinc-alum waste industrial site in Australia, by two authors of this article (Khan *et al.*, 1996) found that established vegetation in the contaminated area was exclusive and included genera belonging to the families Apiaceae, Euphorbiaceae, Asteraceae, Malvaceae and Verbenaceae with *Ricinus communis* (common cater oil plant) being the most dominant. Rhizospheres of all the plants growing on the contaminated site harbored VA mycorrhizal spores , while their roots were also mycorrhizal. In another survey of plants growing on abandoned silver mine in Australia (Khan and Chaudhry, 1996), many endemic mycorrhizal plants, some with nitrogen fixing nodules were reported. Almost all the HM-accumulator plant species known today were discovered on metalliferous soils, often growing together with metal tolerant/metal extruder species. There is an urgent need for primary reconnaissance and screening plants from other parts of the globe, including Pakistan. It is likely, as pointed out by Raskin *et al.* (1994), that many more as yet unidentified HM-accumulator plants growing on natural and man-made metalliferous soils remain to be discovered. It is vital that rare and endemic HM-hyperaccumulator plants are identified and preserved before they become extinct. In employing phytoremediation, not only are the biodiversity and physical structure of Pakistani soils maintained but the technique is environmentally safe, potentially cheaper, visually unobtrusive and offers the possibility of biorecovery of metals. There is an urgent need to prioritize the existing and potential problems associated with soil, plant, animal and human safety and health in Pakistan and preserve biological diversity in Pakistani soils by employing phytoremediation strategy.

REFERENCES

ABDA, M. AND OREN, V., 1993. Removal of Cadmium and Associated Contaminants from Aqueous Wastes by Fibrous Carbon Electrodes. *Water Res.,* **27**: 1535-1544.

ADRIANO, D.C., 1992. *Biochemistry of Trace Metals.* Lewis Publishers, USA.

ATLAS, R.M. AND BARTHA, R., 1993. *Microbial Ecology: Fundamentals and Applications* (3rd ed.). The Benjamin/ Cummings Publishing Company, Inc., CA. USA.

BAKER, A.J.M. AND BROOKS, R.R., 1989. Terrestrial higher plants which hyperaccumulate metallic elements are view of their distribution, ecology and phytochemistry. *Biorecovery*, **1**: 81.

BAKER, A.J.M., MCGRATH, S.P., SIDOLI, C.M.D. AND REEVES, R.D., 1994. The possibility of *in situ* heavy metal decontamination of polluted soils using crops of metal-accumulatingplants.*Resources Conservation and Recycling*, **1**: 41-49.

BROWN, M.T. AND WILKINS, DA., 1985. Zinc tolerance of *Amanita* and *Paxillus. Trans. Br. Mycol. Soc.*, **84**: 367-369.

CACCUATORE, D.A. AND MCNEIL, M.A., 1995. Principles of Soil Bioremediation, *Bio Cycle,* 61-64.

CUNNINGHAM, S.D., BERTI, W.R. AND HUANG, J.W., 1995. Phytoremediation of contaminated soils. *Trends Biotechnol.,* **13**:393-397.

DUECK, T.H.A., VISSER, P., ERNST, W.H.O. AND SCHAT, H., 1986. Vesicular-arbuscular Mycorrhizae decrease zinc-toxicity to grasses growing in zinc-polluted soil. *Soil Biot Biochem.,* **18**(3): 331-333.

EPA. 1990. Draft Report Government of Punjab Lahore. Environment Protection Agency, Government of Pun] ab, Lahore. pp. 51.

FINSTEIN, M.S. AND MORRIS, M.L., 1975. Microbiology of municipal solid waste composting. *Advances in Applied Microbiology,* **19**: 113-151.

GALLI, U., SCHUEP, H. AND BRUNOLD, C., 1994. Heavy metal binding by mycorrhizal fungi. *Physiology Plantarum,* **92**: 364-368.

GRIFFIOEN, W.A.J. IETSWAART, I.H. AND ERNST, W.H.O., 1994. Mycorrhizal infection of an *Agrostis capillaris* population on a copper contaminated soil. Plant Soil, **158**: 83-89.

HANSEN, J.E. AND STEVENS, D.K., 1992. Biological and Physico-chemical Remediation of Mercury- Contaminated Hazardous Waste *Arsenic and Mercury: Workshop on Removal, Recovery, Treatment, and Disposal.* EPA/600/R-92/105. pp.121-125. August.

HORN, I.M., BRUNKE, M., FLECKWER, W.D. AND TIMMIS, K.N., 1992. "Development of Bacterial Strains for the Remediation of Mercurial Waste. *Arsenic and Mercury. Workshop on Removal, RecoveryTreatment, andDkposat* EPA/6001R-92/105. pp. 106-109. August.

HUSSAIN, Z., CHAUDHRY, M.R., ZUBERI, V.A., HUSSAIN, Q. AND SHARIF, M., 1996. Contaminants and the soil environment of Pakistan, *Proc. 1st Australian-Pacific Confrrence on Contaminants and Soil Environment in the Australian-Pacific Region,* Adelaide, Australia, 18-23 February, 1996.

KHAN, A., IBRAHIM. M., AHMED, N. AND ANWAR, SA., 1992. Studies on accumulation and distribution of heavy metals in agricultural soils receiving sewage effluent irrigation. *Proc. 4th National Congr. Soil Sci. Soc. of Pakistan 'Efficient use of plant nutrients'.* May 24-26, 1992, Islamabad, Pakistan. pp. 607-610.

KHAN, A.G., KHOO, C., HAYES, W., HILL, L., FERNANDEZ, R. & GALLARDO, P. AND. CHAUDHRY, T.M., 1996. Biological, Chemical, and Physical characterisation of **a** Zinc-Alum waste site at Port Kembla, NSW, Australia. *Water; Air and Pollution (Manuscript submitted).*

KHAN, A.G. AND CHAUDHRY, T.M., 1996. Effects of metalliferous mine pollution on the vegetation and their mycouhizal association at Sunny Comer-A silver town of the 1880's. *First International Conference on Mycorrhizae,* August 4-9, 1996, University of California, Berkeley, USA.

KUMAR, P.B.A.N., DUSHENKOV, V., MOTTO, H. AND RASKIN, I., 1995. Phytoextraction: The use of plants to remove heavy metals from soil. Environ. Sci. Technol. **29**:1232-1238.

LEYVAL, C., BERTHELIN, J., SCHONTZ, D., WEISSENHORN, I. AND MORE, J.L., 1991. Influence of endomycorrhizas on maize uptake of Pb, Cu, Zn, and Cd applied *as* mineral salts or sewage sluge. In: Heavy metals in the Environment. (Ed.) J.G.F armer. pp 204-207. CEP Consultants Ltd, Edinburgh.

MANAHAN, S.F., 1994. *Environmental Chemistry* (6th ed.). CRC Press, USA.

MARKERT, B., 1994. *Plants as Biomonitors for Heavy Metals Pollution of the Terrestrial Environment* VCH, Weinheim.

MATTISON, P.L., 1993. *Bioremediation of Metals - Putting It To Work* COGNIS. Santa Rosa, CA.

McGRATH, S.P., DUNHAM, S.J., SIDOLI, C.D.M. AND LODICO, F., 1996. Extended Abstracts: *Ist. International Conference on Contaminants and the Soil Environment in the Australia-Pacific Region,* 18-23 February, 1996, Adelaid, South Australia, pp.323. C.S.I.R.O., Division of Soils, Adelaid, Australia.

MEANS, J.L., NEHRING, K.W. AND HEATH, J.C., 1993. Abrasive Blast Material Utilization in Asphalt Roadbed Material. *Third Int. Symp. Stabilization/ Solidification of Hazardous, Radioactive, and Mixed Wastes.* ASTM STP 1240. American Society for Testing and Materials, Philadelphia.

MUBARIK, A. AND JABBAR, A., 1992. Effect of pesticides and fertilization shallow groundwater quality, Pakistan Council of Research in Water Resource, Islamabad. pp.130.

NILMANI, M., LEHNER, T. AND RANKIN, W.J., 1994. Pyrometallurgy for Complex Materials and Wastes Australian Asian Pacific Course and Conference. June 6-8, 1994, Melbourne, Australia.

PALIOURIS, G. AND HUTCHINSON, T., 1991. Arsenic, Cobalt and nickel tolerances in two populations of *Silene vulgaris* (Moench) Uarcke from Ontario, Canada. *New PhytoL,* **117**: 449-459.

PERRY, R.H. AND CHILTON, C.H., 1984. *Chemical Engineers's Handbook,* 6th ed. McGraw-Hill Book Company, New York.

RASKIN, I., NANDA, K.P.B.A., DUSHENKOV, V. AND SALT, D.E., 1994. Bioconcentration of heavy metals by plants. *Current Opinion in Biotechnology. 285-290.*

SHAW, A.J., (ed.), 1990. *Heavy Metal Tolerance in Plants: Evolutionary Aspects.* CRC Press, Boca Raton, Florida, USA.

SHETTY, K.G., HATRICK, B.A.D. AND SCHWAB, A.P., 1995. Effects of mycorrhizae and fertilizer amendments on zinc tolerance of plants. *Environmental Pollution* **88**: 307-314. Stenhouse, F. (1991). The 1990 Australian market basket report. NHMR~FA (Aust. Govt. Printing Ser Canberra).

SMITH, L.A., MEANS, J.L., CHEN, A., ALLEMAN, B., CHAPMAN, C.C., TIXIER, JR., J.S., BRAUNING, S.F., GAVASKAR, A.R. AND ROYER, M.D., 1995. *Remedial Options for Metals-Contaminated Sites.* CRC Press, Inc., USA.

U.S. DEPARTMENT OF HEALTH, EDUCATION AND WELFARE, 1970. *Sanitary Landfill Facts.* SW 41s. Government Printing Office, Washington D.C.

U.S. EPA., 1989. *Technical Guidance Document Final Covers on Hazardous Waste Landfills and Surface Impoundments. EPA/530-SW*-89-047. U.S. Environmental Protection Agency, Office of Solid Waste, Washington, D.C.

U.S. EPA., 1992. *Handbook: Vitrification Technologies for Treatment of Hazardous and Radioactive Waste.* EPA/625/R-92/002. Office of Research and Development. Washington, D.C.

VOGAL-LANGE, R. AND WAGNER, G.J., 1989. Subcellular 1ocalization of cadmium and cadmium binding peptides in tobacco leaves. *Plant Physiol,* **92**: 1086-1093.

WEISSENHORN, L AND LEYVAL, C., 1995. Root colonization of maize by a Cd-sensitive and a Cd-tolerant *Glomus mosseae* and cadmium uptake in sand culture. Plant and Soil, **175**: 233-238.

WILLIS, B.A., 1985. *Mineral Processing Technology* (3rd ed.). Pergamon Press, New York.

WOLF, K., 1989. Contaminated Soils '88. Kluwer Academic Publisher. Dordrecht, Netherlands.

SECTION II

BIODIVERSITY OF PLANTS

TAXONOMIC EVALUATION, DIVERSITY AND DISTRIBUTION PATTERN OF THE GENUS *POTENTILLA* L. (ROSACEAE) IN PAKISTAN AND KASHMIR

MUQARRAB SHAH AND C.C. WILCOCK*

Pakistan Museum of Natural History, Islamabad
**Department of Plant and Soil Sci,. Uni. of Aberdeen, Scotland, U.K.*

Abstract: Taxonomic evaluation of the genus *Potentilla* L. (Rosaceae) has been made from Pakistan and Kashmir. The genus comprises nine subgenera and 15 sections. A total of 63 taxa occur in our area out of which four species and two varieties are new to science. Distribution pattern and diversity of the taxa are discussed in detail, three species are subcosmopolitan and 21 are endemic. The flavonoid composition of 90 samples from 54 species of *Potentilla* together with four species of *Sibbaldia,* two species of *Fragaria* and one species of *Geum* is made. These studies provided extremely useful taxonomic markers at the subgeneric level in the genus *Potentilla.* A scanning electron microscopic study of the achene surfaces of 10 species is made. This investigation is the first of its kind undertaken for the genus.

About one third of the species of *Potentilla* reported from Pakistan and Kashmir are useful and traditionally being utilized for various local treatments. Appropriate conservation measures are required to protect these species, as some are threatened with extinction.

INTRODUCTION

Genus *Potentilla* L. belongs to the family Rosaceae, subfamily Rosoideae and tribe Potentilleae. It is generally confined to the Northern Hemisphere. It is common in high alpine regions, on open sunny sites and a few are woodland plants. The genus is poorly represented in the warmer regions. The main taxonomic studies of *Potentilla* are the revision of Lehmann (1856) and the monographic works of Rydberg (1898) and Wolf (1908). The floristic accounts of Hooker (1878), Juzepczuk (1941) and Schiman-Czeika (1969) have also provided useful information. Besides this, considerable amount of phytochemical studies have also been done by various workers in the past. (Bate-Smith 1961, Challice 1974, 1981; Kohli & Denford 1977 and Harborne and Nash 1984).

The present studies have been undertaken to give a taxonomic account of the genus represented in Pakistan and Kashmir in the first instance as no floristic work of this kind has been done for the genus from this area. Secondly where taxonomic problems related to the genus became apparent during the course of the preparation of the Flora, it necessitated some more intensive surveys such as SEM of achene surface and phytochemical studies. The genus *Potentilla* proved to be particularly amenable to flavonoids and as no previous studies were available from Pakistan and Kashmir, consideration was given to their value as potential taxomonic markers.

MATERIALS AND METHODS

More than five thousand herbarium specimens of *Potentilla* were examined which were borrowed from different herbaria of the world including B,BM,CAL,CGE,CHR,DD,E,G,K,KUH,L,LE,LIV,MICH,O,P, PR,RAW, S,TUR,UC,US,W,WU, and Z besides collections at ABD gathered during field trip to N. Pakistan during summer of 1987. The material was examined using stereoscopic microscope. Quantitative measurements of the vegetative and floral parts were made from comparable positions by ruler and caliper. Hairs were measured by a graticule magnifier (6x). Achenes of 10 critical species were examined under SEM. The mature achenes were fixed to a metal stub (12.5 mm. diam.) with double sided tape. The stubs were coated with gold for two minutes in an Emscope Sc 500 sputter coater. The gold coated specimens were then examined in a Cambridge instrument stereoscan 5600, 7.5 Kv. All measurments are in millmicron.

For flavoniod studies 90 samples from 54 species of *Potentilla* together with four species of *Sibbaldia,* three species of *Fragaria* and one species of *Geum*, were investigated. Herbarium material borrowed from different herbaria of the world mentioned above was used for the investigations apart from field collections from northern Pakistan by one of the authors in 1987. The techniques employed for screening of flavonoids followed those of Harborne (1984). Authentic commerical markers of flavonoids (myricetin, quercetin and kaempferol) were co-chromatographed near the corner of the plates. Chromatographic seperations were made in two dimensions using Forestal (gl. acetic acid, HCl, water 30:3:10) for the first run for four hours and water for the second run for 45 minutes. Identifications of the compounds were done after examination of the profile under UV light.

RESULTS

A. TAXONOMIC:

During the course of studies on the genus *Potentilla*, it has been observed that the genus in composed of a heterogeneous group of species with several distinct subgroups which can best be treated as subgerera. The 63 species of the genus are grouped into nine subgenera and 15 sections (Table I). The two genera *Sibbaldiopsis* Rydb. and *Duchesnea* J.E. Smith have been reduced to subgenera, whereas subgenus *Potentilla* Syme and subgenus *Hypargyrium* (Fourr.) Juz. have been united under subgenus *Potentilla* Syme. A brief account of the subgenera of *Potentilla* is given below:

1. Subgenus Dasiophora (Rafin.) Panigr. & Dixit, Bull. Bot. Surv. India 27 (1-4): 179 (1985) L.Type *Potentilla fruticosa* L.

Dioecious or monoecious shrubs. Stem woody. Leaves articulated, 3-7 foliate, leaflets entire. Petals suborbicular. Anthers monothecous. Stigma 4-lobed, styles subbasal or lateral, clavate. Achenes completely concealed.

2. Subgenus Trichothalamus (Lehm.) Rechb. Consp. Regn. Veget. 167 (1828). L. type *Potentilla lignosa* Willd. ex Schlecht.

Undershrubs or herbs. Leaves 2-3 pinnate, leaflets finely serrulate. Flowers 2-3 on dicotomous peduncles, petals obovate, creamy white to light pink. Anthers dithecous. Styles lateral. Achenes with long soft hairs.

3. Subgenus Lasiocarpa Dixit & Panigr. J. Sci. Club 33-34: 39 (1983). L. type *Potentilla salesoviana* Steph.

Spreading undershrubs. Stem woody. Leaves interrruptedly pinnate, leaflets sharply 7-12 dentate, coriaceous, farinose beneath. Petals white, suborbicular, Stamens monothecous. Styles very long, filiform, lateral. Achenes concealed in long hairs.

4. Subgenus Schistophyllidium Juz. ex Fed. Fl. Armen. 3: 87 (1958). Type *Potentilla bifurca* L.

Perennial dioecious herbs. Stem partly subterranean. Leaves crowded at the base, 3-8 pinnate, leaflets entire. Petals obovate. Anthers monothecous. Styles fusiform. Achenes glabrous.

5. Subgenus Sibbaldiopsis (Rydb.) Shah & Wilcock Edin. J. Bot. 50 (2): 175 (1993). L. type *Potentilla tridentata* Soland.

Basal stem partly subterranean, woody, floral stem herbaceous. Leaves ternate, leaflets tridentate. Petals orbicular, white or yellow.

Anthers monothecous. Styles filiform, lateral or subbasal. Achenes concealed.

6. Subgenus Fragariastrum (Heist. ex Fabr.) Rchb. Consp. Regn. Veg. 167. (1828). L. type *Fragaria sterilis* (L.) Schur. = *Potentilla sterilis* (L.) Garcke.

Caespitose bushy undershrubs. Basal stem partly subterranean. Leaves forming rosettes at the base, leaflets dentate to deeply segmented.Petals obcordate. Anthers with two thecae. Receptacle villose. Styles filiform, subterminal. Stigma not dilated.

7. Subgenus Chenopotentilla (Focke) Juz. in Kom. Fl. USSR 10: 164 (1941). Type *Potentilla anserina* L.

Trailing herbs, stoloniferous. Leaves imparipinnate. Flowers solitary, petals obovate. Anthers with 2 thecae.Styles lateral. Achenes glabrous.

8. Subgenus Duchesnea (J.E.Smith) Shah & Wilcock Edin. J. Bot. 50 (2): 176 (1993). Type *Duchesnea indica* (Andr.) Focke

Trailing herbs rooting at the nodes. Leaves ternate. Flowers yellow, outer sepals trilobed.Stamens dithecous. Styles subterminal. Receptacle much enlarged swollen and spongy, bright red. Fruit bright red, tasteless.

9. Subgenus Potentilla Syme in Smith & Sowerby, English Botany ed.3, 3: 143 (1864). L.type *Potentilla reptans* L.

Perennial herbs, rarely annuals. Stem prostrate ascending or erect. Leaves ternate, quinate or pinnate. Flowers in terminal cymes, rarely axillary and solitary. Anthers dithecous. Styles subterminal, clavate or coniform. Achenes glabrous.

Table I. A checklist of *Potentilla* species in Pakistan and Kashmir

Subgenus	Section	Species
Trichothalamus		*P. sericophylla**
Dasiophora		*P. ochreata**
		P. arbuscula
		P. dryadanthoides
		*P. rigida**
Lasiocarpa		*P. salesoviana*
Schistophyllidium		*P. bifurca ssp. orientalis*
Sibbaldiopsis		*P. cuneifolia*
Fragariastrum	Eriocarpae	*P. eriocarpa*
	Biflorae	*P. biflora var. biflora*
		*P. biflora var. lahulensis**
		P. l atipetiolata

	Curvisetae	*P. curviseta**
		*P. collettiana**
		*P. pteropoda**
Chenopotentilla	Luconotae	*P. peduncularis*
		P. polyphylla
		*P. commutata**
	Anserinae	*P. anserina***
Duchesnea		*P. indica*
Potentilla	Terminales	*P. argentea*
		P. virgata
		*P. kashmirica**
	Tanacetifoliae	*P. bannehalensis**
		P. gerardiana
	Aureae	*P. gelida*
		P. doubjouneana
		P. turczaninowiana ssp. *turczaninowiana*
		P. turczaninowiana ssp. *kuramensis**
	Pensylvanicae	*P. multifida*
		P. ornithopoda
		P. plurijuga
		P. exigua
		P. agrimonioides
		P. pamirica var. *pamirica*
		P. Pamirica var. *pamiroalaica*
		P. hololeuca
		*P. khunjrabensis**
	Rivales	*P. monanthes*
		P. desertorum
		P. sundaica
		*P. supina ssp. paradoxa***
		P. supina ssp. costata
		P. heynii
	Niveae	*P. sinonivea*
		P. evestita
		P. grisea var. grisea
		*P. grisea var. vilijuga**
		P. blanda
		P. saundersiana
	Chrysanthae	*P. chrysantha*
	Persicae	*P. flabellata*
		*P. stewartiana**
	Haematochroae	*P. atrosanguinea*
		*P. Nepalensis**
		*P. cathaclines**
	Potentilla	*P. reptans***
	Hybrid Taxa	*P.x mutabilis**
		*P.x subdigitata**
		*P.x gilgitica**
		*P.x grandiloba**
		*P.x brevicissa**
		P.x aphanes

* endemic; ** subcosmopolitan

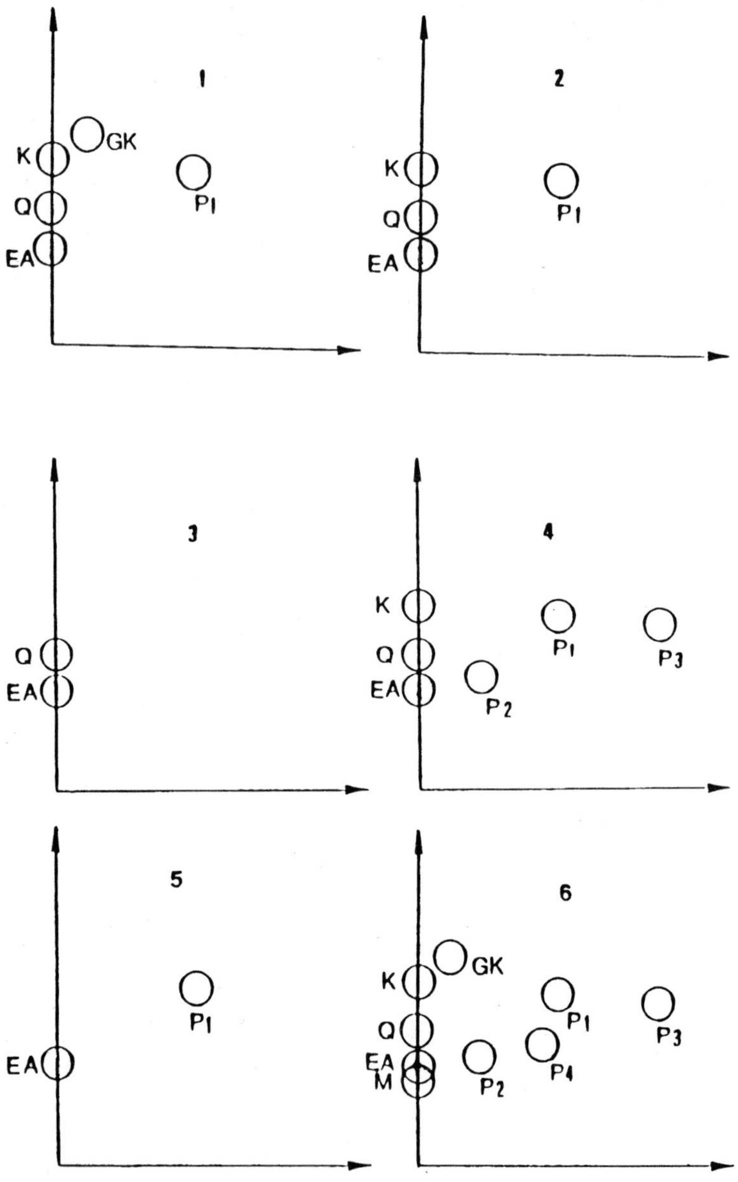

Fig. 1 Six different flavonoid profiles in *Potentilla* and allied genera, EA (Ellagic acid) is present in all chromatograms

DISTRIBUTION PATTERN

The species of the genus *Potentilla* are mostly found in the northern mountain ranges of Pakistan and Kashmir. The main concentration however, is in Kashmir 51 taxa followed by Gilgit 32 taxa, Baltistan 26 taxa and Hazara and Chitral 22 taxa each (Table II). Three recently described new species, *P. stewartiana, P. gilgitica and P. grandiloba* have also been included in the account (Shah & Wilcock 1991 a & b). Three species are subcosmopolitan and 21 are endemic (see Table I).

Table II. Representation of *Potentilla* species in different areas of Pakistan and Kashmir

Area	Representative species
Kashmir	51
Gilgit	32
Baltistan	26
Chitral	22
Hazara	22
Swat	09
Kurrum valley	06
Waziristan	03
Peshawar & Khyber Agency	02
Baluchistan	02
Punjab	02
Sind	01

B. PHYTOCHEMICAL RESULTS

The results are listed in Table III together with the previous data showing the flavonoid profiles of 54 species of *Potentilla* with four species of *Sibbaldia*, three species of *Fragaria* and one species of *Geum*. The following six distinct groups are visible in these chromatograms (Fig.1).

1. Subgenera *Trichothalmus, Dasiophora, Schistophyllidium, Sibbaldiopsis, Fragariastrum, Chenopotentilla* (section *Luconotae), Potentilla* (except sections *Pensylvanicae* & *Haematochroae)* and *Geum* with flavonoids such as Quercetin (Q) Kaempferol (K) and C-glucoside of Kaempferol (GK).

2. Subgenus *Comarum, Potentilla* (section *Pensylvanicae*), *Fragaria and Sibbaldia* with Q & K.

3. Subgenus *Lasiocarpa* with Q only.

4. Subgenus *Potentilla* (section *Haematochroae*) with additional unidentified spots P_2 & P_3.

5. Subgenus *Duchesnea* with absence of both Q & K.

6. Subgenus *Chenopotentilla* (section *Pentaphylloides*) with the presence of an additional flavonoid myricetin (M) apart from Q & K.

Table III. Distribution of principal flavonoids in the leaves of *Potentilla* and allied genera.M=Myricetin,=Quercetin, K= Kaempferol, GK= Kaempferol C-glucoside. - = Undetected, + to +++ = detected at increasing concentration, Tr = Trace, () = not present in all species.

Genus	Subgenus	No. of spp. examined *	M	Q	K	GK	Source of data **
Potentilla	*Trichothalamus*	1 (2)	-	++	Tr	+++	1c
	Dasiophora	3 (8)	-	++	+	+++	1a, 3c
	Lasiocarpa	1 (1)	-	+++	-	-	1c
	Comarum	1 (1)	-	++	++	-	1a, 1c
	Schistophyllidium	1 (4)	-	++	+	++	1c
	Sibbaldiopsis	2 (3)	-	+++	+	+++	1a, 1c
	Fragariastrum	5 (7)	-	++	+	+++	1a, 5c
	Chenopotentilla sect. Luconotae	3 (7)	-	++	++	-	3c
	Chenopotentilla sect.Pentaphylloides	4 (4)	+++	+++	++	+	1a, 4c
	Duchesnea	1 (2)	-	-	-	-	1a, 1c
	Potentilla	34 (c.400)	-	+++	+++	(++)	36a, 3b, 33c
Sibbaldia		4 (8)	-	++	+	-	4c
Fragaria		3 (12)	-	++	+	-	2a, 2c
Geum		1 (65)	-	+++	++	++	1a, 1c

* The first number refers to the total number of species chemically examined,in this and previous studies, and the number in brackets to the total number of species in the taxon.

** a = Bate-Smith (1961), b = Kohli & Denford (1977), c = Shah (1990). The numbers indicate the number of species studied in each work. The total may add to more than the total number of species studied as some species may have been examined more than once.

SEM STUDIES:

SEM studies of Achene surfaces in the genus *Potentilla* showed the following two main types a) Achenes with smooth surfaces b) Achenes with ridged surfaces (Fig.2). Members of the subgenera *Schistophyllidium* and *Fragariastrum* and section *Pensylvanicae (P. multifida* group) of the subgenus *Potentilla* have smooth achene surfaces whereas the rest of the species of the genus have more or less ridged achene surfaces.

DISCUSSION

For the taxonomic evaluation of the genus, various important characters like number of leaves, shape of the leaflets, stamens and carples, have been taken into account. Considerable variation has also been noticed in the type, position and length of the hairs particularly on the petioles and stems, leaflet surfaces and achenes. Infrageneric classification of the genus *Potentila* is based primarily on the following taxonomic characters: i) shape of stamen and carpels ii) Type of indumentum on the achene.

Potentilla is a complex and polymorphic genus. Apomixis and pseudogamy play an important role in formation of complexes of microspecies. Most apomicts are polyploid, where sexual ancestors are known, they are usually cross breeding, often self incompatible and dioecious. Formation of an unreduced embryo and the predominant capacity of egg cells to undergo parthenogenetic development, are the two major factors involved in pseudogamy and apomixis in *Potentilla*. Hybridization is a common phenomenon in the genus. Normally the polyploids are well adapted in contrast to their diploid progenies (Barber 1970). Intergeneric hybrids of *Potentilla* and *Fragaria* have also been reported. An artificially crossed *Potentilla* x *Fragaria* hybrid went on sale in Chelsea Flower show in London under the name "Pink Panda" (report-New Scientist vol. 122 no. 1666, 27 May 1989. p. 29).

Sino-Himalayan mountain ranges are among the primary centres of diversity of the genus *Potentilla* (Shah et al., 1992). All except three species (*P. supina, P. heynii* and *P. sundaica*) are perennials. The perennial species are generally alpine and subalpine and have very brief growing periods. The plants are generally dwarf and caespitose and these conditions are caused by severe winds and frost. Nevertheless they survive in harsh conditions in nature through their very thick and stout rootstock. Mostly the plants are densely hairy

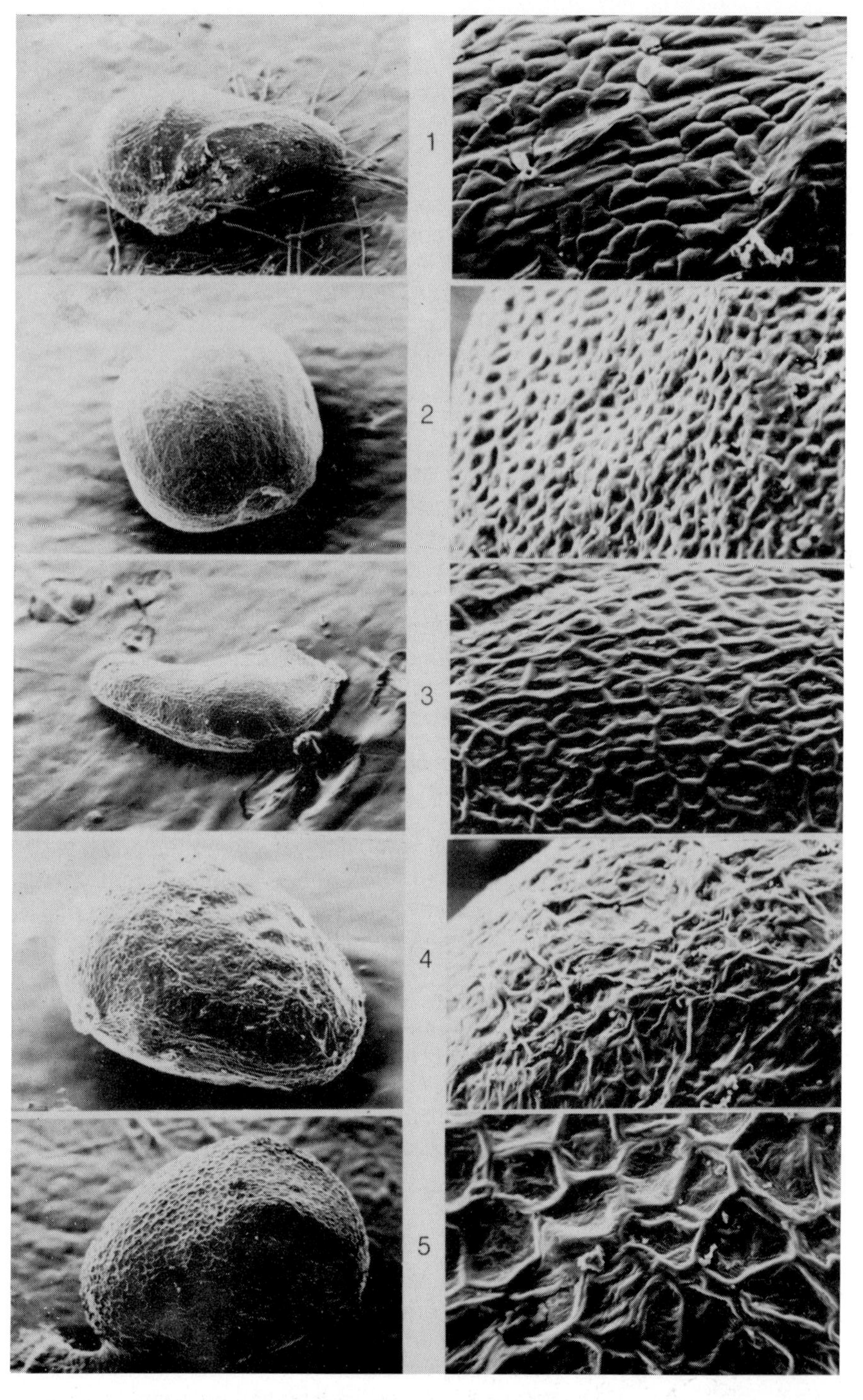

Fig. 2a SEM of achene surfaces of *Potentilla*, subgenus *Dasiophora* (1 *P. ochreata)*, subgenus *Schistophyllidium* (2 *P. bifurca)*, subgenus *Fragariastrum* (3 *P. curviseta)*, subgenus *Chenopotentilla* (4 *P. polyphylla*, 5 *P. anserina)*

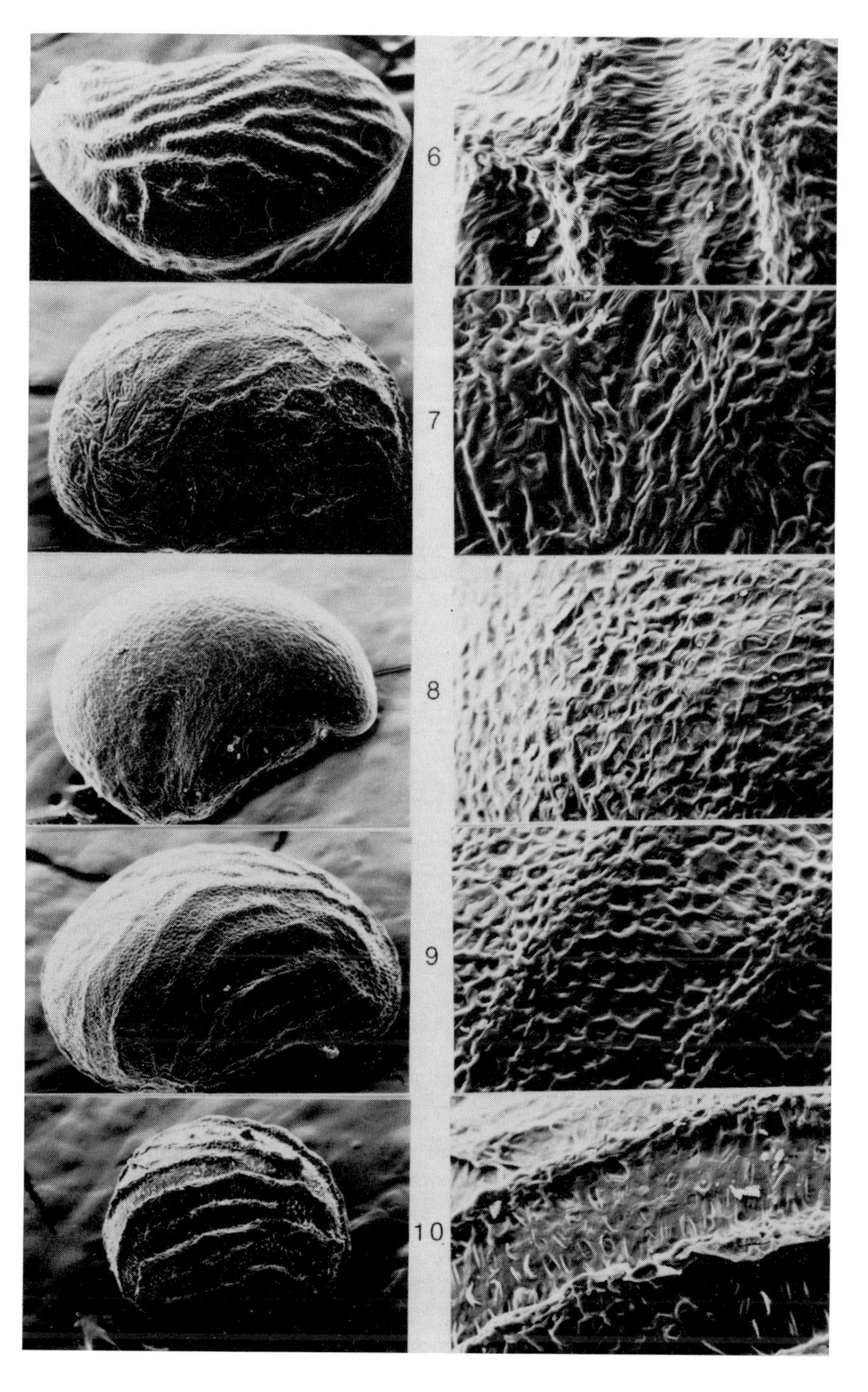

Fig. 2b subgenus *Potentilla* (6 *P. gerardiana*, 7 *P. gelida*, 8 *P. multifida*, 9 *P. desertorum*, 10 P. *sundaica)*

probably an adaptation for protection from high intensity ultraviolet rays. Our northern mountains are famous for high rate of endemism in plants and *Potentilla* is no exception. One third of the species are endemic to our region. Eighteen species are economically important and traditionally being utilized by locals over centuries (Shah & Bano, 1994). Some species have been over exploited like *P. chrysantha* and *P. dryadanthoides*, so that these have become extinct. Preventive measures are required for the conservation of these valuable species.

Phytochemistry of the genus is interesting. Six distinct groups are visible on the chromatograms under UV light (Fig.1). The principal flavonoids of all the primitive subgenera (except subgenus *Lasiocarpa*) are quercetin (Q), kaempferol (K) and Kaempferol c-glucoside (GK) - group 1. In the subgenera *Comarum* and *Potentilla* (section *Pensylvanicae*) and in genera *Fragaria* and *Sibbaldia*, Q and K are present - group 2. In subgenus *Lasiocarpa* only Q is visible - group 3, whereas in subgenus *Duchesnea* both Q and K are absent - group 5. On the other hand flavonoid gain has occurred in subgenus *Potentilla* (section *Haematochroae*) where two additional unidentified purple spots P2 and P3 are visible in addition to Q and K - group 4. In group 6, additional flavonoid myricetin (M) occurs which is interesting because this trihydroxy derivative is not found in any other subgenus of *Potentilla*. Ellagic acid is present in all the subgenera.

ACKNOWLEDGEMENTS

We wish to thank Dr. Bashir Ahmad Sheikh, Chairman, Pakistan Science Foundation, Islamabad for his guidance and Dr. Shahzad A. Mufti, Director General, Pakistan Museum of Natural History, Islamabad for going through the manuscript and offering his valuable comments.

REFERENCES

BARBER, H.N. 1970. Hybridisation and evolution of plants. *Taxon* **19**:154-160.

BATE-SMITH, E.C. 1961. Chromatography and taxonomy in Rosaceae with special reference to *Potentilla* and *Prunus*. *Bot. J. Linn. Soc. London* **58**: 39-54.

CHALLICE, J.S. 1974. Rosaceae - Chemotaxonomy and origin of Pomoideae. *Bot. J. Linn. Soc. London* **69**:239-259.

CHALLICE, J.S. 1981. Chemotaxonomic studies in the family Rosaceae and the evolutionary origin of subfamily maloideae. *Preslia* **53**:289-304.

HARBORNE, J.B. 1984. *Phytochemical Methods*, 2nd ed. Chapman and Hall. London.

HARBORNE, J.B. AND NASH. R.J . 1984. Flavonoid pigments responsible for ultraviolet patterning in petals of the genus *Potentilla. Biochem. System. and Ecol.* **12**: 315-318.

HOOKER, J.D. 1878. *Fl. Brit. Ind.* **2**:345-360.

JUZEPCZUK, S.V. 1941. In Komarov, V.L. *Fl. USSR* **10**:78-223.

KOHLI, B.L. & DENFORD, K.E. 1977. A study of the flavonoids of the *Potentilla pensylvanica* complex in North America. *Canad. J. of Bot.* **55**:476-479.

LEHMANN, C. 1856. *Revisionem Potentillarum Iconibus illustrata Vratislaviae et Bonnae.*

RYDBERG, P.A. 1898. A monograph on North American Potentilleae. *Memoirs from the Department of Botany of Columbia Universtiy,* **2**.

SCHIMAN-CZEIKA, H. 1969. *Potentilla* in K.H. Rechinger *Fl. Iran.* **1**:78-114.

SHAH, M. 1990. Taxonomic studies in the genus *Potentilla* (Rosaceae) from Pakistan and Kashmir. Ph. D. Thesis. University of Aberdeen, U.K.

SHAH, M. AND WILCOCK, C.C. 1991a. Notes on the genus *Potentilla* (Rosaceae) from Pakistan and kashmir. *Willdenowia* **21:** 195-199.

SHAH, M. AND WILCOCK, C.C.1991b. Notes on genus *Potentilla* (Rosaceae) from Pakistan and Kashmir. *Folia Geobot. et Phytotax.* **26**: 193-196.

SHAH, M., SHINWARI, Z.K. LEGHARI, M.K. AND NAKAIKE, T. 1992. A note on centres of diverstiy of the the genus *Potentilla* (Rosaceae) *Bull. Natn. Sci. Mus., Tokyo*, Ser. B, **18**: 117-122.

SHAH, M. AND BANO, F. 1994. Economic importance of *Potentilla* species in Pakistan and Kashmir *Sci. Tech. & Devel.* **13**: 56-58.

STEWART, R.R. 1972. *An annotated catalogue of vascular plants of West Pakistan and Kashmir.* Fakhri Printing Press. Karachi.

WOLF, T. 1908. *Monographie der gattung Potentilla.*

ETHNOBOTANICAL STUDIES OF SWAT DISTRICT (PAKISTAN)

M. RASHID AWAN AND SALEEM AHMAD

Pakistan Museum of Natural History, Garden Avenue, Shakarparian, Islamabad

Abstract: Ethnobotanical studies of Swat District were carried out during 1989-1992, to collect information about plants being traditionally utilized over generations by different ethnic groups. In this paper effort has been made to document the ethnobotanical knowledge of sixty eight important plant species from all available sources including the local names and uses.

INTRODUCTION

Swat district lies between 34^0 09' and 35^0 56' north latitude and 72^0 07' to 73^0 00' east longitude. It is mountainous with fairly broad and almost flat strips which are used for crop cultivation with fertile soil conditions along both banks of the Swat river. The region is highly variable in lithology and pedology, complemented by the existence of different climatic zones like subtropical, temperate and alpine. All these factors produce quite variable and rich flora that includes plants of ethnobotanical importance as well.

Swat valley is inhabited by large tribal populations such as Dalazaks, Swatis, Yousafzai and other ethnic groups such as Awan, Kashmiri, Kohistani, Gujar, Hindki, Sayyed and Sahibzadgan, etc. These groups have their distinct ways of life, beliefs, traditions, and cultural heritage. They have been utilizing local plants for various purposes over generations. Substantial extraction of medicinal plants for many decades has resulted in the scarcity of a number of plants which are considered to be of immense benefits to mankind.

The Flora of Pakistan (edited by Nasir, E. and S.I. Ali, nos. 1-189) alongwith Stewart's Annotated Catalogue (1972), Stewart's Checklist (1967) and Beg and Samad (1974) were major sources of information in this field. Although the area has been extensively surveyed by various workers in the past yet little effort has been made to document the ethnobotanical knowledge of plants of Swat.

MATERIALS AND METHODS

Ethnobotanical information was obtained through general conversation with local inhabitants of Swat. A number of locals belonging to different ethnic groups were interviewed during the field work regarding tradional uses of plants. Collected plant material was labelled, identified and preserved at PMNH Herbarium.

RESULTS AND DISCUSSION

Information about sixty eight plant species of ethnobotanical importace is presented below:

Juglans regia Linn. (Common name : Akhrot, walnut)

It is a deciduous tree, found wild and cultivated from 1000-3300 meters altitude. The wood is excellent for furniture, carving and for gun stocks. The bark is also good for the gums and sold in the local market under the name "Dandasa" which is used for cleaning the teeth. Bark is also used as a dye. The seeds yield an oil used for cooking. The "Kaghzi" variety of walnut is valued for its thin shell and edible fruit.

Ricinus communis L. (Common name: Arund, castor oil)

It is a common shrub. The oil from the seed has many uses such as an illuminant, purgative, a leather-preservative and a lubricant, especially used in delicate machinery; the oil-cake is used as fertilizer and fuel.

Dioscorea deltoidea Wall. ex Kunth (Common name: Yams)

It is a very common perennial herb from plains to hilly areas in Swat. Its tubers are used for washing shawls, woolen clothes and blankets. It is also said to be used for the expulsion of intestinal worms.

Valeriana jatamansi Jones. (Common name: Mushkbala)

It is a very common perennial herb with aromatic underground parts, found in early spring from 1300- 3200 m. in Swat. The rhizome yields an aromatic oil, which is used in the preparation of tranquilizers and a remedy for the suppression of urine and an important ingredient in perfumed powdwers.

Platanus orientalis Linn. (Common name: Chanar)

It is a large tree, its wood is used for small painted boxes and cabinet work and paneling.

Podophylum emodi Wall. ex Royle (Common name: Bankakri)

It is a small herbaceous plant found from 2500-3000 m in a coniferous forest. The rhizomes are used in liver and bile diseases. Due to indiscriminate extraction of roots there is an immediate need to regenerate this species.

Woodfordia fruticosa (Linn.) S.Kurz. (Common name: Dhawi)

Leaves and twigs yield a yellow dye used in printing while its petals yield a red dye. Flowers and leaves are used medicinally as astringent and analgesic.

Ribes himalense Decne. (Common name: Karan)

Berries are edible and are sold as currants for making jams, puddings and cakes etc.

Daphne mucronata Royle (Common name: Kuttilal)

The leaves are poisonous but are tolerated by goats; can be applied for abscesses. The bark is used in diseases of bones and for washing hair. Gunpowder charcoal is said to be made from the wood. The fruit can be eaten and is used as a dye for leather.

Colchicum luteum Baker (Common name : Suranjan talkh)

It is an early flowering, small perennial herb found soon after the snow melts in upper Swat. It is a famous remedy for rheumatism and diseases of liver and spleen.

Quercus incana Roxb. (Common name: Ban, Ringi)

It is a common tree in Marghazar from 1000 to 2700 meters. The wood is used in buildings whereas the leaves are used as fodder.

Q. baloot Griff. (Common name: Breh)

An evergreen small tree or a shrub, 2 to 6 m tall. The wood is used for construction purposes and the bark yields tannin.

Q. dilatata Royle (Common name: Barungi, Moru)

An evergreen tree up to 20 m tall found between 1600 to 2900 meters altitudinal range. The wood is used for fuel and for making charcoal and as timber.

Punica granatum Linn. (Common name: Jungli Anar, Daruna)

It is a small tree or shrub. The fruit is delicious to eat ; the juice is used as a tonic in fever. Dried seeds " Anardana" are used for adding taste to certain foods like " Chatni". Bark of the root and wood is used as a vermifuge for tapeworms; also used for diarrhoea and dysentry; a number of dyes can be obtained from it; black writing ink is also made from it.

Paeonia emodi Wall. ex Royle (Common name: Mamekh)

The underground tubers are useful in nervous disorders. The dried flowers are used for stomach complaints. The seeds are purgatives.

Flacourtia indica (Burm.) Merrill. (Common name: Pachnala)

It is a small tree or shrub; its fruit is edible .

Viola serpens Linn. (Common name: Banafsha)

It is a small prostrate annual herb which is used in cough and cold conditions and in lung diseases.

Melia azedarach Linn. (Common name: Dhrek, Bakain)

It is a fast growing tree of the plains and foot-hills, cultivated as a shady plant along road-sides in Swat valley. Juice of twigs and leaves is used in many skin diseases. The fruit is eaten by goats and sheep and the stony endocarp is used as beads.

Ziziphus nummularia (Burm.f.) Wight & Arn. (Common name: Malla, Jer beri)

The branches are often used for fencing the fields, the leaves are threshed out and used as fodder while its fruit is said to be edible. Leaves also applied in scabies and boils.

Ziziphus jujuba Mill. (Common name: Singli, Unab, Baryan)

Economically an important tree or shrub. Fruit is used for bronchitis, the bark contains tannin. It is a major ingredient of "Joshanda" which is best for cough and cold.

Acacia modesta Wall. (Common name: Phulai)

The wood is hard and durable. It is used for cane crushers, wheels and agricultral implements. It is also used as fuel. The gum is used in local medicine and twigs used as "Maswak" for cleaning the teeth.

Albizia lebbeck (L.) Benth. (Common name: Sharin)

Commonly planted as a roadside tree. Wood is excellent for furniture, picture frames, building etc.

Linum usitatissimum Linn. (Common name: Alsi)

Oil extracted from the seeds has medicinal importance while the seed cake is used for feeding cattle.

Plantago major Linn. (Common name: Isafgol)

It is a perennial herb . Leaves are said to be diuretic.

Plantago lanceolata Linn. (Common name: Danichk, Isabgool)

Seeds are used as purgative and in many stomach diseases while leaves are applied to wounds.

Hippophae rhamnoides Linn.

A small shrub or tree, variable in size and scaliness of the leaves. The wood is used for fuel. The fruit, though acidic, when boiled with sugar can be eaten.

Elaeagnus angustifolia Linn. (Common name: Russiam olive)

It is mostly a cultivated tree. The yellow flowers have a strong but pleasant odour. The fruit is astringent and is edible.

Acer cappadocicum Gleditsch (Common name: Kilpattar)

Trees up to 25 meters tall. It's wood is used for making implements, poles and bedsteads.

Dodonaea viscosa (L.) Jacq. (Common name: Sanatha)

An evergreen shrub. The branches are used as firewood and as a support in the mud roofs in village Kacha houses.

Sapindus mukorossi Gaertn. (Common name: Ritha)

The soap nut tree is cultivated for its fruit, the pericarp is saponaceous and used for washing clothes and hair. It is also planted along roadside for shade.

Cedrela toona Roxb. ex Willd. (Common name: Tun)

Tun is cultivated in the plains and the foot hills up to 1000 m. It yields a reddish timber of good quality used for making furniture, carving and cigarette boxes. The sweet scented flowers yield a dye and the bark is used in medicine.

Myrsine africana Linn. (Common name: Chapra, Bebrang)

It is small and evergreen shrub up to 1 meter tall.It burns quickly, even when green, because of its glandular leaves. The fruit is used as an anthelmic.

Alhagi maurorum Medic. (Local name: Shiz)

Fresh leaves are crushed and juice is collected. This juice is used to relieve soreness of eyes. Paste of the root is also used as an external treatment for swellings.

Asphodelus tenuifolius Cavan (Local name : Piazi)

A weed of wheat crops, leaves are eaten raw as a vegetable.

Boerhaavia procumbens Roxb. (Local name: Itsit, Wasao)

Rare small herb. Favourite fodder of livestock. The roots are said to be medicinal and used as purgative. The juice extract from the plant is used as diuretic.

Fraxinus xanthoxyloides (G.Don) DC. (Common name: Hanuz, Sum)

The wood is hard, white and close-grained; used for tool handles and walking sticks. The foliage is used as fodder.

Calotropis procera (Ait.) Ait. f. (Local name: Ak)

A perennial shrub, 6-18 feet tall.The milky latex is said to be irritant to skin and cause blindeness. The seed floss is used for stuffing pillows.

Capparis decidua (Forsk.) Edgew. (Local name: Karir, Kary)

It is one of the commonest shrubs in the area , it's fruit is eaten by locals when ripe and also used for making pickle. The wood is hard and bitter in taste and resistant to attacks by white ants, it is used for making planks of boats. Flowers are used as vegetable.

Olea ferruginea Royle (Common name: Kahu)

It is a slow growing evergreen tree found from 500 to 2000 meters. The wood is very hard and heavy, used for making ploughs, sticks etc.

Jasminum humile Linn. (Common name: Chambeli)

It is a large shrub, deciduous or evergreen used as ornamental plant due to its fragrat flowers.

Agave cantala Roxb. (Common name: Kantala, Naghband)

Large rhizomatous herb with stolons. Cultivated on a small scale by locals for hedge and its fiber.

Pinus roxburghii Sargent (Common name: Chir, Chil)

It is common from 600 - 1800 meters. It's wood may be used for construction purposes. It is also valuable for its resin extract which is used as tarpentine oil and in varnish etc.

P. gerardiana Sargent (Common name: Chilghoza pine)

Trees up to 18 m tall. Common in Janshey, Uthror and Gabral areas. The seeds " Chilghoza" are edible .

Abies pindrow Royle (Common name: Partal, Paludar)

Silver fir is common and gregarious usually on the north aspect in upper Swat area. It's wood is very useful for building purposes and is also used for making matches and paper pulp.

Picea smithiana (Wall.) Boiss. (Common name: Spruce)

The spruce is fairly common from 250 - 3300 meters. Used as timber.

Cedrus deodara (Roxb. ex D. Don) G. Don. (Common name: Deodar, cedar)

The cedar is commonly gregarious at an altitude from 2000-3000 meters. The wood is of an excellent quality and used for construction and funiture purposes.

Ailanthus altissima (Mill.) Swingle. (Common name: Tree of Heaven)

Cultivated as a roadside tree in middle and lower Swat areas. It is an excellent soil binder.

Achyranthes aspera Linn. (Common name: Puthkanda)

It is a common herb found in dry areas of Buner and Lower Swat. Its roots, leaves and stems are used in various ways. Mixed with water it is used for asthma, cough, stomach pain and skin eruptions.

Chenopodium album L. (Common name: Josag)

It is a small and common herb, its boiled leaves are eaten as a vegetable.

Peganum harmala Linn. (Common name: Harmal, Ispand)

It is a small perennial herb 25 - 60 cm tall, corymbosely branched. Seed powder is used in asthma, colic, jaundice and as an anthelmintic against tapeworms and reducing temperature in chronic malaria. Seeds are used as narcotic. It increases the flow of milk in livestock and is stimulant. The smoke is considered to be antiseptic and wounds are fumigated by burning of seeds and leaves.

Betula utilis D.Don (Common name: Bhojpattra)

Birch is commonly found at the upper limit of trees, from 3000 to 4500 meters. The bark is smooth and white when peeled horizontally and is said to be used for roofing and a subsitute for writing paper.

Parrotiopsis jacquemontiana (Dcne.)Rehder (Common name: Pishor)

A common tree from 1200 to 2800 meters altitude in Swat. The wood is strong and is commonly used for making handles, walking sticks. The twigs are used for making baskets.

Alisma plantago-aquatica Linn. (Common name: Alisma)

Root powder is used as a cure in hydrophobia.

Skimmia laureola (DC.) Sieb. (Common name: Ner)

A small strong-scented shrub, commonly growing as undergrowth in shady coniferous forest in temperate zone in upper Swat. The leaves are used in medicine.

Zanthoxylum armatum DC. (Common name: Timbar)

A fairly common xerophytic small, spiny tree or shrub, found in the foothills up to 1500 meters. The twigs are used as tooth brushes "Maswak" and the stems made into walking sticks. Seeds are also used for stomach pains and piles.

Robinia pseudo-acacia Linn. (Common name: Robinea, Kiker)

Cultivated as a roadside tree for shade and to prevent soil erosion .

Solanum nigrum L. (Common name: Mako, Kach-Mach)

It is a small herb. It's fruit is edible while the young shoots are cooked as a pot herb. Plant parts contain the alkaloid solanine. The juice of this plant is said to be diuretic, and used in the treatment of enlarged livers.

Solanum tuberosum L. (Common name: Alu, Potato)

An erect, unbranched perennial herb. Stem tubers under ground, of various shapes and sizes. The Potato is widely cultivated for its edible tubers in Swat. The green part of the plant contain poisonous alkaloid, solanine, which is soon lost on boiling the tubers; sprouting or green tubers should be avoided for edible purposes. Apart from starch, the potato is also a rich source of protein and vitamin C.

Withania somnifera (L.) Dunal in DC. (Common name: Aksan)

It is a common herb throughout waste places upto 2300 meters. The plant parts have alkaloids with sedative properties. The root is used in rheumatism. The leaves are used in fevers while roots are useful for ulcers, boils etc. The fruit is said to be diuretic.

Nicotiana tabacum Linn. (Common name: Tumbaku, tobacco)

Plants up to 1 meter tall and is widely cultivated on a commercial scale in lower and middle Swat. Commonly used for smoking and for insecticidal purposes for woolen clothes.

Indigofera heterantha Wall.ex Brandis. (Common name: Kanthi)

It is a common undergrowth in coniferous forests. It's branches are used for basket making and in the roof material of mud houses.

Aesculus indica (Wall. ex Camb.) Hook. f. (Common name: Bankhor)

It is found wild and cultivated as an ornamental tree, from 1200-3300 meters. The bark is said to be astringent. The fruit is used for rheumatic pains while leaves are used as fodder.

Berberis lycium Royle. (Common name: Samlu, Kashmal)

It is common shrub in sub tropical chirpine zone of Swat. It's root is said to be used for skin disease, in chronic diarrhoea and for piles .

Foeniculum vulgare Mill. (Common name: Saunf, fennel)

It is commonly cultivated herb in the area. It's fruit is used as condiment, in stomach disease and is said to be good for eye sight.

Coriandrum sativum L. (Common name: Dhania, coriander)

It is a common cultivated herb. Fruit is used as condiment and as carminative.

Pimpinella stewartii R.R. Stewart (Common name: Tarpakhi)

It is a common large herb which is considered as a good remedy for stomach pain. The fruits are also aromatic, carminative and diuretic.

Trachyspermum ammi (L.) Sprag. (Common name: Ajwain, thymol)

It is a cultivated herb. Its fruit is said to be used for stomach diseases. It is also used for kidney stones and for stomach insects.

Cassia fistula L. (Common name: Amaltas, Sundali)

It is a common cultivated ornamental tree. Latex of leaves and seed is used in skin diseases while fruit in rhematism and snakebite.

REFERENCES

BEG, A. R. AND SAMAD, K.A. 1974. Flora of Malakand Division.**1**(B) *Pak. J. For.,* **24**:230-286.

BEG, A.R AND KHAN, M.R. 1980. The present situation and the future of dry oak forest zone at Swat Valley, *Pakistan. Pak. J. For.,* **30**:109-122.

BEG, A.R AND KHAN, M.H. 1984. Some more plant communities and the future of dry oak forest zone in Swat Valley. *Pak.J.For.* **34**: 25 - 35.

CHAMPION, H.G., SETH, S. K. AND KHATTAK, G.M. 1965. *Forest types of Pakistan.* Pak. Forest Inst. Peshawar. 238 pp.

NASIR, E. AND ALI, S.I. (ed.) 1970 - 1993. *Flora of West Pakistan/Pakistan.* Nos. 1- 193, Islamabad / Karachi.

STEWART, R.R. 1967. Check list of the plant of Swat State North West Pakistan. *Pak. J. For.* **17**(4):457-528.

STEWART, R.R. 1972. *An Annotated Catalogue of the Vascular plants of Pakistan and Kashmir*, Karachi.

STEWART, R. R. 1982. History and Exploration of plants of Pakistan and adjoining area. (ed. E. Nasir and S.I. Ali) *Flora of Pakistan.* National Herbarium, Islamabad.

EFFECT OF ALTITUDE, ASPECT AND BIOTIC FACTORS ON THE PLANT DIVERSITY OF DABARGAI HILLS, SWAT, PAKISTAN

FARRUKH HUSSAIN, ABDUL KHALIQ AND IHSAN ILAHI

Department of Botany, University of Peshawar, Peshawar

Abstract: Studies on the vegetation of Dabargai Hills, District Swat indicated that south facing slopes had high air and soil temperature than the north-east facing slopes. However, with increasing altitude the differences in soil and air temperature reduced and at the top (3200 m) they were similar. Water holding capacity of the soil also declined with increasing elevation.

The plant species and their communities similarly differed on south- and north-east facing slopes. Hemicryptophytes, chamaephytes and phanerophytes dominated the top. At lower altitude, therophytes and nanophanerophytes were dominant. The leptophyllous and nanophyllous species increased with the rising altitude especially on south facing slopes. The herbage production also declined with the increasing elevation especially on the north-east facing slopes. The area needs revegetation and protection with the help of local participation.

INTRODUCTION

The vegetation of N.W.F.P., Pakistan, varies due to tremendous altitudinal and climatic changes. Deforestation, overgrazing, terrace cultivation and erosion have decreased the forest cover in the mountainous areas (Hussain and Ilahi, 1991). Chaghtai and Ghawas (1976) observed that the vegetation of Malakand Pass was dominated by shrubs and non palatable plants. Beg and Khan (1980, 1984) reported that deforestation has reduced the *Quercus baloot* forests to scrub in various parts of Swat. Hussain and Shah (1989), Shah *et al.* (1991) and Hussain *et al.* (1992) observed that deforestation and over grazing has changed the *Pinus roxburghii*, *Quercus incana* and *Pinus wallichiana* forests of Docut Hills, Swat into scrubs and grasslands. Hussain *et al.* (1995) reported that *Pinus roxburghii* occurred in isolated form on south facing slopes of Girbanr Hills, District Swat. They concluded that deforestation, overgrazing, uprooting of plants and terrace cultivation have accelerated soil erosion in the area.

Dabargai Hills are situated near Madyan District. The average altitude varies from 1320 m to 3277 m (Margater Sar). The area lies inbetween latitudes 35° 5' to 35° 15' N and 72° 32' to 72° 41' E longitude. Climatically, the lower parts fall into subtropical region while the upper parts lie within the dry temperate zone with mild summer and cold winters (Ahmad, 1951). The maximum rain is received during early spring and then in the winter along with snow. Summers are relatively dry owing to weak monsoon influence. Soils are shallow and eroded of loamy-sand or sandy-loam type. No quantitative data on the phytosociology and habitat of Dabargai hills is available. This study reports the plant diversity and productivity of herbaceous layer. This base line data might help range and wildlife managers in their future planning .

MATERIALS AND METHODS

The data on south and north-east facing slopes was collected at 1600 m (foot hills), 2000 m, 2400 m, 2800 m, and 3200 m (top hill). The density, frequency and circumference of the tree species at breast height (1.5 m) and herbage cover of herbaceous species was determined following Hussain (1989). There were 5, 10x2 m quadrats for trees; 10, 2x4 m and 15, 0.5x0.5 m quadrats for shrubs and herbs respectively. The data was converted to relative values (Hussain, 1989). The importance values for all tree, shrub and herb species was summed up to get total importance value for each of these groups.The communities were named after 3 leading species with the highest importance values. Life form and leaf size spectra were determined after Raunkiaer (1934). Fresh biomass of herbaceous plants was determined by clipping the above ground parts in 5, 0.5x0.5 m quadrats at each of the sites. Plants were separated into forbs and grass/grass-like species.

Duplicate soil samples, collected from each site upto a depth of 15 cm, were mixed together and air dried. Texture, pH, electrical conductivity, $CaCo_3$, organic matter, N, P , K ,total soluble salts and water holding capacity were determined following standard methods (Jackson, 1962; Hussain, 1989). Air and soil temperature was measured with air and soil thermometers by taking 5 observations.The study was conducted during November, 1994. Plants were identified in National Herbarium, National Agricultural Research Center, Islamabad.

RESULTS AND DISCUSSION

Habitat Features

Soils were either of loamy-sand or sandy-loam type (Table I). The % N was almost similar in various sites while $CaCo_3$ contents varied from 0.25 % to 2.56% at various sites and organic matter form 1.15% to 1.96% (Table II). The P contents varied from 1 to 8.4 ppm while K differed from 85 to 225 ppm. The pH, electrical conductivity and total soluble salts did not vary among the sites.The water holding capacity declined with the altitude (Table II).

Table I: Some physical features of soil and air from various sites and aspects of Dabargai Hills during November 1994

Height (m)	Texture		Water Holding Capacity		pH		Ec X10		Temperature C° Air		Temperature C° Soil	
	S	NE	S	NE	S	NE	S	NE	S	NE	S	NE
1600	LS	LS	22	18	7	7	0.02	0.01	24	22	16	14
2000	LS	SL	21	6	7	7	0.01	0.02	22	20	14	12
2400	SL	LS	11	22	7	7	0.02	0.01	16	14	9	8
2800	LS	SL	20	32	7	7	0.01	0.01	12	12	8	7
3200	LS	LS	20	13	7	6	0.01	0.02	7	7	4	4

LS=loamy sand, SL=Sandy loam; S & NE, respectively represent south and north east facing slopes.

Table II: Chemical features of soils from various sites of Dabargai Hills during November 1994.

Alt. (m)	CaCo3 %		Organic matter %		N %		P ppm		K ppm		TSS %	
	S	NE	S	NE	S	NE	S	NE	S	NE	S	NE
1600	0.3	0.3	1.4	1.4	0.1	0.1	4.4	3	85	225	0.006	0.003
2000	1.5	2.5	1.7	1.79	0.89	0.89	3.3	1	122	160	0.003	0.006
2400	1.5	1.3	1.4	1.1	0.07	0.06	0.2	2.3	132	197	0.006	0.003
2800	2.5	2.6	2	1.5	0.09	0.08	8.4	1	188	188	0.003	0.003
3200	1.3	2.5	1.4	1.8	0.07	0.08	2.6	1	179	160	0.003	0.003

Alt.= Altitude; TSS = Total soluble salts

The low nutrient status of various sites might be due to soil erosion. Similar findings regarding low soil fertility due to deforestation and erosion from various parts of Swat have been reported (Shah & Hussain, 1989; Hussain *et al.* 1992). The water holding capacity was relatively high in sites having high organic contents. The low water holding capacity at the top was due to erosion, low organic matter and compaction of the soil.

The air and soil temperature was high on the south facing slopes (Table I). A gradual decrease in the temperature with rising altitude was observed. The top' south- and north-east facing slopes had same air and soil temperatures. Shah *et al.*(1991) also reported that south facing slopes of Docut Hills were warmer than north facing slopes. Hussain and Tajul-Malook (1984) also observed that the north and south facing slopes of Karamar Hills differed in micro climate. The north-east and south facing slopes of Girbanr Hills exhibited differences in temperature and vegetation due to slope exposure (Hussain *et al.* 1995).

Phytosociology

The south facing slopes at 1600 m had *Plectranthus-Cyndon-Themeda* community while *Cynodon-Plectranthes-Chrysopogon* community was present on north-east facing slopes (Table III). There were 33 and 29 species on the south and north-east facing slopes, respectively. The similarity between them was 32.27%. *Abies pindrow, Cedrus deodara, Picea smithiana, Pinus wallichiana* and *Quercus dilatata* were present on the north-east facing slopes. While *Ailanthus altissima, Celtis caucasica, Diospyrus lotus, Ficus carica, Morus alba* and *Zizyphus sativa* were present on the south facing slopes. There were, respectively, 6 and 5 shrub species on the north-east and south facing slopes. The total importance values for tree and shrub species was greater on the south facing slopes (Table IV). Thero-nanophanerophytic species dominated south facing slopes while north-east slopes had thero-hemicryptophytic vegetation (Table V). Both the slopes were dominated by micro-nanophyllous species.

Plectranthes-Cynodon-Themeda community ascended to 2000 m on south facing slopes. It had 19 species (Table III), while north-east slopes had *Viburnum-Fragaria-Chrysopogon* community with 22 species. There was 4.18% similarity between them. *Cedrus deodara, Picea smithiana, Pinus wallichiana* and *Olea ferruginea* were present on the north-east slopes. Both slopes had 5 shrub species. The total importance values for trees, shrubs, and herbs on the north-east facing was respectively 60.4, 65.9 and 173.1 (Table IV). On the south facing slopes shrubs and herbs had a total importance value of 91.3 and 208.3, respectively. Thero-hemi-nanophanerophytic vegetation dominated south facing slopes while north-east facing slopes had hemicryptophytic species (Table V). Micro-nanophyllous species dominated both the slopes (Table V).

Pinus-Chyrysopogon-Fragaria and *Picea-Abies-Viburnum* communities were present on south and north-east facing slopes at 2400 m (Table IV). Both the communities had 41.54% similarity. *Picea smithiana* was confined to north-east facing slopes. *Plectranthus rugosus* and *Sarcococca salgina* occurred on the south facing slopes while *Skimmea* and *Viburnum* were confined to north-east facing slopes. Each community had 18 species. The total importance values for tree species was 90 and 87.4 on south and north-east facing slopes, respectively (Table IV). Each slope had 4 shrub species with a total importance values of 44 and 56.2 on south and north-east facing slopes, respectively (Table IV). The total importance values of herbaceous species was 165.7 and 156.3, respectively on south and north-east facing slopes.

The percentage of hemicryptophytes and nanophanerophytes was similar on both the slopes (Table V). The community was micro-nanophyllous on south and nano-microphyllous on north-east facing slopes (Table V).

Sibaldia-Thymus-Cedrus and Thymus-Sibaldia-Abies communities were respectively present on south and north-east facing slopes at 2800 m (Table IV). They shared 56.74% similarity. *Cedrus deodara* occurred on south facing slopes. The total importance values for tree species was respectively 72.6% and 81.4% on south and north-east facing slopes. The 2 shrub species on north-east slopes had a total importance value of 53.88 while 4 shrub species on south facing slopes had a total importance value of 24.88 (Table IV). Herbaceous species contributed total importance value of 202 on the south and 164.8 on the north-east facing slopes.

Both the slopes had hemi-megaphanerophylic communities (Table V). South facing slopes had lepto-nanophyllous community while north-east facing slopes had micro-leptophyllous community (Table V).

The south and north-east facing slopes at 3200 m had *Sibaldia-Thymus-Achillea* and *Sibaldia-Thymus-Juniper* communities (Table III). They had 79.51 % similarity. South facing slopes had no tree species. While north-east facing slopes had *Abies pindrow* and *Picea smithiana.* They had a total importance value of 46.8 (Table IV). *Juniperus communis* and *Salix* Sp. were the only shrubs on north-east facing slopes. They had a total importance value of 51.3. The only Juniper on south facing slope had an importance value of 27.1. There were 5 and 7 herbaceous species respectively on south and north-east facing slopes with total importance values of 272.4 and 201.6.

Hemicrypto-megaphanerophytic community dominated both the slopes (Table V). Lepto-microphyllous species were abundant on the south facing slopes and micro-nanophyllous species on the north-east slopes (Table V).

Pinus wallichiana, Abies pindrow and *Cedrus deodara* forests extend from 2400 to 2800 m on south facing slopes. However, on the north-east facing slopes they start from 1600 m and ascend to the sub-alpine regions. The association of *Olea*-oak and *Olea-Cedrus/Pinus wallichiana* indicates a dry oak zone. Beg and Khan (1980, 1984) reported that *Quercus baloot* associates with *Olea ferrugnea* or *Pinus roxburghii* in the sub-tropical zone. It may also extend to *Juniperus* and *Cedrus deodara* zone at higher altitude. Kaul and Sarin (1974) also described oak-blue pine forests in-between 1600 to 1900 m in Bhadarawah Hills, India. While the upper parts had *Abies pindrow* and *Cedrus dodara.* Coventry (1929) stated that *Pinus wallichiana* and *Quercus incana* make mixed forests in the temperate zone between 1600 to 2600 m in the Punjab Himalayas. The upper zone between 2600 to 3800 m had *Abies pindrow* and *Picea smithiana* . Champion *et al.* (1965) and Hussain and Ilahi (1991) stated that temperate forests in between 1600 to 1900 m consists of *Pinus wallichiana* and *Quercus incana* while *Cedrus deodara, Abies pindrow* and *Pinus wallichiana* occur from 2600 to 3800 m.

Life form and leaf size spectra are indicators of biotic interaction, climate and habitat deterioration. The vegetation of Dabargai Hills upto 2400 m is dominated by therophytes, hemicryptophytes and nanophanerophytes and thereafter,at higher elevation it shifts to chamaephytic and nanophanerophytic types. Degradation due to deforestation and over-grazing has increased therophytes. Furthermore, soil erosion has hampered the regeneration of plants in the area. The findings of Rajwar and Gupta (1985) support our findings in this respect. Hussain *et al.* (1995) observed a similar situation in the nearby Girbanr Hills. Therophytes and nanophanerophytes gradually decreased while hemicryptophytes and chamaephytes enhance with increasing altitude.. This agrees with Kaul and Sarin (1974) and Gupta and Kachroo (1983) who reported similar trend for the flora of Yasmarg Valley, Occupied Kashmir. Overgrazing and animal trampling increases therophytic species (Cain, 1950; Deschene, 1969). Same might be true in this case as grazing is common in this area. Cain (1950) stated that chamaephytes are important in high altitude and latitudes.

The communities in the lower elevation were mostly micro-nanophyllous while at higher altitude they were lepto-microphyllous. Plants suffer from aridity, strong winds, poor soil development and short growing season at high altitude. They therefore adapt to reduce their size, height, foliage and duration of growth.

The production of fresh herbage was more on the north-east facing slopes (Table VI). There were more forbs than grasses. Furthermore, grazing has decreased palatable plants. At higher altitude leptophylls, chamaephytes and hemicryptophytes dominate which might have also decreased the herbage productivity.

Deforestation and overgrazing have decreased the plant diversity of these fragile ecosystems. Many species of plants and animals have either become threatened or endangered. The extraction of fuel wood, timber wood, medicinal and other plant resources has dramatically changed the overall habitat conditions. There is dire need of restoration of these degraded habitats. For this purpose a participatory approach with the collaboration of the local communities might bring success to reforestation and protection of the area. Introduction of *Acer pentapomicum, Cedrus deodara, Fraxinus excelsior, Pinus wallichiana, Ulmus villosa, Celtis caucasica, Morus alba, Ailanthus, Buxus wallichiana* and *Parrotiopsis jacquemontiana* between 1300 to 2500 m might improve the area. Among the fruit trees *Malus pumila, Cydonia oblonga, Pyrus communis, P.lindleyi, P.pashia, P.bukharensis, Ficus carica* and Vitis vinifera could be tried. *Salix, Populus, Platanus, Juglans* and *Diospyrus* along irrigated places might perform well. Below 1300 m, *Acer, Robinia pseudo-acacia, Acacia modesta, Olea ferruginea, Zizyphus mauritiana, Pistacia integerrima, Dodonaea viscosa, Grewia optiva, Punica granatum* and *Citrus* could be planted. Likewise reseeding and proper management of grassland is required. The biodiversity must survive for the survival of mankind.

Table III. Importance value of Dominants, tree and shurb species (other than dominants) in various plant communities of Dabargai Hills during November (Note: The dominant trees and shrubs have been listed under the dominants)

Altitude	1600m		2000m		2400m		2800m		3200 m	
Aspect	S	NE	S	NE	S	NE	S	NE	S	NE
Communities	PCT	CPC	PCT	VFC	PCF	PAV	STC	TSA	STA	STJ
A. Dominant Species										
Abies pindrow	-	19.1	-	-	16.9	34.3	17.9	28.3	-	23.5
Achillea mellifolia	-	-	-	-	-	-	18.6	8.5	32	20.9
Cedrus deodara		-	4.77	-	10.1	-	-	20.1	-	-
Chrysopogon montanus-	20.2	-	25.9	38.2	22.4	-	14.9	-	-	
Cynodon dactylon	25.1	49.6	39.9	-	-	-	-	-	-	-
Fragaria nubicola	-	19.6	-	31.5	34.9	-	3.11	20.7	-	-
Juniperus communis	-	-	-	-	-	-	8.69	-	27.1	37.7
Picea smithiana	-	9.34	-	22	-	45.4	15.9	36.6	-	23.3
Pinus allichiana	-	5.46	-	24.3	73.1	7.76	18.7	6.52	-	-
Plectranthus rugosus	36.1	23.3	59.8	-	15.7	-	-	-	-	-
Sibaldia cuneata	-	-	-	-	10.4	23.2	86.7	46.3	125	88.1

Themeda anathera	21.9	-	23.3	-	-	-	-	-	-	-
Thymus serpyllum	-	-	-	-	2.03	18.7	76.5	50.8	-	84.6
T. Vulgare	-	-	-	6.25	10.6	13.7	3.11	5.68	106	-
Viburnum nervosum	-	-	-	37.1	-	23.2	-	20.1	-	-
B. Tree species (Other than dominants)										
Ailanthus altissima	3.47	-	-	-	-	-	-	-	-	-
Celtis caucasica	4.63	-	-	-	-	-	-	-	-	-
Diospyros lotus	15.7	-	-	-	-	-	-	-	-	-
Elaeagnus angustifolia	8.8	-	-	-	-	-	-	-	-	-
Ficus carica	16.2	-	-	-	-	-	-	-	-	-
Morus alba	18.2	-	-	-	-	-	-	-	-	-
Olea ferruginea	-	-	-	4.03	-	-	-	-	-	-
Quercus dilatata	-	11	-	-	-	-	-	-	-	-
Salix Sp.	-	-	-	-	-	-	16.2	29.4	-	13.6
Zizyphus sativa	13.2	-	-	-	-	-	-	-	-	-
Z. oxyphylla	6.81	-	-	-	-	-	-	-	-	-
C. Shrub species (Other than dominants)										
Berberis lycium	12.2	1.97	6.19	-	-	-	-	-	-	-
Buxus sempervirense	-	4.37	9.28	-	-	-	-	-	-	-
Cotoneaster microphylla	-	-	9.8	-	-	-	-	-	-	-
Indigafera herterantha var. *herterantha*	-	5.93	6.19	2.22	20.1	20.2	-	-	-	-
Jasminum humile.	-	-	-	5.92	-	-	-	-	-	-
Myrsine africana	1.89	3.95	-	-	-	-	-	-	-	-
Parrotiopsis jacquemontiana	--	3.95	-	15.5	3.19	7.89	-	-	-	-
Rosa Sp.	14.2	-	-	-	-	-	-	-	-	-
Sarcococca saligna	-	-	-	-	4.99	-	-	-	-	-
Skimmea laureola	-	-	-	-	-	4.89	-	1.98	-	-
Spiraea lindleyana	-	-	-	5.24	-	-	-	1.98	-	-

KEY:PCT=Plectranthus-Cynodon-Themeda community; CPC= Cynodon - Plectranthus-Chrysopagon community; VFC = Viburnum-Fragaria-Chrysopogon community; PCF = Pinus-Chrysopogon-Fragaria community; PAV = Picea-Abies-Viburnum community; STC = Sibaldia-Thymus- Cedrus community; TSA = Thymus-Sibaldia-Abies community; STA = Sibaldia-Thymus-Achillea community; STJ = Sibaldia-Thymus-Juniperus community

Table IV: Number and the total importance values (TIV) of of tree, shrub and herb species on different sites of Dabargai Hills

Altitude	1600 m		2000 m		2400 m		2800 m		3200 m	
Aspect	S	NE	S	NE	S	NE	S	NE	S	NE
Community	PCT	CPC	PCT	VFC	PCF	PAV	STC	TSA	STA	STJ
Trees No	6	5	-	4	2	3	4	3	-	2
TIV	71.4	49.9	-	60.4	90	87.4	72.6	81.4	-	46.8
Shrubs, No	7	6	5	5	4	4	2	4	1	2
TIV	83.7	53.3	91.3	65.9	44	56.2	24.88	53.3	27.1	51.3
Herbs, No	20	18	14	13	12	11	8	9	5	7
TIV	144.9	195.8	208.3	173.1	165.7	156.3	202	164.8	272.4	201.6
Total	33	29	19	22	18	18	14	16	6	1

Key for communities is given in Table III; S = South; NE=north-east

Table V: Fresh herbaceous biomass of plants at various sites and slopes of Dabargai Hills during November.

Altitude (m)	Grasses/grass-like plants, g/m2		Forbs, gm/m2		Total, g/m2	
Aspect	South	north-east	South	north-east	South	north-east
1600 m	180.8	227	128	160	308	387
2000 m	118	249	105	220	223	469
2400 m	103	224	200	222	327	446
2800 m	98	180	201	227	278	407
3200 m	2	3	234	235	236	238

Table VI :Life Form and Leaf Size spectra of plant communities at different sites of Dabargai Hills during November

Altitude	1600, m		2000, m		2400, m		2800, m		3200, m	
Aspect	S	NE	S	NE	S	NE	S	NE	S	NE
Community	PCT	CPC	PCT	VFC	PCF	PAV	STC	TSA	STA	STJ
a. Life Form Spectrum										
Thero	30.3	34.5	31.6	22.7	11	5.6	7.2	-	-	9.1
Hemi	18.2	24	31.6	36.4	50	55.6	35.7	50	66.7	45.5
Geo	-	-	5.3	-	5.6	-	14.3	6.3	16.7	91
Cham	12.2	3.4	5.3	-	-	-	-	-	-	-
Nano	21.2	20.7	26	22.7	22	22	14.3	25	16.7	18.2
Micro	15	3.4	-	4.5	-	-	-	-	-	-
Mega	-	10.4	-	9.1	5.6	11	21	12.4	-	18.1
Life form	TN	TH	THN	HT	HN	NH	NG	HN	HG	H
b. Leaf size Spectrum										
Le	12	6.89	5.3	18.2	22.2	22.2	28.6	25	50	18
Na	27.3	37.9	36.8	22.7	22.2	33.3	28.6	18.8	16.6	27.3
Mii	51.5	48.8	42	31.8	44.5	33.2	21	25	16.6	9.1
Me	9.1	6.9	10.5	18	11	11	21	25	16.6	9.1
Mg	-	3.4	5.3	4.5	-	-	-	-	-	-
Pa	-	-	-	4.5	-	-	-	-	-	-

a. Life Form: Thero= Therophyte; Hemi= Hemicryptophyte; Geo= Geophyte; Chame= Chamaephyte; Nano=Nanophanerophyte; Micro=Microphanerophye; Meso= Meso-phanerophyte; Mega=Megphanerophyte; TN= Thero-nanophanero phytic ; TH= Thero-hemicryptophytic; THN= Thero-hemi-nanaphanerophytic; HT= Hemi-therophytic; HN= Hemi-nanophanerophytic; HG= Hemi-geophytic. b.Leaf Size spectra; Le=Leptophyll; Na=Nanophyll; Mi=Microphyll; Me=Mesophyll; Mg=Megaphyll; P=Parasite

REFERENCES

AHMAD, K.S., 1951. Climatic Regions of Pakistan. *Pak. Geog. Rev.*, **6**: 1-35.

BEG, A.R. AND KHAN, M.H., 1984. Some more plant communities and the future of dry oak forest zone in Swat valley. *Pak. J. Forest.*, **34**: 25-35.

CAIN., S.A. 1950. Life form and phytoclimates. *Bot. Revg.*, **16**: 1-32.

CHAGHTAI, S.M. AND GHAWAS, I.H., 1976. The study of the effect of exposure on community set up in Malakand Pass, N.W.F.P., *Pakistan. Sultania*, **2**: 1-8.

CHAMPION, H.G., SETH, S.K. AND KHATTAK, G.M., 1965. *Forest Types of Pakistan.* Pakistan Forest Inst. Peshawar.

COVENTRY, B.O. 1929. The forest vegetation and evidence of denudation. *Ind. For. Rec.*, **14**: 49-60.

DESCHENES, J.M. 1969. Vegetation differences of north and south parts of grazed pastures of Sussex Country, New Jersey. *Bull. N.J. Acad. Sci.*, **11**: 22-29.

GUPTA, V.C. AND KACHROO, P., 1983. Life form classification and biological spectrum of the flora of Yasmarg, Kashmir. *Tropical Ecol.*, **24**: 22-27.

HUSSAIN, F. 1989. Field and Laboratory Manual of Plant Ecology. University Grants Commission, Islamabad.

HUSSAIN, F. AND SHAH, A., 1989. Phytosociology of vanishing subtropical vegetation of Swat with special reference to Docut Hills. Winter Aspect. *Scientific Khyber.*, **2**: 27-36.

HUSSAIN, F. AND TAJUL-MALOOK, S., 1984. Biological spectrum and comparison of coefficient of communities between the plant communities harboring Karamar Hills, District Mardan, *Pakistan. Jour. Sci. and Technol.*, **8**: 53-60.

HUSSAIN, F. AND ILAHI, I., 1991. *Ecology and Vegetation of Lesser Himalayas.* Jadoon Printing Press, Peshawar.

HUSSAIN, F., SALJOQI, A.R. SHAH, A. AND ILAHI, I., 1992. Phytosociology of the vanishing sub-tropical vegetation of Swat with special reference to Docut Hills. II. Spring aspect. *Sarhad J. Agric.*, **8**: 185-191.

HUSSAIN, F., ILYAS, M. AND KIL, B., 1995. Vegetation studies of Girbanr Hills, District Swat, Pakistan. *Korean J. Ecol.*, **18:** 207-218.

JACKSON, M.A. 1962. *Soil Chemical Analysis.* Constable and Co. London. P. 497

KAUL, V. AND SARIN, Y.K., 1974. Studies on the vegetation of the Bhadrawah Hills. I. Altitudinal zonation. *Bot. Notiser.*, **127**: 500-507.

RAJWAR, G.S. AND GUPTA, K.S., 1984. Biological spectrum of Garhwal Swalik. *Ind. Forester*, **110**: 1171-1176.

RAUNKIAER, C. 1934. *Life form of Plants and Statistical Plant Geography.* Clarendon Press, Oxford.

SHAH, A., AYAZ, S. AND HUSSAIN, F., 1991. Similarity indices, biological spectrum and phenology of plant communities of Docut Hills, District Swat during winter. *Jour. Sci. Technol.*, **15**: 15-21.

SHAH, A AND HUSSAIN, F., 1989. Phytosociology of vanishing subtropical vegetation with special reference to Docut Hills. 1. Winter aspect. *Sci. Khyber.*, **2**: 27-36.

SECTION III

BIODIVERSITY OF INVERTEBRATES

BIOGEOGRAPHY & DIVERSITY OF BUTTERFLIES OF NORTHWEST HIMALAYA

SYED AZHAR HASAN

Pakistan Museum of Natural History, Shakarparian, Islamabad

Abstract: Present study is a part of the plan to investigate butterfly diversity of Pakistan so that a conservation strategy could be evolved for them. Eighty species belonging to the families Nymphalidae, Pieridae, Danaidae, Papilionidae, Hesperidae, Satyridae, Lycaenidae, Erycinidae and Libytheidae have been recorded and described from Galiat and Azad Kashmir of which 60 are common and distinctive while 20 are less common.

INTRODUCTION

Butterflies have attracted attention of the environmentalists the world over by having a special function as " indicator species" of the ecosystem health. Several of these splendid insects are exclusively forest and garden dwellers and, being conspicuous, their presence or absence serves to monitor ecological changes in habitat, thus warning us about the deteriorating environment. Butterflies having exact environmental and feeding requirements to survive and reproduce, are the worst sufferers by rapid environmental degradation and the effects of insecticides.

Pakistan, being at the junction of Palaearctic, Oriental and Ethiopian zoogeographical regions represents interesting butterfly diversity which owes much to varied origin as well as to the routes that different species were forced to undertake to overcome geographical barriers especially Himalaya to the north of it. The Himalaya stretching from Burma in the east terminates in the west at the bend of the River Indus just around Mount Nangaparbat. Its width varies from 80 Km. to over 300 Km and its mean elevation is higher than the highest mountains of Europe and comparable only to the South American Andes. Though the Himalaya is usually divided into Eastern, Central, Western and Northwestern regions but these divisions are rather arbitrary. The more natural division of the Himalaya based on sound geological, geographical, ecological and biogeographical grounds appears at the defile of the River Sutlej from the Himalaya to the east (Mani, 1986).
Altitudinal and climatic differences between these two Himalayan regions not only affect the type of vegetation but also its associated

butterfly fauna. The prominent habitats in Northwest Himalaya includes Sub-tropical Pine Forest, Himalayan Moist Temperate Forest, Dry Temperate Coniferous Forest and Tropical Deciduous Forest.

Despite a large volume of literature dealing with the butterfly fauna of Oriental region (Elwes, 1882; Doherty, 1886; Watkins, 1927; Puri, 1931; Evans, 1923, 1932; Gough, 1935; Peile, 1937; Talbot, 1939, 1947; Malik, 1970 & 1973; Ahsan and Iqbal, 1975; Iqbal, 1978; Lewis, 1987; Mani, 1986; Antram, 1986; Mohyuddin, 1987; Gay et. al., 1992; Hasan, 1994) only occasional description of butterflies from the Northwest Himalaya included now in Pakistan have been made, consequently our knowledge about this important biological resource is inadequate and fragmentary. Many species known to have existed within historic times have become extinct or are endangered.

Present study is a part of the plan to investigate butterfly diversity of Pakistan so that a conservation strategy could be evolved for them. Eighty species belonging to the families Nymphalidae, Pieridae, Danaidae, Papilionidae, Hesperidae, Satyridae, Lycaenidae, Erycinidae and Libytheidae have been recorded and described from Galiat and Azad Kashmir of which 60 are common and distinctive while 20 are less common.

NYMPHALIDAE

This is by far the largest family of butterflies represented the world over. These are robust bodied and of beautiful colours. Forelegs in nymphalids, unlike other butterflies, are reduced, bearing no claws for perching or walking, but instead are provided with brush-like tufts. The adults are mostly eager flower visitors, but many also feed at rotten fruits.

Aglais cashmirensis (Kollar)

Description: Also known as *Vanessa cashmirensis*; upperside orange-red with black markings, basal half of the hindwing pale with long dark hairs; underside mostly brown; has two broods in a year; males territorial, spent much of their time in chasing away other butterflies.

Wingspan: 55-65 mm.

Range: Murree Hills, Azad Kashmir

Argyreus hyperbius (Johanns.)

Description: The nymphalid mimics *Danaus chrysippus*, commonly found from April to October; upperside orange-yellow mottled with black spots and markings; Underside of the hindwings is greenish- brown variegated with ochraceous, brown and white markings; female differs from the male by having apical half of the forewing black with four white spots and a white band.

Wingspan: 65-82 mm.

Range: From Murree foothills upto 7000 ft. extending eastward of the Himalayas to Burma, China, Japan, Malaysia.

Argynnis kamala (Moore)

Description: Upperside dark brown with jet black spots. Hind- wing dentate. Underside of forewing red, apex pale yellow and green with a silver loop and black spots. Hindwing metallic green with white bands.

Wingspan: 50-55 mm.

Range: Shogran, Neelum Valley

Argynnis lathonia (Linn.)

Description: Fairly common on open hillsides at 7000 ft., especially in May and June. Upperside is orange-yellow, speckled with black markings. Underside orange-yellow but with four silver spots on the apex of forewing and several large silver spots on the hindwing.

Wingspan: 46-50 mm.

Range: Besides Kaghan Valley in N.W. Himalayas also occurs in Europe.

Argynnis aglaia (Linn.)

Description: Upperside tawny yellow with black markings; underside of the forewing pale yellow with black markings as on upperside; basal and sub-terminal areas of the hindwing suffused with green.

Wingspan: 72-80 mm.

Range: Azad Kashmir

Ariadne merione (Cram.)

Description: Upperside reddish-brown, hindwing with black markings and stripes; underside with dark brown bands.

Wingspan:45-60 mm.

Range: Besides Kaghan valley in N.W Himalaya also found in India, Malaysia and Sri Lanka.

Nymphalis xanthomelus (Esper)

Description: Also· known as *Vanessa xanthomelus* and commonly found in April and May; upperside yellowish-orange with few black spots on the forewing and a narrow irregular blue band on the hindwing.

Wingspan: 55-60 mm.

Range: Besides N.W. Himalaya widely distributed in Central and Southern Europe through Asia to China.

Phalanta phalantha (Drury)

Description: More common in the foothills of Murree and remains active from July to October; upperside bright ochraceous with black spots and markings; underside pale-ochraceous with faint markings.

Wingspan: 55-65 mm.

Range: Besides Murree foothills also occurs in the whole Indo-Malayan & Indo- Australian region.

Junonia almana (Linn.)

Description: A common nymphalid of the plains, occurs from April to October, also known as *Precis almana*; upperside orange-yellow with prominent eye-spots; underside yellowish-brown with faint markings.

Wingspan: 60-65 mm.

Range: Widely distributed in the plains and Murree foothills. Also occurs in India, Malaysia, Japan, Philippines, Hong Kong, Andaman & Nicobar Islands.

Junonia hierta (Fabr.)

Description: Also known as *Precis hierta*; upperside bright yellow with a brilliant blue oval spot on the hindwing; found all the year round except in extreme winter.

Wing span: 45-55 mm.

Range: Common in the plains and Murree foothills. Also occurs in India, Sri Lanka, Burma, through Cambodia to W. & S. China.

Junonia orithya (Linn.)

Description: Common nymphalid occurring almost throughout the year but more abundant in July and August, neither rises high nor covers long distances and usually found resting on the ground. Basal two thirds of the forewing velvety black and the apex pale-brown with white band and rings; hindwing shining blue with a basal black patch and orange rings.

Wingspan: 40-50 mm.

Range: Widely distributed in the plains and foothill of Pakistan, also ranges from Africa to Asia and Australia.

Vanessa cardui (Linn.)

Description: Found almost throughout the year from sea level up to 15000 ft. in the Himalayas; upperside is pinkish-red with golden brown base, apical half and termen of the forewing dusky black having white spots, hindwing mottled with olive brown and ochraceous; underside paler than the upperside and the hind wings sandy brown with white markings.

Wingspan: 54-62 mm.

Range: Widely distributed in Pakistan including Murree and Azad Kashmir. Also occurs throughout the world except S. America, Arctic and Antarctic regions.

Vanessa indica (Herbst.)

Description: Common from March to November, upperside dark brown with white spots, apex of the forewing bears a red bend and a large black spot, termen of the hindwing red with black spots.

Wingspan: 52-60 mm.

Range: Ranges from Galiat and Azad Kashmir to India, China, Burma and Thailand.

Vanessa canace (Johanns.)

Description: Common at high altitudes in May and June; upperside deep indigo blue with a light blue band on both the wings and a series of black dots on the hindwing; underside brownish-black with a white spot on the hindwing; outer margin of the forewing distinctly concave.

Wingspan: 62-68 mm.

Range: Found in the N.W. Himalayas including Chitral and Murree Hills.

Neptis hylas (Linn.)

Description: Frequently encountered in the foothills of Islamabad and Murree Hills from mid April to mid October; upperside black speckled with white bands and spots, forewing with an elongate triangular white spot; underside golden ochraceous and the white markings on it defined in black.

Wingspan: 50-60 mm.

Range: Besides Murree and Azad Kashmir, widely distributed throughout the Indo- Australian region.

Neptis mahendra (Moore)

Description: Fairly common in May and June, resembles *Neptis hylas* in colour but differs from it by having an elongated discal streak well separated from a conical blunt spot on the forewing and postdiscal series of spots not margined by a black line.

Wingspan: 56- 60 mm.

Range: Ranges from the Murree Hills to Kashmir and India.

Limenitis trivena (Moore)

Description: Fairly common in June; upperside dull black with broad white band on both wings, a narrow transverse band in the cell and two white spots at the apex of the forewing; margins of both wings distinctly dentate.

Wingspan: 60-67 mm.

Range: Found in N.W. Himalayas including Murree Hills and Azad Kashmir.

Euthalia garuda (Moore)

Description: Frequents mango and chestnut trees from April to October; upperside olivaceous brown, forewing with five discal and two preapical white spots; underside is ochraceous.

Wingspan: 60-67 mm.

Range: Besides the Murree Hills also occurs in India, Burma and Sri Lanka.

Euthalia patala (Koll.)

Description: Upperside pale olive green with an oblique whitish discal band on the forewing and a large quadrate white spot on the hindwing; underside pale.

Wingspan: 80-100 mm.

Range: Occurs from Murree Hills to Nepal

Cyrestis thyodamas (Bsd.)

Description: Male on the upperside usually white, female often pale yellow; forewing with a series of dusky rings with yellow tinge, hindwings ochraceous with black and brown markings; underside similar to the upperside.

Wingspan: 60-70 mm.

Range: Azad Kashmir

Neptis yerburyi (Btlr.)

Description: Commonly found from April to October; upperside of forewing with an elongated and acutely pointed triangular spot; underside pale, triangular spot much closer to the apex of the discoidal streak.

Wingspan: 55-65 mm.

Range: Azad Kashmir, Murree, Nathia Gali

Neptis zaida (Dbdy.)

Description: Upperside of the forewing brownish-black with pale ochraceous markings, discal stripe broad, elongate and acute, not notched subapically; hindwing with sub-basal and postdiscal ochraceous and narrow bands.

Wingspan: 65-70 mm.

Range: Murree and Ghora Gali.

Polygonia egea (Cram.)

Description: Upperside tawny with black spots; underside of hindwing with a silver comma-shaped mark.

Wingspan: 50-65 mm.

Range: Nathia Gali, Tret

Sephisa dichroa (Koll.)

Description: Less common, found high up on Oak trees from May to September; upperside black with four large yellow markings, hindwing with conspicuous black veins; underside of the forewing with less conspicuous markings.

Wingspan: 62-74 mm.

Range: Murree, Chirtal

PIERIDAE

These butterflies constitute a large family with members almost everywhere, even in the extreme arctic. As their name implies, they usually have a preponderance of white or yellow pigment. These pigments are derived from the common excretory waste i.e., uric acid. Like the 'Swallowtails' but unlike most other butterflies, the adult of both sexes have fully developed and functional forelegs. The pierids due to their remarkable habit of migration have attracted the attention of naturalists the world over.

Anaphaeis aurota (Fabr.)

Description: Usually occurs throughout the year, migrate in large swarms; upperside white with black streaks on the apex and at the end of the cell of the forewing; dimorphic with dry and wet season forms. In the dry season form black markings narrower than the wet season form.

Wingspan: 40-55 mm.

Range: Widely distributed in the plains, Murree foothills. Also occurs in Asia and Africa, particularly well adapted to semi- desert regions.

Aporia leucodice (Evers.)

Description: Frequents flowers of horsechestnut in May and June; upperside white having black veins; underside of the forewing white with apex and costal edge pale-yellow; hindwing pale yellow with precostal area chrome-yellow.

Wingspan: 50-70 mm.

Range: Islamabad and Murree foothils

Catopsilia pomona (Fabr.)

Description: A powerful flier and a well known migrant, colour varies from pale yellow to white; in males base of the wings sulphur-yellow, silvery red ringed spots present at the end of the cell.

Wingspan: 55-65 mm.

Range: Islamabad, Murree Hills. Also found in Indo-Malayan and Indo-Australian regions.

Catopsilia pyranthe (Linn.)

Description: Common from July to September; upperside white tinted with green; underside darker in both sexes; female differed from male by having a postdiscal black band on the forewing and black edging along termen on the hindwing.

Wingspan: 50-60 mm.

Range: Besides Islamabad and Murree Hills, widely distributed throughout Indo-Malayan,and Indo- Australian regions and Africa.

Catopsilia crocale (Cram.)

Description: Common from July to September, fast flier, rises high in the air and can cover long distances; male greenish white with yellow base and black terminal angle, female yellow with termen and tornus of hindwing black. Underside of both sexes is dark.

Wingspan: 55-65 mm.

Range: Islamabad and Murree Hills.

Pieris brassicae (Linn.)

Description: A serious pest of crucifers, found upto 2000 m.; upperside creamy white with two black spots, apex of forewing also black; underside of the hindwing dirty yellow dusted with black spots; females are larger than males and have more prominent black spots on the upperside of the forewing.

Wingspan: 57-66 mm.

Range: Widely distributed in Pakistan including Galiat area and Azad Kashmir. Also occurs in Asia, Northern Africa and Europe.

Pieris canidia (Sparr.)

Description: Found almost throughout the year except in extreme winter, prefers damp spots and the edges of terraced fields higher up; upperside of forewing white with black, dentate border and two black spots; hindwing white and has marginal black veins and spots; underside without marginal black borders and spots.

Wingspan: 50-55 mm.

Range: Besides Chitral, N.W. Frontiers and Murree Hills, occurs throughout Indo-Malayan and Indo-Chinese regions.

Pieris napi (Linn.)

Description: Upperside white, veins, apex and terminal border black and a black spot in the outer half of interspace. Hindwing with a black subcostal spot; underside of forewing apically black, hindwing tinged with yellow.

Wingspan: 45-50 mm.

Range: Occurs in Murree and Kaghan Valley in N.W. Himalayas and also found in Sikhim and Bhutan

Eurema hecabe (Linn.)

Description: One of the most common pierids having low fluttering flight, usually seen flying over grass patches; upperside dark yellow terminally; underside of both wings with reddish-brown markings.

Wingspan: 25-30 mm.

Range: Occurs in the plains and Murree foothills.

Pontia daplidice (Linn.)

Description: Upperside white with black spots on the apex and in the middle of the forewing; underside mostly green with large white patches; has two generations in a year, one in March- April and the other in July-August.

Wingspan: 40-50 mm.

Range: Widely distributd in Pakistan including Islamabad and Murree Hills.

Gonopteryx rhamni (Linn.)

Description: Remarkably different from other pierids in the shape of the wings and even saffron yellow colour of the male and pale green colour of the female; hibernates over the winter months and survives longer than most other butterflies.

Wingspan: 60-70 mm.

Range: Islamabad, Murree Hills, Neelum Valley. Also widespread in most of Asia, North Africa and Western Europe.

Gonopteryx aspasia (Men.)

Description: Common from March to June at grassy slopes. In male forewing yellow and hindwing cream-white on the upperside and in female both pair of wings cream-white; discal spots small, orange and dark ringed; costa of forewing concave medially.

Wingspan: 50-55 mm.

Range: Murree and Azad Kashmir

Colias erate (Esper.)

Description: Upperside lemon yellow, costa of the forewing, base of both wings and a posterior half of the hindwing dusted with black, a prominent black spot on the forewing and an orange yellow discocellular spot on the hindwing; occurs in dry and wet season forms.

Wingspan: 45 - 55 mm.

Range: From 5000 to 9000 ft. in Galiat area, ranges into E. Europe and across temperate Asia to Korea.

Colias crocea (Fourcroy)

Description: A powerful flier, has several broods a year; upperside orange-yellow, forewing marginally black with a black spot, hindwing with an orange spot.

Wingspan: 45-58 mm.

Range: Besides Islamabad, Murree Hills and Azad Khamir also occurs in most of Asia, North Africa, Southern Europe.

Polyommatus stoliczkana (Moore)

Description: Also known as *Colias stoliczkana*; upperside orange-yellow with black margins; underside more or less dusted with green, a prominent red spot on the discal area of the hindwing

Wingspan: 38-42 mm.

Range: Chitral, Murree Hills, Azad Kashmir in N.W. Himalaya to West Himalaya

DANAIDAE

This family is confined almost entirely to the tropical regions. A peculiar characteristic of the danaids is the toxic substance with which caterpillars, chrysalis as well as adults, are endowed. These substances are derived from the caterpillar foodplants of the milkweed family. The danaids are very strong butterflies and have the ability to survive in adverse conditions. They fly with a rather slow, almost flapping motion and favour the sunny edges of the forest or roadways. They are normally found in the plains.

Danaus chrysippus (Linn.)

Description: One of the commonest butterflies found in abundance from July to September; upperside tawny, apex of forewing black with white spots, hindwing of the male has three and the female has two black spots.

Wingspan: 70-80 mm.

Range: Occurs in the plains of Punjab upto 3000 m., also ranges to India, China, Japan, Malaysia, N. Australia, N. Africa.

Danaus limniace (Cram.)

Description: Also known as *Tirumala limniace*, occurs more frequently in sheltered cool forests or plantations from March to September; upperside black with translucent bluish- white spots and bars; part of the forewing and the whole hindwing olive brown underside.

Wingspan: 90 - 100 mm.

Range: Common in Islamabad and Murree Hills, also ranges to India, Burma, Bangladesh, Sri Lanka and Hong Kong.

Danaus genutia (Cram.)

Description: Prefers edges of the scrub forest; upperside tawny with broad black veins, speckled with white spots and bars; hindwings in dry season forms paler than the wet season forms.

Wingspan: 80-90 mm.

Range: Occurs upto 2,700 m. in Galiat area and Azad Kashmir, also found in India, Malaysia, West and Central China, and Australia.

PAPILIONIDAE

This family comprises the most beautiful and largest butterflies which flutter on the flowers in gardens and in the forest. The common name 'Swallowtails' has been given to the group as most of them possess tailed-hindwings. Bright colours predominate, with patterns of yellow, white, orange, red and iridescent blue and green. The papilionids are represented everywhere in the world except in the extreme north and south, and in desert areas.

Papilio polyctor (Boisd.)

Description: Commonly found from mid April to mid September; upperside dull black having a golden green band on the forewing and a bright blue band on the hindwing; underside chocolate- brown, hindwing produced in the form of a long and spatulate tail.

Wingspan: 100-115 mm.

Range: From Murree Hills, Azad Kashmir to Afghanistan, N. India and West China.

Papilio demoleus (Linn.)

Description: A fast flying butterfly, commonly seen in gardens where there are lemon and orange trees; upperside black, having yellow specks and spots which turn to deep orange as the butterfly ages, hindwing with a brick-red spot on its margin.

Wingspan: 80-100 mm.

Range: Widely distributed in Pakistan including Islamabad, Galiat. Also occurs in Iran, India, China, Malaysia and N. Australia.

Papilio polytes (Linn.)

Description: Usually found in gardens and open woods upto 6000 ft., upperside black having a series of yellowish-white spots on the margin of the forewing and a white band on the hindwing; males more active than females.

Wingspan: 90-100 mm.

Range: Found in Murree Hills, Kaghan Valley and Shogran. Also occurs in the whole Indo-Malayan and Indo-Chinese regions.

Papilio philoxenus (Gray)

Description: Common at Murree Hills in April to August; upperside black and red, forewing black above with paler streaks between the veins, hindwing broadly scalloped with two white spots and crimson crescent patterns; underside with black spots.

Wingspan: 95-110 mm.

Range: Ranges throughout N.W. Himalaya including Azad Kashmir.

Papilio machaon Linn.

Description: A Palaearctic species, of which subspecies asiatica occurs at higher elevation from May to October; upperside yellow with black veins and markings, a broad black band from apex of forewing to tornus of hindwing; tail of hindwing black.

Wingspan: 75-90 mm.

Range: Kaghan Valley, Neelum Valley and Shogran.

Graphium cloanthus (West.)

Description: Frequent the tops of hills; forewing on the upperside transparent green with black margin along costa and termen; hindwing with tail.

Wingspan: 70-90 mm.

Range: Azad Kashmir.

HESPERIDAE

Hesperids are considered as the most primitive butterflies. They occur in all geographical regions of the world. They resemble moths more than butterflies. The swelling on their antennae is characteristic of nocturnal moths but their habits are essentially diurnal. Generally these are considered a connecting link between moths and butterflies. The majority are not brightly coloured. Usually they are dull brown or orange-brown.

Gomalia albofasciata (Moore)

Description: Also known as *Alenia elma*; upperside blackish-brown, forewing with a white narrow streak on the apex and a black streak in discal area, hindwing white marginally; underside pale with white bands.

Wingspan: 18-20 mm.

Range: Widely distributed in Pakistan, India and Sri Lanka.

Parnara guttata (Bremer & Grey)

Description: Commonly found from July to September; upperside dark brown with 5-6 white spots on the forewing and 4-5 white spots on the hindwing, margins of both wings white; underside similar to the upperside.

Wingspan: 35-38 mm.

Range: Murree Hills, Kashmir, India, Sri Lanka and Burma

Gegenes nostrodamus (Fabr.)

Description: Upperside brackish-brown and underside ashy gray; male without any spots, female with white apical spots and a row of pale yellow spots between the veins.

Wingspan: 30-34 mm.

Range: Besides Pakistan widely occurs in Asia Minor and S. Europe.

Badamia exclamationis (Fabr.)

Description: A strong flier and has a rapid, bounding flight, commonly occurs in jungle; upperside dark brown, forewing long and narrow and has three white spots in the middle of the cell; underside resembles the upperside.

Wingspan: 50-55 mm.

Range: Widely distributed in Pakistan, India, Sri Lanka, China and Australia

Aeromachus stigmata (Moore)

Description: Upperside black; underside of hindwing with a row of dark spots.

Wingspan: 25-30 mm.

Range: Murree, Nathia Gali

SATYRIDAE

This family occurs in all geographical regions of the world. These are small to medium-sized butterflies often associated with grassy or heathland areas. They are usually brown, but sometimes gray. Most species have eye-spot markings. The satyrids frequent shady undergrowth and have slow, jerky flight close to the ground.

Satyrus parisatis (Koll.)

Description: Common from April to September; upperside black above with black notches on its inner side continued along the veins to the margins, marginal dentitions black; underside with large spots.

Wingspan: 50-56 mm.

Range: N.W. Himalayas including Murree Hills and Kashmir.

Lethe rohria (Fabr.)

Description: 'The Common Tree Brown' is fairly common in open country as well in jungle from April to October. It frequents animal dung, overripe fruits and sugary tree sap. Upperside is brown with large black spots on the hindwing. Underside is pale, shaded with dark brown.

Wingspan: 58-70 mm.

Range: Distributed from Murree Hills to Kashmir and further eastwards.

Lethe confuse (Auriv.)

Description: A common satyrid; upperside brown with white bands across forewing and two apical white spots, hindwing uniform; underside brown with a discal white band on the forewing and a series of yellow ringed ocelli on the hindwing.

Wingspan: 50-55 mm.

Range: Whole of Himalayas, China, Tibet, Malaya, Thailand, Burma

Aulocera swaha (Koll.)

Description: Frequents open, grassy hillsides from August to October; upperside black with creamy bands on both wings; underside of the forewing tinged pale-yellow and the base of the hindwing bronze-green.

Wingspan: 70-75 mm.

Range: N.W. Himalayas including Chitral and the Murree Hills.

Aulocera padma (Koll.)

Description: Fairly common from May to October, occasionally settles on the ground or bushes; upperside of the forewing black with a broad white discal band, hindwing dark with a series of postdiscal black markings; underside silky brown.

Wingspan: 72-78 mm.

Range: N.W. Himalaya to China.

Ypthima sakra (Moore)

Description: Common among under-growth in shady places from May to October; upperside amber-brown and underside ochraceous-brown with black eye-spots.

Wingspan: 45-50 mm.

Range: Occurs in N.W. Himalayas including Murree Hills.

Pararge schakra (Koll.)

Description: Fairly common from April to October at grassy hillsides; upperside brown with a prominent black eye-spot at the apex of the forewing and three smaller marginal eye-spots on the hindwing; underside of forewing orange-yellow and the hindwing pale gray with yellowish-brown lines.

Wingspan: 55-60 mm.

Range: Occurs in N.W. Himalayas including Chitral and the Murree Hills.

Callerebia nirmala (Moore)

Description: Common from May to September; upperside dark brown with an oval black ocellus on the forewing and subtornal ocellus and one or more postdiscal ocelli on the hindwing; the ocellus on the underside of the forewing more prominent and with yellow ring.

Wingspan: 50-55 mm.

Range: N.W. Himalayas: Azad Kashmir, Murree

LYCAENIDAE

This is a family of the smallest butterflies commonly found in tropical countries. They fly in grassy places at the edges of woodland and in gardens. Their hindwings are tailed or lobed and have prominent eyespots. They generally feed on plants in the family Leguminosae, clovers, and vetches. The caterpillars of certain lycaenids have a very interesting relationship with ants. The caterpillars through specialized organs on their last body segment exude droplets of a sweet liquid on which the ants feed. In return the ants protect the caterpillars against their indefatigable enemies especially parasitic wasps and flies which lay their eggs inside or on the caterpillars. This type of relationship in which both parties are benefited is called symbiosis.

Aphnaeus ictis (Hew.)

Description: Also known as Spindasis ictus; upperside of the forewing dark brown with a pale pre-apical triangular area invaded by three dark brown bands, subcostal of the hindwing brown, medial blue and subdorsum , pale brown; underside pale brown with silver bands.

Wingspan: 28-30 mm.

Range: From Murree Hills through Kashmir to N. India.

Azanus ubaldus (Cram.)

Description: Active from April to December; being dimorphic the sexes differ in colour, the male purplish-blue above and grayish-brown underneath, the female silky brown, purplish at base with marginal black lines; two broods occur in a year, markings on underside more conspicuous in autumn brood than in the spring brood.

Wingspan: 18-20 mm.

Range: From Murree Hills through India to Sri Lanka

Azanus uranus Butler

Description: Upperside blue in female and dull blue in male; underside dark brown in female than male with grayish-white markings; has two broods, one in winter and the other in the rainy season.

Wingspan: 24-26 mm.

Range: Besides Murree Hills, also found in India and Bangla Desh.

Lampides boeticus (Linn.)

Description: Frequents gardens, open spaces and grasslands almost throughout the year but most abundant from March to June; upperside of both sexes differ in colour: violet blue in male and brown in female, blue only at the base of the forewing; underside grayish-brown, crossed by pale brown bands, two tornal sides of the hindwing are black.

Wingspan: 30-36 mm.

Range: Occurs in Murree foothills. Also widely distributed in Asia, Africa, Europe through Malaya to Australia and Hawaii.

Prosotas nora (Fldr.)

Description: Also known as *Nacaduba nora*; common from May to October; upperside violet-blue with bright blue sheen on forewing in male, dark purple in female; underside brownish-gray with dark bands.

Wingspan: 22-24 mm.

Range: Besides Murree foothills also ranges to India, Burma and Sri Lanka.

Castalius rosimon (Fabr.)

Description: More abundant in December and January; besides perching on flowers, can also be seen on bird droppings and dead insects. upperside white with black spots and streaks, bases of wings are metallic blue.

Wingspan: 24-32 mm.

Range: Ranges from Nathia Gali and Azad Kashmir to India, Burma, and Thailand.

Tarucus callinara (Butl.)

Description: Upperside violet to clear dark blue; underside with postdiscal band broken up into spots and markings, forewing with a prominent spot.

Wingspan: 18- 20 mm.

Range: Whole Himalayan range from N.W. to East.

Zizeeria knysna (Trim.)

Description: Upperside purple in the forewing and dull brown in the hindwing; underside pale with a series of terminal black spots.

Wingspan: 16-20 mm.

Range: Besides Murree foothills also widespread in Asia, Africa, S. Europe and Australia.

Zizeeria maha (Kollar)

Description: A tailless lycaenid, more abundant in August and September; upperside pale blue with dark borders in male and dark brown with more or less basal blue scaling in female; underside brownish gray with black discal spots.

Wingspan: 26-30 mm.

Range: Occurs in Margalla Hills, Murree foothills. Also widely distributed in Iran, India, Tibet, China and Japan.

Heliophorus bakeri (Evans.)

Description: Common from May to September; upperside deep blue, nonmetallic, with very wide black borders in male, small orange band on forewing and hindwings in female.

Wingspan: 68-74 mm.

Range: From Murree Hills through Kashmir to N. India.

Lycaenopsis vardhana (Moore)

Description: Also called as *Celastrina vardhana*, a common from April to October; upperside grayish-blue, speckled with dark brown markings; underside sliver white with a large black spot on the forewing and minute dark brown spots on the hindwing.

Wingspan: 36-44 mm.

Range: N.W. Himalayas including Murree Hills and Kashmir.

Aricia agestis (Bgst.)

Description: Found at 8000 ft. in May and June; upperside dark brown with marginal row of black spots.

Wingspan: 25-30 mm.

Range: Murree, Chitral

Chaetoprocta odata (Hew.)

Description: Upperside shining purple with apex and margin black; underside white with ochraceous bands especially at the end of cell.

Wingspan: 30-35 mm.

Range: Galiat area, Azad Kashmir, Chitral

Chrysozephyrus syla (Koll.)

Description: Fairly common from May to October; upperside metallic green in male, brownish black in female; underside light gray.

Wingspan: 40-45 mm.

Range: Murree, Shogran, Neelum Valley

Euaspa milionia (Hew.)

Description: Common in May and June; upperside light blue with a white discal patch and black apex on the forewing and a filiform tail on the hindwing; underside amber brown basally and brown marginally.

Wingspan: 30-34 mm.

Range: Murree, Azad Kashmir

Rapala melampus (Cram.)

Description: Upperside orange red with dark brown costal-discal borders; underside ochraceous brown with a hair streak.

Wingspan: 30-36 mm.

Range: Occurs in Murree foothills and also in India, Sri Lanka

Rapala selira (Moore)

Description: Upperside dark brown, tinged violet with orange- yellow spot on forewing and hindwings; underside brownish gray.

Wingspan: 26-30 mm.

Range: Occurs in Murree, Chitral

ERYCINIDAE

This is a small family which comprises some very attractive butterflies. The front pair of legs is imperfect in male and perfect in female.

Dodona durga (koll.)

Description: Active in the lower Murree Hills from April to September, usually seen settled on sand and roadside plants; upperside dark brown with prominent dark yellow spots, a bar at the end of cell, hindwing lobed; underside ochraceous brown.

Wingspan: 30-40 mm.

Range: Kashmir, Tibet, W. & C. China.

LIBYTHEIDAE

This family is treated by some workers as a subfamily of the Erycinidae. Due to the presence of a number of distinguishing characters the group is now considered as a separate family. The most remarkable feature of this family is the head which is unusually long and protruding like a beak. This is a small family consisting of a few species usually occurring in the tropical climate.

Libythea lepita (Moore)

Description: Fairly common in May and June; upperside dark brown with an orange-yellow streak and dark spots on costal margin of forewing and whole of hindwing; underside of forewing brown, and hindwing variable, grayish.

Wingspan: 35-43 mm.

Range: From Murree Hills to Kashmir, India, China and Japan.

CONSERVATION OF BUTTERFLIES

Biological diversity, which comprises every form of life from the tiniest microbe to the mightiest beast and the ecosystem of which they are a part, is one of the most pressing environmental issues. Over exploitation of natural resources by man and concentration on short term gain, has led to a global crisis. Loss of biodiversity- the habitats, species and genes that make up life on earth- is proceeding at an ever accelerating rate.

Although the disappearance of species is a natural process, there is growing concern that it is now being fastened dramatically by changing natural habitats; cutting them down, plowing them up, overgrazing them, paving them over, damming and diverting water, flooding or draining areas, spraying them with pesticides and acid rain, pouring oil into them, changing their climate, exposing them to increased ultraviolet radiation etc. etc. How fast is this diversity now disappearing? Although it is impossible to say with precision, the answer clearly is 'frighteningly fast'.

In a developing country like Pakistan, with limited natural resources and a rapidly multiplying population, the environmental picture seems even more gloomy. Its natural systems such as rivers, mountains, plains and forests are under tremendous pressure. If these resources are depleted or destroyed or are not sustained, the ecological support system may erode and eventually collapse. For many centuries primitive mankind existed in this region in a state of relative ecological equilibrium. But in the last five or six decades of this century Pakistan has experienced profound and accelerated negative ecological changes which have eroded the marvelous diversity of butterflies. Many species known to have existed within historic times have become extinct or are endangered.

As butterflies are primarily dependent on plants as food source both as caterpillars and adults, the broad-scale disruption of habitats which reduces or eliminates plant species or communities is harmful. Conservation of butterflies is essential, since the most insignificant species play a crucial role in the ecosystem to which they belong. Besides pollinating flowers, feeding song birds and delighting the eye, butterflies function as 'indicator species' of ecosystem health. Today many species of butterflies are being closely watched for sign of climate change as well as other habitat disturbance. While flagging population may indicate trouble, healthy butterfly population can mean that the ecosystem's various organisms and cycles are still in balance. Though other insects might be equally good indicators, butterflies being conspicuous and beautiful are much preferred over other insects. Since butterfly species are handy "umbrella species" to represent an entire ecosystem, these are being used by conservationists in selecting site for " Wildlife Park/ Protected Area".

BUTTERFLY GARDENS which are becoming popular the world over might have a role in Pakistan in developing appreciation for butterflies in nature and adopting hobbies like butterfly watching and photography. If we love the butterflies well enough to save them, we have to preserve whole community of plants and animals with which the future of humanity is bound.

REFERENCES

AHSAN, M. AND LQBAL, J., 1975. A contribution to the butterflies of Lahore with the addition of new records. *Biologia,* **21** (2): 143-158.

ANTRAM, C.B., 1986. *Butterflies of India.* 1-226, Periodical Expert Book.

DOHERTY, W., 1886. List of Butterflies taken in Kashmir. *J. Asiatic Soc. Bengal.* **55**: 103-140.

ELWES, H.J., 1882. On a collection of butterflies from Sikhim. *Proc. Zool. Soc.; London,* 398-407.

EVANS, W.H., 1932. The Butterflies of Baluchistan. *J. Bombay nat. Hist. Soc.,* **36**: 195-209.

EVANS, W.H., 1923 The identification of Indian butterflies (Papilionidae, Pieridae). J. *Bombay nat. Hist. Soc.,* **29**: 230-260.

GAY, T., KEHIMKAR, I.D. & PUNETHA, J.C., 1992. *Common Butterflies. of India.*1-67. W.W.F. India.

GOUGH, W.G.H., 1935. Some Butterflies of Nepal. *J. Bombay nat. Hist. Soc.,* **38:** 258-265.

HASAN, S.A., 1994. *Butterflies of Islamabad and Murree Hills.*1-68+ Fig. 49. Asian Study Group.

IQBAL, J., 1978. A preliminary Report on butterflies of Rawalpindi and Islamabad. *Biologia,* **24**: 237-247.

LEWIS, H.L., 1987. *Butterflies of the world.* 1- 312. Harrison House, New York.

MALIK, J.M., 1970, Notes on the butterflies of Pakistan in the collection of the Zoological Survey Department. *Part 1. Rec. Sur. Pakistan* , **2**: 25-54.

MALIK, J.M., 1973, Notes on the butterflies of Pakistan in the collection of the Zoological Survey Department. *Part 2. Rec. Sur. Pakistan,* **5**: 11-28.

MANI, M.S., 1986. *Butterflies of the Himalaya..* 1- 181. Dr. W. Junk publisher.

MOHYUDDIN, A., 1. 1987. *A catalogue of insects and mites in the reference collection of PARC- CIBC Station up to 1986,* **1**: 34. PARC-CIBC Publication.

PEILE, H.D., 1937. *A guide to collecting Butterflies of India.* 1- 360. John Bale, Sons and Danielsson, Ltd. London.

PURI, D.R., 1831. Fauna of Lahore: Butterflies. *Bull. Dept. Zool. Punjab University,* 1: 1-61.

TALBOT, G., *1939.The Fauna of British India including Ceylon and Burma* 1: 1-600. Taylor and Francis, London.

TALBOT, G., 1947. *The Fauna of British India including Ceylon and Burma* **2**: 1-23 1. Taylor and Francis, London.

WATKINS, H.G.T., 1927. New Himalayan Butterflies. *Entomologist,* 151.

A PRELIMINARY SURVEY OF DIVERSITY AND DISTRIBUTION OF BUTTERFLIES OF NORTHERN PAKISTAN: GILGIT TO KHUNJERAB

DAVID SPENCER SMITH & S. AZHAR HASAN*

Department of Biological Sciences, Florida International University, University Park, Miami, FL33199, USA
** Pakistan Museum of Natural History, Garden Avenue, Shakarparian, Islamabad*

INTRODUCTION

The foundations of knowledge of the butterfly fauna of the Indian subcontinent were established by the works of Marshall and DeNiceville (1882-1890), Moore and Swinhoe (1890-1913), Bingham (1905, 1907), Evans (1932) and others. Mani (1986) has provided an account of early work on the butterflies of the Himalaya/Karakoram. Of this vast area, the present Northern Areas of Pakistan have been less well documented than many other regions that have, in the past, been more accessible. Comprehensive accounts of the butterflies of the adjoining countries have recently been published: for Afghanistan by Sakai (1981) and for China by Chou (1990).

Northern Pakistan includes, in the Hindu Kush and Karakoram ranges, an extensive portion of the meeting point between the Palaearctic and Oriental zoogeographical regions, extending eastwards along the Himalayan chain. Evans (1932) recognized this, attributing the butterflies of Chitral, Hunza, Baltistan and Ladak to a "central Asian section of the Palaearctic region with a strong infiltration from the Himalayas". Of these four areas, the valley system from Gilgit through Nagar and Hunza and surrounding areas, to the border with Badakshan (Afghanistan) and Sinkiang/Xinjiang province (China) is most poorly documented. The descriptions of new taxa from India and Burma by Tytler (1926) included subspecies from Chitral, and a few from the Gilgit Agency including Chilas and Astor, from the Yasin valley, and from Misgar on the approach to the Mintaka Pass, west of Khunjerab. Evans (1927) gave an account of 27 species collected by Visser-Hooft, between May and October, 1925, in the western Karakoram, during a mountaineering expedition that passed through Hunza to Shimshal, and to the then remote and inaccessible valley of Khunjerab.

In 1994, the Pakistan Museum of Natural History and the University Museum, Oxford, started a collaborative project designed, firstly, to document the diversity of the butterfly fauna between Gilgit and Khunjerab, along the main valley of the Indus, in selected side valleys, and in the high altitude Khunjerab National Park. The area passes through a wide altitudinal range, from ca. 4500-16000 ft (ca. 1500-4800m), and a further aim of the survey is to investigate species distribution in the context of ecological preference and altitudinal range.

RESULTS AND DISCUSSION

This preliminary summary primarily covers field data obtained in the study area during the first two weeks of July, 1994 and 1995, augmented by records made at other times by Gulam Naseer, Murtazabad, Hunza. The area around Gilgit is now almost entirely cultivated, and the vicinity of the Indus along the Karakoram Highway through Nagar and lower Hunza, and continuing through upper Hunza (Gojal) to Sost, includes wide stretches of arid mountain slopes, with vegetation most obvious along the edges of glacier-fed streams, and in the irrigated orchards and fields surrounding villages. Numerous side valleys are confluent with the main Gilgit-Nagar-Hunza course of the Indus, providing access to high pine/juniper pasture. As is mentioned further below, the last habitat type was sampled briefly (1995) in the Naltar and Minapin valleys, respectively in Gilgit and Nagar districts. Along the main valley, beyond the entrance to Khunjerab National Park at Dih, the arid terrain gives rise relatively abruptly, above ca. 12500 (ca. 3800m), to low, alpine ground cover, flowering briefly and unpredictably.

To date, we have recorded over 50 species from the area studied. A precise figure cannot be given at this point, since taxonomic determination is as yet incomplete. Many of the butterflies present in the area have evolved local geographical races or 'subspecies' through genetic isolation between populations, across their sometimes very extensive range. While some subspecific taxa are readily recognized, others are much more difficult to assign, and/or the subject of current revisionary work, and in this preliminary account trinomials are used only where taxonomic work on the material is complete.

The Oriental component:

The great majority of species recorded to date from the Gilgit-Hunza region have Palaearctic affinities. In a general account of the butterflies of Himalaya, Mani (1986) stressed the great richness of the eastern fauna, particularly of mountain forest species, compared with that of the west. He suggests that an abrupt fall in diversity occurs at the Sutlej Defile, passing from Tibet through the Zascar Himalaya at the Shipki La, at the transition to geologically less ancient, and far more arid mountains; a region some 700 km SE of Khunjerab.

The tropical montane forest of the east is absent from northern Pakistan, and the Oriental component in the study area is represented by ingress of plains species. The commoner butterflies of the Islamabad/Murree area have recently been discussed by Hasan (1994), and most of the Oriental representatives found commonly along the Indus valley in Kohistan, 250-350km south of Gilgit, for example *Danaus chrysippus* (Linn.) and *Papilio polyctor* (Boisd.) do not penetrate into the arid terrain further north. The lycaenids *Pseudozizeeria maha* (Kollar), *Heliophorus sena* (Kollar) and *Everes indicus* (Evans) are common around Gilgit, and the latter extends at least to 7000 ft (2100m) at Aliabad, lower Hunza. *Eurema hecabe* (Linn.), abundant in the plains, has been found north of Gilgit only to Joglot at 4600 ft (1400m), while another Oriental pierid *Catopsilia pyranthe* (Linn.) reaches the 10,000ft (3000m) level at Kamris, above Gulmit in Gojal.

Altitudinal distribution:

It is already possible to recognize altitudinal preferences in many of the Palaearctic species of the area. The present picture will be refined as records are extended during each period of field work, and the ecological factors underlying this stratification will emerge as the botanical structure of habitats, and the phenology and larval food plant choices become more adequately known. The altitudinal groupings are to some extent arbitrary, but in part reflect the position of field sites with significant vegetation, separated by often extensive arid stretches, virtually devoid of plants and butterflies, alike.

(a) High altitude species:

Throughout the entire Himalayan region, a characteristic high altitude ('hypsobiont'; Mani, 1986) butterfly fauna is present, entirely above the timberline, or equivalent altitude as, in our area, the full ca. 2500 km^2 extent of Khunjerab National Park. In the eastern Himalaya the snowline lies at ca. 15,000ft (4500m) but rises to ca. 16,000-17,000ft (4800-5200m) in the arid Karakoram (Mani, 1986). The international boundary at Khunjerab Pass lies at about 16,000ft, and even within the Park, terrain below 12,500ft (3800m) is largely occupied by rock and scree slopes, with very sparse vegetation and few butterflies on the wing. It is already clear that this altitudinal zone (12,500-16,000ft) possesses an admixture of strictly hypsobiont species, and the highest point in the range of a small number of very widely ranging species, notably *Vanessa cardui* (Linn.) and *Pieris brassicae* (Linn.). In July 1994, we recorded four species restricted to the open terrain at the highest elevation, with low, alpine ground cover, representing the first butterflies listed from Khunjerab since the Visser-Hooft expedition of 1925: the lycaenid *Albulina asiatica* (Elwes), the pierids *Colias cocandica hinducucica* (Verity) and *C. eogene eogene* (C & R Felder), and one of several subspecies of *Parnassius epaphus* (Oberth.). At somewhat lower elevation (11,500-12,500ft), along the Indus near Barkhoom, other localised species were noted: a nymphalid (*Melitea* sp.), and *Pieris deota* (DeNicev.). This sparsely vegetated locality also marked the lowest elevation, from our data, for the pierid *Pontia callidice kalora* (Moore) and the highest elevation for the swallowtail *Papilio machaon asiaticus* (Men.), elsewhere widely distributed at all altitudinal levels to Gilgit.

A comment may be inserted at this point, noting the present total lack of understanding of the biology of these high altitude butterflies, in Pakistan. On 5 July 1994, the high pass was in full sun, 20°C (69°F) at midday and carpeted with plants in full flower: purple *Astragulus*, the yellow blooms of *Potentilla pamirica*, *Ranunculus pulchellus*, *Papaver nudicaule,* a white *Aster* sp., *Artemisia* and others. In these conditions, butterfilies and other insects were active but on 4 July 1995 flowering had scarcely commenced; no insects were seen and we encountered a snowfall. The question of how the phenology and reproductive biology of butterflies and other insects are adapted to meet such year-to year fluctuations remains to be answered. This is merely a dramatic instance of the need for research on the biology of butterflies in the area, in general.

(b) Distribution patterns at lower altitudes:

Of the majority of species occurring below the National Park, some are widely distributed, others so far recorded only from a relatively narrow altitudinal zone. Of the former group, for example, five pierids are generally common in all vegetated localities from Gilgit up to the altitude of Sost (9000ft; 2700m): *Pontia daplidice moorei* (Rober), *P. chloridice alpina* (Verity), *Pieris rapae* (Linn.), *Colias erate* (Esper) and *C.fieldii* (Men.). The lycaenid *Lyceides samudra* (Moore) occupies a similar range, becoming more abundant as the valley is ascended. Several species recorded within the main Gilgit-Nagar-Hunza valley have been found more locally, in a relatively narrow altitudinal range. For example, the satyrids *Pararge m. menava* (Moore) and *Aulocera padma* (Kollar) have been noted in Aliabad and Murtazabad (Lower Hunza) at 7000-7500ft (2100-2300m), while *Pieris kreuperi* (Staud.) has so far been recorded only from the vicinity of Ghulmet-Nagar (6000ft; 1800m).

(c) Diversity and Isolation:

The above examples of butterfly distribution in the main valley of the Indus lay the groundwork for investigating the ecological basis of habitat selection. However, the area traversed by this valley is set in a vast and complex terrain of valleys, separated and often isolated by high peaks and glaciers. Thus the species list from the areas readily accessible from the KKH is not only just a small part of the entire region, but moreover on that has been substantially modified by human habitation and cultivation. An important part of the present survey is to document at least a sample of the 'secondary' valleys. It cannot be assumed that valleys, isolated by extremely high mountains and glaciers, across which butterfly dispersal seems improbable, will maintain the same species structure and diversity.

Our preliminary data suggest that this view is correct. In July 1994, we worked, albeit briefly, in two well separated sites: the Minapin valley in Nagar district and Naltar valley in Gilgit district. With the proviso that the details of this picture are likely to be modified by each future visit, our first data showed at least six species recorded from Minapin (but not from the main valley or from Naltar), and a similar number so far found, in our survey, only in Naltar. The former group includes an *Aulocera*, a subspecies of *Parnassius charltonius* (Gray), and the lycaeenids *Cupido buddhista* (Alperaky) and *Aricia eumedon privata* (Staud.). Taxa so far found in Naltar include two satyrids, and the lycaenids *Albulina chitralensis* (Tytler) an *Aricia allous nazira* (Moore). Isolation of butterfly populations in the mountain ranges of Central Asia has led to a rich evolutionary radiation of taxonomically recognized species, and of geographical races or subspecies; the key to documentation of biodiversity of the area of study lies in comparison between its component faunas. Our plans for field work in 1996 include revisiting the Naltar and Minapin sites, and commencing a survey of two more valleys leading to the Indus. In the context of evolutionary diversification through isolation, it may be noted that phenotypic similarity between two or more geographically isolated populations may mask genetic discontinuity. For example, we recorded the *lycaenid Polyommatus devanicus* (Moore) from Naltar, Minapin and above Gulmit, each locality at ca. 9000-10,000ft (2700-3000m), but further comparison must await application of techniques of molecular genetics.

In this brief summary, we have illustrated some general distribution patterns by examples. Some taxa remain to be determined, and others, based as yet on single records, are insufficient as reliable guides to distribution. Moreover, it must be stressed that the data obtained must be viewed in the context of knowledge of the butterfly faunas of the Palaearctic/Oriental regions in general, and of adjacent regions of Central Asia, in particular. We suggest that a systematic, 'high-resolution' view of this part of the Northern Areas of Pakistan will not only fill a significant gap in knowledge of the country's insect fauna, but contribute more widely to montane butterfly biogeography, ecology and evolution.

ACKNOWLEDGEMENTS

We are grateful to Dr. Shahzad A. Mufti, Director General of Pakistan Museum of Natural History, for approving the project and for his encouragement. We thank Mr. Yusuf Ali, Conservator of Forests/Northern Areas, for permission to obtain records within Khunjerab National Park. The vehicle provided by PMNH has been in the care of Mr. Gulam Mustafa, whose wide knowledge of the area has been invaluable throughout our field work. The able help of Mr. Fiaz Ahmed, Field Assistant of PMNH, is acknowledged with gratitude, as is the contribution to records made by Mr. Ghulam Naseer, Murtazabad. DSS thanks the Commonwealth Science Council for their generosity that allowed his participation in the 1995 International Symposium on Biodiversity of Pakistan, Islamabad, where this preliminary account was presented. He also records his appreciation for the valuable help of Mrs. Malgosia Atkinson, the Hope Entomological Collections, University Museum, Oxford, in preparing and identifying material under study, and to Father Alan Bean for helpful discussions. Our colleague Dr. Muqarrab Shah, PMNH, kindly identified plants collected during the survey. Dr. Zsolt Bálint, the Hungarian Natural History Museum, Budapest, has advised us on the taxonomy of many of the Central Asian lycaenids, a group in which he has unique expertise. Following the preliminary work described here, the project was supported by NSF grant INT-9700669.

REFERENCES

BINGHAM, C.T., 1905. *The Fauna of British India including Ceylon and Burma. Butterflies*, Vol. **1**. Taylor & Francis, London. pp. 511.

BINGHAM, C.T., 1907. *The Fauna of British India, including Ceylon and Burma. Butterflies*, Vol. **2**. Taylor & Francis, London. pp. 480.

CHOU, I., 1990. *Monographia Rhopalocerorum Sinensium (Monograph of Chinese Butterflies),* Vols. 1 and 2, Henan Scientific & Technological Publishing House. pp. 854 (in Chinese).

EVANS, W.H., 1927. Lepidoptera-Rhopalocera obtained by Mme J.Visser-Hooft of the Hague (Holland) during an exploration of previously unknown country in the Western Karakorum, N.W. India. *Tijdschrift voor Entomologie*, **70**: 158-162.

EVANS, W.H., 1932. *The identification of Indian butterflies.* 2nd edit., Diocesan Press, Madras, pp. 454.

HASAN, S.A., 1994. *Butterflies of Islamabad and the Muree Hills.* Asian Study Group, Islamabad, pp. 68.

MANI, M.S., 1986. Butterflies of the Himalaya. *Series Entomologica* (K. A. Spencer, edit.) 36. Junk, Dordrecht, pp. 181.

MARSHALL, G.F.L. and DE-NICEVILLE, L., 1882-1890. *The Butterflies of India, Burma and Ceylon.* Vols. 1-3, Central Press Co., Calcutta, pp. 327, 332, 503.

MOORE, F. and SWINHOE, C., 1890-1913. *Lepidoptera Indica.* Vols. 1-10, Lovell, Reeve, London.

SAKAI, S., 1981. *Butterflies of Afghanistan.* pp. 271 (in Japanese).

TYTLER, H.C., 1926. Notes on some new and interesting butterflies from India and Burma. *Journal of the Bombay Natural History Society*, **31**: 248-260.

SOME FEATURES OF ZOOGEOGRAPHICAL INTEREST IN THE BIODIVERSITY OF TERMITES OF PAKISTAN

M. SAEED AKHTAR AND MUZAFFER AHMAD

Department of Zoology, University of the Punjab, Lahore

INTRODUCTION

Termite fauna of Pakistan consists of 49 species, belonging to 17 genera and four families. The most interesting feature of zoogeographical importance of termites of Pakistan is their biodiversity. Though most of Pakistan forms part of the Indo-Malayan region, generic components of its termite fauna reveal an interesting composition of Ethiopian, Palaearctic, Indo-Malayan and even Neotropical elements.

According to Emerson (1955), the Indo-Malayan/Oriental region incudes the tropical Asian continent from Western India, Southern China just north of Burma and Southern coast of China, through the tropical Islands of Formosa, the Philippines, and East Indies to Weber Line west of Moluccas. Although Emerson (1955) pointed out that Pakistan south of Himalayas has an Indo-Malayan fauna clearly related to the more southern tropics, its present status, which is discussed in this paper, reveals some interesting features.

RESULTS AND DISCUSSION

Family Kalotermitidae

In Pakistan, this family is represented by four genera, namely *Postelectrotermes, Bifiditermes, Epicalotermes* and *Cryptotermes.* Interestingly, none of these genera originated in the Indo-Malayan zoogeographical region.

Two species of the genus *Postelectrotermes* have been recorded from Pakistan, both from the Palaearctic part of the country (Ahmad, 1955; Akhtar, 1974). The eastern distribution of *Postelectrotermes* extends upto Thailand from where one species, *P. tongyii,* was reported by Ahmad (1965). The evidence points out to *Postelectrotermes* possible origin in the Palaearctic region (Buillon,

1970). Its occurrence in the Palaearctic part of Pakistan and Iran (Raven and Akhtar, unpublished) lends support to Bouillon's view (1970). From the Palaearctic region, the genus invaded Ethiopian region, from there to Malagasy region and east to the Indo-Malayan region. The genus shows a great diversity in its adaptation to ecologically very different areas, extremely cold alpine vegetation to tropical rain forest.

The genus *Bifiditermes* has been recorded from Ethiopian, Indo-Malayan and Australian zoogeographical regions. It has largest number of species in the Ethiopian region which suggests its origin in this region. Bouillon also opines that *Bifiditermes* originated in the Ethiopian region. Only one species, *B. beesoni* (Gardner) has been reported from Indo-Pakistan sub-continent. Its occurrence in the Ethiopian, Indo-Malayan and Australian regions suggests that it had a wide continuous distribution initially and when the Gondwanaland split up in various continents, the genus *Bifiditermes* species also got divided.

The genus *Epicalotermes* was till recently considered as endemic to Africa. With the discovery of *E. pakistanicus* by Akhtar (1972), its range extended east to Indo-Malayan region.

The drywood termite genus *Cryptotermes* has a worldwide distribution having been reported from the Australian, Papuan, Indo-Malayan, Ethiopian, Malagasy, Nearctic and Neotropical regions. The genus is primarily tropical in distribution with the exception of one species occurring in the warmer parts of Florida. Its wide distribution is due to introduction. The introduced species have established themselves in the coastal areas. The only species occurring in Pakistan, *C. karachiensis* Akhtar, is also a case of introduction. Its origin in the Ethiopian region at the end of the Cretaceous or the beginning of the Eocene is suggested by Bouillon (1970).

Family Termopsidae:

This family is represented in Pakistan by a single monotypic genus, *Archotermopsis*, which is Palaearctic in origin. This is the only termite inhabiting high altitude (5,000 to 11,000 ft.) in the Indo-Pakistan sub-continent.

Family Hodotermitidae:

The family Hodotermitidae is also represented by a single genus, *Anacanthotermes* in Pakistan. It is a subtropical desert genus extending from North Africa to India. This genus presumably originated in southern Palaearctic steppes and deserts and dispersed east to tropical Indo-Malayan region (Emerson, 1955). In Pakistan,

Anacanthotermes is represented by four species, ranging from Palaearctic to subtropical parts of the country.

Family Rhinotermitidae

Three Rhinotermitidae genera - *Psammotermes, Coptotermes* and *Heterotermes* - occur in Pakistan. Genus *Psammotermes* originated in the Ethiopian region. The first case of its invasion to Indo-Malayan region was reported by Roonwal and Bose (1964) who described a new species from Rajisthan desert. The second case of its occurrence in the Indo-Malayan region was reported by Akhtar (1972).

The genus *Coptotermes* has a worldwide distribution, having been reported from all the zoogeographical regions except Nearctic. In Pakistan, 2 species of *Coptotermes* have been recorded. These are *C. heimi* and *C. formosanus*. The latter species has been introduced in many parts of the world (Ahmad, 1953; Gay, 1969). Judging from the number of its species occurring in a particular region, genus Coptotermes probably originated in the Indo-Malayan region.

There occurs a single species of *Heterotermes* in Pakistan. Like *Cryptotermes*, the genus *Heterotermes* has also been introduced in many countries. Recently, it has been reported from Miami, Florida by Scheffran and Sue (1995). Its continuous occurrence in Indo-Malayan, Papuan and Australian region suggests that it evolved before the separation of Australia from the Indo-Malaya in the Cretaceous.

Family Termitidae

Two of the fungus growing termite genera occur in Pakistan. These are *Odontotermes* and *Microtermes* which are confined to the Ethiopian and Indo-Malayan regions. In Pakistan, most of these termites occur east of Kirther range and are very scantily present in Balochistan west of Kirther range. The pattern of distribution of the fungus growers in Pakistan suggests areas west of Kirther range as forming Palaearctic region (Akhtar, 1974). The fungus growing termite genera originated in the Ethiopian region and some of them migrated east to the Indo-Malayan region. Five of 12 fungus growing genera are represented in the Indo-Malayan region also, while the remaining genera are confined to the Ethiopian region except the genus *Microtermes* which occurs in the Malagasy region also. The distributional pattern of the subfamily Macrotermitinae which comprises the fungus growing genera clearly indicates its origin in the Ethiopian region (Emerson, 1955; Krishna, 1970).

The Apicoterminid genus *Speculitermes* is very interesting from zoogeographical point of view. It is found in the Neotropical (South America) and the Indo-Malayan regions. It was first described by Wasmann (1902). Till 1960 it was regarded as soldierless termite and was considered closely related to *Anoplotermes* which is also soldierless. Roonwal and Chhotani (1960) reported for the first time the occurrence of soldier caste in *Speculitermes.* Ahmad (1965) collected *S. macrodentatus* from Thailand with a queen, 2 soldiers and workers. In Pakistan there occurs one species of *Speculitermes* i.e., *S. cyclops.* Roonwal (1970) reported ten species from the Indo-Malayan region. In South America, 6 species have been reported by Emerson (1955). *Speculitermes* is the only termite genus occurring in Neotropical (South America) and Indo-Malayan (Pakistan, Thailand and India) regions. The authors are of the opinion that this termite genus had a very wide distribution at the time when the continents in the southern part were together, as the genus occurs in widely separated Neotropical and Indo-Malayan regions. Its absence from the Ethiopian region between Neotropical and Indo-Malayan regions is surprising. It could possibly be due to lack of collecting.

The only nasute termite occurring in Pakistan is *Trinevritermes biformis* (Wasmann) recorded by Akhtar (1972) from Dalwal Rakh Forest and Fort Munro. Since it has been reported from interior of Pakistan, the possibility of its being an introduced species is excluded. The genus *Trinevritermes* is of Ethiopian origin from where it invaded Indo-Malayan region.

Three genera belonging to the subfamily Termitinae are recorded from Pakistan. These are *Microcerotermes, Amitermes* and *Angulitermes.*

Microcerotermes is represented in Pakistan by 11 species. It occurs in all the zoogeographical regions except Nearctic. It originated in Africa during the Cretaceous and dispersed through the Oriental region to the new world through Bering land Bridge (Emerson, 1955). Although it reached the new world through the Bering Strait, its absence in North America is surprising.

The genus *Eremotermes* has been recorded from Palaearctic and Indo-Malayan regions. In Pakistan, 3 species of this genus occur. Since majority of the *Eremotermes* species occurs in Indo-Malayan region, its origin in this zoogeographical region is postulated.

Amitermes occurs in all the zoogeographical regions except Papuan region. Since it is most abundant in the Australian region next to Ethiopian region, its absence in the adjoining Papuan region could be due to lack of collection. Its origin in the Ethiopian region is indicated by its being most abundant in Africa.

The termite species belonging to the genus *Amitermes* exhibit a marked degree of biodiversity in their nest building behaviour dependant upon ecological conditions in which they occur. In Pakistan, the *Amitermes* species build subterranean nests. In Thailand rain forests, some of the species live in the nests of other termites. In Australia, *Amitermes* builds large mounds which form a conspicuous feature of the landscape. The most remarkable of these nests is the compass nest build by *A. meridionalis*. Invariably, all the nests wherever they are built have their long axis pointing north and south. The orientation of the nest results into maintaining a relatively stable internal temperature in the hot days. In sharp contrast to the compass nest, *A. exullens* in the tropical rainforests of South America constructs nests on the trees with finger-like projections which shed the rain water during heavy tropical downpour.

Angulitermes in the only genus occurring in Pakistan with snapping type of mandibles. It includes 14 species distributed as follows; Ethiopian region (4); Indo-Malayan region (9) and Palaearctic region (1). From the pattern of distribution it is postulated that it originated in the Indo-Malayan region.

Place of origin of the termite genera and their occurrence in different zoogeographical regions is indicated in Table I. The most interesting element is *Speculitermes* which originated in the Neotropical region and later on migrated to the Indo-Malayan region. This is the only termite genus which has this kind of distribution i.e., occurring in the Neotropical and Indo-Malayan regions. Although the species occurring in the Indo-Malayan region are greater in number, it is nevertheless presumed that *Speculitermes* originated in the Neotropical region, because genus *Anoplotermes*, of which it was regarded as a sub-genus, has a Neotropical origin.

Maximum resemblance at the generic level has been noticed with the Ethiopian region (Table II); as 53 % of the termite genera found in Pakistan originated in the Ethiopian region. Pakistan received 18.64% of the genera from the Palaearctic region and 6 % from the Neotropical region. Only 23% of the genera of Pakistan belong to the Indo-Malayan/Oriental region to which Pakistan belongs. This means that the Ethiopian elements found in Pakistan had a wider distribution to begin with and then because of climatic changes disappeared from several localities. Same can be true of the genus *Speculitermes* found in the Neotropical region and the Indo-Malayan region, or the Wegener's theory of continental drift is the only appropriate answer for this type of distribution.

Table I: Zoogeographical distribution of the termite genera of Pakistan

Genus	Place of origin	Present distribution
1. *Postelectrotermes*	Palaearctic	Indo-Malayan, Ethiopian, Malagasy, Palaearctic.
2. *Bifiditermes*	Ethiopian	Australian, Indo-alayan, Ethiopian.
3. *Epicalotermes*	Ethiopian	Ethiopian, Palaearctic, Pocket of Pakistan.
4. *Cryptotermes*	Ethiopian	Australian, Indo-Malayan, Ethiopian,Malagasy, Nearctic, Neotropical.
5. *Archotermopsis*	Palaearctic	Palaearctic, Palaearctic Pocket of Indo-Malayan.
6. *Anacanthotermes*	Palaearctic	Ethiopian, Palaearctic, Indo-Malayan.
7. *Psammotermes*	Ethiopian	Ethiopian, Palaearctic, Indo-Malayan.
8. *Coptotermes*	Indo-Malayan	Ethiopian, Palaearctic, Indo-Malayan,Australian, Papuan, Malagasy,Neotropical.
9. *Heterotermes*	Indo-Malayan	Palaearctic, Indo-Malayan,Ethiopian, Malagasy, Neotropical.
10. *Odontotermes*	Ethiopian	Ethiopian, Indo-Malayan.
11. *Microtermes*	Ethiopian	Ethiopian, Indo-Malayan, Malagasy.
12. *Speculitermes*	Neotropical	Indo-Malayan, Neotropical.
13. *Trinervitermes*	Ethiopian	Ethiopian, Indo-Malayan.
14. *Microcerotermes*	Ethiopian	Cosmopolitan except Nearctic
15. *Eremotermes*	Indo-Malayan	Indo-Malayan, Palaearctic
16. *Amitermes*	Ethiopian	Cosmopolitan
17. *Angulitermes*	Indo-Malayan	Indo-Malayan, Ethiopian, Palaearctic

Table II: Relationship of the termite genera of Pakistan (based on place of origin) with other Zoogeographical regions

Family	No. of Genera	Common with (% of Total)			
		Ethiopian	Palaearctic	Indo-Malayan	Neotropical
Kalotermitidae	4	3(17.64%)	1(5.88%)	-	-
Hodotermitidae	2	-	2(11.76%)	-	-
Rhinotermitidae	3	1(5.88%)	-	-	-
Termitidae					
Macrotermitinae	2	2(11.76%)	-	-	-
Apicotermitinae	1	-	-	-	1(5.88%)

Nasutitermitinae	1	1(5.88%)	-	-	-
Termitinae	4	2(11.76%)	-	2(11.76%)	-
Total No. of genera (with % of Total)	17	9(53 %)	3(18 %)	4(23 %)	1(6 %)

REFERENCES

AHMAD, M., 1953. Two new cases of introduction of Termites. *Spolia Zeylan.,* **37**: 35-36.

AHMAD, M., 1955. Termites of West Pakistan. *Biologia* **1**: 202-264.

AHMAD, M., 1965. Termite (Isoptera) of Thailand. *Bull. Am. Mus. Nat. Hist.,* **131**: 1-113.

AKHTAR, M.S., 1972. *Studies on the taxonomy and zoogeography of termites of Pakistan.* Ph.D.Thesis, University of the Punjab, Lahore.

AKHTAR, M.S., 1974. Zoogeography of termites of Pakistan. *Pakistan J. Zool.,* **6**: 84-104.

BOUILLON, A., 1970. Termites of Ethiopian region. In: *Biology of Termites* (eds. K. Krishna and F.M. Weesner), **2**:154-273, New York.

EMERSON, A.E., 1955. Geographical origins and dispersions of termite genera. *Fieldiana Zool.*, **37:** 465-521.

GAY, F.J., 1969. Species introduced by man. In: *Biology of Termites* (eds. K. Krishna and F.M. Weesner),**1**: 459-491, New York.

HUANG, F., LI G.X., ZHU, SHI-M.O., 1989. *The Taxonomy and Biology of Chinese Termites,* 1-605.pp., Tianze Press, China.

KRISHNA, K. 1970. Taxonomy, phylogeny and distribution of termites. In: *Biology of Termites* (eds. K. Krishna and F.M. Weesner), **2**: 127-150, New York.

ROONWAL, M.L., 1970. Termites of Oriental region, In: *Biology of Termites* (eds. K. Krishna and F.M. Weesner), **2:** 315-384, New York.

ROONWAL, M.L. AND BOSE, G., 1964. Termites of Rajasthan, India, *Zoologia Stuttg,* **40:** 1-58.

ROONWAL, M.L. AND CHHOTANI, A.B., 1960. Soldier cast found in the termite genus *Speculitermes*, *Sci. Cult.* (Culcutta), **26**: 143-144.

SCHEFFRAN, R.H. AND SUE, N.Y. 1995. *Heterotermes* in Miami, Florida. *Isoptera Newsletter*, **5:** 1-8.

WASMANN, E., 1902. Termiten, Termito-Philen und Myrmecophilen. Gesammelt auf Ceylon von Dr. W. Horn. *Zool. Jb.*, **17**: 99-164.

MARINE WORMS (ANNELIDA: POLYCHAETA) OF PAKISTAN

JAVED MUSTAQUIM

Centre of Excellence in Marine Biology, University of Karachi, Karachi

Abstract: An account of the polychaete fauna of Pakistan coast (Northern Arabian Sea) is given. A total of 109 species of polychaetes have been reported from the coastal waters during the last about 71 years (since 1924). The classification and nomenclature of some of the species have been changed during this period, resulting in synonyms. The zoogeographical affinities of the fauna are analysed.

INTRODUCTION

The polychaete fauna of the northern Indian ocean is well documented with 883 species (Hartman, 1974). Of these more than 200 species are known to occur in Persian Gulf and Gulf of Oman (Mohammad, 1973). However, the coast of Pakistan , which borders the northern Arabian Sea and lies adjacent to Gulf of Oman remained neglected until recently.

The first worker who reported polychaetes from Karachi was Aggarwala (1924). He collected "seventeen different kinds of polychaetes" from the rocky shore of Manora and Oyster islands. These seventeen kinds of polychaetes belong to families Hesionidae, Aphroditidae, Nereidae, Amphinomidae, Eunicidae, Glyceridae and Terebellidae. Aggarwala (1924) was however, unable to identify them and in the absence of adequate description and figures it is almost impossible to determine the species. Bindra (1927) published a monograph describing five species of the genus *Eurythoe* from Karachi coast. Of these, two species, namely *E. karachiensis* and *E. matthaii* were new to science. However, Fauvel (1953) considered *E. karachiensis* a synonym of *E. complanata.* In 1938, Aziz published a more comprehensive account and described 34 species of polychaetes from Karachi. Aziz (1938) named four new species which are *Eunice manorae, Perineries matthaii, Dasychone gravelyi* and *D. kumari. Eunice manorae* Aziz is now considered as indeterminable by Fauchald (1992), who reviewed the genus. A characteristic feature of these early named worms from Karachi is the fact that they are large, conspicuous, brilliantly coloured in life and come from intertidal zone.

More recently Hasan (1960) described 18 species, out of which one species *Aricidea alisdairi* is new to science. Ahmad (1969) described 17 species from the intertidal zone of Karachi coast. Other workers who published reports on small collections or individual families includes Bhatti and Soofi (1949), Ashraf (1968), Jawed and Khan (1974), Siddiqui and Mustaquim (1988), Habib and Mustaquim (1988), Rehana and Mustaquim (1989), Mustaquim (1990, 1991a, b), Swaleh (1993) and Swaleh and Mustaquim (1993).

COMPOSITION OF THE POLYCHAETE FAUNA

The number of polychaete species reported from Pakistan is 109 (Table I). Of this total, 45 species (41.3 %) belong to group Errantia, whereas 64 species (58.7 %) belong to Sedentaria. The errantiate families Nerelidae (15) and Eunicidae (13) are well represented and in total account for 62 % of the errant species. Sedentary species are greater in number (64 out of 109) with the families Spionidae (14), Terebellidae (11), Sabellidae (10), and Serpulidae (10) containing most of the species. There are seven families, namely Phyllodocidae, Polydontidae, Ampharetidae, Orbiniidae, Paraonidae, Pectinariidae, and Trichobranchidae, which are represented by one species in each family.

There are four endemic species namely, *Eurythoe matthii* Bindra, *Glycera manorae* Fauvel, *Perineries matthaii* Aziz and *Aricidea alisadairi* Hasan.· Five species, *Perineries cultrifera* (Grube), *Platyneries dumerilii* (Aud. & M.Edw.), *Eulalia viridis* (Muller), *Chaetopterus variopedatus* Renier, and *Hydroides norvegica* Gunnerus are cosmopolitan in their distribution.

Table I: Polychaete species reported from Pakistan

ERRANTIA

Amphinomidae
Euphrosyne myrtosa
Eurythoe complanta
E. indica
E. macrotricha
E. matthaii

Aphroditidae
Harmothoe imbricata
H. dictyophora
Lepidonotus hedleyi
L. tenuisetosus

Eunicidae
Aglaurides fulgida
Diopatra neapolitana
Eunice antennata
E. australis
E. monorae
E. siciliensis
Lumbericonereis heteropoda
Lumbrinereis impatiens
Lysidice collaris
Marphysa corallina
M. macintoshi
M. mossambica
M. sanguinea

Glyceridae
Glycera alba
G. manorae

Hesionidae
Hesione pantherina
Leocrates claparedii

Nereidae
Ceratonereis marmorata
Leonnates jousseaumei
Nemalycastis indica
Neanthes capensis
Nereis coutieri
N. persica
Perinereis cultrifera
P. matthaii
P. nigropunctata
P. nuntia
P. vancaurica
Platynereis dumerilii
Pseudonereis anomala
P. variegata
Tyloneresis bogoyawlenskyi

Phyllodocidae
Eulalia viridis

Polydontidae
Polyontes melanonotus

Syllidae
Syllis exilis
S. variegata

SEDENTARIA
Ampharetidae
Melinna aberrans

Arenicolidae
Arenicola brasiliensis
A. cristata

Chaetopteridae
Chaetopterus variopedatus
Mesochaetopterus capensis
Phyllochaetopterus elioto

Chloraemidae
Stylariodes bengalensis
S. parmatus

Cirratulidae
Audouinia anchylochaeta
Cirratulus cirratus

Mageloniidae
Magelona sp.
Magelona papillicornis

Nephthydidae
Nephtys sp.
Pectinariidae
Pectinaria capensis

Sabellariidae
Pallasia pennata
Sabelleria spinulosa

Sabellidae
Amphiglena mediterranea
Branchiomma cingulata
B. gravely
B. kumari
B. serratibranchis
Hypsicomus phaeotaenia
Potamilla ehlersi
P. leptochaeta
Sabella melanostigma
Sabellastarte sanctijosephi

Serpulidae
Serpula vermicularis
Hydroides albiceps
H. exaltatus
H. heteroceros
H. norvegica
H. tuberculata
Pomatoleios kraussii
Protula tubularia
Spirobranchus tetraceros
Vermiliopsis glandigerus

Spionidae
Apoprionospio sp.
Boccardia polybranchia
Paraprionospio pinnata
Paraprionospio sp.
Polydora armata
P. ciliata
P. flava
P. giardi
P. hoplura
P. spondylana
Prionospio sp.
Pseudopolydora antennata
P. paucibranchiata
Nerine cirratulus

Terebellidae
Eupolymnia nebulosa
Loimia annulifilis
L. medusa
Nicolea venustula

Orbiniidae
Phylo foetidae

Paraonidae
Aricidea alisdairi

Pista quadrilobata
Streblosoma atos
S. minutum
S. persica
Terebella ehrenbergi

T. pterochaeta
Thelepus setosus

Trichobranchidae
Terebellides stroemi

ECONOMICALLY IMPORTANT SPECIES

Polychaetes constitute an important group of marine animals. Because of their wide distribution and abundance, they play a significant role in the food chain of the oceans. Some of these are primary consumers and others eat mud and thereby convert organic debris into protoplasm. Polychaetes in turn are taken as food by other animals such as fish, shrimps, crabs and lobsters.

Many species of lug-worms such as *Arenicola* are also used as bait in other parts of the world. The common British lug-worm *Arenicola marina*, is extensively used as bait in Britain for flat fish and haddock and it has been shown by Ashworth (1904) that *A. marina* is most successful bait for flat fish as compared to other baits, like clams, mussels, and limpets. There are two species of lugworms namely *A. cristata* and *A. brasiliensis,* which have been reported by Bhatti and Soofi (1949) and Ashraf (1968) from Pakistani coast. These worms are yet to be exploited by bait diggers.

Polychaetes are also important because they are fouling organisms and they can damage coastal installations and breakwater walls, either by their boring activity or by settling in masses. They usually attach to boats and ship hulls along with other fouling organisms. Haq *et al.,* (1978) mentioned that "considerable damage is reported to have been caused to the sea water cooling system of Karachi Nuclear Power Plant (KANUPP) by the settlement and growth of the larvae of fouling organism". The organisms found to be foulers include such species of polychaetes as *Hydroides norvegica* and *Dasychone* spp. (now *Branchiomma* spp.). *Hydroides norvegica* is a cosmopolitan species and seems to have been recorded from all the seas that have been thoroughly investigated. It lives in a calcareous tube, usually attached to hard objects. It prefers floating objects such as buoys and ship hulls and in particular the shaded areas. *Branchiomms* spp. which belong to family Sabellidae are also tubicolous. The tube is lined with mucoprotein and covered with sand and mud. *Branchiomma cingulata* is quite abundant on our rocky shores and it usually meaures 2-8 cm in length.

Several species of polychaetes, especially those belonging to family Spionidae, are found in waters polluted with oil or thermal and industrial effluents. They can survive in such growth inhibitory environment and for this reason have been the subject of studies involving chemical effects of such effluents on the populace. For the same reason, they are regarded as pollution indicator species.

ZOOGEOGRAPHICAL AFFINITIES

At this stage zoogeographical affinities of Pakistan fauna with adjacent localities should be regarded only as provisional since many areas of Pakistan, especially Balochistan coast, still remain to be investigated and further records, as well as taxonomic revisions, will modify the present analysis. The data for this analysis are given in Table II. In compiling the distributional data for each species the following references have been consulted: (1) for Persian Gulf and Gulf of Oman - Wesenberg-Lund,1949; Fauvel,1953; Ten-Hove, 1990 and (2) for west coast of India - Fauvel, 1932, 1953 and Hartman, 1974.

Table II shows that the fauna of Pakistan has a small overlap with the fauna of Persian Gulf and Gulf of Oman (22%) and even smaller overlap with the west coast of India (13.5%). Although the coast of Pakistan lies between the coasts of Iran and the north west of India, the faunal overlap is not large. This may be due to the fact that polychaete fauna of Pakistan has not been investigated thoroughly. None of the planktonic polychaetes are known to occur in the coastal waters of Pakistan nor any serious attempt has been made to study subtidal polychaetes. Furthermore, the entire Balochistan coast is yet to be investigated.

Table II : Zoogeographical affinities of polychaete fauna

	Persian Gulf & Gulf of Oman	**West Coast of India**
Total No. of species in common (Pakistan total = 109)	26	19
No. of cosmopolitan species in common (Pakistan = 5)	03	05
No. of non-cosmopolitan species in common (Pakistan = 104)	23	14
Percentage	22	13.5

REFERENCES

AGGARWALA, A.C., 1924. Polychaeta and Gephyrea from Karachi. *Proc. Lahore Phil. Soc.*, **3**: 69-70.

AHMAD, S.A., 1969. On a collection of polychaetes in the Zoological Survey Museum, Karachi. *Rec. Zool. Sur. Pakistan,* **1**:17-25.

ASHRAF, S.A., 1968. Occurrence of *Arenicola brasilensis* a polychaete from Karachi. *Pakistan J. Sci.,* **20**: 285-286.

ASHWORTH, J.H., 1904. *Arenicola* (the lugworm). *LMBC Mem.,* **11**:1-18.

AZIZ, N.D., 1938. Fauna of Karachi, 2 - Polychaetes. *Mem. Dept. Zoology, Punjab University, Lahore.*, **2**: 19-52

BHATTI, H. K. AND SOOFI, M., 1949. *Arenicola* a polychaete from Karachi. *Pakistan J. Sci. Res.*, **1**: 76

BINDRA, S.S., 1927. Fauna of Karachi, 1-A Study of the genus *Eurythoe* Family:Amphinomidae). *Mem. Dept. Zool., Punjab Univ., Lahore.,* **1**: 1-18

FAUCHALD, K., 1992. A review of genus *Eunice* (Polycheata: Eunicidae) based upon type material. *Smithsonian Contr.Zool.*, **523**: 422.

FAUVEL, P., 1932. Annelida Polychaeta of the Indian Museum Calcutta. *Mem. Indian Mus.*, **12**: 1 - 262.

FAUVEL, P., 1953. *The fauna of India including Pakistan, Ceylon, Burma and Malaya. Annelida: Polychaeta.* Allahabad, **12**: 1-507.

HABIB, F. AND MUSTAQUIM, J., 1988. New records of polychaete annelids from coastal waters of Pakistan. *Pakistan J. Zool.,* **20**: 304-306.

HAQ, S.M., MOAZZAM, M. AND RIZVI, S.H.N., 1978. Studies on the marine fouling organisms from Karachi coast I. Preliminary studies on the intertidal distribution and ecology of fouling organisms at Paradise Point. *Pakistan J. Zool.,* **10**: 103-115.

HARTMAN, O., 1974. Polychaetous annelids of the Indian Ocean including an account of species collected by members of the International Indian Ocean Expedition 1963-1963,and a catalogue and bibliography of the species from India. Part-2. *J. mar. biol. Ass. India.*, **16**: 540-582.

HASAN, S.A., 1960. Some polychaetes from the Karachi coast. *Annals. Magz.Natur. Hist.,* **3**: 103-112.

JAWED, M. AND KHAN, M.A., 1974. Zonation in the macrofauna inhabiting the mud flats of Baba Island with special reference to *Lingula murphians* King. *J.Sci. Univ. Karachi,* **3**: 78-83.

MOHAMMAD, M.B.M. 1973. New species and records of Polychaete Annelids from Kuwait, Arabian Gulf. *Zool. J. Linn. Soc.*, **52**: 23-44.

MUSTAQUIM, J., 1990. The occurrence of the fan-worm *Amphiglens mediterrana in* Pakistan. *Pakistan J. Zool.,* **22**: 402-403.

MUSTAQUIM,J., l991a. *Harmothoe dictyophors* (Grube)(Polychaeta: Annelida) a new record for Pakistan. *Pakistan J. Zool.*, **23**: 362-363.

MUSTAQUIM, J., l991b. *Polydontes melanonotus* (Grube) (Polychaeta: Polydontidae) a new distributional record for Pakistan. *Pakistan J. Sci. Ind. Res.*, **34**: 40-41.

REHANA, K. AND MUSTAQUIM, J., 1989. New additions to polychaete fauna of Pakistan. *Pakistan J. Zool.,* **21**: 393-394.

SIDDIQUI, N.N. AND MUSTAQUIM, J., 1988. Four new records of nereid worms (Polychaeta:Annelida) from Karachi. *Pakistan J. Zool.*, **20**: 306-309.

SWALEH, R., 1993. Morphological and biochemical systematics of spionids worms (Polycaheta: Annelida) from Karachi coast. M. Phil. Thesis (unpublished). Karachi Univ. 155 pp.

SWALEH, R. AND MUSTAQUIM, J., 1993. New records of *Pseudopolydors* species (Polychaeta: Spionidae) from Pakistan. *Pakistan J. Sci. Ind. Res.*, **36**: 203-204.

TEN-HOVE, H.A., 1990. Description of *Hydroides bulbosus sp. nov.* (Polychaeta:Serpulidae) from the Iranian Gulf, with a terminology for opercula of *Hydroides.* Beaufortia, **41**: 115-120

WESENBERG-LUND, W. 1949. Polychaetes of the Iranian Gulf. *Danish Scientific Investigations in Iran,* 400 pp.

DIVERSITY, DISTRIBUTION AND CLADISTICS OF ANTESTIINES (HETEROPTERA: PENTATOMIDAE) FROM INDO-MALAYAN REGION

IMTIAZ AHMAD AND S. AZHAR HASAN*

Department of Zoology, University of Karachi, Karachi
**Pakistan Museum of Natural History, Garden Avenue, Shakarparian, Islamabad*

Abstract: Stinkbug tribes Pentatomini Leach and Antestiini Stål are distinguished and the latter keyed to its three genera viz *Antestia* Stål *Plautia* Stål and *Neoplautia* Ahmad and Rana with their major distinguishing features highlighted. Ten species of *Antestia* Stål from Indo-Pakistan sub-continent and Malayan subregion are keyed, cladistically analysed with their distributional ranges given and on this basis cladistics of *Antestia* Stål is discussed within the tribes Antestiini Stål and Pentatomini Leach from the above areas.

INTRODUCTION

The representatives of Antestiini Stål and Pentatomini Leach are mostly green coloured shield/stink bugs feeding legumes and vegetables (Ahmad, 1996) and are easily confused with each other. Gross (1976) split the tribe Antestiini Stål into *Antestia* group and *Pentatoma* group, the latter in addition to Pentatomine genera included an antestiine *Plautia* Stål.

Ahmad et. al. (1995) in his taxometric studies analysed seventeen antestiine genera and Ahmad (1996) revised the tribe Pentatomini Leach from Indo-Pakistan subcontinent. Ahmad et al. (1996) taxometrically analysed the pentatomine and antestiine genera confirming his (Ahmed and Rana, 1989) earlier belief contrary to that of Gross (opcit), that *Antestia* and *Plautia* are indeed related.

Presently the two stinkbug tribes Pentatomini and Antestiini (representing Gross is *Antestia* and *Pentatoma* groups) are distinguished with a key and major distinguishing features of the three genera viz *Antestia, Plautia* and *Neoplautia* given. A key to the species of *Antestia* Stål is given from Indo-Pakistan subcontinent and Malayan subregion based upon the studies of Ahmad and Rana (1989) and Hasan (1991). In this light a cladistic analysis of *Antestia* spp. with three antestiine genera viz *Antestia*, *Plautia* and *Neoplautia* and six pentatomine genera from the above areas is also briefly discussed.

PENTATOMINI

Pentatomini Leach, 1815: 120; Ahmed, 1996: in press (Revision, key to six Indo-Pakistani and Malayan and three exotic genera with their cladistics).

Distinguishing features: Body predominently green or pale; scutellum triangular, distinctly longer than broad, with narrow apical lobe; venter of abdomen basally at least with a tubercle or usually armed with a prominent spine reaching or reaching much beyond metacoxae, some times reaching to procoxae; connexiva with acute exposed tips. Spermathecal bulb with usually two finger-like asymmetrical processes, when processes absent, bulb elongate with distal duct leading into an expanded portion before proximal flange or bulb medially constricted or at least with another much larger bulbous portion before proximal flange and distal duct connecting this portion markedly twisted. Paramere irregular, straight, "T"_ or "Y"_ shaped.

Species examined: As shown in Table I.

ANTESTIINI

Antestiini Stål, 1872: 31

Distinguishing features: Body brightly coloured, green with pinkish tinge, sometimes with black spots on head, pronotum, scutellum, hemelytra and on venter all over ; scutellum usually broader than long or when slightly longer than broad, broadly rounded at apex; venter of abdomen basally unarmed, never spinose; connexiva with round concealed tips; spermathecal bulb with "T" or inverted "L" shaped apical finger-like processes, sometimes additional proximal processes present, if processes absent length of bulb distinctly longer than pump region; paramers "F"-, "J" or broadly "C"-shaped.

KEY TO THE GENERA

1. Pronotum with several black spots, basal antennal segments at least reaching or reaching beyond head apex --------------*Antestia* Stål

-. Pronotum without black spots, basal antennal segments much shorter than head apex ----------------------------------- ----------------- 2

2. Pronotum slightly more than 2X broader than long, ostiolar peritreme short and broad, hardly reaching half distance of metapleuron -------------------------------- *Neoplautia* Ahmad and Rana
-. Pronotum distinctly 2.5X or more as broad as long, ostiolar

peritreme much narrow and elongate reaching 3/4 distance of metapleuron -- *Plautia* Stål

Neoplautia

Ahmad and Rana 1981: (description, relationships of new genus and new species *N. pakistanica* Ahmad and Rana).

Type: *N. pakistanica* Ahmad and Rana

Distinguishing features: Body brightly green, dorsum with some pinkish tinge; basal antennal segment much shorter than head apex; labium reaching posterior margin of third abdominal venter; pronotum slightly more than 2X broader than long; scutellum distinctly broader than long; ostiolar peritreme short and broad, hardly reaching 1/2 distance of metapleuron; pygophore distinctly broader than long; parameral blade "f" - shaped; inflated aedeagus with ventral thecal appendages; sprmathecal bulb very small, round with two short tubular processes with remarkably long and broad pump region.

Species examined: As shown in Table 1.

Plautia

Plautia Stål 1865: Ahmad and Rana 1996: 22; Hasan 1988, Hasan and Kitching 1993:61 (revision key, descriptions and cladistics of Indo-Malayan species).

Type: *P. fimbiata* (F.)

Distinguishing features: Body generally brightly green with pinkish tinge; basal antennal segments always shorter than head apex; labium distinctly reaching beyond metacoxae but rarely reaching posterior margin of third abdominal venter; pronotum atleast 2.5X as broad as long; scutellum longer or as long as broad; ostiolar peritreme elongate, reaching 3/4 distance of metapleuron; pygophore as long as or longer than broad; parameral blade with apex remarkably broad with sinuate inner margin; inflated aedeagus without ventral thecal appendages; spermathecal bulb "T" or inverted "L"-shaped, sometimes with processes.

Species examined: As shown in Table I.

Antestia

Antestia Stål 1865: 200; Ahmad and Rana, 1989:349 (description of Indo-Pakistani species), Hasan, 1991: 205 (Description and key to species of Malayan subregion) *Otantestia* Breddin, 1900: 324

Type: *Antestia lyamphata* Kirkaldy

Distinguishing features: Body brightly coloured with large black spots; basal antennal segment as long as or longer than head apex; labium usually not reaching beyond metacoxae; pronotum atleast 2.5x as broad as long; scutellum longer or as long as broad; ostiolar peritreme of variable length; pygophore as long as or longer than broad; parameral blade usually broadly "C"-shaped; inflated aedeagus without ventral thecal appendages; spermathecal bulb usually with symmetrical apical processes, sometimes one process reduced or altogether lost, when bulb without processes bulb longer than pump region and distal duct much longer than combined lengths of bulb and pump region).

Species examined has shown in Table I.

Table I: Species examined

Acrostrnum arabicum Wangner
Acrostrnum bactrianum Kritschenko
Acrostrnum breviceps (Jakovlev)
Acrostrnum graminea (Fabricius)
Acrostrnum heegeri (Fabricius)
Acrostrnum prunasis (Dallas)
Acrostrnum pusniensis Ahmad and Rana
Bathycoelia indica (Dallas)
Chinavia acuta (Dallas)
Chinavia pallidoconspersa (Stål)
Glaucias albomaculatus (Distant)
Glaucias beryllus (Fabricius)
Glaucias dorsalis (Dohrn)
Grazia tincta (Distant)
Nezara antennata Scott
Nezara viridula (Linnaeus)
Pausias ++ (Linnavuorius) leprieuri (Signoret)[x]
Pausias++ (Pausias) martini (Puton)
Pausias (Linnavuorius) pulverosus Linnavuori
Piezodorus guildinii (Westwood)
Piezodorus hybneri (Gmelin)*
Piezodorus lituratus (Fabricius)
Piezodorsus pallescens (Germar)
Piezodorus purus (Stål)
Piezodorus sindellus Ahmad and Rana
Rhaphigaster nebulosa (Poda)
+++*Neoplautia pakistanica* Ahmad and Rana
++*Plautia fimbriata* (F.)
Plautia assamensis Ahmad and Rana
Plautia kaptaiensis Ahmad and Rana
Plautia pictorata Distant

Antestia+++++ anchora (Thunberg)
Antestia cameronica Hasan
Antestia cruciata (F.)
Antestia korinchiensis Hasan
Antestia major Ahmad and Rana
Antestia megaviini Hasan
Antestia partita Hasan
Antestia pulchra (Dallas)
Antestia unicolor Hasan

+Orian (1965), Linnavuori (1972 and 1982) have described and illustrated the pygophore, paramere and spermatheca of most of the Ethiopian and Rolston (1983) has revised the new world taxa.

+Freeman (1940) has revised and illustrated the genus and Linnavuori (1982) the Ethiopian taxa.

++Linnavuori (1982) has illustrated the pygophore, paramere and spermathecae of most of the species and recently Ahmad (1995) has reviewed the world species.

+*Pausias elegans* Vidal was synonymised with *Pausias leprieuri* (Signoret) by Linnavuori (1982)

+++ Ahmad and Rana (1981) described with phylogenetic relationships

++++ Ahmad and Rana (1996) described Indo-Pakistani spp. with cladistics of Antestiini.

+++++ Ahmad and Rana (1989) described the Indo-Pakistani species. Hasan (1991) keyed and described the species of Malayan region.

Key to the species of Antestia

1. Anteocular distance equal to or shorter than remainder of head (Fig. 1f) spermathecal bulb without processes, (Fig. 10f), pygophore longer than broad,·dorsolateral lobes small, conical (Fig. 5h) ------- 2

-. Anteocular distance longer than remainder of head, spermathecal bulb always with finger-like processes, pygophore always broader than long, dorsolateral lobes larger, never conical (Figs 1a, 10a & 5a)-- ------------- 3

2. Anteocular distance shorter than remainder of head (Fig. 1i), pronotum more than 2.5x broader than long (Fig. 2i), scutellum as long as broad (Fig. 3i), peritreme short, spout-like, evaporative area very small (Fig.4i), first gonocoxae rounded at apices (Fig. 9i),

spermathecal bulb remarkably elongate, much longer than pump region (Fig. 10i) ------------------ *pulchra* Dallas (Malayan subregion).

-. Anteocular distance equal to remainder of head, pronotum 2.5x broader than long, (Fig.1f & 2f), scutellum longer than broad (Fig. 3f), peritreme long, pointed (Fig. 4f), evaporative area not as above, first gonocoxae acutely pointed at apices (Fig. 9f), spermathecal bulb ovate, in length about equal to pump region (Fig. 10f) ------------------- ---------------------------------- *major* Ahmad and Rana (Karachi, Sindh).

3. Small species, length about 7.5 mm in male and less than 8.5 mm in female, anterior half of pronotum with four spots or none (Fig. 2d)- ---4.

-. Large species, length about 9.0 mm in male and at least 10.0 mm in female, anterior half of pronotum always with two spots (Fig. 2a)--- -- 6.

4. Anterior half of pronotum without black spots (Fig. 2h), scutellar spots continuous with lateral or anterior margins (Fig. 3h), first gonocoxae with triangular patch continuous with apex, (Fig. 9h), dorsolateral lobes of pygophore triangular (Fig. 6g), vesica thick, stout, reaching dorsal membranous conjunctival appendge (Fig. 8g) -- -- *partita* (Walker) (Indonesia)
-. Anterior half of pronotum with four and posterior half with six black spots, scutellar spots not as above, first gonocoxae without triangular patch as above, dorsolateral lobes of pygophore broadly round, vesica weak, not reaching dorsal membronous conjunctival appendage.--------- -- 5

5. Anterior pronotal spines laterally directed (Fig. 2j), black spots on posterior half of pronotum much narrower and more elongate, rectangular (Fig.2j), lateral angles of scutellum without black markings (Fig. 3j) meso-, and metapleuron adjacent to metathoracic scent auricle without black spots (Fig. 4j), first gonocoxae each with a black spot near apex (Fig. 9j), spermathecal bulb with a large T-shaped process, one limb of "T" slightly smaller (Fig. 10j), parameral blade truncate at apices (Fig. 7i)----------- *unicolor* Hasan (Malaysia).

. Anterior pronotal angles without spines (Fig. 2d) black spots on posterior half of pronotum much wider and more or less oval or globular (Fig. 2d), lateral angles of scutellum with elongate black markings (Fig. 3d), first gonocoxae without black spots as above (Fig. 9d), spermathecal bulb with a long finger-like and a short knob-like process (Fig. 10d), parameral blade acute at apex (Fig. 7d), ------ -------------- *cruciata* (F.) (India, Srilanka, Malaysia and Indonesia).

6. Posterior half of pronotum with six spots, black marking at each basal angle of scutellum representing long curved, distinct pointed

triangle (Fig. 2e & 3e), posterior margin of peritreme more or less entirely convex (Fig.4e), spermathecal bulb with a long curved process, the other one very short, triangular (Fig. 10e), dorsolateral lobe of pygopnore very small, conical, ventral margin medially shallow boat-shaped (Fig.5e & 6e)------------ *korinchiensis* Hasan (Indonesia)

-. Posterior half of pronotum with always four spots, black marking on each basal angle of pronotum never representing a long curved distally pointed triangle, posterior margin of peritreme atleast outwardly concave, spermathecal bulb always with two long finger-like processes, dorsolateral lobe of pygopnore ventrally broadly round or bilobed but never as above, ventral margin medially never shallow boat-shaped as above-- 7

7. Black marking on each basal angle of scutellum dot-like (Fig. 3b), posterior margin of peritreme bilobed (Fig. 4b), spermathecal pump region very stout, longer than the bulb (Fig. 10b), ventromedian margin of pygophore deeply and narrowly concave (Fig. 6b), vesica very short, not at all reaching dorsal membranous conjunctival appendage (Fig. 8b)------------- *arlechino* (Vollenhoven). (Indonesia).

-. Black marking on each basal angle of scutellum larger, never dot-like as above, posterior margin of peritreme usually markedly concave, not as above, spermathecal pump region much shorter and narrower, never as stout or as long as bulb as noted above, ventromedian margin of pygophore variable but never as above, vesica never as short as above, at least reaching dorsal membranous conjunctival appendage--------- -- 8

8. Fused posterior margin of second gonocoxae bilobed (Fig. 9g), ninth paratergites triangular (Fig. 9g), ventrally dorsolateral lobes of pygophore bilobed, ventral margin medially markedly convex (Fig. 6f), vesica thin, slender only reaching apex of dorsal memboranous conjunctival appendage (Fig. 8f) --------- *mcgavini* Hasan (Malaysia).

-. Fused posterior margin of second gonocoxae more or less convex, ninth paratergites broadly round at apices, ventrally dorsolateral lobes of pygophore broadly round, ventral margin medially never as above, vesica long, curved, extending much beyond apex of dorsal membranous conjunctival appendage --------- -------------------- ------- 9

9. Ocelli not surrounded by black spots (Fig. 1a), first gonocoxae conical at apices, fused posterior margins of triangulin broadly round, ninth paratergites much longer than fused posterior margins of eighth paratergites (Fig. 9a), distal spermathecal duct much longer than combined lengths of spermathecal bulb and pump region (Fig. 10a), pygophore with ventroposterior margin medially, only slightly concave (Fig. 6a), parameral blade leaf-like with acute apex, at base with inner

conical projection (Fig. 7a) ---
-------*anchora* (Thunberg). (India, Sikkim, Bangladesh and Malaysia).

-. Ocelli surrounded by black spots (Fig. 1c), first gonocoxae with broadly round apices, fused posterior margin of triangulin conically projected, ninth paratergites as long as fused posterior margin of eighth paratergites (Fig. 9c), distal spermathecal duct much shorter than combined lengths of bulb and pump region (Fig. 10c), ventroposterior margin of pygopnore medially V-shaped (Fig. 6c), parameral blade razor-like with round apex, inner basal projection absent (Fig. 7c)-------------- ------------- *cameronica* Hasan (Malaysia)

ILLUSTRATION OF FIGURES

Figs 1-9 *Antestia* species, aI, *anchora*; aII, same on lower scale showing variation; b, *arlechino*; c, *cameronica*; dI, *cruciata*; dII, same on lower scale showing variation; e, *korinchiensis*; f, *major*, g, *mcgavini*; h, *partita*; i, *pulchra*; j, *unicolor*, 1, head (dorsal view); 2, pronotum (dorsal view); 3, scutellum (dorsal view); 4, metathoracic scent auricle (ventral view); 5, pygophore (dorsal view); 6, same (ventral view showing posterior margin; 7, paramere (inner view); 8, inflated aedeagus (dorsal view); 9, female terminalia (ventral view); 10, spermatheca (dorsal view showing apical bulb, proximal and distal flanges and distal duct.

Male of *major* (figs 5-8) was not available.

Illustrations modified from Ahmad and Rana (1989) and Hasan (1991).

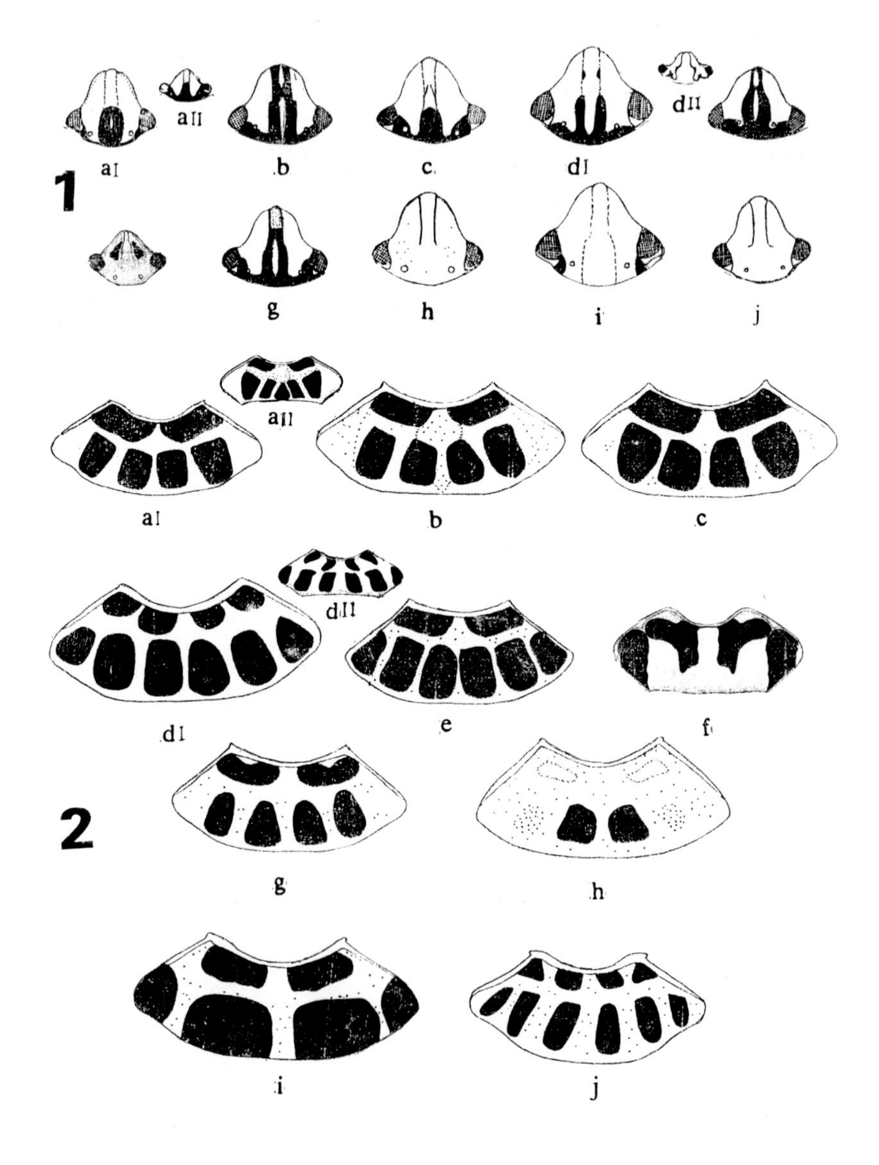
1
aI
aII
b
c
dI
dII
g
h
i
j
aII
aI
b
c
dII
dI
e
f
2
g
h
i
j

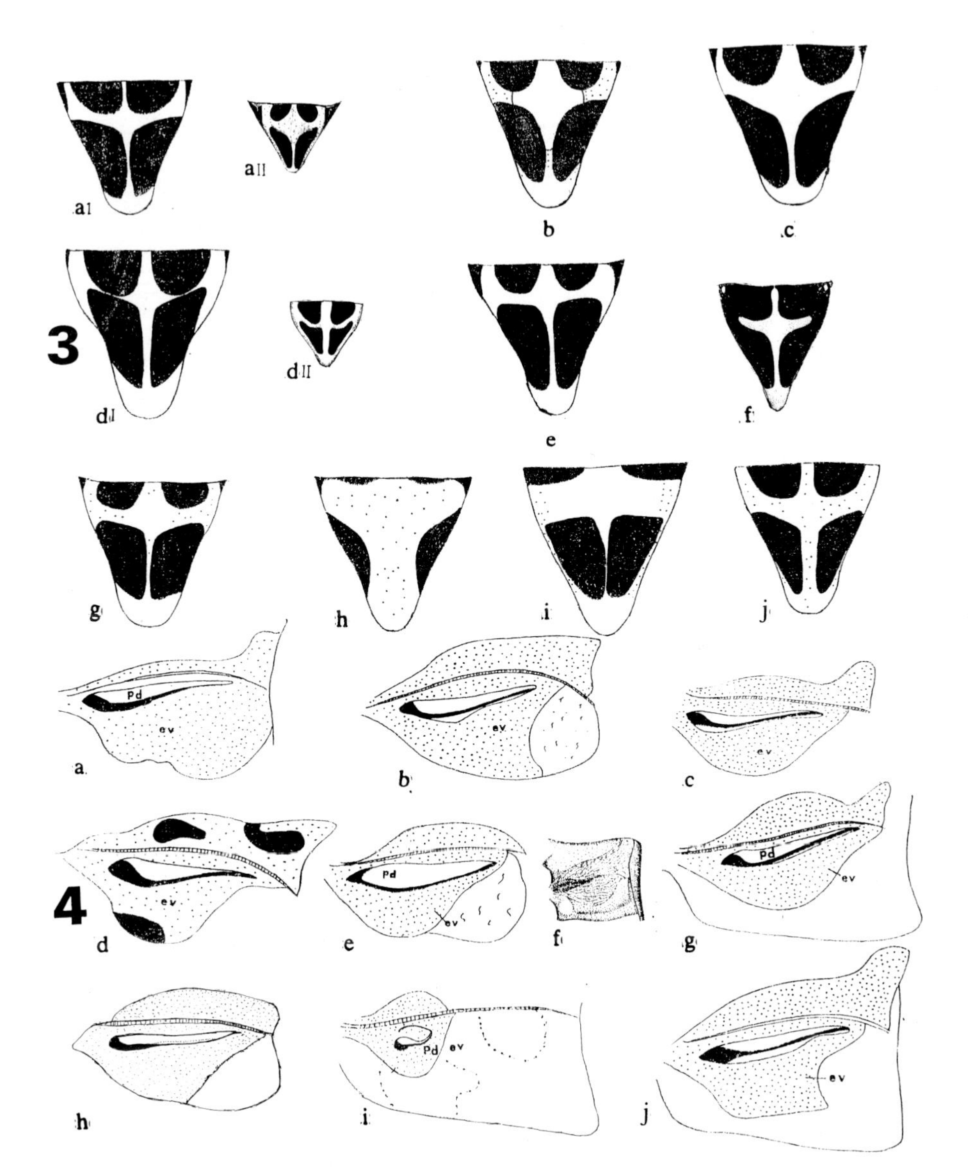

aI
aII
b
c
3
dI
dII
e
f
g
h
i
j
Pd
ev
a
b
c
4
d
e
f
g
h
i
j

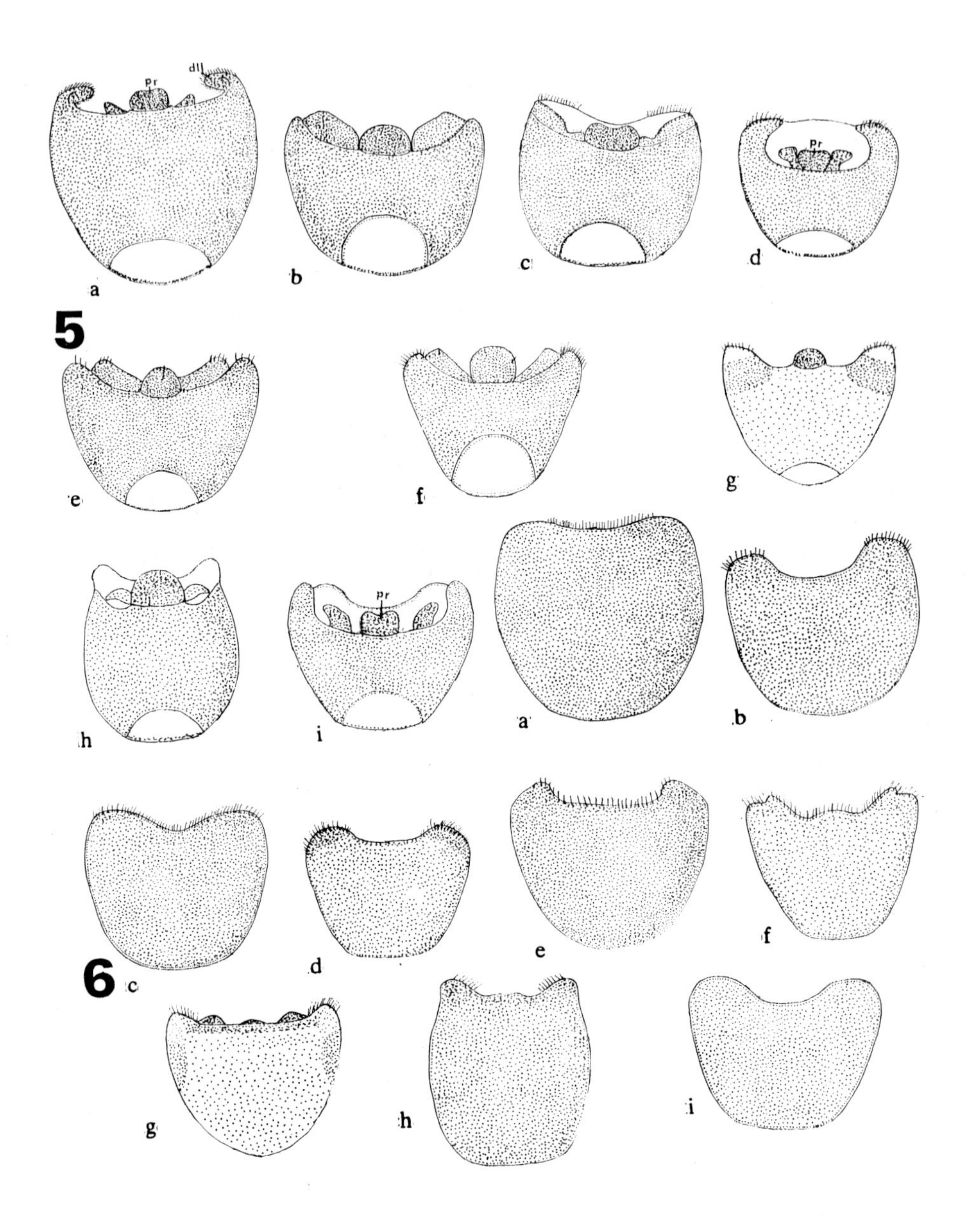
pr
dl
pr
pr
5
6

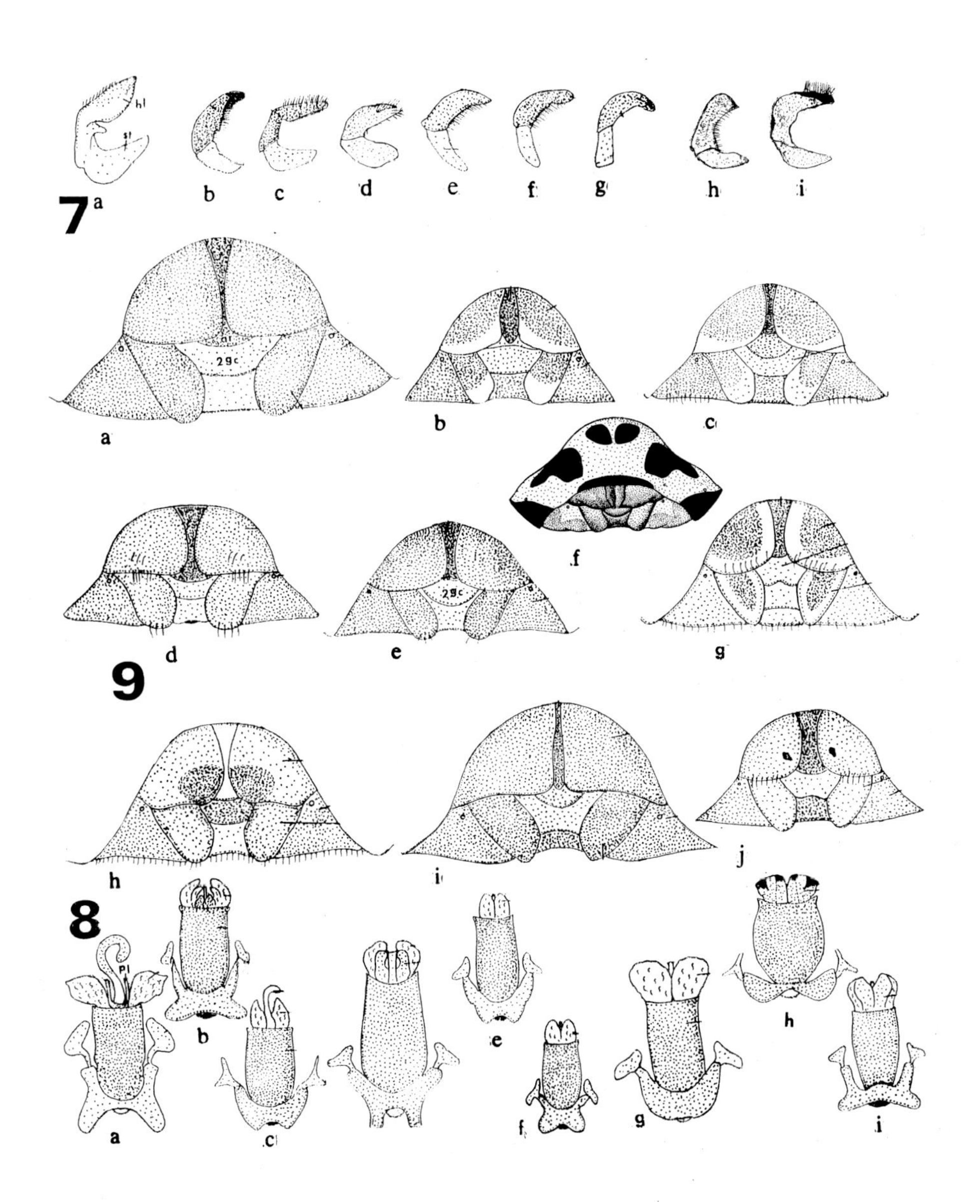

7
a
b
c
d
e
f
g
h
i
hl
st
a
b
c
d
e
f
g
9
h
i
j
8
a
b
c
e
f
g
h
i
Pl
2gc

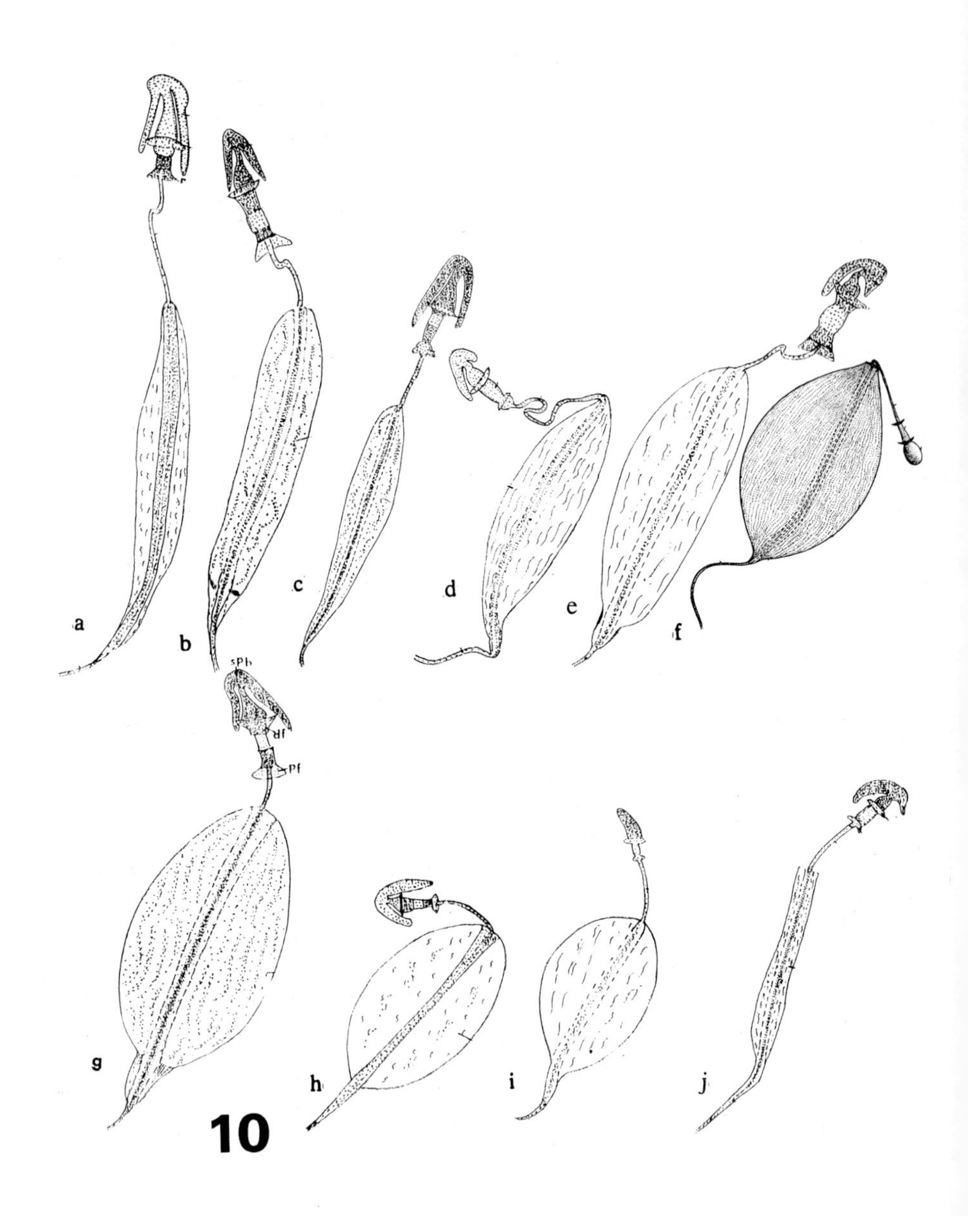
a
b
c
d
e
f
sPh
Rf
Pf
g
h
i
j
10

CLADISTIC ANALYSIS

In all 82 characters with their different states were randomly taken from different parts of the body of both sexes. "a_o", "b_o", "c_o" etc. in the "Table II" indicate plesiomorphic states whereas "a_1", "b_1", "c_1" etc. represent the apomorphic states. "a_1", "a_2", "a_3" indicate the derived, more derived and further derived states of the same trait. Each taxon is defined by its apomorphic state. The characters of antestiine genera and *Antestia* species were compared by their closest allies in their subfamily Pentatominae and indeed in their family Pentatomidae at large. No homoplasies had to be invoked. The cladogram was constructed following the principle of parsimony after Ahmad (1987), Schaefer and Ahmad (1987) and Ahmad (1995 and 1996), Hasan and Kitching (1993).

Table II: Characters and character states in the species of the tribes Antestiini and Pentatomini.

Colouration:

"a_o", body green or pale green.
"a_1", body brilliantly coloured (antestiine species).
"b_o", body without black spots, more or less of uniform colour.
"b_1", body with black spots (Antestia species).
"c_o", ocelli exposed not surrounded by black spots.
"c_1", ocelli surrounded by black spots (*cameronica*).
"d_o", pronotum without black spots.
"d_1", anterior half of pronotum with two black spots (most of presently treated Antestia species except *cruciata, unicolor* and *partita*).
"d_2", anterior half of pronotum with two large and medially closely associated black spots (*korinchiensis, arlechino, mcgaviini, cameronica* and *anchora*).
"d_3", anterior half of pronotum with two and posterior half with four black spots (*arlechino, mcgaviini, cameronica* and *anchora*).
"d_4", anterior half of pronotum with two transversely narrow black spots (*arlechino*).
"d_5", anterior half of pronotum with two transversely bilobed and medially distant black spots (*mcgaviini*).
"d_6", lateral spots on posterior half of pronotum on humeral angles and central spots very large close to spots on anterior half of pronotum (*pulchra and major*).
"d_7", central spots on posterior half of pronotum fused with those on anterior half, of irregular shape (*major*).
"d_8", posterior half with six spots (*korinchiensis*).
"d_9", anterior half with four and posterior half with six spots (*cruciata* subgroup i.e. *cruciata, unicolor* and *partita*).

"d_{10}", spots on anterior and posterior half of pronotum much narrower and elongate, more or less rectangular (*unicolor*).

"d_{11}", spots on anterior half of pronotum lost, posterior half with only two centrally placed spots, all other spots lost (*partita*).

"e_o", basal scutellar angles without markings and scutellar disc without spots.

"e_1", basal scutellar angles with markings and basal scutellar disc with two anteriolateral and two posteriolateral spots (presently treated *Antestia* species).

"e_2", basal scutellar markings and two anteriolateral and two posteriolateral spots on scutellar disc separately and independently placed (all presently treated *Antestia* species except *pulchra* and *major*).

"e_3", basal scutellar angles with markings pointing towards posteriolateral spots on scutellar disc (*kroinchiensis, arlechino, mcgaviini, cameronica* and *anchora*).

"e_4", basal scutellar markings short triangular or narrow, straight, rod-like (*arlechino, mcgaviini, comeronica* and *anchora*).

"e_5", basal scutellar markings long, curved and triangular (*korinchiensis*).

"e_6", basal scutellar markings fused and continuous with anteriolateral spots on scutellar disc (*pulchra* and *major*).

"e_7", basal scutellar markings fused and continuous with not only anteriolateral but also with posteriolateral spots on scutellar disc (*major*).

"e_8", basal scutellar markings very small with anteriolateral spots closely associated with posteriolateral ones, markings and the spots continuous with each other or the markings absent (*partita, cruciata* and *unicolor*).

"e_9", basal scutellar markings entirely absent (*unicolor*).

"e_{10}", anteriolateral spots on the scutellar disc reduced to transversely elongated patch (*partita*).

"f_o", meso-, and metapleura adjacent to scent auricle without spots.

"f_1", meso-, and metapleura adjacent to scent auricle with spots (*cruciata*).

"g_o", seventh abdominal venter without spots.

"g_1", seventh abdominal venter with several spots (*major*).

"h_o", first gonocoxae without spots.

"h_1", first gonocoxae with a dot at its apex (*unicolor*).

"i_o", first gonocoxae without pigmented patch.

"i_1", first gonocoxae each with a triangular pigmented patch at its apex (*partita*).

Size:

"j_o", size moderate, 9.0-10.0 mm long.
"J_1", size reduced, male 7.5 mm, female 8.5 mm (*partita, cruciata and unicolor*).

Head and its appendages:

"k_o", Length anterocular distance and remainder of head equal.
"k_1", length anteocular distance distinctly more than remainder of head (all the treated *Antestia* species except *pulchra* and *major*).
"k_2", anteocular distance reduced subequal (to or shorter than remainder of head (*pulchra* and *major*).
"k_3", anteocular distance distinctly shorter than remainder of head (*pulchra*).
"l_o", basal antennal segment approaching head apex.
"l_1", basal antennal segment at least reaching or reaching beyond head apex (Most of the presently treated *Antestia* species).
"l_2", basal antennal segment reduced in length, much shorter than head apex (*Neoplautia* and *Plautia* species).
"m_o", labium of moderate size, reaching or reaching slightly beyond mesocoxae
"m_1", labium long, always passing distinctly beyond metacoxae (presently treated *Neoplautia* and *Plautia* species).
"m_2", labium at least reaching posterior margin of third abdominal venter (*Neoplautia pakistanica*).
"m_3", labium usually reduced, restricted to metacoxae (most of the treated *Antestia* species).

Thorax:

"n_o", anterior pronotal angles unspinose.
"n_1", anterior pronotal angles at least slightly produced or conical (most of the treated *Antestia* species).
"n_2", anterior pronotal angles spinose, laterally directed (*unicolor*).
, scutellum longer than broad with narrow apex.
"o_1", scutellum as long as or broader than long with much broader apical lobe (most of the presently treated antestiine species).
"p_o", metathoracic ostiolar peritreme developed, reaching 0.5x distance of metapleuron.
"p_1", metathoracic ostiolar peritreme much narrow and elongate reaching at least 0.75x distance of metapleuron (presently treated *Plautia* species).
"p_2", posterior margin of peritreme markedly concave (*anchora, cameronica* and *mcgaviini*).
"p_3", posterior margin of ostiolar peritreme bilobed (*arlechino*).

"p_4", peritreme reduced not quite reaching 0.5x distance of metapleuron (*Neoplautia pakistanica*).
"p_5", peritreme further reduced not reaching 0.5x distance of metapleuron (*major* and *pulchra*).
"p_6", peritreme remarkably reduced, spout-like, with evaporatoria drastically reduced (*pulchra*).

Abdomen:

"q_o", abdomen basally unspinose.
"q_1", basal abdominal spine on 3rd abdominal venter developed (pentatomine species).
"q_2", basal abdominal spine lost, sometimes with a tubercle (antestiine species).
"r_o", connexival tips round, unspinose.
"r_1", connexival tips markedly pointed, spinose (pentatomine species).

Male genitalia:

"s_o", pygophore as long as broad.
"s_1", pygophore longer than broad (*pulchra, major*).
"s_2", pygophore at least slightly broader than long (*Antestia* species except *pulchra* and *major*).
"s_3", pygophore much broader than long (*Neoplautia pakistanica*).
"t_o", dorsolateral lobes of pygophore round.
"t_1", dorsolateral lobes of pygophore bilobed (*mcgavini*).
"t_2", dorsolateral lobes small, conical (*partita).*
"t_3", dorsolateral lobes very small, conical (*korichiensis*).
"t_4", dorsolateral lobes reduced triangular (*pulchra* and *major*).
"u_o", paramere without inner lobe at base of blade.
"u_1", paramere with a prominent inner lobe at base of blade (antestiine species).
"u_2", inner lobe at base of parameral blade reduced or absent (presently treated *Antestia* species).
"u_3", inner lobe at base of parameral blade reduced, conical.
"v_o", apex of parameral blade narrow, round.
"v_1", apex of parameral blade remarkably broad (*Plautia* species).
"v_2", apex of parameral blade leaf-like (*anchora*).
"w_o", vesica slender approaching dorsal membranous conjunctival appendage.
"w_1", vesica stout, reaching dorsal membranous conjunctival appendage (*partita*).
"w_2", vesica long, reaching much beyond dorsal membranous conjunctival appendage (*cameronica* and *anchora*).
"w_3", vesica short not reaching dorsal membranous conjunctival appendage (*pulchra* and *major*).

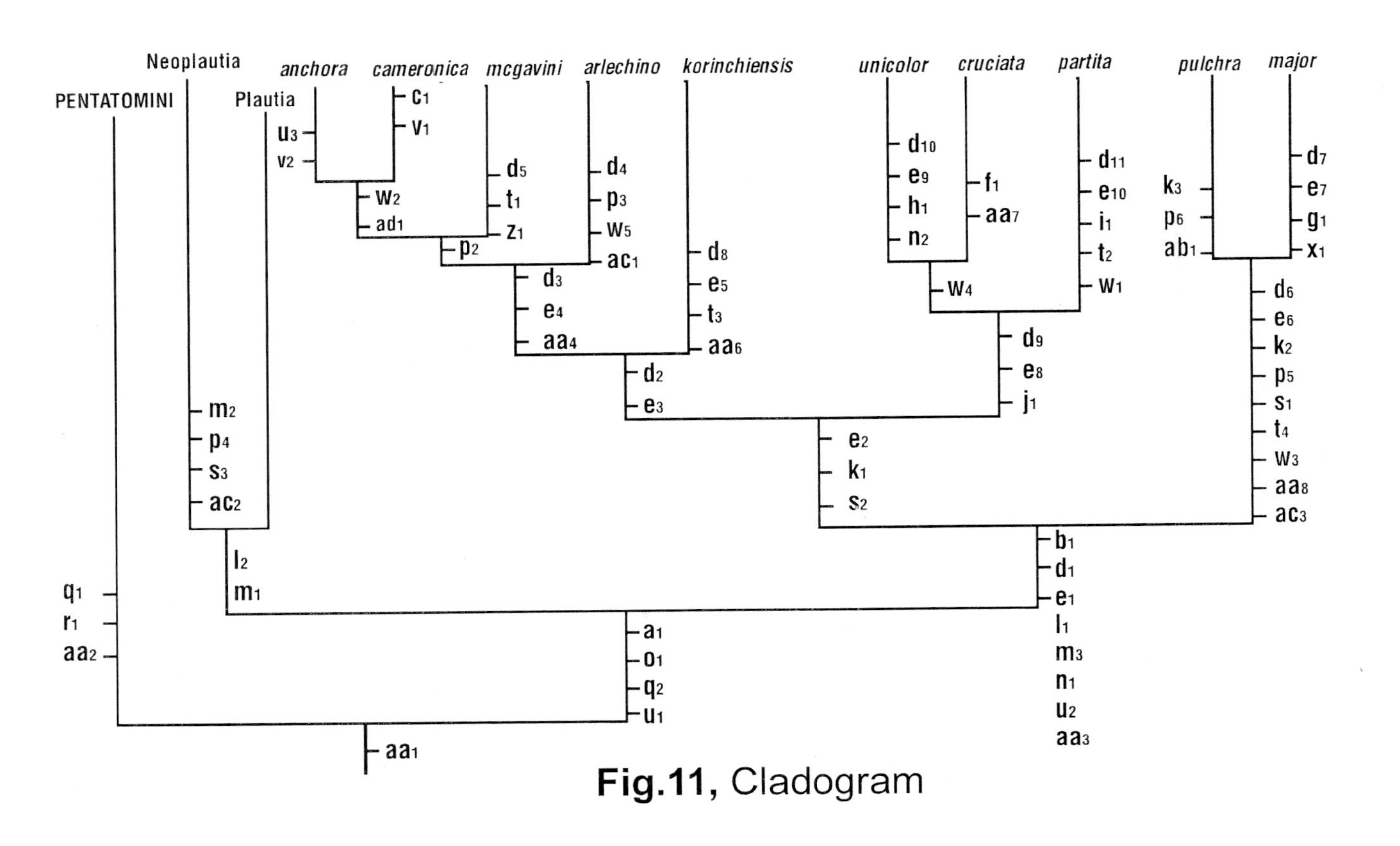

Fig.11, Cladogram

"w_4", vesica reduced not reaching dorsal membranous conjunctival appendage (*cruciata* and *unicolor*).
"w_5", vesica very short not at all reaching dorsal membranous conjunctival appendage (*arlechino*).

Female genitalia:

"x_o", first gonocoxae round at apices.
"x_1", first gonocoxae acutely pointed (*major*).
"y_o", fused posterior margin of triangulin, straight.
"Y_1", fused posterior margin of triangulin conically projected (*cameronica*).
"z_o", fused posterior margin of second gonocoxae straight.
"z_1, fused posterior margin of second gonocoxae bilobed (*mcgavini).*
"aa_o", spermathecal bulb round.
"aa_1", spermathecal bulb elongately ovate (antestiine and pentatomine species).
"aa_2", constriction in the bulb present (pentatomine species).
"aa_3", processes on the bulb usually present (antestiine species).
"aa_4", finger-like processes more or less symmetrical, long (*arlechino, mcgavini, cameronica* and *anchora*).
"aa_5", proximal portion of spermathecal bulb with more processes (*Plautia* species).
"aa_6", one of the two finger-like processes reduced to a triangular lobe (*korinchiensis*).
"aa_7", one of the two finger-like processes reduced into a knob (*cruciata*).
"aa_8", processes on spermathecal bulb lost (*majora* and *pulchra*).
"ab_o", bulb globular or ovate.
"ab_1", bulb much elongate (*pulchra*).
"ac_o", pump region slender, shorter than length of bulb.
"ac_1", pump region stout, longer than bulb (*arlechino*).
"ac_2", pump region long and broad, much longer than distal spermathecal duct (*Neoplutia pakistanica*).
"ac_3", pump region reduced (*majora* and *pulchra*)
"ad_o", distal spermathecal duct of moderate size.
"ad_1", distal duct much longer than bulb and pump region combined (*anchora* and *cameronica*).

DISCUSSION ON THE CLADOGRAM

The group of nine pentatomine genera including six from Indo- Pakistan subcontinent and Malayan subregion cladistically treated by the present first author (1996) in his review of the legume bug tribe Pentatomini appears to be outgroup of the presently cladistically treated tribe Antestiini. Indeed the present cladogram (Fig. 11) confirms the earlier belief of Ahmad and Rana (1981 and 1996), Ahmad et al. (1995 and 1996) and of Ahmad and Kamaluddin (1996) that Neoplautia and Plautia are remarkably closely related and are sistergroups. However the two are neatly defined by their autapomorphies as shown in the present cladogram.

Contrary to Gross (1976) and in accordance with Ahmad et al. (1995, 1996), Ahmad and Kamaluddin (1995), Ahmad and Rana (1996) and finally Ahmad (1996), *Antestia* and *Plautia* are indeed closely related as is confirmed by the present cladogram and the two do not fall into two different tribes Pentatomini and Antestiini (*Pentatoma* and *Antestia* groups of Gross, op. cit.).

In the species of *Antestia* from Indo-Pakistan subcontinent and from Malayan subregion *major* and *pulchra* appear to be most closely related playing sistergroup relationship with each other and outgroup relationship with the rest of the treated *Antestia* species. The remaining *Antestia* species appear to fall into two distinct subgroups i.e. *anchora* subgroup and *cruciata* subgroup playing sistergroup and outgroup relationships with each other.

In the *cruciata* subgroup *partita* appears to play outgroup relationship with the two sistergroup species *cruciata* and *unicolor.*

In the *anchora* subgroup *anchora* and *cameronica* appear to be the most closely related species of which *mcgavini* is their outgroup. Similarly *arlechino* plays outgroup relationship with this group of species and *korinchiensis* plays the outgroup relationship with the latter larger subgroup comprising *anchora, cameronica, mcgaviini* and *arlechino.*

REFERENCES

AHMAD, I. 1987. A cladistic analysis of Tricentric (Homoptera: Auchenorrhyncha. Membracidae) with a note on their origin, distribution and food plants. *Proc. 6th Auchen. Meeting, Turin,* Italy.

AHMAD, I. 1994. A review of taxa of pentatomine genus Pausias Jakovlev (Hemiptera: Pentatomidae. *Pakistan j. entomol. Karachi* **9**: 29-42.

AHMAD, I. 1995. A review of pentatomine legume bug genus Piezodorus Fieber (Hemiptera:Pentatomidae: Pentatominae) with its cladistic analysis. *Proc. Pakistan Congr. Zool.* **15**: 329-358.

AHMAD, I. 1996. A revision of the green stinkbug tribe Pentatomini Leach (Hemiptera: Pentatomidae: Pentatominae) from

Indo-Pakistan subcontinent attacking cereals and vegetables with special reference to their cladistics. *Proc. Pakistan Congr. Zool.* **16**: in press.

AHMAD, I. AND KAMALUDDIN, S. 1996. A cladistic analysis of the seventeen genera of the tribe Antestiini Stål (Hemiptera: Pentatomidae: Pentatominae) from Oriental and Ethopian regions. *Proc. Pakistan Congr. Zool.* **16:** in press.

AHMAD, I. AND RANA, N.A. 1981. A new genus and a new species of Antestiini (Pentatomidae: Pentatominae) from Pakistan and their relationships. *Pakistan J. Zool.* **13** (1&2): 45-50.

AHMAD, I. AND RANA, N.A. 1989. The Indo-Pakistani species of *Antestia* (Heteroptera:Pentatomidae). *Mitt.Schweiz.Entomol. Ges.* **62:** 349-358.

AHMAD, I. AND RANA, N.A. 1996. A review of Antestiini from Indo-Pakistan subcontinent with a revision of the genus Plautia Stål (Hemiptera: Pentatomidae: Pentatomini) with their cladistics. *Pakistan j. entomol. Karachi* **11**(1&2): in press.

AHMAD,I., SHAUKAT, S.S. AND KAMALUDDIN, S.1995.Taxometric studies on antestiine genera (Hemiptera: Pentatomidae: Pentatominae). *Proc. Pakistan Congr. Zool.* **16:**

AHMAD,I., SHAUKAT, S.S.AND KAMALUDDIN,S.1996.Taxometric studies on pentatomine genera of Indo-Pakistan subcontinent along with three most closely related exotic genera and four genera of related groups. *Proc. Pakistan Congr. Zool.* **16**: in press.

FREEMAN, P. 1940. A contribution to the study of the genus Nezara Amyot and Serville (Hemiptera: Pentatomidae). *Transactions of the Royal Ent. Soc. Lond.* **90:** 351-374.

GROSS, E.G. 1976. *Plant-feeding and other bugs (Hemiptera) of South Australia, Heteroptera. Part II.* A.B. James, Government Printer, South Australia.

HASAN, S.A. 1988. *The morphology and cladistics of the Pentatomidae, with a revision of three tribes from the Malayan subregion.* Ph.D. Thesis, University of Oxford, Oxford, U.K.

HASAN, S.A. 1991. A revision of the genus Antestia Stål (Heteroptera: Pentatomidae) from the Malayan subregion with description of four new species. *Annot. zool. bot. Bratislava* **205**: 1-23.

HASAN, S.A. AND KITCHING, I.J. 1993. A cladistic analysis of the tribes of the Pentatomidae (Heteroptera). *Japan J. Ent.* **61**(4): 651-669.

LEACH, W.E. 1815. *Hemiptera in Brewster (S.D.).* The Edinburgh Encyclopedia conducted by S.D. Brewster, pp. 120-124.

LINNAVUORI, R.E. 1982. Pentatomidae and Acanthosomatidae of Nigeria and the Ivory coast with remarks on species of the adjacent countries in West and Central Africa. *Acta Zool. Fenn.* **163**: 1-176.

SCHAEFER, C.W. AND AHMAD, I. 1987. A cladistic analysis of the genera of the Lestonocorini (Hemiptera: Pentatomidae: Pentatominae). Proc. Entomol. Soc. Wash. 89 (3): 444-447.

STÅL, C. 1865. *Hemiptera Africana* 1-3: 1-200. Heteroptera, Stockholm.

STÅL, C. 1872. *Enumeratio Hemipterorum 2 Kongl.* Svensta Vet. Akad. Handl. 1&4: 1-159.

THE BIOGEOGRAPHY OF THE LAND SNAILS OF PAKISTAN

KURT AUFFENBERG

Division of Malacology, Florida Museum of Natural History, University of Florida, Gainesville, Fl 32611, U.S.A.

Abstract: Recent surveys of the land snails of Pakistan have increased the known fauna from 50 species to almost 100. Our current knowledge of land snails is still very preliminary, however, and in Pakistan many additional taxa remain to be collected and described. Several areas of rich biodiversity are especially apparent, among them the Ziarat area of Balochistan. The high level of species richness of land snails in Ziarat and the western ranges of Pakistan is similar to the high levels of biodiversity and endemism found in small mammals of the same zone.

INTRODUCTION

The land snail fauna of Pakistan is one of the least known in Asia. Although some of the earliest well-localized species of Asian land snails were originally described from Pakistan (Hutton, 1849), the fauna has remained poorly collected until recently. Early land snail collecting in Pakistan was undertaken by members of the British colonial forces and staff members of the Indian Museum, Calcutta. Historically, there has been a relatively high degree of difficulty in travel due to both the harsh terrain and adamant tribal land protection. Much of the country is inaccessible except by foot and large portions of some provinces remain off-limits to foreign travelers. Biodiversity surveys conducted jointly by staff members of the Florida Museum of Natural History and the Zoological Survey Department of Pakistan during 1989-1994, representing approximately 350 field stations from throughout much of the country, have added greatly to our understanding of the diversity and distribution of terrestrial mollusks in this geologically complex region. Previously, only about 50 species (11 families) of land snails had been recorded from Pakistan. Most of these early records were from single collections or unlocalized areas such as 'Sind'. These recent surveys have increased the known fauna to 96 species, representing 23 families (Table I). Many of the unrecorded species were known extralimitally,

Fig. 1 The Biogeographic subprovince of Pakistan based on land snail distributions

while others represent undescribed taxa. The preliminary biogeographic analysis presented below may require revision vision after a thorough systematic examination of these specimens is complete.

Biogeographic Provinces in Pakistan

Historical Perspective - Previous discussions of the biogeographic provinces in Pakistan have been largely anecdotal (Annandale and Rao, 1923; Hutton, 1849; Nevill, 1878; Theobald, 1881; Woodward, 1856). This was due mainly to the general lack of available field material on which to base tenable hypotheses. Although a few collections were made along trade routes and at hill stations throughout northwestern India, most of the land snail collections in Pakistan were scattered and paltry in comparison. Capt. Thomas Hutton (1849) described the first well-localized land snails known from Pakistan which he collected at the western terminus of the Bolan Pass, Balochistan "during the advance of the Army of the Indus into Afghanisthan in 1839". In this report he stated that some of the species collected were also encountered earlier in western India (Hutton, 1834). Ferdinand Stoliczka added greatly to our knowledge of the high elevation fauna in northeastern Pakistan through comprehensive collections made during the Yarkand Expeditions. Unfortunately, he succumbed to the effects of altitude sickness at Karakoram Pass during his return trip to India (Godwin-Austen, 1900). Stoliczka's observations and a taxonomic review of his collections were presented by Nevill (1878). Later work by Theobald (1881) in the Muree area more clearly elucidated the geographical limits of the Himalayan land snail fauna in Pakistan. Blanford and Godwin-Austen (1908), Godwin-Austen (1882 - 1920) and Gude (1914) made numerous contributions to the understanding of the biogeography of land snails in southern Asia, including Pakistan. Annandale and Rao (1923) discussed the land snail fauna of the Salt Range. The biogeographical provinces of the entire Indomalayan region, including those of Pakistan, was provided by Udvardy (1985).

Table I. List of the land snails of Pakistan and their occurrence by biogeographic subprovince. The few species reported from Lower Chitral is probably an artifact of inadequate sampling. * = possible introduction

Taxon	Thar Desert	Ziarat Highlands	Himalayan Foothills	Lower Chitral	Alpine	Pamir-Karakoram
Carychiidae						
Carychium sp. A					X	
Cochlicopidae						
Cochlicopa lubrica						X
Pupillidae						
Pupilla muscorum		X			X	X
Pupilla signata		X				
Pupoides coenopictus	X		X			
Pupoides sp. A		X	X			
Valloniidae						
Vallonia asiatica		X				
Vallonia ladacensis		X			X	X
Vallonia pulchella		X*			X	X
Vertiginidae						
Boysia boysii	X					
Gastrocopta huttoniana		X	X		X	X
Gastrocopta sp. A	X					
Gastrocopta sp. B			X			
Truncatellina sp. A		X				
Truncatellina sp. B			X			
Truncatellina sp. C					X	X
Vertigo antivertigo					X	
Vertiginidae undet.			X			
Pyramidulidae						
Pyramidula humilis			X			
Chondrinidae						
Granaria lapidaria		X				

Taxon	Thar Desert	Ziarat Highlands	Himalayan Foothills	Lower Chitral	Alpine	Pamir-Karakoram
Enidae						
Jaminia continens		X				
Pseudonapaeus sp. A				X		
Subzebrinus beddomeana			X			
Subzebrinus candelaris			X			
Subzebrinus coelebs			X			
Subzebrinus coelocentrus			X			
Subzebrinus dextrosinister			X			
Subzebrinus dominus			X			
Subzebrinus eremita		X				
Subzebrinus hazarica			X			
Subzebrinus kayberensis			X			
Subzebrinus longstaffi			X			
Subzebrinus mainwaringiana			X			
Subzebrinus nevilliana			X			
Subzebrinus pretiosa			X			
Subzebrinus rufistrigata			X			
Subzebrinus salsicola			X			
Subzebrinus sindhicus			X			
Subzebrinus smithei			X			
Subzebrinus tandianiensis			X			
Subzebrinus tipperi				X		
Subzebrinus spp. A - J			X			
Styloptychus sp. A				X		
Clausiliidae						
Cylindrophaedusa cylindrica			X			
Hemiphaedusa farooqi			X			
Hemiphaedusa waageni			X			
Ferussaciidae						

Taxon	Thar Desert	Ziarat Highlands	Himalayan Foothills	Lower Chitral	Alpine	Pamir-Karakoram
Caeciliodes spp. A - B	X		X			
Coilostele scalaris		X	X			
Subulinidae						
Allopeas gracile	X*					
Zootecus insularis	X					
Streptaxidae						
Gulella bicolor	X*					
Punctidae						
Punctum pygmaeum						X
Toltecia sp. A			X			
Succineidae						
Succinea sp. A	X					
Succinea sp. B			X			
Helicarionidae						
Bensonies jacquemonti			X			
Bensonies monticola			X			
Bensonies wynnii			X			
Euaustenia monticola			X			
Euaustenia sp. A					X	
Euconulus fulvus					X	X
Girasia sp. A			X			
Kaliella bullula			X			
Kaliella fastigiata			X			
Kaliella nana			X			
Khasiella hyba			X			
Macrochlamys vesicula			X			
Parvatella flemingi			X			
Rhadella tandianensis			X			
Syama promiscua			X			

Taxon	Thar Desert	Ziarat Highlands	Himalayan Foothills	Lower Chitral	Alpine	Pamir-Karakoram
Syama prona			X			
Syama theobaldi			X			
Vitrinidae						
Vitrina pellucida		X			X	
Zonitidae						
Hawaiia sp. A		X			X	X
Oxychilus sp. A		X				X
Zonitoides nitidus						X
Parmacellidae						
Parmacella rutellum			X			
Limacidae						
Deroceras laeve	X*		X		X	X
Hygromiidae						
Euomphalia bactriana		X	X			
Helicellidae						
Monacha sp. A			X			
Xeropicta candaharica			X			
Bradybaenidae						
Cathaica chitralensis				X		
Cathaica phaeozona					X	
Cathaica sp. A						X
Arionidae						
Anadenus altivagus			X			
Anadenus lahorensis	X*					
Total Number of Species = 96	11	16	64	4	13	12

Taxon	Thar Desert	Ziarat Highlands	Himalayan Foothills	Lower Chitral	Alpine	Pamir-Karakoram
Restricted to one subprovince in Pakistan, may also be extralimital, N (% subprovincial fauna)	7(64)	6(38)	55(86)	4(100)	5(39)	4(33)
Endemic, N(%subprovincial fauna	2(1)	(13)	19(30)	4(100)	(15)	1(8)

Blanford (1870, 1901) developed a scheme of the biogeography of India, including much of present-day Pakistan, based on observations of vertebrate and invertebrate distributions. Prashad (1942) reviewed and added to Blanford's original discussion. Prashad (1942) divided much of the land mass now in Pakistan into two subregions, the Western Frontier Territory and the Himalayas. Although the exact geographical limits of the Western Frontier Territory was not given, Prashad agreed with Blanford (1870) that this area should be associated with the Mediterranean subregion of the Palaearctic. Prashad informally subdivided the Himalayas altitudinally based on a distinct faunal change occurring at tree line. The lower elevations were associated with the Mediterranean Palaearctic, while the higher elevations, including intervening forested valleys, were associated with the Central Asian Palaearctic fauna. Prashad further subdivided the lower elevations into a western portion related to the Central Asian Palaearctic fauna and an eastern zone more closely associated with the Indo-Malayan fauna. Blanford and Prashad were unable to clearly delineate these regions because of the severe dirth of adequately localized material available for their examination. Although it is clear, based on the specimens collected during our recent surveys, that this overly simplified scheme must be revised, their observations were remarkably accurate. Reviews of the biogeography of India, including much of Pakistan, were presented by Mani (1974a, b) and Udvardy (1985).

The biogeographic subprovinces of Pakistan are relatively clear with little overlap (Fig. 1), particularly those at low elevations. The subprovinces in the northern one third of the country become more mosaic in character with the primary distinction being elevation. A brief overview of the physical geography, tectonic activity, geology and climate of the country is necessary for the following discussion.

Physical Geography

Pakistan (Fig. 1) ranges from the Arabian Sea in the south to some of the highest mountains in the world along the borders with Afghanistan and China. The lowlands consist mostly of the lower Indus River valley and its' numerous tributaries, most notably the Jhelem, Chenab, Ravi, Sutlej and Beas rivers which join the Indus in Punjab Province. This area, comprising about one third of the land mass of Pakistan, is generally subdesertic or desertic and *Acacia* savanna. A lowland desertic region also extends from the Kirthar Mountains west to the Iranian border. The western uplands of Balochistan Province is closely associated with the Iranian Plateau and adjacent physiographic features in southern Afghanistan. This area consists of deserts and several isolated mountain ranges. An extremely important zoogeographic feature of western Pakistan is a fan-like series of mountain ranges, some of which attain relatively high elevations. These include the Bruhai, Sulaiman and Kirthar Mountains. This series of ridges extends south and then west to form the Makran Ranges and others which are generally lower in altitude. The zoogeographically important Salt Range is separated from the main mountain ranges of the west by the Indus River. The Potwar Plateau connects the Salt Range to the Himalayan Foothills to the north. The Himalayan Mountains enter Pakistan from the southeast and terminate at Nanga Parbat. The associated Pir Panjal Mountains extends through the Murree Hills to the Indus River. The lower elevations of the Pir Panjal Mountains and the series of ranges extending to the west of the Indus River comprises another of the most important areas in Pakistan in regards to land snail diversity. This region is dissected by several major tributaries of the Kabul River, including the Swat and the Panjkora Rivers, forming serial linear mountain ranges with relatively high altitudes separated by valleys of comparatively low elevation. The complex highlands of Chitral, Gilgit, Skardu and the Deosai Plateau provide varied habitats between the foothills of the Himalayas and mountain ranges further to the north. The northwestern and northeastern borders of Pakistan are guarded by the massive Hindu Kush and Karakoram Mountain

Ranges, respectively. These two mountain ranges contain most of the highest mountains in Asia and form an effective boundary between the steppes of Central Asia and the Indus River valley.

Tectonic Activity - Much of the physical geography of Pakistan is a direct result of the Indian Plate - Eurasian Plate collision about 50 - 55 million years ago (Treloar and Coward, 1992). This major tectonic event had several major pulses, uplifting sediments throughout the Central Asian region (Krishnan, 1974). The collision forced the closure of the Tethys Sea in the region (Krishnan, 1974) and the rapid uplifting of the Tibetan Plateau and Himalayan Mountains (Patriat and Achache, 1984). The 'twisting' of the Indian Plate to the northwest along the northwest Indian syntaxes caused sediments to be 'squeezed out' to the west and southwest (Treloar and Coward, 1991). The uplifting caused by these 'excess' sediments formed the Salt Range, Sulaiman and Kirthar Mountains and other associated ranges in southwest and south central Pakistan (Jadoon, et al., 1994). This entire mountain range complex was referred to as the Balochistan Arc and the uplift was partly attributed to a deeply situated 'Kashmir Wedge' of the Indian Plate by Krishnan (1974).

The remarkable uplifting caused by the collision of the Indian and Eurasian Plates drastically altered the regional and local monsoonal patterns (Mani, 1974c). Major watersheds such as those of the Indus, Ganges and Brahmaputra were greatly affected by the uplifting and unroofing of the Himalayas and associated mountain ranges, however debate continues concerning the degree of influence (see Mani, 1974d for discussion).

Geology - Due to the extensive tectonic activity in this region, geological features are quite complex and poorly understood. However, a very simplified discussion is provided only as a guide to explain observed land snail distributions and diversity presented below. Most of the eastern portion of Pakistan south of the Himalayan Foothills is made up of alluvial deposits. The northern portions of the country are metamorphic and intrusive in origin. The Himalayan Foothills and the western mountain ranges, including those fanning out to the south and west into western Balochistan, are composed mostly of Mesozoic and Tertiary sedimentary deposits. Ophiolites are also present on these ranges. Portions of the Lower Swat District are comprised of Paleozoic and Pre-Cambrian sedimentary deposits (Krishnan, 1974). Land snails generally prefer substrates of sedimentary (usually limestones) origin. The limestone outcropping and calcareous soils of portions of the Himalayan Foothills provide ideal land snail habitat.

Climate - Very few places on earth exhibit the range of temperatures and climatic conditions present in Pakistan. Air temperature in the south may exceed 50° C, while a relatively short distance to the north lies a region with possibly the greatest number of glaciers outside Antarctica. Most of the country has a tropical to subtropical continental climate. The cooler subtropical to temperate climates of the Himalayan Foothills are optimal for land snail diversity. The high mountains of the north and west have cooler temperate and/or arctic climates. Rainfall is greatest during the southwest monsoon (June - September, depending on locality). Annual rainfall varies greatly from the south (100 - 200 mm) to the Himalayan Foothills (500 - 1000 mm). Rainfall in the far north is once again low (100 - 200 mm) (Mani, 1974c; Ramdas, 1974). Although daily temperature variation may be high and precipitation is quite seasonal, the Himalayan Foothills provide the most optimal land snail habitats in Pakistan.

The Subprovinces - In the following discussion of the biogeographic subprovinces of Pakistan, taxonomic problems have been largely simplified. For instance, *Subzebrinus* (s. l.) is utilized instead of the numerous enid genera deployed by Schileyko (1984) and Muratov (1992), generally based on anatomical characters, which have not yet been examined in the Pakistan material. The detailed examination of these problematic taxa may provide a higher degree of resolution, but the general biogeographic trends outlined below are not expected to be greatly altered. It should also be noted that large areas of the country were not sampled during these surveys. Most of Balochistan Province remains unexplored. Only a few highly localized areas were sampled, i.e. Quetta, Ziarat, Khuzdar and the coastal area from the Hab River to Gwadar. Undoubtedly, many exciting discoveries await in the isolated mountain ranges of southwestern and eastern Balochistan. The higher elevations of the northern provinces are also poorly sampled. Many of these areas can be reached only on foot or the roads are but of marginal utility. Records from northern Chitral are almost entirely lacking, while those from the uplands of Gilgit, Skardu and the Deosai Plateau are scant.

Six biogeographic subprovinces can be recognized in Pakistan based on present land snail distributions (Fig. 1). Additionally, large desertic and glacial areas, as well as the high elevation barren mountains of the north, appear to be devoid of mollusks. Highly eroded land snail shells (*Zootecus* sp.) are found occasionally in blowing sand desert areas, but these probably indicate former subdesertic populations overrun by advancing sand dunes.

1) **The Thar Desert Subprovince** (0 - 250 m) extends in a narrow band east from the Iranian border to the western Indian border and northward in the Sindh and Punjab Provinces to the Salt Range and the southern foothills of the Himalayas. Diversity is extremely low.Taxa such as *Zootecus insularis* (Ehrenberg, 1831) and *Pupoides coenopictus* (Hutton, 1834) dominate this region of *Acacia* savanna, subdesertic and desertic habitats. Both of these species belong to the Ethiopian Region fauna and occur from northern Africa through the Middle East and north central India to xeric habitats in central Burma and Java. The possibly introduced species, *Allopeas gracile* (Hutton, 1834) and *Gulella bicolor* (Hutton, 1834), also occur throughout this subprovince. A few anomalies are present, such as *Pupoides karachiensis* Peile, 1929, an endemic species known from the Karachi area to the southern Kirthar Mountains. The small, unusually shaped, *Boysia boysii* (Pfeiffer, 1846) was previously known only from along the Nerbudda River in India (Gude, 1914), but was found at one locality about 20 km west of Karachi.

2) **The Ziarat Highlands Subprovince** (2200 - 2700 m) is quite unusual in several aspects. Additional collecting throughout this region is required. A mixed array of species occur in the arid juniper forests in the vicinity of Ziarat. *Vallonia ladacensis* Nevill, 1878, *Vallonia asiatica* Nevill, 1878 and *Gastrocopta huttoniana* (Benson, 1849) were previously known from localities far to the north of Ziarat, although at similar or higher elevations (Gude, 1914). *Pupilla muscorum* (Linne, 1758), *Pupilla signata* (Mousson, 1873), *Vitrina pellucida* (Muller, 1774) and *Euconulus fulvus* (Muller, 1774), previously known from Afghanistan north and east to western China (Gude, 1914; Likharev and Starobogatov, 1967, Schileyko, 1984), are common members of this fauna. One species of *Jaminia* was collected at the highest elevations around Ziarat. This species is virtually indistinguishable from *Jaminia continens* (Rosen, 1892) or its several subspecies occurring on the Iranian Plateau (Forcart, 1959). Species of the zonitid genera, *Oxychilus* and *Hawaiia*, probably also have affinities with species found to the west and north (Likharev and Starobogatov, 1967). Similar relationships of the small mammals found in the Ziarat area and those of more northern localities were observed by Woods (1997).

3) **The Himalayan Foothills Subprovince** (250 - 2700 m) has the greatest land snail diversity and local endemism in Pakistan (see Climate and Geology sections). The area encompasses the lower elevation mountains and hills from Azad Kashmir Province, through the northern Pir Panjal Range and westward to the Salt Range and the Afghanistan border west of Peshawar and into lower Chitral. Previously, the species occurring in this region, particularly the eastern portion, had been associated with the Indo-Malayan fauna because of a high diversity of apparently tropical helicarionids. This viewpoint may be accurate, but necessary dissections in some groups have not been made. *Macrochlamys* (s. l.) and *Kaliella* (s. l.) are reasonably diverse and presumed congenerics are known from localities in the Himalayan foothills to western India (Blanford and Godwin-Austen, 1908).The clausiliids, *Hemiphaedusa waageni* (Stoliczka, 1872), *H. farooqi* Auffenberg and Fakhri, 1995 and *Cylindrophaedusa cylindrica* (Pfeiffer, 1846) belong to the Phaedusinae, a diverse group in southern and eastern Asia (Zilch, 1959 - 1960). The large helicarionid, *Bensonies monticola* (Hutton, 1838), is known from several localities in the Pir Panjals, but was not collected west of the Galis of Pakistan. *Bensonies* (s. l.) *jacquemonti* (Martens, 1869), *B.* (s. l.) *wynnii* (Blanford, 1880) and *Syama* spp. represent adaptations to the drier and hotter climate of the lowland foothills (Solem, 1979), and are probably more closely related to Indo-Malayan taxa than Central Asian groups. The large, helicarionid semi-slugs, *Girasia* spp. and *Euaustenia* spp., are well-documented in the Himalayan foothills of India (Blanford and Godwin-Austen,1908). *Parvatella* spp. is well-known from Kashmir westward to Afghanistan and north to Tadjikistan (Solem, 1979). Two undescribed vertiginid taxa also appear to be related to groups found to the east. Despite this seemingly large number of species with Indo-Malayan affinities, many species, even in the high, moist Galis, are more closely related to Palaearctic or Central Asian species. In fact, the Himalayan Foothills Subprovince is best characterized by the occurrence of the enid genus, *Subzebrinus* (s. l.),which occurs throughout South Central Asia to Afghanistan and West China (Gude, 1914; Solem, 1979; Zilch, 1959 - 1960). The group possibly reaches its greatest diversity in Tadjikistan and Pakistan. In Pakistan, *Subzebrinus* (s. l.) occurs from most of the Jhelum River

valley (Rajagopal and Rao, 1972) and the moist, forested slopes of the Gali Hills (Gude, 1914) to the drier conifer forests of Chitral (Prashad, 1927) and the arid lowlands of the Potwar Plateau and the Salt Range (Annandale and Rao, 1923). Several additional taxa also have Central Asian (Turkmenian), Palaearctic or Holarctic affinities, i.e. *Pyramidula humilis* (Hutton, 1838), *Toltecia* spp., *Succinea* spp., and the giant arionid slugs, *Anadenus* spp. A few lowland species, particularly the helicoids, have more truly Mediterranean affinities, i.e. *Monacha* spp., *Xeropicta candaharica* (Pfeiffer, 1846) and *Euomphalia bactriana* (Hutton, 1849).

4) **The Lower Chitral Subprovince** (1200 - 1800 m) is an extremely interesting area. Unfortunately, very few specimens are available for study. Despite this lack of material it is clear that although the area has affinities to the western Himalayan Foothills (i.e. similar *Subzebrinus* spp.) this part of Chitral is also quite unique in Pakistan. In the vicinity of Drosh, members of the Zoological Survey of Pakistan collected several specimens of a large *Pseudonapeaus* and one species of *Styloptychus*. If the tentative identifications are correct, these are the only records for these genera in Pakistan and clearly indicate a substantial faunal change above the Lowari Pass in Lower Chitral District.

5) **The Alpine Subprovince** (2800 - 3600 m) is unique in that it is not continuous, but is divided into numerous smaller parcels of land on the tops of the north - south ridges in northern Pakistan. These high ridgelines are alpine in appearance with conifer forests and meadows. The upper limits of this subprovince may be barren except for isolated groves of trees and shrubs. Although each of these areas in Chitral, Swat and Gilgit may have its own unique fauna, these areas can be loosely united by their similar climates, elevations and floras (although some areas are quite dry and a mesic conifer forest may be replaced by juniper). These areas are characterized by the family Bradybaenidae. This group, particularly *Cathaica* (s. l.) spp., is quite diverse at high elevations in Central Asia, apparently becoming more speciose in West China (Zilch, 1959 -1960). Some species, such as *Cathaica phaeozona* (Martens, 1874), probably occur in scattered deems throughout the Central Asian region. Other species of *Cathaica* (spp.) occur in the Dras Valley of northern India

(Nevill, 1878) and high forests of Chitral (Odhner, 1963), but apparently not in intervening areas.

6) **The Pamir-Karakoram Subprovince** (1300 - 4700 m) ranges from the forested elevations of the Gilgit region (as far south as the Chilas vicinity) northward through Hunza and Karimabad and east to the Skardu valley and the Deosai Plateau. In the far north this Holarctic and Palaearctic fauna is restricted to forested valleys and the vicinity of vegetated seeps and springs. Tundra at Khunjerab Pass (4730 m) also yielded a suite of Holarctic species, although depauperate. No land snails were found on the intervening arid, barren mountains north of Karimabad or those in the Skardu region. Land snails found in this subprovince include *Vallonia* spp., *Cochlicopa lubrica* (Muller, 1774), *Carychium* sp., *Punctum pygmaeum* (Draparnaud, 1801), *Pyramidula* sp., *Vertigo antivertigo* (Draparnaud, 1801), *Pupilla muscorum* (Linne, 1758) and *Euconulus fulvus* (Muller, 1774).

DISCUSSION

The biogeographic subprovinces of Pakistan are fairly clear with minimum overlap, geographically or altitudinally (Fig. 1). Delineations are somewhat blurred in the Himalayan Foothills area which represents the contact zone for all but the Ziarat Highlands (further collecting in the western mountain ranges north to Peshawar may extend this subprovince northward to the Himalayan Foothills). The Himalayan Foothills provide the most optimal land snail habitats in Pakistan. The air temperatures in this region are generally more tolerable and the precipitation is relatively high. Combined with extensive limestone outcropping, these climatic factors provide ideal habitats for land snail dispersal and local speciation. The influence of the Palaearctic Region is generally lessened in the more southern areas, but a substantial number of the species in the Ziarat Highlands are of Palaearctic origin, while others are of more Iranian Plateau/Transcausasian affinity. The species composition of the northern subprovinces is almost entirely Palearctic. The Himalayan Foothills are also greatly influenced by the Palaearctic (Turkmenian), since *Subzebrinus* (s. l.) must be considered a Central Asian element. Indo-Malayan influence is minimal occurring only in the Himalayan Foothills and rapidly decreasing in diversity westward toward the Afghanistan border. The land snail fauna of the Thar

Desert Subprovince is a very depauperate Ethiopian fauna extending eastward to localized areas in central India, Burma and Java.

Similar biogeographic patterns were observed by Woods (1997) for the small mammals of Pakistan, particularly those of the Ziarat Highlands,Alpine, and Pamir-Karakoram subprovinces. The Ziarat area is especially rich in endemic species of small mammals, and voles of the genus *Hyperacrius* are fragmented in distribution in alpine habitats (Alpine biogeographical subprovince).

Endemism is high considering the generally depauperate fauna (Table 1). The number of endemic species ranges from 1 - 19 (avg. = 5) per biogeographic subprovince (8 - 100% of subprovincial fauna, avg. = 31%). The Lower Chitral Subprovince has the highest percentage of endemic species, but this is probably an artefact of inadequate collecting.The Himalayan Foothills has the highest number of endemic species (N = 19), but also has the greatest land snail diversity, undoubtedly due to the optimal substrate and climatic conditions which prevail in this region. The Pamir-Karakoram Subprovince has the lowest endemism due to the high percentage of widespread Palearctic species found at these higher elevations.

Much of the Pakistan land snail fauna may occur extralimitally, but is restricted to a single subprovince (range 4 - 55 species, avg. = 13.5, 33 - 100% of subprovincial fauna, avg. = 60%). The northern subprovinces of the Ziarat Highlands, Alpine and Pamir-Karakoram have low subprovincial endemism due to the sharing of numerous taxa and the high number of widespread Palearctic species. Lower Chitral is not adequately sampled to reveal realistic conclusions. The Thar Desert Subprovince has seven species (64% of subprovincial fauna) which may be found extralimitally, but do not occur in other subprovinces in Pakistan. Approximately 86% of the Himalayan Foothill fauna is restricted to that area in Pakistan, but may also occur extralimitally. Most of these non-endemic species are *Subzebrinus* (s. l.) spp. and the helicarionids of Indo-Malayan affinities.

Several points of biogeographic interest require discussion. The land snail fossil record in Pakistan is very poor (Prashad, 1925). Without additional information it is impossible to even speculate on the origins and timing of speciation and/or dispersal for the land snail faunas of Pakistan. Somewhat surprisingly, no terrestrial operculates (Prosobranchia) have been collected in Pakistan. Two major groups occur in southern Asia, the Pomatiasidae and the Cyclophoridae. The minor family, Cochlostomidae (considered by some workers as a subfamily of the Cyclophoridae), occurs in portions of the Middle East and Europe. The Pomatiasidae has two subfamilies, the larger being the Pomatiasinae, which occurs from eastern Africa, Madagascar, Europe and Eastern Asia south to the Arabian Peninsula. The Cyclotopsinae has several species in the Mauritius, Seychelles and unexpectedly at Poona, Maharastra State and the Kathiawar Peninsula of Gujarat State in India (Wenz 1938 - 1944). The Cyclophoridae is an extremely diverse tropical family widely deployed from New Guinea and through eastern and southern Asia. The family apparently reaches its western distributional limits in the vicinity of Bombay and mesic habitats to the northeast (Gude, 1921).

The truly Palaearctic/Holarctic component of the northern provinces possibly dispersed to the Pakistan highlands prior to the uplifting of the Hindu Kush and Karakoram Mountains or entered the region through the few existing passes and valleys. Further southerly dispersal by a few taxa (*Pyramidula*, etc.) into the Himalaya Foothills is evident. The occurrence of *Vallonia* spp. and *Pupilla* spp. in the Ziarat Highlands is problematic. Until the higher mountain regions between Ziarat and the western Himalayan Foothills are adequately sampled, it cannot be determined if these indicate continuous or relict distributions. Nevertheless, the occurrence of these typically Palaearctic genera at such low latitudes in Asia is remarkable. The widely deployed Holarctic family, Zonitidae, is found as far east as the Indus Valley in the vicinity of Skardu and as far south as the Ziarat region of Balochistan, but is rapidly replaced by the tropical Helicarionidae in the eastern Himalayan Foothills.

The pulmonate family, Clausiliidae, is extremely speciose throughout Europe and much of southern Asia. The European and Middle Eastern groups extend as far east as the Transcaucasus region. The eastern subfamily, Phaedusinae, is very diverse in southeast Asia to Assam, Nepal and the Himalayan Foothills in Pakistan (an outlier group also occurs in the Transcaucasus area) (Zilch, 1959 - 1960). Two of the three species found in Pakistan appear to be related to species in Nepal and Assam. A third species is widespread from the Muree Hills in Pakistan to Darjiling, India (Auffenberg and Fakhri, 1995).

The distribution of the few helicoids of Mediterranean affinities (i.e. *Euomphalia, Xeropicta* and *Monacha*) is somewhat disturbing. These species are not found in eastern or southern Pakistan. Eastward dispersal of these taxa may have been hindered by the overtly harsh desertic conditions in this region, but other members of these groups have adapted to extreme environments in the Middle East and northern Africa.

The distribution of the Turkmenian enids (*Subzebrinus* (s. l.), etc.) in Pakistan is also noteworthy. This very diverse family ranges from Europe east to China and south into Africa, the Middle East and northwestern India (Zilch, 1959 - 1960). In Pakistan this group is extremely variable in shell form and habitat preference, filling most of the limited niches in this comparatively harsh environment. The numerous species and forms are quite ubiquitous from elevations of about 250 - 2500 m, as far south as Khuzdar, Balochistan and north to central Chitral. In the Indus Valley species occur as far north as the Besham area. Preliminary examination of the survey material at hand indicates that the Pakistan fauna is polyphyletic. The origin of this fascinating radiation is unclear. Whether these species or their ancestors existed in the Pakistan region before or during the massive mountain-building events associated with the collision of the Indian and Eurasian Plates is unknown. The diversity of the enid fauna of Tadjikistan, eastern Turkistan, extreme western China, etc. is equal to or greater than that of Pakistan. Faunal similarities between the enids of Pakistan and these more northern locales are overwhelming, but due to the lack of a fossil record and/or anatomical information, further discussion is mere speculation. The enid fauna of northwestern India is relatively high, but diversity is rapidly reduced south and east of the Kashmir region. The presence of *Jaminia* sp. in the Ziarat Highlands is interesting and indicates a probable relationship with the Iranian Plateau fauna.

The Indo-Malayan fauna has relatively little influence on the land snail diversity of Pakistan. This fauna is represented in the higher mountains of the eastern Himalayan Foothills by fewer than twenty species of helicarionids, some of which are widely distributed to the east, while others are endemic. The semi-slugs, *Euaustenia* spp.and *Girasia* spp. extend westward to the Swat District. *Parvatella* and the xeric-adapted genera, *Bensonies* and *Syama*, occur as far west as Afghanistan (Solem, 1979).Two of the minute vertiginids (*Gastrocopta* sp. and one species which can not be readily assigned at genus level) found in the Himalayan Foothills as far west as the Swat District probably have Indo-Malayan affinities.

The land snail fauna of the southern xeric habitats of Pakistan is extremely depauperate and is of Mediterranean/Ethiopian descent. The few species are closely related to or conspecific with species occurring from the Cape Verde Islands and/or northern Africa eastward to xeric portions of north central India, central Burma and Java. A few unexpected distributions occur in the southern Kirthar Mountains and portions of Sindh Province (i.e., *Pupoides karachiensis* (Peile, 1929) and *Boysia boysii* (Pfeiffer, 1846). *Anadenus lahorensis* Bhatia, 1926 is undoubtedly an introduction (the type locality is a botanical garden in Lahore) or an improper generic assignment.

Much more intensive investigation is required before the biogeography of the land snails of Pakistan can be fully understood. Large portions of the country are imperfectly sampled or completely unknown. Extensive surveys in the western mountain corridors, the isolated mountain ranges of western Balochistan, Chitral and further collections on the Deosai Plateau will undoubtedly lead to many new discoveries. It is hoped that these preliminary results will initiate further investigations in Pakistan so this fascinating, and largely ignored, fauna can be fully documented.

ACKNOWLEDGMENTS

Many persons provided specimens and assistance during fieldwork in Pakistan, particularly Muhammad Farooq Ahmed, Hafeez Rehman, S. M. Shamin Fakhri, Aleem Khan and Muhammad Zubair of the Zoological Survey of Pakistan. I thank the Pakistan Museum of Natural History and the Pakistan Science Foundation for their support of the Biodiversity of Pakistan Project, in particular Dr. Shahzad Mufti and Dr. Azhar Hasan. Dr. Charles Woods, Florida Museum of Natural History, provided enjoyable companionship during a survey of the southern Kirthar National Park and collected very interesting specimens during his fieldwork on the Deosai Plateau and Chitral. Dr. Walter Auffenberg, Professor Emeritus, Florida Museum of Natural History, assisted with the collection of specimens throughout the country. He introduced me to malacology, a passion I have enjoyed for over twenty years. Mr. Dan Cordier, formerly of the Florida Museum of Natural History, provided many enlightening discussions on geological topics. Ms. Florence Sergile, Florida Museum of Natural History, assisted me greatly in manuscript preparation. Funding for the fieldwork in Pakistan was provided by the U. S. Fish and Wildlife Service and the Thomas LaDue McGinty Fund, University of Florida Foundation.

REFERENCES

ANNANDALE, N. AND RAO, H. S. 1923. The molluscs of the Salt Range. *Records of the Indian Museum,* **25**:387-397, pl. 9.

AUFFENBERG, K. AND FAKHRI, S. M. S. 1995. A new species of land snail from Pakistan (Gastopoda: Pulmonata: Clausiliidae: Phaedusinae). *Archiv fur Molluskenkunde,* **124**(1/2):89-92.

BLANFORD, W. T. 1870. Notes on some reptilia and amphibia from central India. *Journal of the Asiatic Society of Bengal,* **39**(2):335-381.

BLANFORD, W. T. 1901. Distribution of vertebrate animals in India, Ceylon and Burma. *Philosophical Transactions of the Royal Society of London, (B)* **194**:335-436.

BLANFORD, W. T. AND GODWIN-AUSTEN, H.H. 1908. *The fauna of British India including Ceylon and Burma, Mollusca. Testacellidae and Zonitidae.* Taylor and Frances, London, 311 pp.

FORCART, L. 1959. Two new subspecies of *Jaminia* (*Euchondrus*) *continens* (Rosen) from Iran. *Journal of Conchology,* **24**(9):315-317.

GODWIN - AUSTEN, H. H. 1882 - 1920. Land and freshwater mollusca of India, including South Arabia, Balochistan, Afghanistan, Kashmir, Nepal, Burmah, Pegu, Tenasserim, Malay Peninsula, Ceylon, and other islands of the Indian Ocean, pts. 1 - 3, London.

GODWIN-AUSTEN, H. H. 1900. Address of the President. *Proceedings of the Malacological Society of London,* **3**:241-262.

GUDE, G. K. 1914. *The fauna of British India including Ceylon and Burma, Mollusca. Pt. 2 (Trochomorphidae - Janellidae).* Taylor and Francis, London, 520 pp.

GUDE, G. K. 1921. *The fauna of British India, including Ceylon and Burma, Mollusca. Pt. 3 Land Operculates (Cyclophoridae, Truncatellidae, Assimineidae, Helicinidae).* Taylor and Frances, London, 386 pp.

HUTTON, T. 1834. On the land shells of India. *Journal of the Asiatic Society of Bengal,* **1**:81-93.

HUTTON, T. 1849. Notices of some land and freshwater shells occurring in Afghanisthan. *Journal of the Asiatic Society of Bengal,* 2nd series, **18**:649-661, 967.

JADOON,I.A. K., LAWRENCE, R.D. AND LILLIE, R.J.1994.Seismic data, geometry, evolution, and shortening in the active Sulaiman Fold-and-Thrust Belt of Pakistan, southwest of the Himalayas. *AAPG Bulletin*, **78**(5):758-774.

KRISHNAN, M. S. 1974. III. Geology. pp. 60-98. In: Ecology and Biogeography in India, ed. M. S. Mani. *Monographiae Biologicae*, Dr. W. Junk b. v. Publishers, The Hague, vol. **23**, 773 pp.

LIKHAREV, I. M. AND STAROBOGATOV, Y.I. 1967. On the molluscan fauna of Afghanistan. *Trudy Zoologicheskogo Instituta Akademii Nauk SSSR*, **42:**159-197.

MANI, M. S.1974a.XXII. Biogeography of the Western Borderlantds. pp. 682-688. In: Ecology and Biogeography in India, ed. M. S. Mani. *Monographiae Biologicae*, Dr. W. Junk b. v. Publishers, The Hague, vol. **23**, 773 pp.

MANI, M. S. 1974b. XXIV. Biogeographical Evolution in India. pp. 698-722. In: *Ecology and Biogeography in India,* ed. M. S. Mani. Monographiae Biologicae, Dr. W. Junk b. v. Publishers, The Hague, vol. 23, 773 pp.

MANI, M. S. 1974c. V. Limiting Factors. pp. 135-158. In: *Ecology and Biogeography in India*, ed. M. S. Mani. Monographiae Biologicae, Dr. W. Junk b. v. Publishers, The Hague, vol. **23**, 773 pp.

MANI, M. S. 1974d. II. Physical Features. pp. 11-59. In: *Ecology and Biogeography in India,* ed. M. S. Mani. Monographiae Biologicae, Dr. W. Junk b. v. Publishers, The Hague, vol. **23**, 773 pp.

MURATOV, I. V. 1992. New taxa of Pseudonapaeinae (Gastropoda, Pulmonata, Enidae). *Ruthenica*, **2**(1):37-44. [in Russian]

NEVILL, G. 1878. Scientific results of the Second Yarkand Mission; based upon the collections and notes of the late Ferdinand Stoliczka PhD, pts. 1 & 2, Calcutta. 22 pp.

ODHNER, N. H. 1963. *Cathaica (Pseudiberus) chitralensis* n. sp. *Proceedings of the Malacological Society of London*, **35**(4):151-153.

PATRIAT P. AND ACHACHE, J. 1984. India-Eurasia collision chronology has implications of crustal shortening and driving mechanisms of plates. *Nature*, **311**:615-621.

PRASHAD, B. 1925. On a collection of land and freshwater fossil molluscs from the Karewas of Kashmir. *Records of the Geological Survey of India*, **56**(4):356-361, pl. 29.

PRASHAD, B. 1927. Notes on molluscs in the collections of the Zoological Survey of India (Indian Museum), Calcutta. *Records of the Indian Museum*, **29**:229-232, pl. 22.

PRASHAD, B. 1942. Zoogeography of India. *Science and Culture,* Calcutta, **7**.:421-429.

RAJAGOPAL, A. S. AND RAO, N.V.S. 1972. Some land molluscs of Kashmir, India. *Records of the Zoological Survey of India*, **66**(1-4):197-212.

RAMDAS, L. A. 1974. IV. Weather and Climatic Factors. In: *Ecology and Biogeography in India*, ed. M. S. Mani. Monographiae Biologicae, Dr. W. Junk b. v. Publishers, The Hague, vol. 23, 773 pp.

SCHILEYKO, A. A. 1984. *Land mollusks of the Suborder Pupillina of the fauna of the USSR (Gastropoda, Pulmonata).* Fauna SSSR, new series, t. III, 3 Leningrad Nauka, 399 pp. [in Russian].

SOLEM, A. 1979. Some mollusks from Afghanistan. Fieldiana, Zoology, new series, no. 1, 89 pp.

THEOBALD, W. 1881. List of Mollusca from the hills between Mari and Tandiani. *Journal and Proceedings of the Asiatic Society of Bengal,* **50**(1):44-49.

TRELOAR P. J. AND COWARD, M.P. 1991. Indian Plate motion and shape: constraints on the geometry of the Himalayan orogen. *Tectonophysics,* **191**:189-198.

UDVARDY, M. D. F. 1985. Biogeographical Provinces of the Indomalayan Realm, Western Part and Eastern Part. Revised maps for a classification of the Biogeographical Provinces of the World (UNESCO Man and the Biosphere Program, Project no. 8.

WENZ, W. 1938 - 1944. Gastropoda, Teil 1: Allgemeiner Teil und Prosobranchia. *Handbuch der Palaozoologie, Berlin,* **6**(1): 1639 pp.

WOODS, C. A. AND KILPATRICK, C.W. 1997. Biodiversity of small mammals in the mountains of Pakistan (this volume).

WOODWARD, S. P. 1856. On the land and freshwater shells of Kashmir and Tibet, collected by Dr. T. Thomson. *Proceedings of the Zoological Society of London*, **24**:185-187.

ZILCH, A. 1959 - 1960. *Gastropoda, Euthyneura.* Handbuch der Palaozoologie, Berlin. 834 pp.

DISTRIBUTION AND RANGE EXTENSION OF PELAGIC SHRIMP *SERGESTES* BELONGING TO *CORNICULUM* SPECIES GROUP

NAUREEN A. QURESHI AND NASIMA M. TIRMIZI

Centre of Excellence in Marine Biology, University of Karachi, Karachi

Abstract: The systematics and distribution of Indian Ocean sergestid shrimps collected on research vessel INS KISTNA during International Indian Ocean Expedition (1962-65) belonging to species group '*corniculum*' is studied. Three species belonging to *corniculum* group, *Sergestes curvatus* Crosnier & Forest 1973, *S. henseni* (Ortmann, 1893)*, and S. paraseminudus* Crosnier & Forest 1973 are described and are new to the Indian Ocean. The frequency distribution of the postlarval stages belonging to *corniculum* species group is also presented.

INTRODUCTION

Planktonic shrimps comprise a significant portion of biomass of the world ocean. They link nekton with the palnkton in the marine food chain and are considered as important transporter of organic mattter from upper producing layer to the great depth. Our knowledge is limited due to lack of basic information on the systematics, biology, and ecology of these planktonic shrimps in the Indian Ocean.

There are about 2,000 species of shrimps in the world of which , at least 200 pass their life in the pelagic phase. There are seven known species of the family Sergestidae viz: *Sergestes* H. Milne Edwards 1830, *Sergia* Stimpson 1860, *Peisos* Burkenroad 1946, *Petalidium* Bate 1881, *Acetes* H. Milne Edwards 1830, and *Lucifer* V. Thompson 1830. Of these three genera *Sergestes, Sergia, Acetes* of sub-family Sergestinae and one genus *Lucifer,* of subfamily Luciferina have been reported from the Indian Ocean. Literature concerning the distribution of these Indian Ocean representative is very poor. Important contribution on the systematics of this group are by Alcock 1901, Kemp 1913, Bernard 1947, 1953, Yaldwyn 1957, Omori 1969, 1974 Kensley 1971, 1981, Crosnier and Forest 1973 and Judkins 1978.

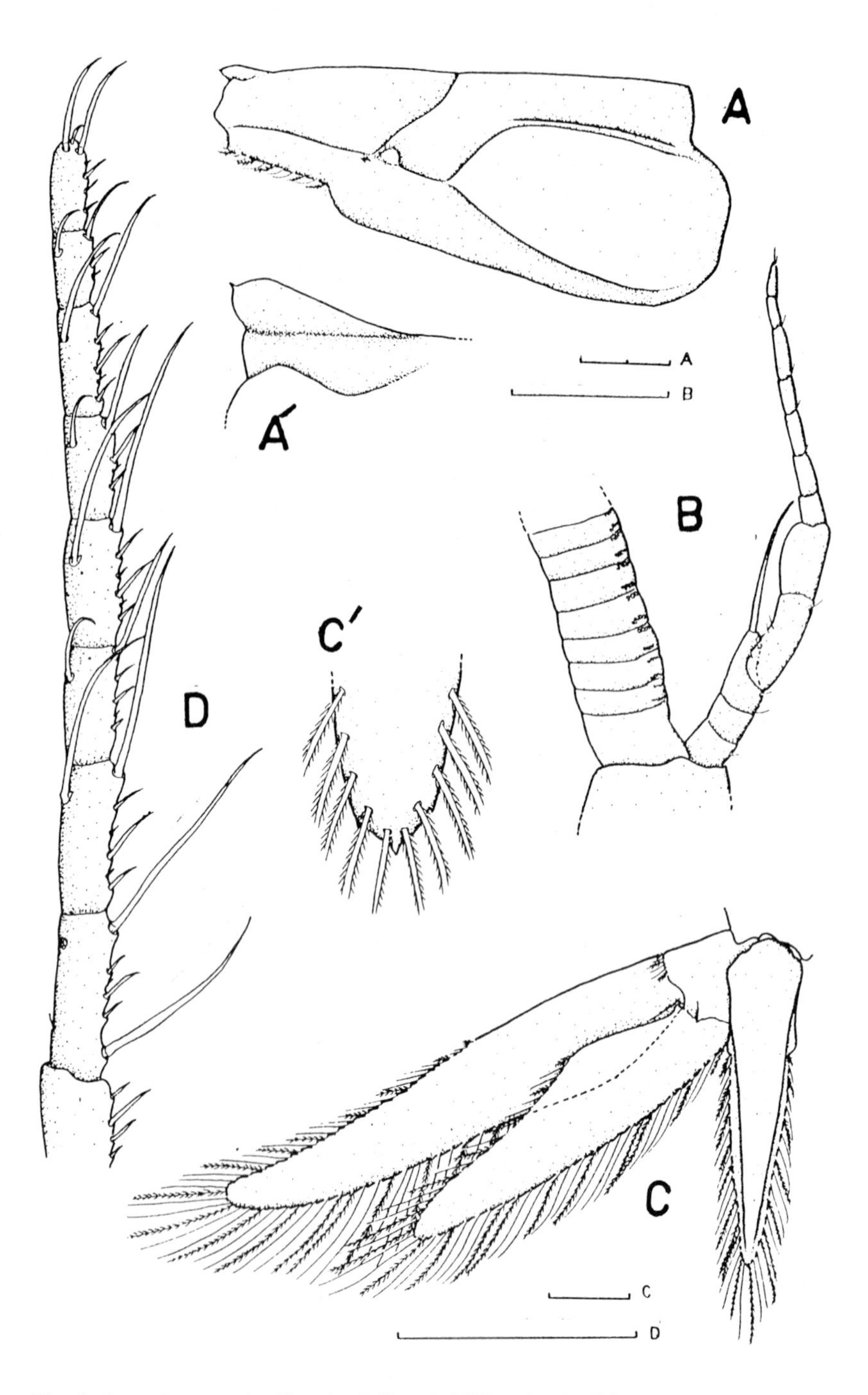

Fig. 1. *Sergestes curvatus* Crosnier & Forest, 1973 male cl. 12.5 mm. A, carapace, lateral view; A', rostrum of the same, enlarged; B, inner flagellum of the right antennule, mesial view; C, telson, and left uropod enlarged, dorsal view; D', tip of telson, enlarged.

The genus *Sergestes* is easily divisible into six distinct species groups: *arcticus, corniculum, atlanticus, sargasi, vigilax, and edwardsi* (Yaldwyn 1957) and these groups appear to be natural phyletic units (Judkins 1978).

The *corniculum* group includes eleven species. The group *corniculum* is characterized by the size of the third maxilliped equal in size with that of the third walking leg, and the fifth walking leg being setose on both margins. The morphological separation of the species of the *corniculum* group is based on the petasma, the male copulatory organ. Species differ to a much lesser degree in the other characters of morphological importance, as numbers of sub segments of dactylus of the third maxilliped and its spination, in the shape of the female genital coxa and in the carapace length and its features.

The present paper reports the distribution of three species, *Sergestes curvatus* Crosnier & Forest 1973, S. *henseni* (Ortmann 1893), and *S. paraseminudus* Crosnier & Forest 1973, belonging to *corniculum* species group. These species were not previously recorded from the Indain Ocean so this provides an extension in the range of the distribution of these species.

MATERIALS AND METHODS

The collection at hand was obtained on board of a research vessel INS Kistna during the International Indian Ocean Expedition. A total of 306 samples collected from different stations during 20 cruises were analysed. The samples were collected by Indian Ocean Standard (IOS) net with mouth diameter of 0.5 m and mesh size of 20 mm. All specimens were examined, identified, and illustrated using wide field and narrow field microscope, using eye piece graticules. The frequency distribution larvae of belonging to the *corniculum* species group is plotted.

RESULT AND DISCUSSION

Sergestes curvatus Crosnier & Forest, 1973 (Figs. 1-2)
Sergestes (Sergestes) curvatus, Crosnier & Forest, 1973 :315-318,
Sergestes corniculum, Hansen, 1922 :126; Kensley, 1968 :307.
Sergestes (Sergestes) corniculum, Kensley, 1971a :236; 1972, :26.

Material and Measurements: Sta 342; Cruise 4; 14.9.63; 10°47'N; 90°55'E; IOS net; 03:05; INS Kistna; 1 male cl. = 12.5 mm.

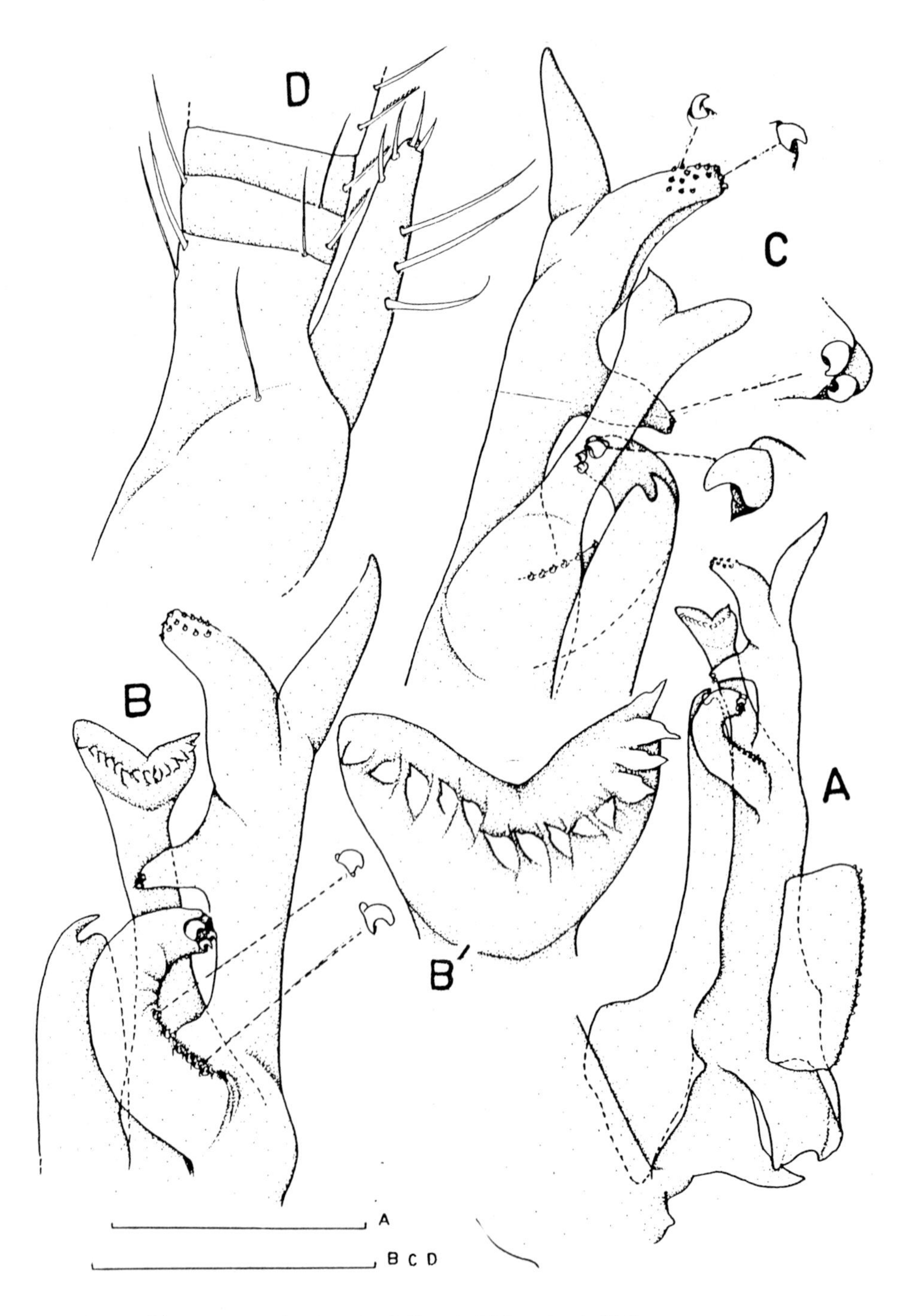

Fig. 2. *Sergestes curvatus* Crosnier & Forest, 1973 male cl. 12.5 mm. A, left petasma, *in situ*, ventral view; B, capitulum of left petasma, ventral view; B', distal part of processus ventralis, enlarged; C, capitulum of left petasma, dorsal view; D, appendix masculina of right side.

Descriptive Remarks: *S. curvatus* is relatively longer (79.5 mm) than the longest recorded specimen (measured as 75 mm by Crosnier and Forest 1973). The body is elongated, slender and fragile. The carapace is also long, with slightly elevated rostrum. The rostrum is convex at the upper edge, sinous at lower edge with a small apical spinule (Fig. 1A & A'). The supra-orbital spine is absent. The supra-orbital ridge is prominent. The hepatic spine is small and acute. The gastro-hepatic groove deep, at about level of hepatic spine the carapace is elevated into hump-shaped anteriorly directed sub-conical projection. The cervical groove is very distinct, the antennal and branchio-cardiac carinae well developed and the supra branchial grove is deep.

The abdomen is long, dorsally rounded and with sixth segment as long as the combined length of the fourth and fifth segments and ends in a dorsal spinule. The telson is three-fourth the length of the sixth segment, tapers into a small terminal spine often flanked by a pair of small fixed lateral teeth (Fig. 1C & C'). The uropod (Fig. 1C) is with the exo- and the endopod long and slender. The exopod is five times as long as wide, with proximal one and half of its external margin non-setose.

The eyes are with a low ocular tubercle, similar to those of *S. henseni.* The antennular peduncle (Fig. 1B) is also like that *of S. henseni.* In males, the inner flagellum is with ten to twelve annuli (Fig. 1B). Third annulus sends off a long spine from the outer margin, extending as far as fifth annulus, the distal half of the spine is striated. The fourth annulus is with the inner margin slightly convex, outer margin notched and bears a few setae in the middle (Fig. 1B).

The third maxilliped is slender extending beyond the third walking leg by the length of dactlyus. The dactylus is subdivided into seven segments. Cronier and Forest (1973) recorded 6-7 sub-segment in their collection. The proximal sub-segment is longest equals the combined length of the preceeding two sub-segments. The sixth and seventh sub-segments are sub equal. The outer margin of the first to sixth subsegments, each bear a long distal spine. The inner margin of all subsegments bears a long distal spine, and 4-7 small spines. The last or seventh sub segment terminally bears three spines of which middle spine is small (Fig. 1D).

The organs of Pesta are present. They are not clear in preserved specimens at hand, however, usuaslly eight distinct organs are present. An antero-lateral pair, a single antero-median and one pair on the lateral mid gastric region. The antero-lateral organs are lobed, whereas, postero-lateral organs are fringed and made up of a single postero-median and a pair of postero-lateral organs (Foxton 1972).

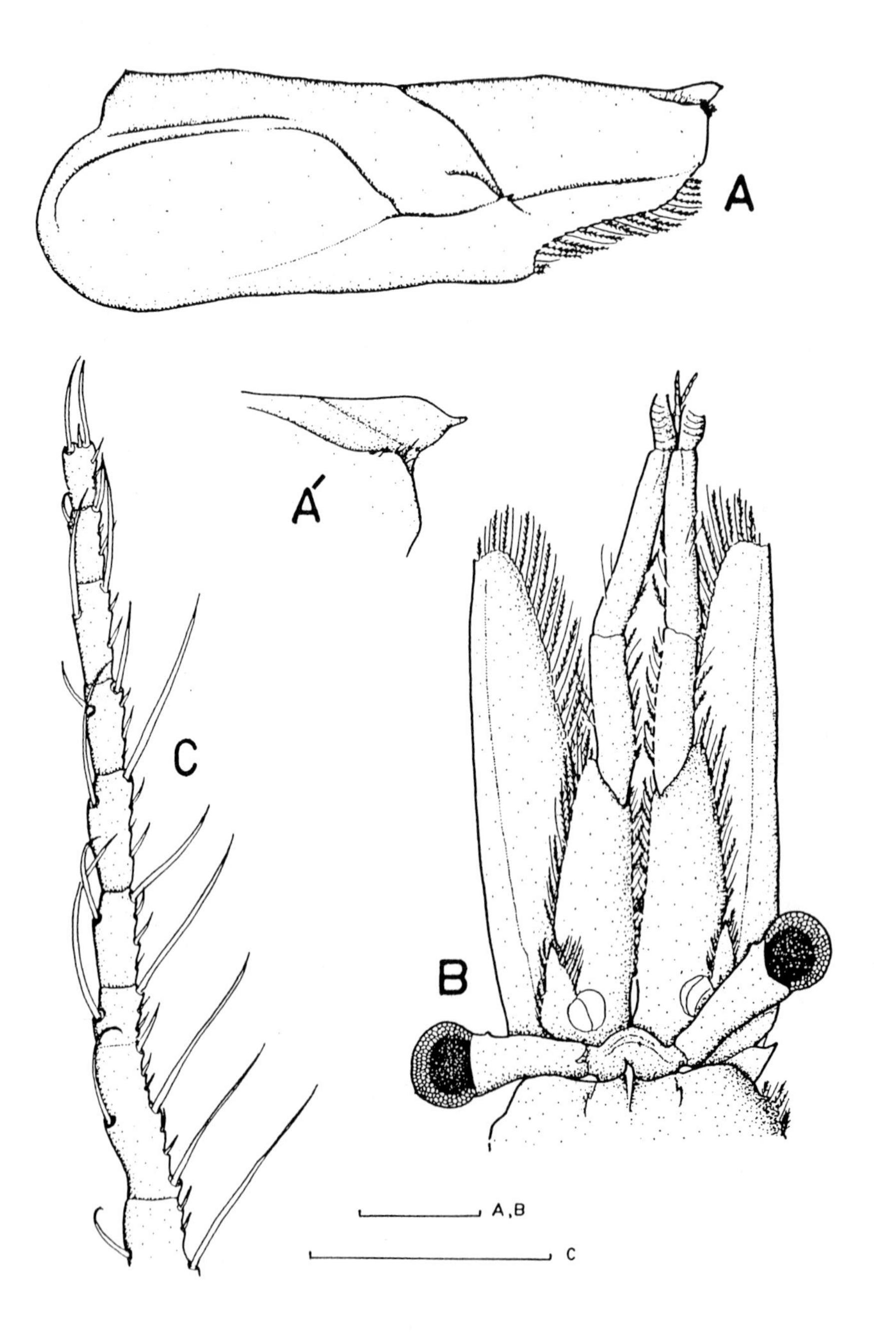

Fig. 3. *Sergestes henseni* (Ortmann, 1893) female cl. 9.0 mm. A, carapace, lateral view; A', rostrum, enlarged; B, anterior part of the body, dorsal view; C, dactylus of left third maxilliped, lateral view.

The petasma (Figs. 2A-C) is with the lamina externa distinctly shorter than processus uncifer. The processes uncifer is long, robust and extends to the tip of lobus armatus, it bears a mesially directed distal hook. The processes basalis is directed postero-mesially, and is modrately long, conical, and acute. The pars media or the capitulum is slender and long with long processus ventralis extending up to the base of the lobus inermis, distially expanding to form Y-shaped structure instead of pallette shape as described by Cronier and Forest (1973). It is much opened with external lobe little longer than the inner lobe and bears 14-15 acute papillae (Fig. 2C'). The shape of processus ventralis resemble more with *S. henseni* (Cronier and Forest 1973). The lobus armatus is stout, strongly curved, and bends backward to direct mesially. It bears distally 1-4 retractile hooks, a big distal hook at its outer margin, and 4-6 hooks are present mesially on the poximal half. The lobus connectens is slender, sub conical directed laterally, apically bears two hooks. The lobus inermis is slender, apically acute and forwardly directed. The lobus terminalis is a little more massive than the lobus inermis, outwardly curved with a cluster of antero-distal hooks.

The second pleopod in the male is biramous and bears a tooth on the proximo-mesial margin on the protopod and a low tubercle on the mesio-lateral margin (Fig. 2D). The endopod bears a slipper-shaped appendix masculina which is distally armed with 4-5 spines, and three long spines in the middle of the inner margin.

Distribution: Atlantic and Indian Ocean. It is known from the east central Atlantic between 37°38' N and 28°11' N and largely off South Africa from south west Indian Ocean between 25° S - 41° S (Crosnier & Forest 1973) and now from the Bay of Bengal.

Sergestes henseni (Ortmann, 1893)(Figs. 3-4)

Sergia henseni Ortmann, 1893 :38.
Sergestes corniculum, Holthuis & Gottlieb, 1958 :13, 111; Zariqviey Alvarez, 1968 :62.
Sergestes (Sergestes) henseni, Crosnier & Forest, 1973 :310-312.

Material and Measurements: Sta. 336; Cr. 14; 13.9.63; 10° N; 93°30'E; IOS net; 01:08; INS Kistna; 1 female cl = 9.0 mm.

Descriptive Remarks: The body is elongated. The carapace (Fig. 3A) is long, with slightly elevated rostrum (Fig. 3A'). The upper edge of the rostrum is convex ending in an acute apical spine. The supra-orbital spine is small and acute; the supra-orbital ridge is prominent. The hepatic spine is also acute; the antennal carina is well developed; the

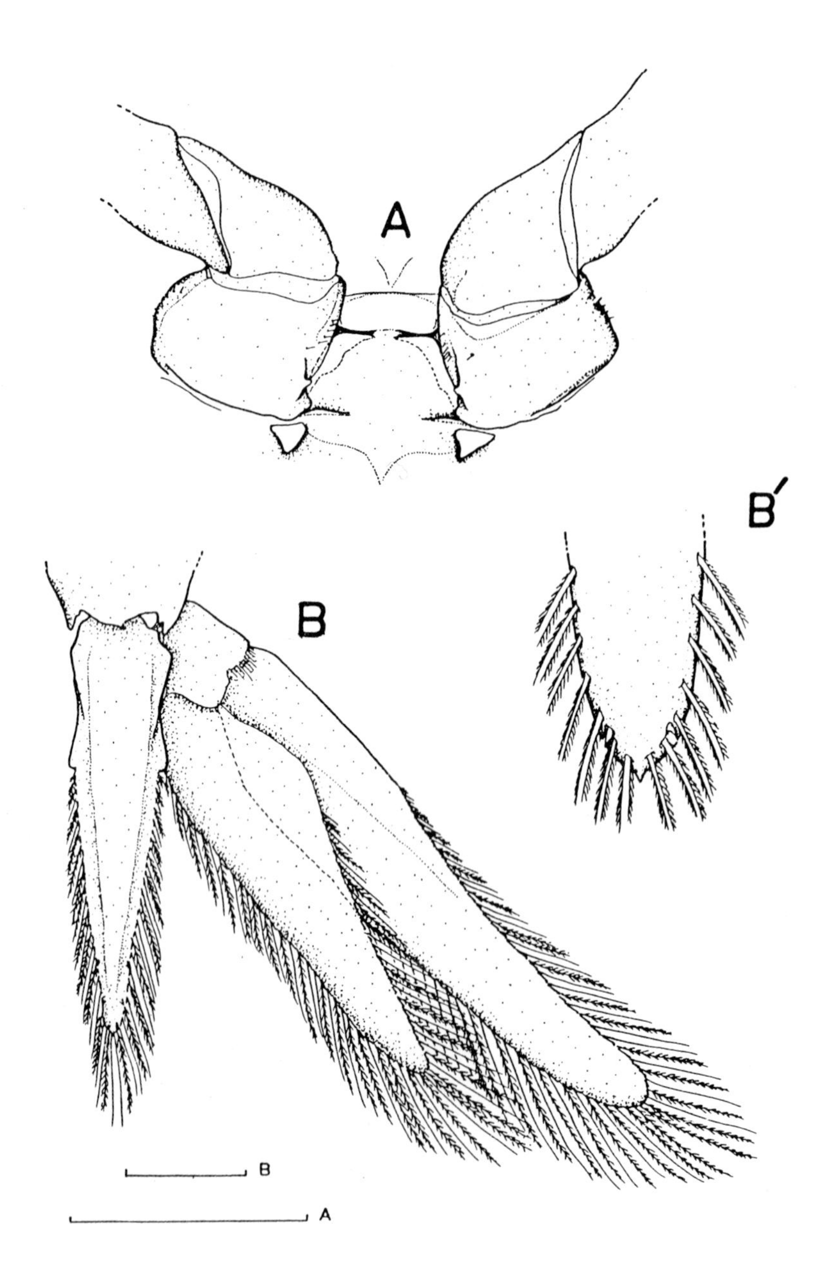

Fig. 4. *Sergestes henseni* (Ortmann, 1893) female cl. 9.0 mm. A, thelycum and coxae of third walking legs, postero-ventral view; B, telson and right uropod, dorsal view; B', tip of telson, enlarged.

gastro-hepatic groove is deep; at about the level of the hepatic spine the carapace is elevated into a hump-shaped anteriorly directed sub-conical projection. The cervical groove is deep; the branchio-cardiac and supra-branchial ridge are well developed with a deep supra-branchial groove.

The abdomen is long, dorsally rounded, with the sixth segment twice as long as high, the minute spine at the tip which is present in *S. curvatus* is absent. The telson (Fig. 4B&B') is as in *S. curvatus*. The uropod (Fig. 4B) has slender, exo- and endopod. The exopod is more than four times as long as wide, with external margin proximally non-setose. The setose portion is twice as long as non setose portion.

The eyes (Fig. 3B) are with long eye stalk which extend a little beyond the middle of the basal segment of the antennular peduncle, the ocular tubercle is present disto-mesially at the base of the cornea. It is more prominent here than observed in *S curvatus* and *S parasiminudus*.

The antennular peduncle (Fig. 3B) is slender, slightly more than half the length of the carapace. The basal segment is longer than the third segment. In female, the third segment terminally bears a long outer flagellum and a slender simple inner flagellum with 8-10 annuli.

The third maxilliped is slightly longer than the third walking leg, with the dactylus (Fig. 3C) sub-divided into seven sub-segments. The size and the armature is similar to *S. curvatus*.

The walking legs are long and slender. The first walking leg reaches to the distal half of the third segment of the antennular peduncle, and is non-chelate. The propodus is twice as long as the carpus and bears numerous stiff barbed setae at the carpo-propodal articulation. The ischia of the first and second walking legs are, each armed with a small tooth disto-laterally. The second and the third walking legs are long and extend well beyond the antennular peduncle, each with a minute chelae. The chela is with the fixed finger half the length of the movable one, terminally bearing tuft of setae. The chela is similar to *S. verpus* which is included in *S. sargassi* species group (Tirmizi and Aziz 1988). Burken road (1937) has also mentioned the similarity and resemblance of certain features, especially the chela, of *S. sargassi* group to that of *S. corniculum* species group.

The organ of Pesta, not visible in the preserved sample, are present. However, Foxton (1972) described ten distinct organs instead od eight organs observed in the related species of *S. corniculum* species group, consisting of an antero-lateral pair, a lateral and mid-gastric pair, two posterior pairs and a single antero-median and postero-median organs.

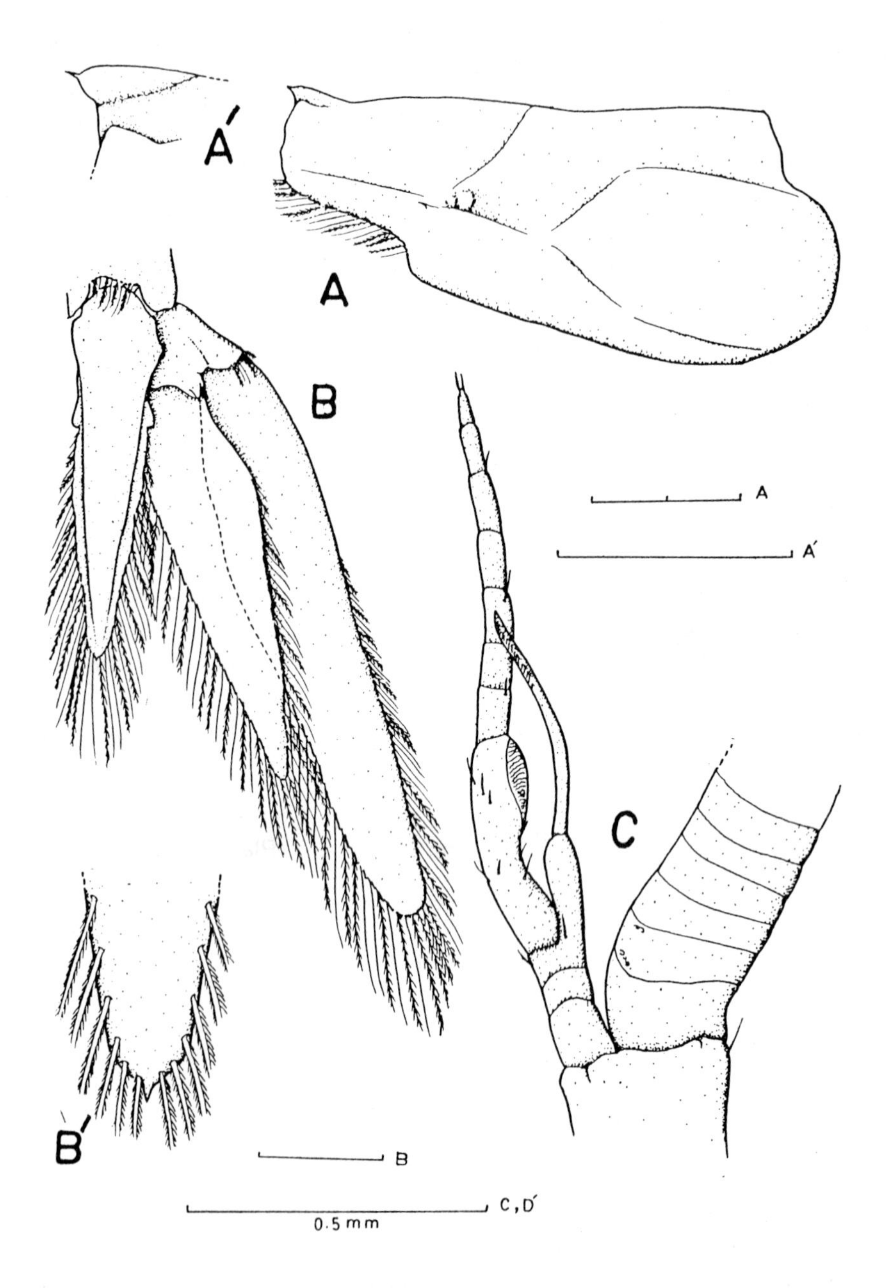

Fig. 5. *Sergestes paraseminudus* Crosnier & Forest, 1973 male cl. 9,0 mm. A, carapace, lateral view; A' rostrum enlarged; B, telson and left uropod, dorsal view; B' tip of the telson, enlarged; C, inner flagellum of right antennule, dorsal view.

The thylecum of female (Fig. 4A) has well developed coxae of the third legs each armed with a tooth mesially on its proximal half. The tooth is small, blunt, curved and slightly forwardly directed, such that the margin proximal to it is concave. The operculum projects posteriorly over the slit like aperture of the sperm recepticles.

Distribution: Atlantic and Indian Ocean. North and South Atlantic Ocean, Mediterranean and now from the Indain Ocean (Bay of Bengal).

Remarks: *S. henseni* has been recorded under various names, however, its systematic position has been thoroughly described by Crosnier and Forest (1973). The specimen at hand, a single female, is referred to *S henseni.* Dr M. Omori agrees with the identification (personal communication); however, it is difficult to distinguish *S. corniculum* from *S. henseni*, when only female is available. A careful examination of the thylecum, shows that the specimen at hand is referable to *S. henseni.* In addition, the characteristic character of *S. henseni* is the presence of sharp and well developed supra orbital spine (Fig. 3B), on which Ortmann (1893) has established this species. This spine is usually not present in the adults of other species and if present it is not well developed (Crosnier & Forest 1973).

Sergester paraseminudus Crosnier & Forest, 1973(Figs. 5-7)

Sergestes corniculum Form C, Gurney & Lebour, 1940 :44.
Sergestes (Sergestes) paraseminudus 1973 :313.
Material and Measurements: Sta. 301; cruise 13; 23.8.63; 00°00'N; 74°00'E; IOS net; 22:11; INSKistna; 1 male cl. = 9.00 mm

Sta 330; cruise 14; 11.9.63; 11°30'N; 95°00'E; IOS net; 23:15; INS Kistna; 1 male (juvenile) cl. = 8.5 mm.

Descriptive Remarks: The body is slender fragile and laterally compressed. The carpace is relatively long and attenuated (Fig. 5A). The rostrum is elevated slightly produced forward, relatively smaller than *S. henseni.* It is compressed and concaves anteriorly, ending in an acute tip or spinule (Fig. 5A'). The ventral margin is concave. The supra-orbital spine is absent, but the supra-orbital ridge is prominent. The hepatic spine is small, but distinct situated at the base of the cervical groove which runs forward from about the dorsal midpoint of the carapace. The cervical groove is distinct. The gastro hepatic groove is deep, the branchio-caridiac and the antennal carinae are well developed. The inferior carina is weakly developed and the groove extends downwards and backwards, at about the level of the hepatic spine. The carapace is elevated into humped shaped anteriorly directed sub conical projection.

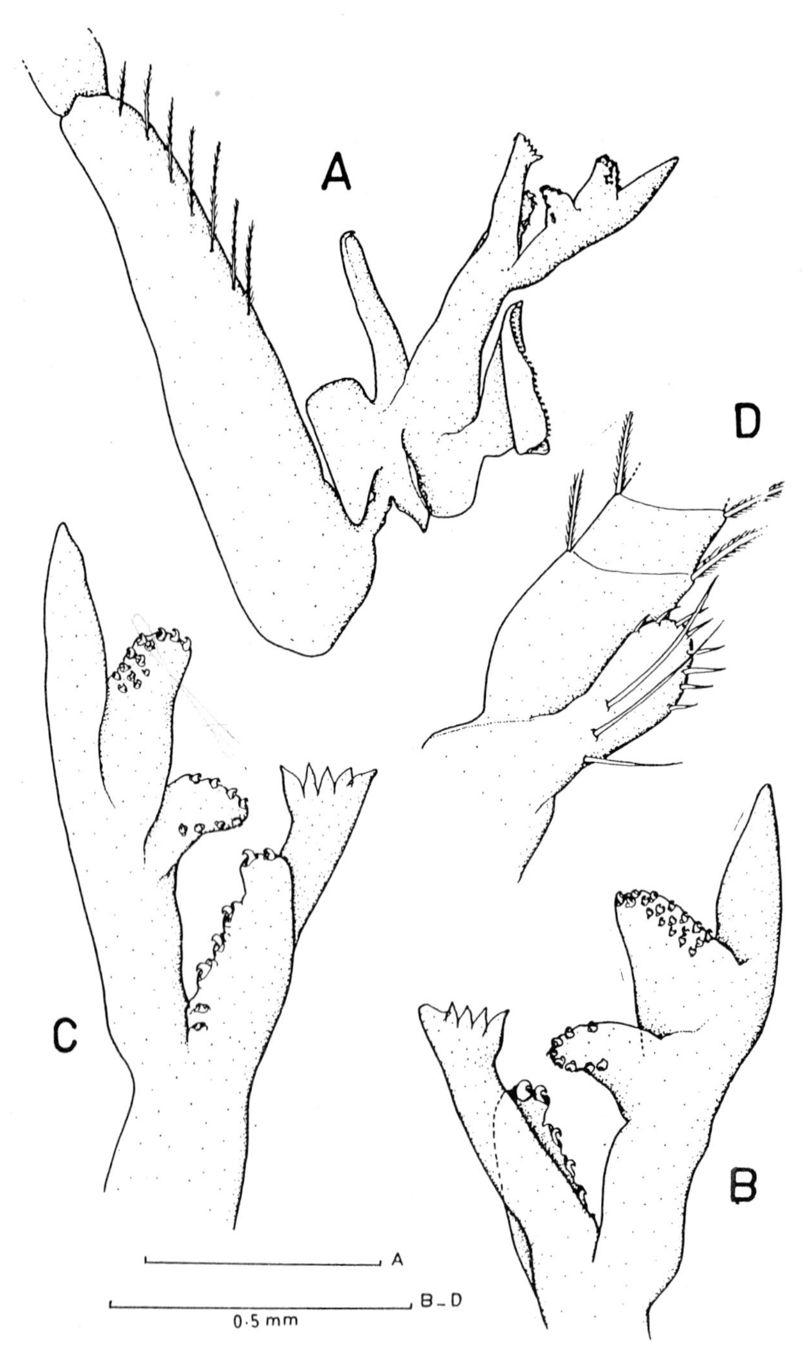

Fig. 6. *Sergestes paraseminudus* Crosnier & Forest, 1973 male cl. 9.0 mm. A, first left pleopod and petasma, *in situ*, ventral view; B, capitulum of left petasma, enlarged, ventral view; C, same capitulum, dorsal view; D, left appendix masculina, mesial view.

The eyes are somewhat small extending as far as the middle of the basal segment of the antennular peduncle. A low ocular tubercle is present at the base of the cornea. The cornea is slightly wider than the eyestalk and is brown in color. The uropod with slender and long exo- and endopod. The endopod six-times as long as wide, with proximally non-setose external margin, half as long as setose portion (Fig. 5B).

The antennule is relatively slender about two-third the length of the carapace. The basal segment is larger than the third, which is itself longer than the second segment. In males, the inner flagellum is slender, have 10-12 annnuli (Fig. 5C). The third annulus sends out long slightly incurved spine from the outer margin. The spine extends as far as the middle of the seventh annulus of the flagellum and is slightly striated on its inner distal half. The fourth annulus with inner margin convex, notched outer margin with few setae in the middle and a striated distal lobe.

The third maxilliped of the adult specimen is missing. The third maxilliped of immature female is equal in length to the third preopod. The propodus and dactylus are equal in length. The dactylus divided into 7 sub-segments (Fig. 7A). The organs of Pesta present, but not clear in the preserved material are similar to *S. curvatus*.

The petasma of adult male (Figs. 6A-C) has lamina externa proximally conical on its outer margin, and shorter than processus uncifer. The processus uncifer is long, robust, hooked and directed mesially. The processus basilis is moderately long, posteriorly directed conical and acute. The pars media is long and slender. The processus ventralis is also long, reaching beyond the distal half of the lobus terminalis, distially expanded forming an open Y, which bears 5-6 papillae. The lobus armatus is short, stout reaches as far as two third of processus ventralis, straight with inner margin bearing row of 6-7 retractile hooks. The lobus connectens is conical curved outwards with 7-8 hooks at the distal half of the lateral margins. The lobus terminalis is little more massive (reverse of *S. disjunctus*), not extends beyond lobus inermis and armed with a cluster of hooks on its antero-distal margin.The lobus inermis longer than lobus terminalis, directed forwards and inwards distally with sub acute distal end.

The appendix masculina (Fig. 6D) is disto-laterally armed with 8-9 spines and three long spines on the proximal half of the inner margin.

Remarks On Sub-Adult Specimen:A sub-adult specimen was recognised from the well developed petasma. The carapace, uropodal endopod and all appendages are in close resemblance to that of adult.

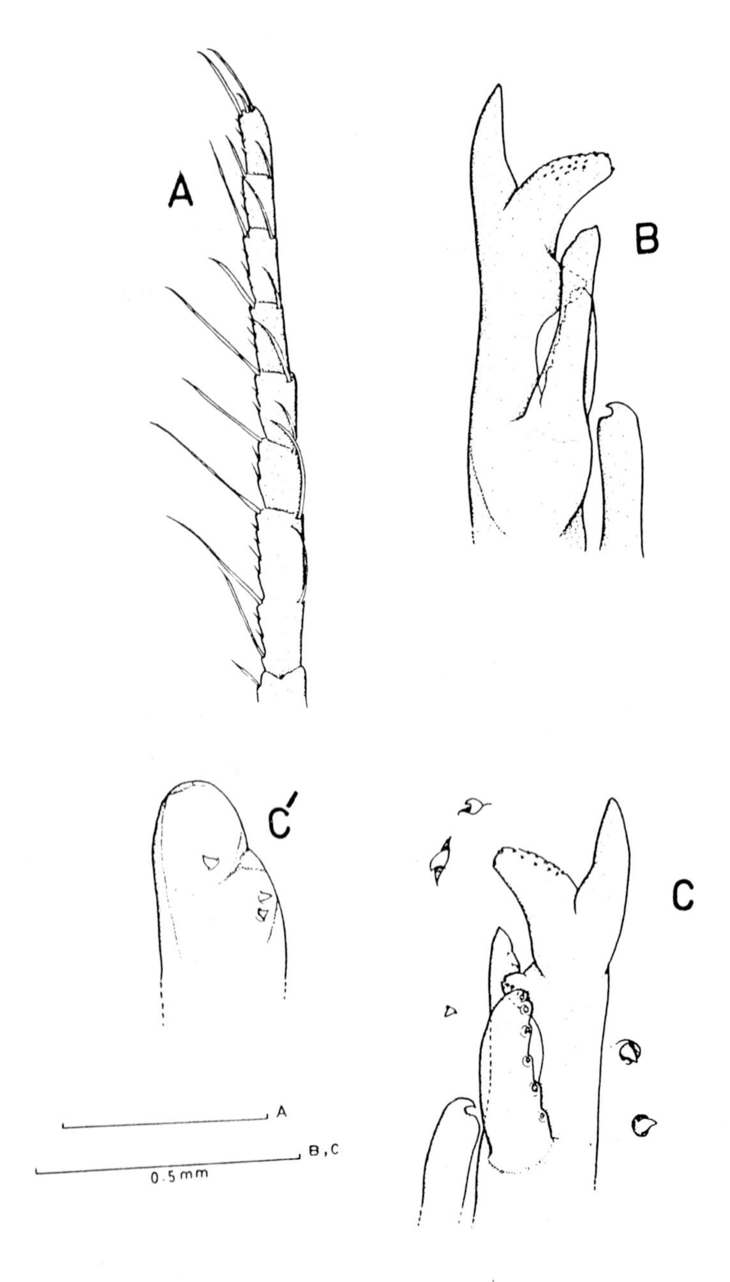

Fig. 7. *Sergestes paraseminudus* Crosnier & Forest, 1973 male (juvenile) cl. 8.5 mm. A, dactylus of right third maxilliped, lateral view; B, capitulum of left petasma, ventral view; C, same capitulum, dorsal view; C' distal part of processus ventralis, enlarged.

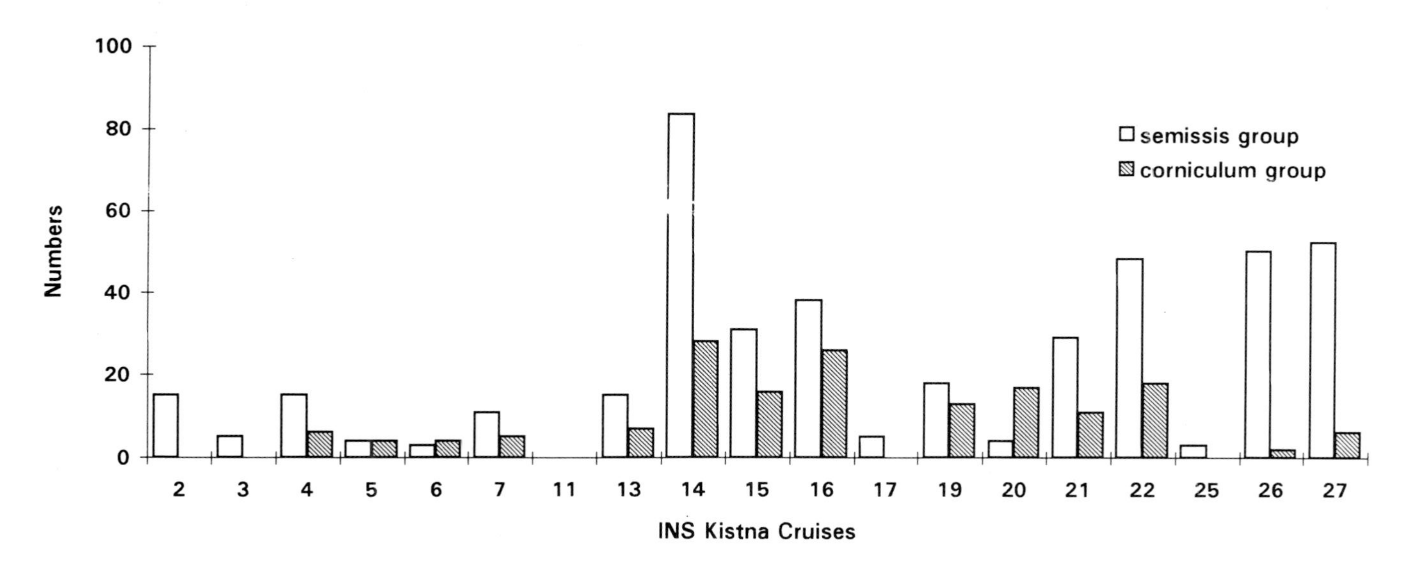

Fig. 8. Frequency distribution of *Sergestes* larvae collected on board INS Kistna during International Indian Ocean Expedition (1962-65).

The petasma (Fig. 7B & C) was well developed but not joined medially by cincinnulli, which is though not rigid criteria but has provided useable criterion for identification of adult specimens (Genthe 1969). The shape of processus basilis and uncifer are similar to that of adult. The processus ventralis reaches almost the middle of lobus terminalis and is not flared fully to form a Y, and bears only 2-3 papillae. The lobus armatus similar to adult with six instead of seven retractile hooks. The lobus terminalis equals to lobus inermis, distally outwardly curved with group of small hooks antero-laterally.

Distribution: South Atlantic, and Indain Ocean. It was recorded from South Atlantic Ocean off Gabon, Congo, and Angola, and now extends its range in Andaman Sea, the Bay of Bengal and south west of India.

Comments on the Distribution

The *corniculum* group is one of the six species groups that comprise the holopelagic penaeid shrimps *Sergestes* (Judkins 1978). Crosnier and Forest (1973) reported their distribution in the upper 0- 130 m depth. On the basis of the collection examined in this study, the vertical distribution of species belonging to the *corniculum* group cannot be described in detail. However, due to presence of few specimens of each species, it appears these species inhabit deeper water and were occasionally caught in the upper shallower layer (200 m) at night hauls where they were present for their diel vertical migration. The larvae appear to have shallower distribution than the adults. More numbers of larvae belonging to the corniculum group were collected (Fig. 8). All larvae of group corniculum even though they belong to different species were placed in *S. corniculum* group and resembled the form B larvae as identified by Gurney and Lebour (1940). The earlier records suggest the epipelagic distribution (few tens of meters) of early developmental stages of genus *Sergestes* (Gurney & Lebour 1940, Omori 1974, Judkins 1978). Judkins (1978) also suggested that the geographical distribution of this group can be profoundly affected by the conditions in the surface layers, therefore, a better understanding of ecology and behavior of these species is needed to understand the distribution patterns.

REFERENCES

ALCOCK, A., 1901. A descriptive catalogue of the Indian deep-sea Crustacea Decapoda Macrura and Anomala, in the Indian Museum. Being a revised account of the deep-sea species collected by the Royal Indian Marine Survey Ship Investigator. Indian Museum, Calcutta. 286 pp., 3 pls.

BARNARD, K.H., 1950. Descriptive catalogue of South African Decapod Crustacea. *Ann. S. Afri. Mus.,* **38**: 1-837.

BATE, C.S., 1888. Report on the Crustacea Macrura collected by H.M.S. Challenger during the years 1873-1876. *Rept. Voy. Challenger, Zool.*, **24**: 1-942.

BURKENROAD, M.D., 1934. Littoral Penaeidae chiefly from the Bingham Oceanographic Collection. *Bull. Bingham Oceanogr. Coll.*, **4**: 1-109.

BURKENROAD, M.D., 1940. Preliminary descriptions of twenty-one new species of pelagic Penaeidea (Crustacea, Decapoda) from the Danish Oceanogrphic Expeditions. *Ann. Mag. Nat. Hist.*, **6**: 35:54.

CROSNIER, A., AND FOREST, J., 1973. Les Crevettes profondes del' Atlantique oriental tropical. *Faune Tropical.,* **19**: 305-345.

FOXTON, P., 1972. Further evidence of the taxonomic importance of the organs of Pesta in the genus *Sergestes* (Natantia, Panaeidea). *Crustaceana,* **22**: 181-189.

GENTHE, H.C. Jr., 1969. The reproductive biology of *Sergestes similis* (Decapoda, Natantia). *Mar. Biol.*, **2**: 203-217.

GORDON, I., 1935. On new or imperfectly known species of Crustacea Macrura. *J. Linn. Soc. Lond. Zool.*, **39**: 307-351.

GURNEY, R. AND LEBOUR, M. V., 1940. Larvae of decapod, crustacea part VI. The genus *Sergestes. Discovery Rept.* **22**: 1-68.

HANSEN, H. J., 1896. On the development and the species of the crustaceans of the genus *Sergestes. Proc. Zool. Soc. London,* 936-970.

JUDKINS, D.C., 1978. Pelagic shrimps of the *Sergestes edwardsii* species group (Crustacea: Decapoda: Sergestidae). *Smith. Contrb. Zool.*, **256**: 1-33.

KEMP, S., 1913. Pelagic Crustacea Decapoda of the Pearcy Sladen Expedition in H.M.S. 'SEA LARK'. *Trans. Linn. Soc. Zool.*, **16**: 5-57.

KENSLEY, B.F., 1971. The family Sergestidae in the waters around southern Africa (Crustacea, Decapoda, Natantia). *Ann. S. Afr. Mus.*, **57**: 215-264.

KENSLEY, B.F., 1981. The South African Museum's Meiring Naude Cruises. Part 12:Crustacea Decapoda of the 1977, 1978, 1979 Cruises". *Ann. S. Afr. Mus.*, **83**: 49-78.

OMORI, M., 1969. Weight and chemical composition of some important oceanic zooplankton in the north Pacific Ocean. *Mar. Biol.* **3**: 4-10.

OMORI, M., 1974. The biology of pelagic shrimps in the ocean. In: *Advances in Marine Biology* (eds. F.S. Russell and M. Yonge), **12**: 233-324. London:Academic Press.

OMORI, M., 1977. Distribution of warm water epiplanktonic shrimps of the genera *Lucifer* and *Acetes* (Macrura, Penaeidea, Sergestidae) Proc. Symp. Warm water Zoopl. Spl. Publ. UNESCO/NIO:1-12.

ORTMANN, S. 1893. Decapoden und Schizopoden. *Ergebn. Atlant. Planktonexped* (G. b.) **2**: 1-120..

TIRMIZI, N.M., AND GHANI, N.A., 1982. New distributional records for three species of *Acetes* (Decapoda, Sergestidae). *Crustaceana*, **42**: 44-53.

TIRMIZI, N.M., AZIZ, N. AND QURESHI, W. M., 1987. Distribution of plankonic shrimp *Sergestes semissis* Burkenroad, 1940 (Decapoda: Sergestidae) in the Indian Ocean with notes on juveniles. *Crustaceana* **53**: 15-28.

TIRMIZI, N. M. AND AZIZ, N., 1988. Rediscovery of *Sergestes verpus* Burkenroad, 1940 (Sergestidae, Crustacea), from the Indian Ocean with description of female. *J. of Nat. Hist., London,* **22**: 199-207.

TIRMIZI, N. M. AND AZIZ, N., 1988. The study of relative growth in the planktonic shrimp, *Sergestes semissis* Burkenroad 1940, collected from the Indian Ocean. *In*:: Marine Science of the Arabian Sea (eds. Thompson Mary-Frances & N. M. Tirmizi). Proceedings of an International Conference March 1986. American Institute of Biological Sciences. 193-205 pp. Washington, D.C.

YALDWYN, J.C., 1957. Deep water crustacea of the genus *Sergestes* (Decapoda, Natantia) from Cook Strait, New Zealand. Zool. *Publ. from Victoria University, Wellington*, **22**: 1-27.

PELAGIC COPEPODS OF NORTH ARABIAN SEA

QUDDUSI B. KAZMI & FARIHA MUNIZA

Marine Reference Collection & Resource Centre, University of Karachi, Karachi

Abstract: This study is based on North Arabian Sea Ecological and Environmental Research samples of two cruises (January 1991, and May 1994) in different seasons. The distribution of subsorted taxa is discussed. During January 1992 and May 1994 different species compositions in pelagic copepods in the north Arabian Sea were noticed. The abundance and distribution did not show much change among the two seasons except in calanoids and poecilostomatoids.

INTRODUCTION

Zooplankton samples collected by the National Institute of Oceanography (NIO) on board R/V 'Behr Paima' for a joint Pak-US cooperative project 'Northern Arabian Sea Ecological and Environmental Research' as NASEER I during January 7 - 22, 1992 and after two and half years in May 10 - 21, 1994, repeated as NASEER IV, both starting from Karachi, covering 1200 nautical miles in the northern and central part of the Arabian Sea and ending up at the Makran coast (Fig. 1) (between 22° 51 to 24° 58N and 60° 05 to 65° 59E). For NASEER I, 32 samples from 18 stations and for NASEER IV, 7 samples from 4 stations are available for analysis, both sets of observations have time series stations (St. 8, 27, 33, 45 & 57).

Distribution and abundance of copepods collected during NASEER-I have already been published (Muniza and Kazmi, in press; Kazmi & Muniza, 1994 and 1995). Now a comparison of analyses made from stations common in both sets of observations is attempted, these two sets were collected in two different seasons i.e., early south west monsoon and early north east monsoon.

MATERIALS AND METHODS

The sampling procedure as communicated by the NIO tells that two times were uesed, as 10 minutes for NASEER I and 5 minutes for NASEER IV in horizontal hauls using bongo net. The data provided to us is date, time, depth, position and flowmeter reading during Cruise I. he sample of Cruise IV are provided only with dates and positions.

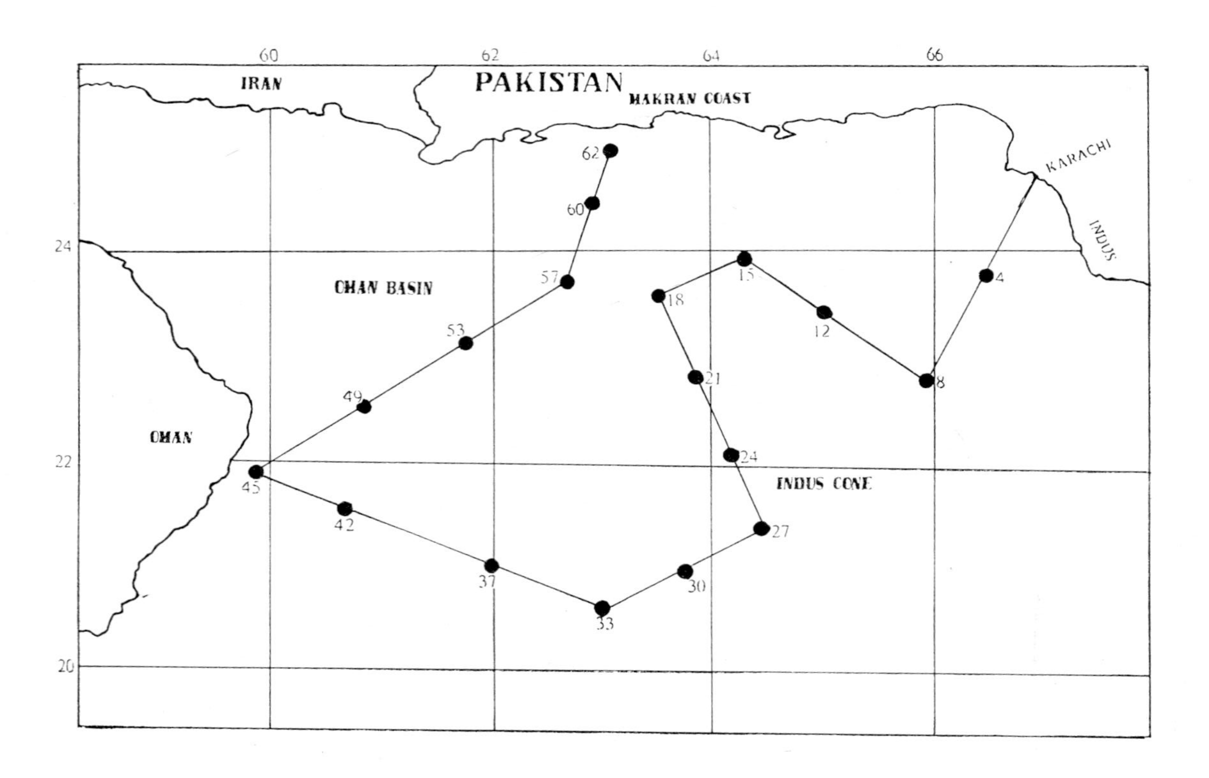

Fig. 1 Map for NASEER cruise.

For the time being the samples are deposited in Marine Reference Collection and Resource Centre (MRC).

Aliquots of 30 ml were taken for counting analysis, and identified upto generic and species levels as far as possible.

In the ensuing account the major groups refer to the four generally recognized orders of the copepods and the dominant taxa; abundance (% age) of common genera from each group is calculated for both the cruises, compared and shown by histograms. Since few genera and some species are still not determined, species richness of the two collections is not comparable.

RESULTS AND DISCUSSION

The copepods were picked and counted, as four general groups - calanoids, harpacticoids, cycopoids and poecilostomatoids. The species were identified as far as possible. Following copepod genera/species were determined from the two sets:

Calanoida: *Eucalanus crassus, Eucalanus pileatus, Eucalanus subtenius, Eucalanus attenuatus, Eucalanus subcrassus, hincalanus nasutus, Rhincalanus cornutus, Acartia* sp., *Paracalanus* sp.,*Calanus* sp., *Cosmocalanus* sp.,*Temora* sp., *Pontella* sp., *Candacia* sp., *Euchaeta* sp., *Undinula vulgaris, Calocalanus pavo*

Harpacticoida: *Microsetella norvegica, Clytemnestra* sp., *Micracia afferata, Macrosetella gracilis*

Cyclopoida: *Oithona* sp.

Poecilostomatoida: *Oncaea venusta, Oncaeamedia,Copilia mirabilis, Sapphirina* sp., *Corycaeus* sp., *Farranula* sp.

The comparison made for the four general group stations indicates (Table I) that the calanoids dominated (up to 90%) in all areas except at St. 27B and 33B of NASEER IV where poecilomatoid (p to 70%) were dominating other groups. The harpacticoid which is the least represented group in NASEER I reached far above the cyclopoids at St. 18, 33B and 37, NASEER IV.

Table I: Copepods present in Naseer Samples I and IV

Sta.	Groups	NASEER I		NASEER IV	
18		12-01-92, 15:00		(18-05-94, 08:55)	
		Total no.	%age	Total no.	%age
	Calanoids	2224	74.00	1905	55.26
	Harpacticoids	2	0.06	267	7.74
	Cyclopoids	86	2.86	25	0.72
	Poecilostomatoid	6932	23.06	1250	36.26
27A		14-01-92, 00:42		(16-05-94,N.A.)	
		Total no.	%age	Total no.	%age
	Calanoids	2655	84.90	1608	62.58
	Harpacticoids	1	0.03	40	1.55
	Cyclopoids	174	5.56	102	3.98
	Poecilostomatoid	297	9.49	820	31.90
27B		14-01-92, 06:25		(17-05-94, 00:30)	
		Total no.	%age	Total no.	%age
	Calanoids	1700	82.16	799	41.48
	Harpacticoids	2	0.09	18	0.93
	Cyclopoids	133	6.42	108	5.60
	Poecilostomatoid	234	11.30	1001	51.97
33B		15-01-92, 22:53		(15-05-94, 16:30)	
		Total no.	%age	Total no.	%age
	Calanoids	1985	79.14	113	22.46
	Harpacticoids	1	0.03	29	5.76
	Cyclopoids	112	4.46	5	0.99
	Poecilostomatoid	410	16.34	356	70.77
33C		16-01-92, 07:15		(15-05-94, 22:30)	
		Total no.	%age	Total no.	%age
	Calanoids	3219	90.88	405	62.69
	Harpacticoids	1	0.02	5	0.77
	Cyclopoids	38	1.07	18	2.78
	Poecilostomatoid	284	8.01	218	33.74
33D		16-01-92, 15:27		(16-05-94, 07-50)	
		Total no.	%age	Total no.	%age
	Calanoids	3350	61.76	2384	62.83
	Harpacticoids	2	0.03	84	2.21
	Cyclopoids	25	0.46	102	2.68
	Poecilostomatoi	2047	37.73	1224	32.26
		17-01-92, 03:18		(14-05-94, 23:45)	
		Total no.	%age	Total	%age
37	Calanoids	3561	74.66	1638	56.32
	Harpacticoids	5	0.10	238	8.18
	Cyclopoids	723	15.16	24	0.82
	Poecilostomatoi	480	10.06	1008	34.66

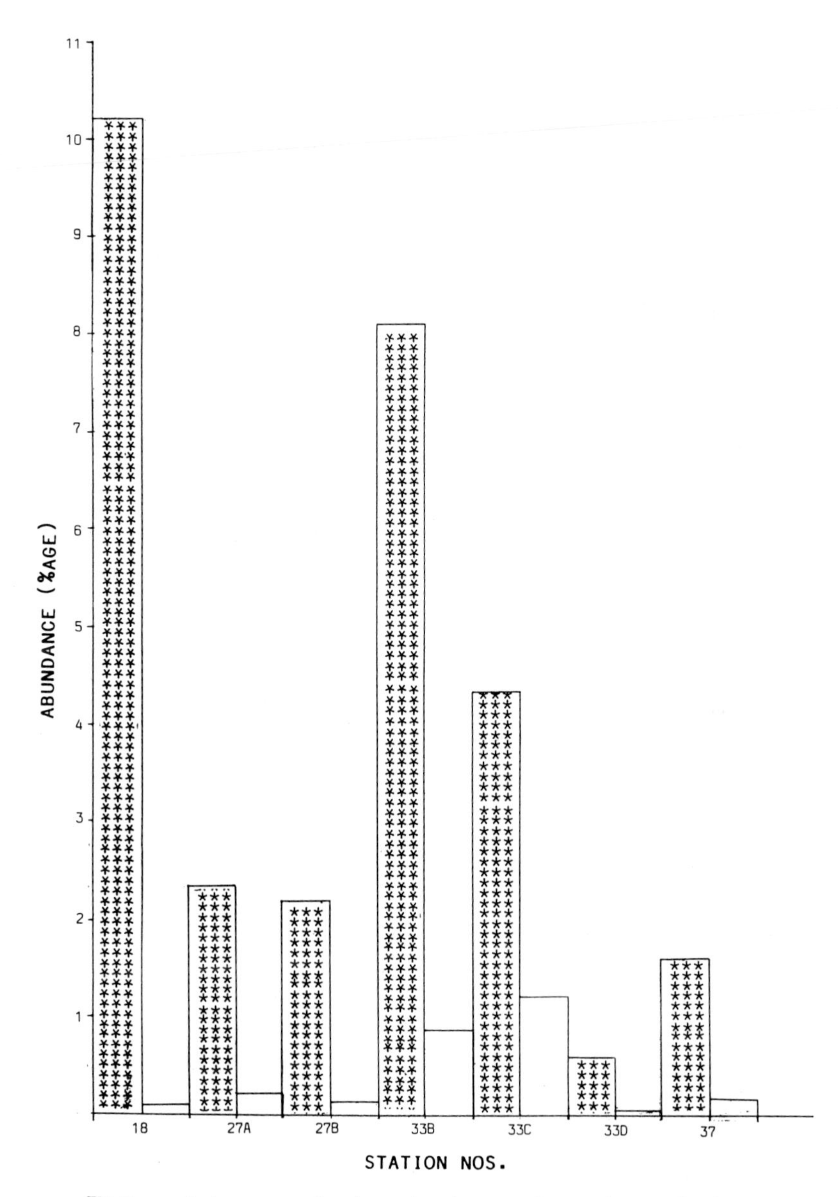

Fig. 2 Histograms showing abundance of _Eucalanus_ during NASEER 1 & NASEER IV.

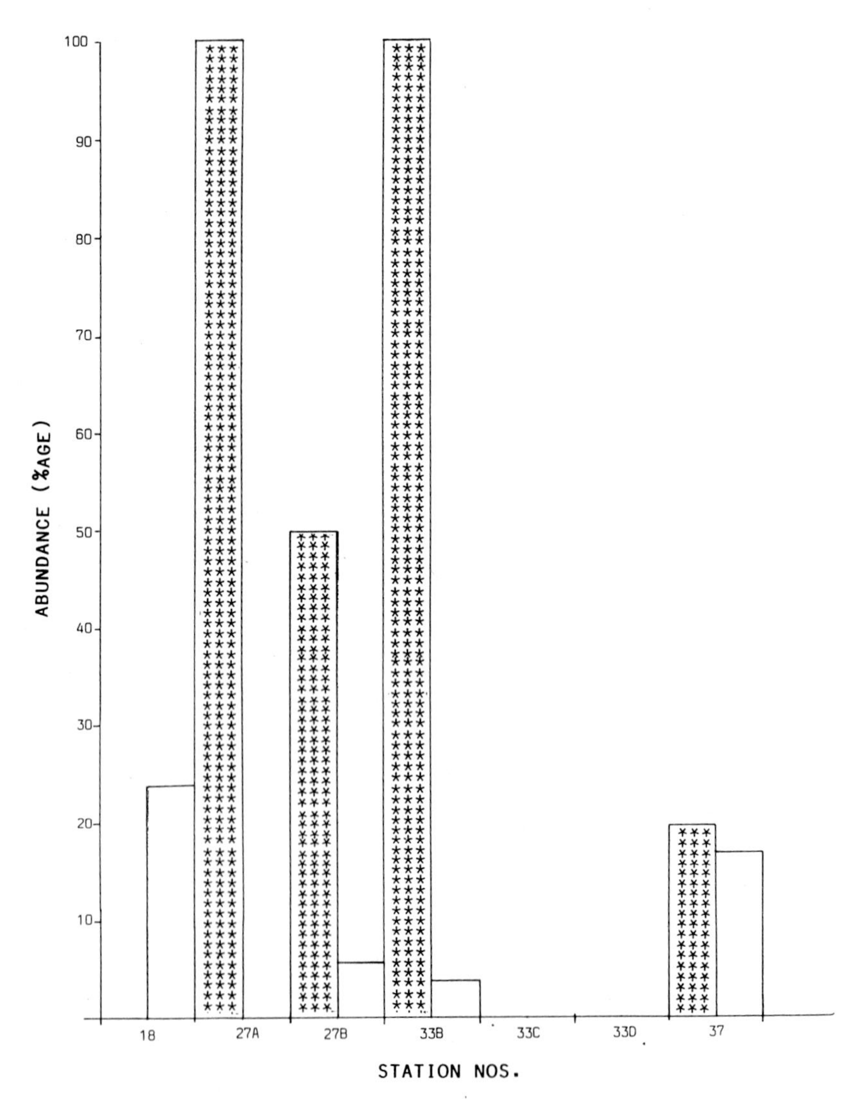

Fig. 3 Histograms showing abundance of <u>Clytemnestra</u> during NASEER 1 & NASEER IV.

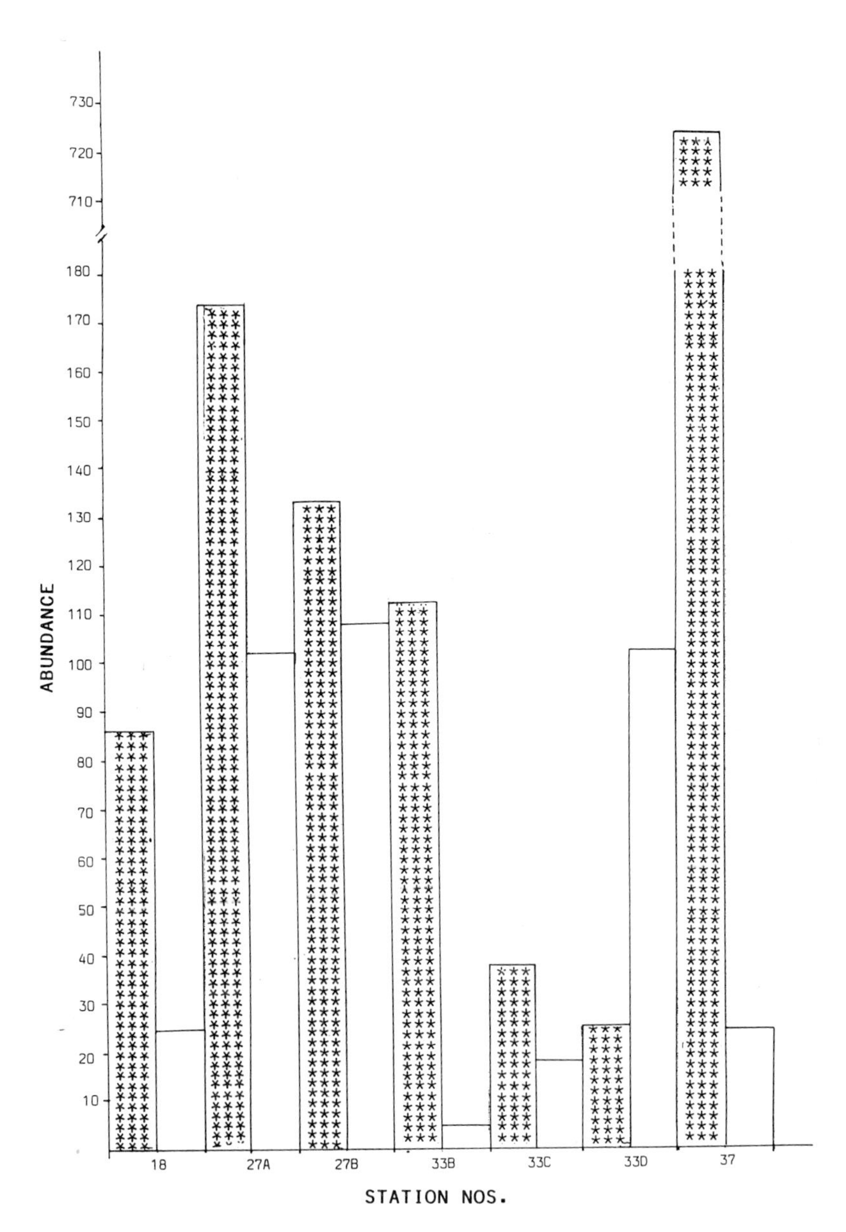

Fig. 4 Histograms showing abundance of <u>Oithona</u> during NASEER 1 & NASEER IV.

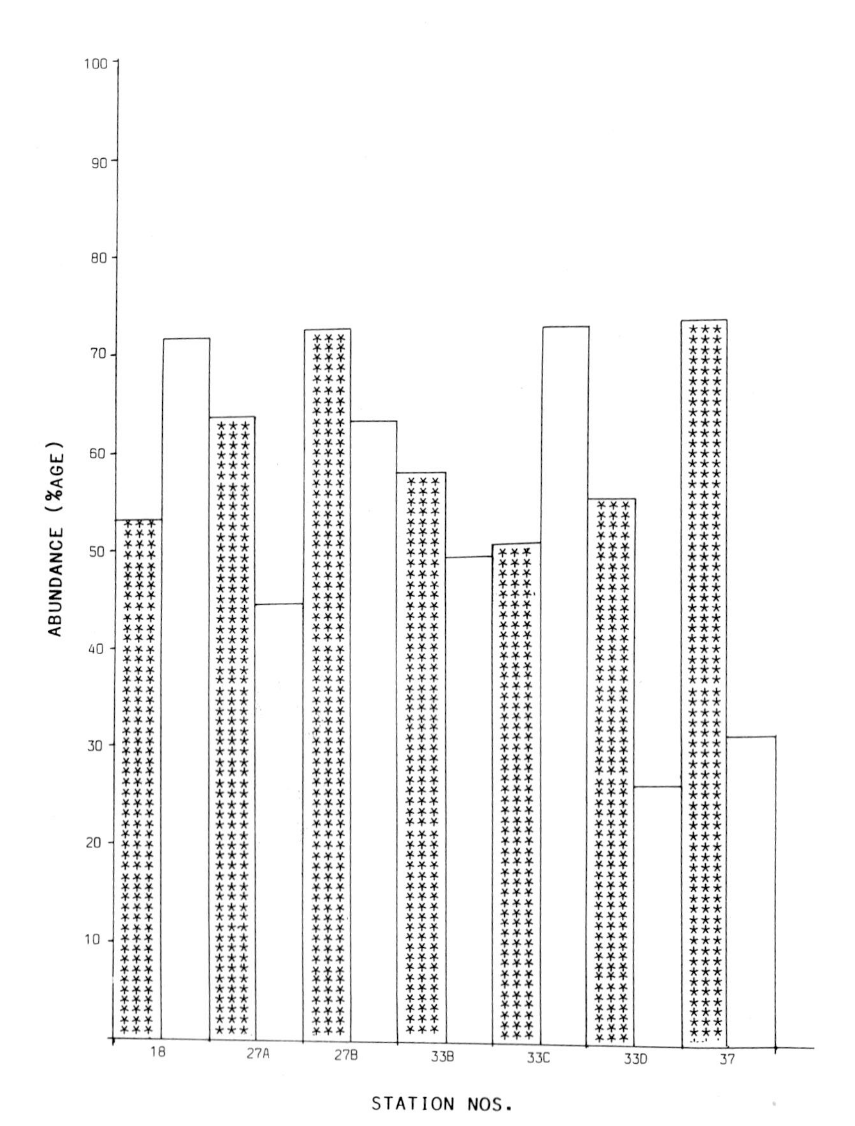

Fig. 5 Histograms showing abundance of <u>Oncaea</u> during NASEER 1 & NASEER IV.

The genera *Rhincalanus, Eucalanus, Acartia, Paracalanus, Calanus, Cosmocalanus, Undinula, Temora, Pontella, Candacia, Euchaeta* and *Calocalanus* were not uniformly abundant in all stations and in both seasons. The genus *Paracalanus* was small sized but most abundant calanoid in both the sets. It was noted that among the calanoids the genus *Rhincalanus* is thinly represented in NASEER I and completely absent among the NASEER IV samples.

Common taxon like *Eucalanus* was upto 10.2% of total calanoids at station 18 during NASEER I, did not attain more than 2% at any of the 7 stations during NASEER IV (Fig. 2). The total count of *Clytemnestra*, a harpacticoid, was far greater in number during NASEER IV than in NASEER I but the percentage among the order is lower due to the presence of other harpacts in the aliquot, whereas total count in NASEER I is low while percentage is high (Fig. 3). For *Oithona* no trend could be set being a sole genus representing cyclopoids (Fig. 4). For *Oncea* over all percentage was lower in NASEER IV than NASEER I. (Fig. 5).

ACKNOWLEDGEMENTS

The samples were kindly made available to the Director, Marine Reference Collection and Resource Centre, University of Karachi through National Institute of Oceanography. Financial assistance was provided by US Office of Naval Research through Grant No. N00014-86-G-0229.

REFERENCES

KAZMI, Q.B. AND MUNIZA, F., 1994. Notes on harpacticoids copepods (Crustacea) gathered from plankton collection during NASEER Cruise I January, 1992 in the Arabian Sea. *Proc. Pakistan Congr. Zool.,* **14**: 151-156.

KAZMI, Q.B. AND MUNIZA, F., 1995. Distribution and abundance of poecilostomatoid population of copepods in NASEER Cruise I January, 1992. 15th Zoological Congress of Pakistan

MUNIZA, F. AND KAZMI. Q.B., Occurrence and abundance of two genera: *Eucalanus* and *Rhincalanus* of the family Eucalanidae (Copepoda: Calanoida) in the samples of NASEER I, January 1992 *Science International, Lahore* (in press).

BIODIVERSITY OF GASTROPODS IN THE EOCENE TIME DURING THE CLOSURE OF TETHYS SEA IN THE CENTRAL SALT RANGE, PAKISTAN

S. R. H. BAQRI, S. AZHAR HASAN, SANJEEDA KHATOON, NAYYER IQBAL

Pakistan Museum of Natural History, Garden Avenue, Shakarparian, Islamabad

Abstract: Gastropod fauna of Palaeo-Tethys Sea from the basal calcareous shales to argillaceous limestones/marls of the Bhadrar Formation represents 14 species of 8 genera, of 8 families and of 3 orders. Six species, *Conus mahmoodi, Euspira cosmanni, Gosavia fatmi, Harpa archiaci, Harpa muftii and Tibia nurpurensis,* have been recognized as new to science. *Gosavia humberti, Tibia nurpurensis nov., Crommium roualti,* and *Gisortia murchisoni* are the most common species observed in these rocks.

INTRODUCTION

The Salt Range (Fig. 1) is known as a "geological museum" of Pakistan due to the preservation of systematic geological records with special reference to the complete stratigraphical sections and palaeo-fauna. The rocks range in age from Cambrian to Pleistocene carrying well preserved fossils of various ages. The Bhadrar Formation, also called Chorgali Formation, of Eocene age is exposed in the central Salt Range. The formation lies on the Sakesar limestones with a local unconformity marked by a conglomeratic bed (Fig. 2). It consists of mostly limestone with calcareous clays, shales and argillaceous limestone. The type locality is exposed at the Khaire-Murat while the principal reference section is exposed at Bhadrar (50 feet thick) in the central Salt Range.

The Bhadrar Formation has been studied by several workers to understand its stratigraphic position and regional extension. Pinfold (1918) called the Bhadrar Formation as the passage beds on the basis of the lithology. Gee and Evens (1937 in Davies and Pinfold) carried out the geological studies in the Salt Range and called these rocks as Bhadrar beds. Fatmi (1973) described the Bhadrar Formation at its principal reference section where it is 50 feet thick and may be divided into two units. The lower unit consists of calcareous shales of greenish

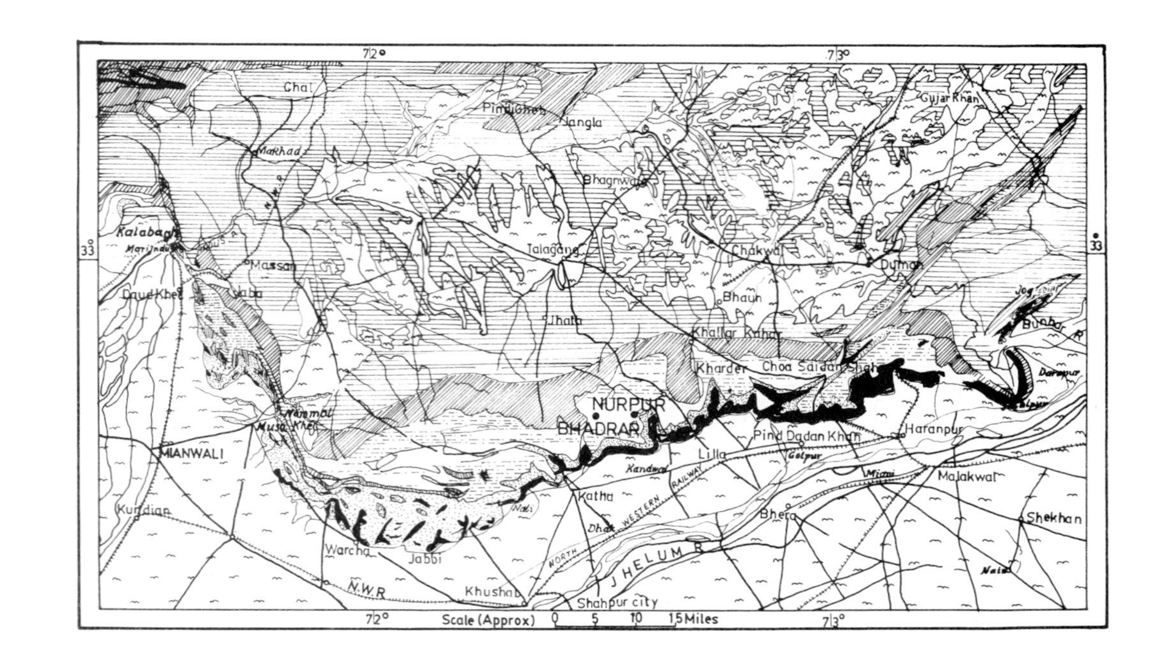

Alluvion (Recent and Plestocene)

Upper and Middle Siwalik

Lower Siwalik and Murree

Eocene (Kohat Saline series shown)

Mesozoic

Upper Palaeozoic

Cambrian or Pre-cambrian

Cambrian or Pre-cambrian

Fig.1. GEOLOGICAL MAP OF THE PUNJAB SALT RANGE SHOWING THE LOCATION AREA OF NURPUR.

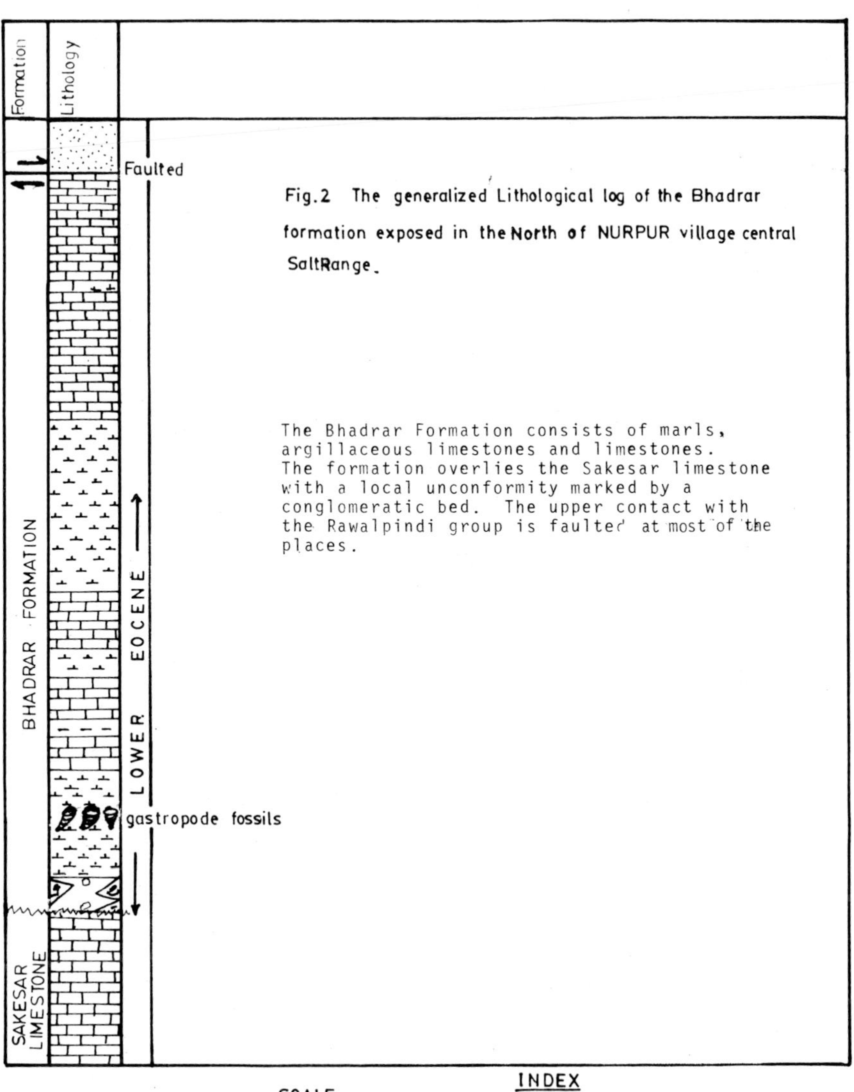

Fig.2 The generalized Lithological log of the Bhadrar formation exposed in the North of NURPUR village central SaltRange.

The Bhadrar Formation consists of marls, argillaceous limestones and limestones. The formation overlies the Sakesar limestone with a local unconformity marked by a conglomeratic bed. The upper contact with the Rawalpindi group is faulted at most of the places.

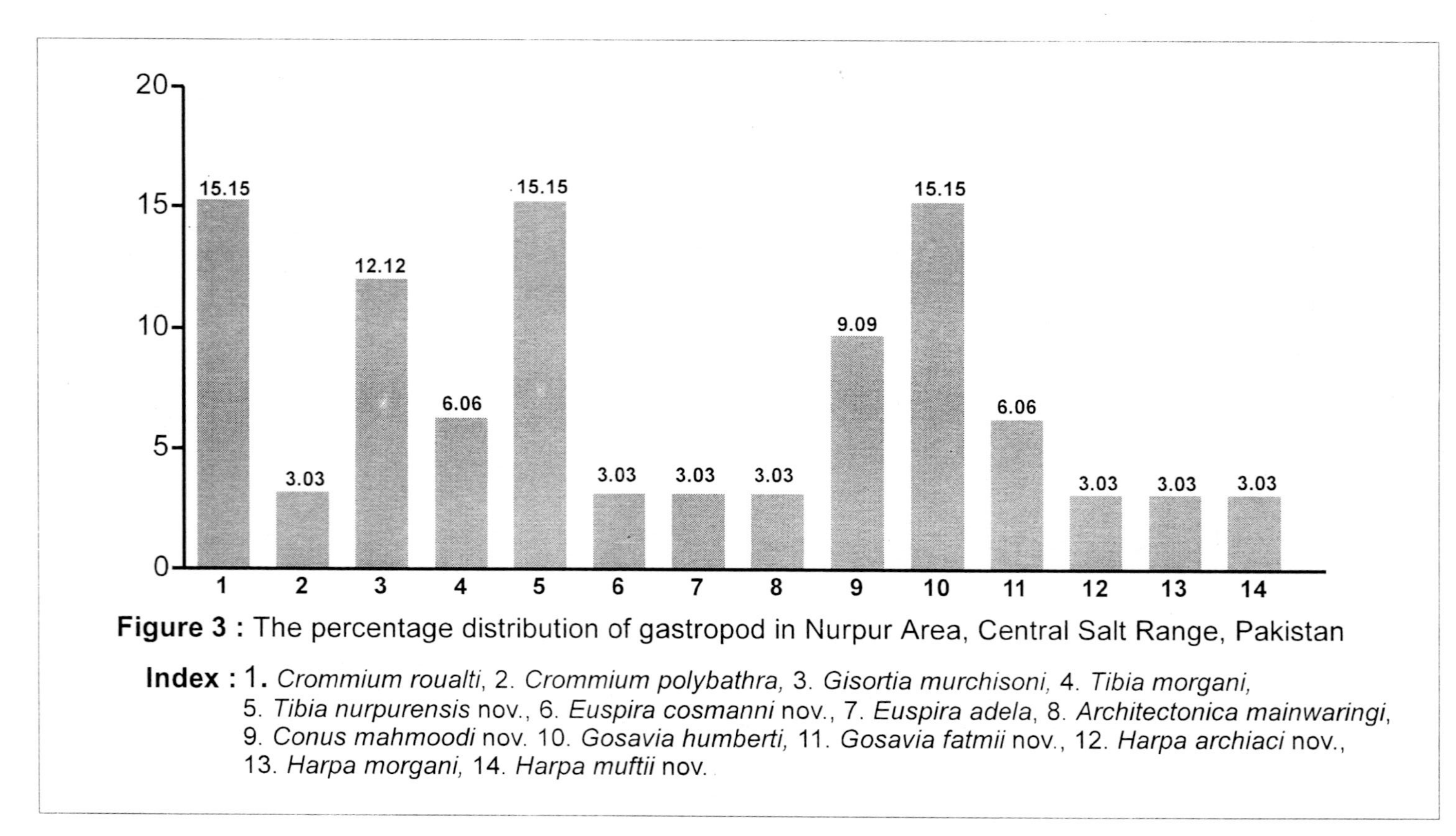

Figure 3 : The percentage distribution of gastropod in Nurpur Area, Central Salt Range, Pakistan

Index : 1. *Crommium roualti*, 2. *Crommium polybathra,* 3. *Gisortia murchisoni,* 4. *Tibia morgani,* 5. *Tibia nurpurensis* nov., 6. *Euspira cosmanni* nov., 7. *Euspira adela*, 8. *Architectonica mainwaringi*, 9. *Conus mahmoodi* nov. 10. *Gosavia humberti,* 11. *Gosavia fatmii* nov., 12. *Harpa archiaci* nov., 13. *Harpa morgani,* 14. *Harpa muftii* nov.

gray green or buff colour and argillaceous limestones. The upper unit is mainly composed of limestones which are white to cream coloured and well bedded. He quoted Gill (1953) and further explained that the lower unit gradually displays the development of calcareous facies from east to west. Fatmi (1973) described its stratigraphic position in the Salt Range and other parts of the Potwar. He stated that the formation confirmably rests on Sakesar limestone in the Salt Range. The upper contact with the Murree Formation is unconformable. Eames (1951, 1952) reported 96 species of gastropods from the Zindapir and other localities and two species from the Charat group exposed in the Shekhan Nala in the district of Kohat. Iqbal (1969) studied the Tertiary gastropod fauna from Drug, Zindapir, Vidor (Distt. D. G. Khan), and Chharat areas and described 37 species of gastropods. He also reported four gastropod species which are restricted to lower Eocene. Same author in 1972 studied the Mega fauna from the Nammal Gorge of the Salt Range and reported the presence of gastropods in the Lockhart Limestone of the Paleocene age. Fatmi (1973) reported the presence of Mollusc fauna such as *Companile gigantum, Euspirocromium oweni, Gosavia humberti, Velates perversus, Vicetia vredenburgi, Discors sp. Euphenax coxi, Trachycadium cotteri, Deltoidnautilus* sp., *Nautilus labechiin* the rocks of the lower Eocene age. Shah (1977) stated that the Bhadrar Formation displays the Mollusc fossils of Early Eocene age.

The present studies were conducted to investigate the Palaeo-environements during the deposition of Bhadrar Formation and the biodiversity of the gastropods from the Nur Pur area of the central Salt Range (Fig. 1). Thirty three gastropods were collected for detailed studies and identifications from the basal argillaceous/marly beds of the Bhadrar Formation (Fig. 2).

RESULTS

The fossils were compared with other works and identified upto species level. Species percentage distribution (Fig. 3) shows *Cromium rouaulti, Gisortia murchisoni* and *Tibia morgani* as the most abundant species. The stratigraphic distribution which varies between Paleocene to Miocene, was compared with the presently identified gastropod fossils of other workers (Table I). Species identified were as:

Phylum : Mollusca
Class : Gastropoda
Order : Ctenobranchiata
Family : Apullosridae
Genus : *Crommium*
Species: *Crommium polybathra* Cossmann & Pissarro, 1909 (Fig. 4)

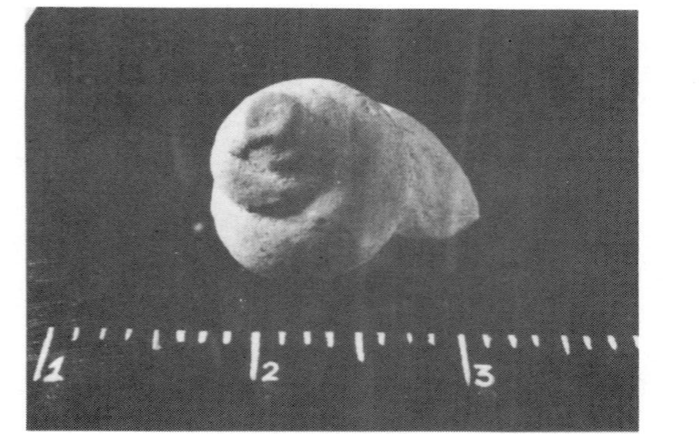

Fig. 4 *Crommium polybathra*

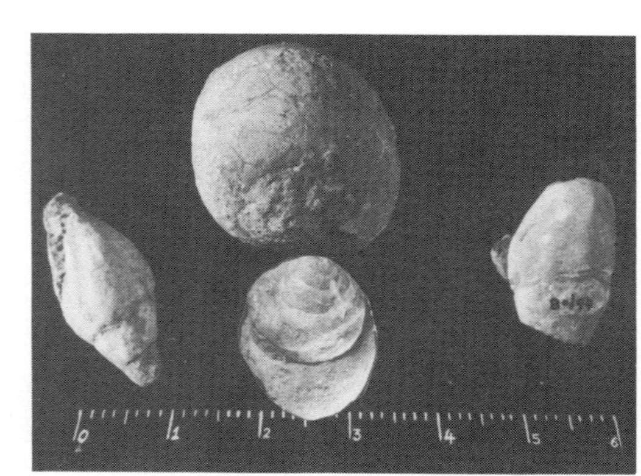

Fig. 5 *Crommium roualti* (LM) , *Tibia morgani* (L & R), *Gistoria murchisoni* (UM)

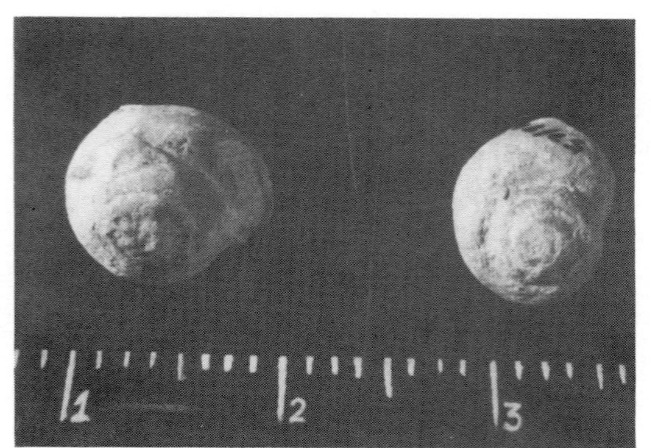

Fig. 6. *Euspira cosmanni nov.* (Left), *Euspira adela* (Right)

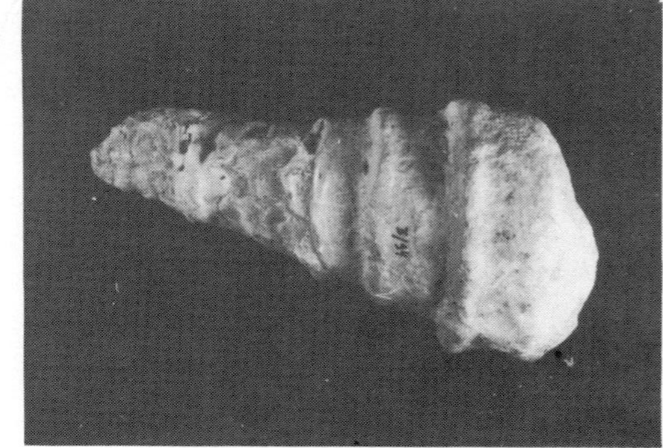

Fig. 7 *Tibia nurpurensis* nov.

Ampulina (*Crommium*) *polybathara* Cossmann & Pissarro,1909

Mus. No. 64/94

Stratigraphic Distribution: Ranges in Pakistan from Paleocene to lower Eocene

Species: *Crommium rouaulti* (Archiac & Haime) 1854 (Fig. 5)

(Mus. No. 34/94, 43/94, 60/94, 82/94, 90/94)

Stratigraphic Distribution: Ranges in Pakistan from Paleocene to Oligocene

Genus : *Gisortia*

Species: *Gisortia murchisoni* (Archiac & Haime), 1854. (Fig. 5)

Mus. No. 29/94, 30/94, 111/94, 126/94 of variable dimensions

Stratigraphic Distribution: Ranges in Pakistan from Paleocene To Lower Eocene

Family : Strombidae

Genus: *Tibia*

Species: *Tibia morgani* (Cosmann & Pissarro),1909 (Fig. 5)

Mus. No. 76/94, 80/94

Stratigraphic Distribution: Ranges in Pakistan from Paleocene to Middle Eocene

Species: *Tibia nurpurensis* nov. (Fig. 7)

Mus. No. 2/94,3/94,109/94,132/94,78/94

Stratigraphic Distribution: Eocene to Recent In India in Miocene In other parts of the world (Miocene to Eocene)

Tibia nurpurensis differs from the other species of the genus by being largest is size (227 mm in height and 90 mm in diameter).

Family: Naticidae

Genus: *Euspira*

Species: *Euspira adela* (Cossmann & Pissarro) 1909 (Fig. 6)

Mus. No 57a/94
Stratigraphic Distribution: Paleocene to lower Eocene.

Species: *Euspira cosmanni* nov. (Fig. 6)

Mus. No. 57 b/94

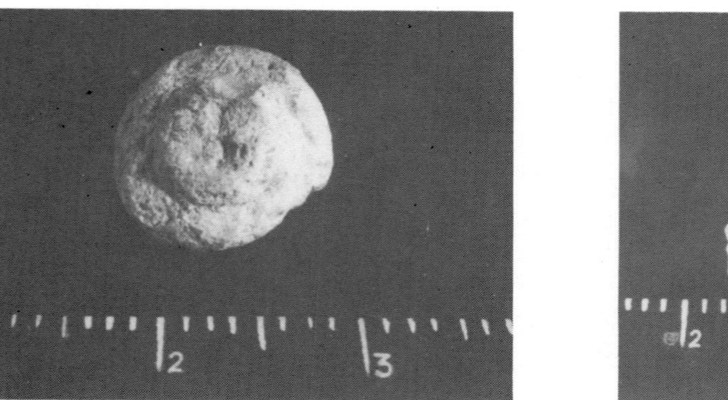

Fig. 8 *Architectonica mainwaringi*

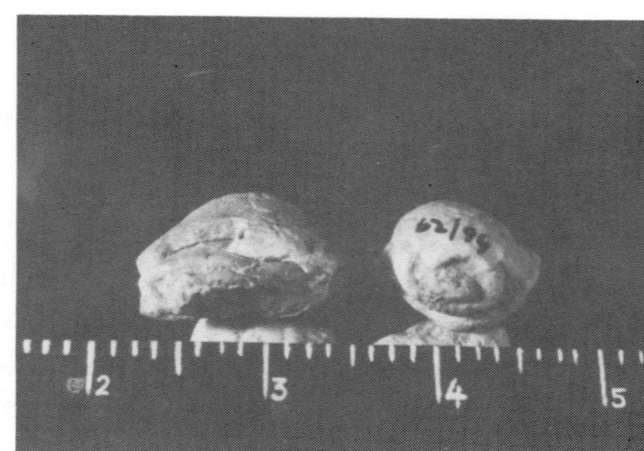

Fig. 9 *Gasavia humberti*

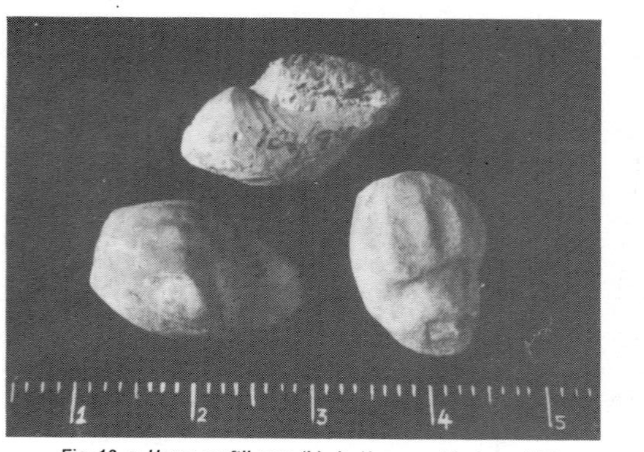

Fig. 10 a. *Harpa muftii* nov. (L), b. *Harpa archiaci nov.* (M), c. *Harpa morgani* (R)

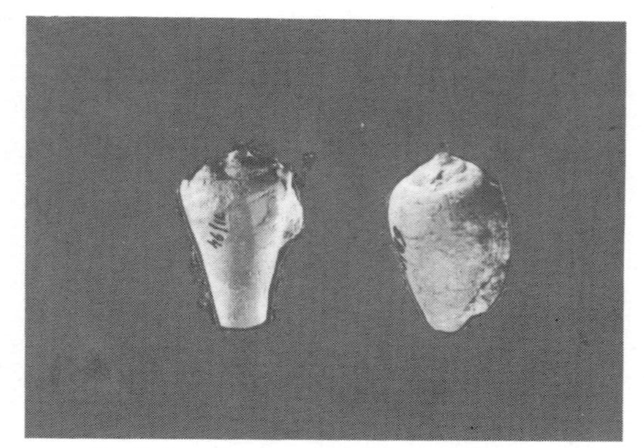

Fig. 11 *Gosavia fatmi* nov. (L), *Conus mahmood nov.* (R)

Stratigraphic Distribution: Restricted to lower Eocene.

Euspira cosmanni differs from its closely related species, *Euspira adela,* in the shape and size of the whorls. The species was named after renowned Palaeontology, Dr. Cossmann, who has described a number of species.

Family : Architectonidea

Genus: *Architectonica*

Species: *Architectonica mainwaringi* Cox 1930 (Fig. 8)

Mus. No. 45/94
Stratigraphic Distribution: Lower Eocene to Paleocene

Order: Stenoglossa

Family: Conidae

Genus: *Conus*

Species: *Conus mahmoodi* nov. (Fig. 11)

Mus. No. 47/94,68/94,87/94
Stratigraphic Distribution: Restricted to lower Eocene

(Note: *Conus mahmoodi* is medium sized with the apical whorl oblique and apex pointed. It differs from the closely related species of the genus in size and shape. The species was named to honour Dr. Khalid Mahmood, Chairman, Pakistan Science Foundation, for his contribution in the field of science.

Order: Rachiglossa

Family: Volutidae

Genus: *Gosavia*

Species: *Gosavia humberti* (Archaic & Haime) 1854 (Fig. 9)

Mus. No. 32/94, 84/94, 88/94, 95/94, 96/94
Stratigraphic Distribution: Ranges in Pakistan from Paleocene to lower Eocene.

Species: *Gosavia fatmii* nov (Fig. 11)

Mus. No. 44/94, 71/94
Stratigraphic Distribution: Restricted to lower Eocene

(Note: The species differs from *Gosavia humberti* by having a blunt apex, straight apical spiral and invisible transverse striations. It was named to honour Dr. A.N. Fatmi, a renowned Palaeontology.

Table I : Distribution of Gastropods species in Central Salt Range and in other parts of Pakistan

Species	1	2	3	4	5	6	7	8	9	10	11	12	13
Architectonica mainwarigi	-	-	-	-	Pu+	Pu+	-	-	-	-	Pu+	El+	PuEL
Conusmahmoodi sp. nov	-	-	-	-	-	-	-	-	-	-	-	El+	El+
Cromium polybathra	-	El+	-	-	Pu+	PEl+	-	-	-	-	-	El+	PuEl
Cromium rouaulti	-	Em+	-	-	Pu+	-	O+	P+	-	-	P+	El+	PuElmO
Euspira cosmani sp. nov.	-	-	-	-	-	-	-	-	-	-	-	El+	El+
Euspira adela	-	-	-	-	Pu+	-	-	-	-	-	-	El+	PuEl
Gosavia humberti	Elm	Elm+	Elm+	Em +	Pu+Elm	PElm+	-	Pl+	El+	El +	El+	El+	PuElm
Gosavia fatmii sp. nov.	-	-	-	-	-	-	-	-	-	-	-	El+	El+
Gisortia murchisoni	-	El+	El+	-	Pu+ El	PEl+	-	Pl+	-	El +	-	El+	PuEl
Harpa morgani	-	-	-	-	Pu+	Pu+	-	-	-	-	-	El+	PuEl
Harpa archiaci sp. nov.	-	-	-	-	-	-	-	-	-	-	-	El+	El+
Harpa muftii sp. nov.	-	-	-	-	-	-	-	-	-	-	-	El+	El+
Tibia morgani	-	Em+	-	-	Pu+	-	-	-	-	-	-	El+	PuElm
Tibianurpurensis sp.nov.	-	-	-	-	-	-	-	-	-	-	-	El+	El+

Index : Drug= 1, Zinda Pir = 2, Rakhi Nala =3, Baluchistan =4, Sind = 5, Salt Range Punjab =6, Vidor D.G. Khan = 7, Jhalar =8, Panoba = 9, Shekhan nala = 10, Hangu NWFP Kohat 11, Central Salt Range Nurpur Area (Present work) 12, General Distribution = 13, Present = +, Absent = -, Lower = L, Middle = M, Upper = U, Eocene = E, Oligocene = O, Paleocene = P.

Genus: *Harpa*

Species: *Harpa* (*Eocithara*) *morgani* Cossmann & Pissarro, 1909 (Fig. 10c)

Mus. No. 49/94

Stratigraphic Distribution: Ranges in Pakistan from Paleocene to Eocene.

Species: *Harpa archiaci* nov. (Fig. 10 b)

Mus. No. 163/94

Stratigraphic Distribution: Restricted to Lower Eocene

(Note: The species differs from the other species of the genus by having a much shorter and depressed apical whorl than the last whorl.

Species: *Harpa muftii* nov. (Fig. 10 a)

Mus. No. 46/94

Stratigraphic Distribution: Lower Eocene in Pakistan

(Note: The species differed from the closely related species by being large in size, having rounded aperture and the last whorl with six longitudinal costa. It was named to honor Dr. Shahzad A. Mufti who has immensely contributed in biodiversity research.

DISCUSSION

The Palaeo Tethys sea during the deposition of the Bhadrar Formation in the Central Salt Range was suddenly uplifted and changed to shallow water conditions due to tectonic movements leading to its ultimate closure. This is indicated by the presence of a local conglomeratic bed at the base of the Bhadrar Formation. The animal life present in the Tethys sea suffered drastic environmental changes as the marine waters were suddenly changed to brackish and fresh water conditions. The changing environmental conditions due to uplifting, tectonic movements and faster sedimentation gradually became hazardous for the gastropods. The clean warm waters of the sea turned to muddy waters as the silt and clays entered into the basin and created environmental pollution for the marine life. Most of the marine life suddenly vanished, buried into the sediments and was finally converted into the fossils. These fossils appear to be buried in such a number that the area appears to be a graveyard of the gastropods.

REFERENCES

DAVIES, L.M. AND PINFOLD, E. S., 1937. The Eocene beds of the Pakistan Salt Range. *India Geol. Surv., Mem. Palaeontology. Indica, New Series,* **24** : 1- 79.

EAMES, F. E., 1951. A contribution to the study of the Eocene in western Pakistan and western India: The contribution of Lamellibranchia from standard sections in the Rakhi Nala and Zinda Pir areas of the western Punjab and in the Kohat District. Royal Soc. London, *Phil. Trans. Ser. B,* **235**: 311-482.

EAMES, F. E., 1952. A contribution to the study Eocene in western Pakistan and western india: The geology of standard sections in the western Punjab and in Kohat District. *Geol. Soc. London, Quart. Jour.*, **107:** 159-172.

FATMI, A. N., 1973. Lithostratigraphic units of the Kohat Potwar pronvince, Indus Basin, Pakistan: *Pakistan Geol. Surv. Mam.*, **10:** 1- 80.

GILL, W. D., 1952. Facies and fauna in the Bhadrar beds of the Punjab Salt Range. *Pakistan: Jour, Palaeontology.*, **27:** 824-844.

IQBAL, M. W. A., 1969. The Tertiary Pelecepods and Gastropods Fauna from Drug, Zindapir, Vidor (District D. G. Khan) Jhallar and Charat (District Campbellpore), West Pakistan. Mem. Geo. Survey Pakistan, *Palaeontology Pakistanica,* **6**:1- 94.

IQBAL, M. W. A., 1972. Bivalve and gastropod from Jherruk Lakhra - Bara Nai (Sind), Salt Range, Punjab and Samana Range (NWFP), Pakistan.

MOORE, P.C.,1964. Treatise on Invertebrate Palaeontology. Part K. Mollusca - Cephalopoda - General features. Endoceratoidea - Actinoceratoidea-Nautiloidea-Bactritoidea. Geo. Soc. Ame. Inc. Univ. Kansas.

MOORE, P.C., 1971.*Treatise on Invertebrate Palaeontology.* Part N. Vol. **3.** Mollusca - Bivalvia. Geo. Soc. Amer. Inc. Univ. Kansas.

MURRAY, J. W., 1984. *Atlas of Invertebrates Macrofossils.* Longman Palaeontology Association.

PINFOLD E.S., 1918. Notes on the structure and stratigraphy in the west Punjab, *India geol. Surv., Res.*, **49:**137-160.

SHAH, S. M. I., 1977. Stratigraphy of Pakistan .*Memoires of GSP,* **12** : 1-138.

WOOD, H., 1967. *Palaeontology Invertebrates.* Cambridge University Press. **8:** 1-477.

BIODIVERSITY OF PLANT PARASITIC NEMATODES IN PAKISTAN

MOHAMMAD A. MAQBOOL

National Nematological Research Centre, University of Karachi, Karachi

Abstract:According to recent calculations micro-organisms and invertebrates together constitute 88% of the species on earth and are crucial to the maintenance of biodiversity as components of community structure. Nematodes are second only to insects in the number of species in the animal kindom. In Pakistan systematic studies carried out on soil and plant parasitic nematodes, has revealed a total of 355 nematode species belonging to 58 genera, 36 subfamilies, 21 families, 9 superfamilies, 3 suborders and 3 orders.

INTRODUCTION

It is a matter of great satisfaction that concepts of Biodiversity and biosystematics are receiving due attention by the scientific community of the world. It is only due to the fact that the implications of declining biodiversity for sustainable agricultural production and environmental protection have now been recognized. However, while justifiable concern is expressed at the need to conserve and prevent from extinction the larger flora and fauna of the world, the importance of micro-organisms and invertebrates in the stable functioning of ecosystems has attracted less attention. Nevertheless, this subject is now recognized as of major significance for a number of issues, such as maintenance of soil fertility and provision of natural enemies for the biological control of pests and pathogens.

Micro-organisms and invertebrates together constitute 88% of the species on earth and are crucial to the maintenance of biodiversity as components of community structure, keystone predators, herbivores, pests and multualists, and environmental bioindicators (Hawksworth & Ritchie,1993). They can contribute to sustainability in nitrogen fixation, mycorrhiza formation, maintaining soil fertility, waste utilization, pollination, pest management, human and animal diseases, natural enemies; biocontrol, bio-products, and quarantine. Only 5-10% of the species in these groups have been recognized, with microorganisms, being proportionately half as well known as the invertebrates.

It is quite interesting that existing resources for microorganisms and invertebrates in developing countries are inadequate to support the demands of biodiversity studies, whereas in developed countries, biosystematics is in a crisis in terms of human resources and its ability to serve identification needs. Reference collections in the UK are almost unparalleled worldwide and it is pre-eminent in the compilation and delivery of biosystematic information. A need for more collaboration and networking between developed and developing country institutions is quite essential.

NEMATODES AND THEIR ENVIRONMENTS

Nematodes are diverse group of roundworms that occur worldwide in virtually every environment; Nematodes are second only to insects in the number of species in the animal kindom. However, only about 3 percent of all nematode species have been studied and identified (Ferris *et.al.,*1972). The thousands of different species include parasites of plants, insects, other animals, and humans, and many that feed on bacteria, fungi, algae, and other nematodes. Nematodes, which are nonsegmented roundworms, have complete nervous, digestive, excretory, muscular, and reproductive systems. These animals range in size from microscopic inhabitants of certain mushrooms, to 27-foot-long parasites that live in the placentas of sperm whales (Hussey, 1994).

Because of the diversity and importance of nematodes, their study is interrelated to other sciences, including invertebrate zoology, medicine, parasitology, plant pathology, microbiology, ecology, marine biology, and environmental toxicology. Research in these fields has revealed that nematodes can directly and indirectly affect our very standard of living.

IDENTIFICATION OF NEMATODE SPECIES

Efficient and accurate identification of nematode species is essential for effective nematode management and regulatory programs. As with all organisms, nematodes vary in shape and size, which can make accurate identification difficult. Biotechnology is beginning to provide tools that can dramatically improve our ability to detect, identify, and quantify nematodes in the developed countries.

It is necessary to make biosystematic studies of the nematode species which may be either harmful to agricultural crops or can play beneficial role in the environment. One of the greatest challenges is understanding and reaching to the roles nematodes play in determining the success or failure of agricultural crops. In fact, their impact is probably even greater because the plant damage and low yields caused by parasitic nematodes frequently go unrecognized or are attributed to other causes (Van Gundy, 1980). Furthermore, new nematode species that cause crop damage are continually being discovered. Nematodes also cause severe damage to ornamental plants, lawns, golf courses, turfgrasses, greenhouse plants, fruit trees, forest trees, and home garden plants (Webster, 1980).

Most nematodes in agricultural and natural habitats are beneficial and contribute to soil processes that enhance plant growth (Ferris *et. al.* 1972). Beneficial nematodes feed on the bacteria and fungi that decompose organic matter and thus influence the balance of carbon in the soil and the release of nutrients used by plants. Because sustainable crop production will increase the food base for these nematodes, studies from the microorganism to ecosystem could clarify which nematodes influence plant nutrient cycling and how these organisms can be managed. In general, the biodiversity of nematodes reflects the amount of human and natural disturbances that alter soil, freshwater, and marine ecosystems. For this reason, nematodes can be sensitive indicators of ecosystem health. By understanding beneficial nematodes in agricultural and natural habitats, we can promote the sustained use of those habitats and monitor our success at achieving beneficial nematode use.

Future nematode management must employ sustainable agricultural practices that take into account beneficial, detrimental, and other nematode species associated with plant roots and soil. Much can be learned from the "biological balance" in natural ecosystems that have minimal changes in the biotic and physical environment. Still, research is required to identify, select, and adopt cropping systems, including cover crops, antagonistic crops, green manure crops, inter-planting, rotations, organic amendments, and minimal tillage, that would enhance populations of beneficial nematodes and other fauna.

As the need for research, education and extension work in plant nematology is unavoidable, the available resources and support for these activities are negligible but if agricultural productivity is to be increased to meet present and future demands, nematology research must be intensified.

NEW NEMATODE SPECIES DESCRIBED FROM PAKISTAN

Quinisulcius solani	Maqbool, 1982
Paurodontella sohailai	Maqbool, 1982
Nothotylenchus goldeni	Maqbool, 1982
Hemicriconemoides ghaffari	Maqbool, 1982
Boleodorus zaini	Maqbool, 1982

Paktylenchus tuberosus	Maqbool, 1983
Merlinius niazae	Maqbool *et al.*, 1983
Aglenchus mardanensis	Maqbool *et al.*, 1984
Basiroides sindhicus	Maqbool *et al.*, 1984
Basiroides citri	Maqbool et al., 1984
Leipotylenchus amiri	Maqbool *et al.*, 1984
Orientylus karachiensis	Maqbool *et. al.*,1984
Scutylenchs quettensis	Maqbool *et al.,* 1984
Cephalenchus sacchari	Maqbool *et al.,* 1984
Dolichorhynchus tuberosus	Maqbool *et al.,* 1984
Criconemella anastomoides	Maqbool & Shahina, 1985
Malenchus labiatus	Maqbool & Shahina, 1985
M. pyri	Maqbool & Shahina, 1985
Pararotylenchus microstylus	Maqbool *et al.*,1985
Scutylenchs baluchiensis	Maqbool *et al.*,1985
Karachienema elongatum	Maqbool & Shahina, 1985
Ottolenchus azadkashmirensis	Maqbool & Shahina, 1985
O. longicauda	Maqbool & Shahina, 1985
Rotylenchs fragaricus	Maqbool & Shahina, 1985
R. pakistanensis	Maqbool & Shahina, 1985
R. alli	Maqbool & Shahina, 1985
Helicotylenchus obliquus	Maqbool & Shahina, 1985
Heterodera pakistanensis	Maqbool & Shahina, 1986
Aulophora karachiensis	Maqbool *et al.*, 1986
Hemicycliophora veechi	Maqbool *et al.*, 1986
Boleodorus arachis	Maqbool & Ghazala, 1986
Cephalenchus longicaudatus	Maqbool & Ghazala, 1986
Tylenchorhynchus quaidi	Golden *et al.,* 1987
T. tritici	Golden *et al.,* 1987
Nagelus saifulmulukensis	Maqbool & Shahina, 1987
Merlinius montanus	Maqbool & Shahina, 1987
Tlenchus naranensis	Maqbool et al., 1987
T. skarduensis	Maqbool & Shahina, 1987
T. bhitai	Maqbool & Shahina, 1987
Heterodera bergeniae	Maqbool & Shahina, 1988
Neothada major	Maqbool & Shahina, 1988
Pratylenchoides maqsoodi	Maqbool & Shahina, 1989
Heterodera cynodontis	Shahina & Maqbool, *1989*
Hoplolaimus tabacum	Firoza *et al.,* 1990
Amplimerlinius parbati	Zarina & Maqbool, 1990
Ogma sadabhari	Shahina & Maqbool, 1990
O. multiannulata	Shahina & Maqbool, 1990
O. qamari	Shahina & Maqbool, 1990
Neopsilenchus peshawarensis	Shahina & Maqbool,1990
N. (Neopsilenchus) curvistylus	Shahina & Maqbool, 1990

N. *(Acusilenchus) bilineatus*	Shahina & Maqbool, 1990
Bolenodorus azadkashmirensis	Maqbool *et al.,* 1990
Rotylenchus capsicumi	Firoza & Maqbool, 1991
Tylenchorhynchus swatensis	Nasira et al., 1991
Helicotylenchus verecundus	Zarina & Maqbool, 1991
Tylenchorhynchus rosei	Zarina & Maqbool, 1991
Tylenchus pakistanensis	Farooqi *et al.,*1991
Xiphinema karachiense	Nasira *et al.,* 1991
Quinisulcius quaidi	Zarina & Maqbool, 1992
Deladensus pakistanensis	Shahina & Maqbool, 1992
Merlinius pistaciei	Fatima & Farooq, 1992
M. pyri	Fatima & Farooq, 1992
Filenchus sindihicus	Shahina & Maqbool, 1993
Gracilacus musae	Shahina & Maqbool, 1993
Criconemoides afganicus	Shahina & Maqbool, 1993
Helicotylenchus microtylus	Firoza & Maqbool, 1993
H. discocephalus	Firoza & Maqbool, 1993
Macroposthonia curvata alpina	Shahina & Maqbool, 1993
Pakcriconemoides anastomoides	Shahina & Maqbool, 1993
Paralongidorus lemoni	Nasira *et al.,* 1993
Rotylenchus goldeni	Firoza & Maqbool, 1993
Tylenchorhynchus tuberosus	Zarina & Maqbool, 1993
Paratrichodorus psidii	Nasira & Maqbool, 1994
P. faisalabadensis	Nasira & Maqbool, 1994
Helicotylenchus meloni	Firoza & Maqbool, 1994
H. striatus	Firoza & Maqbool, 1994
Xiphinema cynodontis	Nasira & Maqbool, 1994
Radopholus allius	Shahina & Maqbool, 1995
R. brassicae	Shahina & Maqbool, 1995
Longidorus trapezoides	Nasira & Maqbool, 1995
Merlinius indicus	Zarina & Maqbool, 1995
Tylenchorhynchus gossypii	Nasira & Maqbool, 1999

Cyst nematodes of Pakistan

Globodera	*Skarbilovich,* 1959
G. pallida	(Stone, 1973*) Behrens,* 1975
G. rostochiensis	(Wollenweber,1923*)*Behrens, 1975
Heterodera	*Schmidt, 1871*
H. avenae	Wollenweber, 1924
H. bergeniae	Maqbool & Shahina, 1988
H. Cajani	Koshy, 1967
H. cruciferae	Franklin, 1945
H. cynodontis	Shahina & Maqbool, 1989
H. fici	Kirjanova, 1954
H. mani	Mathews, 1971
H. mothi	Khan & Husain, 1965

H. oryze	Luc & Berdon Brizuela, 1961
H. pakistanensis	Maqbool & Shahina, 1986
H. sacchari	Luc & Murny, 1963
H. schachtii	A. Schmidt, 1871
H. zeae	Koshy, Swrup & Sethi, 1971

Root-knot nematodes:

Meloidogyne incognita	(Kofoid & White, 1919) Chitwood, 1949
M. Javanica	(Treub, 1885) Chitwood, 1949
M. hapla	Chitwood, 1949
M. Arenaria	(Neal, 1889) Chitwood, 1949

Survey and taxonomy of plant parasitic nematodes:

Total no.of plantations surveyed = 194 (Agricultural and Non-Agrigultural), Cereals = 13; Fruits = 40; Vegetables = 42; Fiber crops=5; Ornamental plants = 54 ; Fodder crops = 7; Oil seeds = 6; Pulses = 11; Forest trees = 6; Grasses & weeds = 10

Losses in crops of economic importance due to nematodes

Crops	% losses (Sasser, 1986-87)	Area under cultivation in Pakistan (m ha)	Annual Production (m tons)	Annual losses (m tons)
1. Banana	9.7	0.023	0.200	0.040
2. Citrus	14.2	0.170	0.160	0.022
3. Corn	10.2	0.800	1.200	0.120
4. Cotton	10.7	2.600	8.400	0.880
5. Potato	12.2	0.640	0.650	0.079
6. Rice	10.0	2.100	3.200	0.320
7. Sugarcane	15.3	0.880	5.600	0.370
8. Tobacco	14.7	0.040	0.074	0.010
9. Wheat	7.0	7.800	14.500	1.000

Some nematode diseases of economically important crops of Pakistan

Crop Type	**Common Name**	**Nematode species**
VEGETABLES		
Potato	Golden cyst nematodes	*Globodera rostochiensis, G. pallida*
	Lesion nematodes	*Pratylenchus pratensis, P. zeae*
	Potato rot nematodes	*Ditylenchus destructor*
	Root-knot Nematodes	*Meloidogyne hapla*
Tomato	Lesion nematodes	*Pratylenchus pratensis, P. zeae*
	Root-knot nematodes	*Meloidogyne incognita,*

		M. Javanica M. Hapla
Sugarbeet	Sugarbeet cyst nematodes	*Heterodera schachtii*
		Pratylenchus sp.
Okra	Root-knot nematodes	*Meloidogyne incognita, M. javanica, M. hapla*
Cauliflower	Root-knot nematodes	*Meloidogyne incognita, M. javanica, M. hapla*
	Cyst nematodes	*Heterodera schachtii*
	Lesion nematodes	*Pratylenchus sp.*
CEREAL CROPS		
Wheat	Oat cyst nematodes	*Heterodera avenae*
	Wheat gall nematodes	*Anguina tritici*
Rice	Rice rot nematodes	*Hirschmanniella oryzae*
	Rice stem nematodes	*Ditylenchus angustus*
	Rice white tip nematodes	*Aphelenchoides besseyi*
Maize	Cyst nematodes	*Heterodera zeae*
	Lesion nematodes	*Pratylenchus zeae*
FRUITS		
Apple, Pear	Lesion nematodes	*Pratylenchus pratensis*
Plum, Peach	Stunt nematodes	*Merlinius brevidens, Tylenchorhynchus brassicae*
	Lance nematode	*Hoplolaimus indicus*
	Dagger nematodes	*Xiphinema basiri, X. Index*
Banana	Spiral nematodes	*Helicotylenchus multicinctus*
	Reniform nematodes	*Rotylenchulus reniformis,*
	Lance nematodes	*Hoplolaimus columbus*
	Root-knot nematodes	*Meloidogyne incognita, M. javanica, M. hapla*
	Burrowing nematodes	*Radopholus similis*
Papaya	Root-knot nematodes	*Meloidogyne incognita, M. javanica*
Mango	Sheath nematodes	*Hemicriconemoides mangiferae*
	Spiral nematodes	*Heliocotylenchus indicus*
Citrus	Citrus nematodes	*Tylenchulus semipenetrans*
	Lesion nematodes	*Pratylenchus pratenis*
Grapes	Dagger nematodes	*Xiphinema index*
	Lesion nematodes	*Pratylenchus penetrans*
OTHER CROPS		
Sugarcane	Spiral nematodes	*Helicotylenchus dihystera*
	Stubby root nematodes	*Trichodorus obtusus*
	Stunt nematodes	*Tylenchorhynchus annulatus*
	Lance nematodes	*Hoplolaimus indicus*
Tobacco	Cyst nematodes	*Heterodera tabacum*
	Lesion nematodes	*Pratylenchus pratensis, P. zeae*
	Root-knot nematodes	*Meloidogyne incognita, M. javanica, M. hapla*
	Stunt nematodes	*Tylenchorhynchus annulatus*
	Spiral nematodes	*Helicotylenchus indicus*
Cotton	*Stunt nematodes*	*Tylenchorhynchus mashhoodi*

REFERENCES

FERRIS, V.R., FELDMESSER, J., HANSEN,E., LEVINE,N. AND TRIANTAPHYLLOU, A.C. 1972. The importance of discoveries in nematology to human welfare. *Bioscience,* **22**: 237-239.

HAWKSWORTH, D.L. RITICHIE, J.M. 1993. *Biodiversity and biosystematic priorities: Microorganisms and invertebrates.* Wallingford, U.K., CAB International 120 pp.

HUSSEY, R.S. 1994. *Plant and soil Nematodes.* Societal impact and Focus for the Future. Department of Plant Pathology, University of Georgia, Athens, G.A. USA. 11 pp.

VAN GUNDY, S.D. 1980. Nematology status and prospects: let's take off our blinders and broaden our horizons. *Journal of Nematology.* **12:** 158-163.

WEBSTER, J.M. 1980. Nematodes in an overcrowded world. *Revue de Nematologie,* **3**: 135-143.

SECTION IV

BIODIVERSITY OF VERTEBRATES

BIODIVERSITY OF FISHES IN THE RIVER INDUS AND ITS TRIBUTARIES BETWEEN KALABAGH AND TARBELA

M. RAMZAN MIRZA

Department of Zooloay, Government College, Lahore

Abstract: The river Indus between Tarbela and Kalabag is under active consideration for various hydroelectric projects like Ghazi-Barotha Hydroproject and the Kalabagh multipurpose project. So the biodiversity of this river is likely to be affected in the near future. It is, therefore, necessary to study the biodiversity of this part of the Indus and its tributaries. The present paper deals with the biodiversity of fishes of this part of Indus and its major tributaries (the river Kabul and Kohat Toi on the right bank and river Haro and Soan on the left bank). The paper is based on the study of fishes done mostly by the author, his students and colleagues in various institutions. There have been found about 101 species, belonging to 61 genera and 19 families, 9 orders of the teleostean fishes. Of these, 7 species are exotic i.e. *Salmo trutta fario*, *Carassius auratus*, *Cyprinus carpio*, *Hypophthalminthys molitrix*, *Oreochromis aureus*, *Oreochromis massambicus*, *Oreochromis niloticus*. One species, *Cyprinion watsoni* is West Asian. Seven species i.e. *Racoma labiata*, *Schizopyge curvifrons*, *Schizopyge esocinus*, *Schizothorax plagiostomus*, *Triplophysa microps*, *Triplophysa naziri*, *Triplophysa yasinensis* are High Asian. The remaining species are South Asian. Most of these species are widely distributed in the South Asian countries. It is remarkable that some peripheral freshwater species, i.e. *Gudusia chapra*, *Xenentodon cancila*, *Glossogobius giuris*, *Chanda nama*, *Parambassis baculis*, and *Parambassis ranga* are distributed in this part of the Indus and its tributaries more than 1750 Km upstream from the mouth of the Indus. Most of the South Asian species are not found in the river Indus upstream of Tarbela except a few species recorded from the river Siran and Unar on the left bank and the river Brandu on the right bank of the Indus. The High Asian species are mostly restricted upstream of Attock Khurd, only two or three species being found upto Kalabagh. The West Asian *Cyprinion watsoni* is restricted downstream of Attock Khurd. The biodiversity of fishes increases gradually downstream of Tarbela.

INTRODUCTION

The river Indus originates from the Western Tibet on the northern flanks of the Kailas Range according to the Times Atlas of the World (1986), the source of the Indus is the lake Manasarowar (according to Fairley, 1993) (31.75 N, 83.00 E). The total length of the Indus according to this Atlas is 3180 Km (1975 miles). It covers about 700 Km before entering into Pakistan in Baltistan at the Line of control. The length of the Indus from the Line of control to its mouth is 2480 Km (Wapda Reports, 1984). It flows between Kailas and eastern Ladakh ranges from southeast to northwest. Near Gol it receives river Shyok, one of its major tributaries in this area. At Jaglot river Gilgit falls into the Indus from the north. A few Km downstream, river Astor joins the Indus from the south. Flowing towards south and west, Indus enters the Northwest Frontier Province receiving many small streams from the east as well as from the west. Downstream of Tarbela, it enters the vale of Peshawar after covering about 1430 Km from its source. At Attock Khurd it receives river Kabul from the west. From Attock to Kalabagh, it flows through a gorge. It receives rivers Haro and Soan from the left and Kohat Toi and Banda Toi from the right. A few Km downstream of Kalabagh, about 1575 Km from the mouth of the Indus, it enters the Indus Plain. Near Mithankot, it receives the Panjnad formed by the union of the five great rivers of the Punjab, the Jhelum, Chenab, Ravi, Beas and Sutlej. Below Mithankot Indus flows slowly and passes into the Sind province, where it falls into the Arabian Sea (24.00 N and 67.11 E).

The fish fauna of river Indus between Tarbela and Kalabagh (about 1775 Km to 1575 Km from its mouth) is of great interest as there is a gradual change from the High Asian genera to the South Asian genera from north to south. It is the part of the river that will be affected by the Ghazi-Brotha Hydropower Project and the Kalabagh Dam. Hence, a thorough survey of the fish fauna in this part of the Indus is of immense importance. Fortunately, there are some preliminary reports covering this area (Mirza, 1973a, b, 1975, 1976; Mirza and Awan, 1976; Mirza and Kashmiri, 1973; Mirza and Omer, 1974; Omer and Mirza, 1975; Butt and Mirza, 1981; Ali et al., 1980; Mirza et al., 1981; Rafiq and Janjua, 1983; Qureshi et al., 1988; Butt, 1986, 1992; Razaq and Mirza, 1992; Mirza and Jan, 1993; Mirza et al., 1995). The present paper is based on the collections of fishes done by the author, his students and friends during the last twenty five years (1970 to 1994). This paper will provide basic information for further research. There are 101 species of fishes belonging to 61 genera, 19 families, 9 orders, 5 superorders and three cohorts of the teleostean fishes, as recognized by Mirza and Alam (1994), so far recorded from this region.

RESULTS

There are 94 species of freshwater fishes native to the river Indus and its tributaries between Tarbela and Kalabagh. In addition, there are about half a dozen of exotic species.

Of the native species, there are 7 high Asian species:

1. *Racoma labiata* Mc Clelland, 1842
2. *Schigopyge curvifrons* (Heckel, 1838)
3. *Schizopyge esocinus* (Heckel, 1838)
4. *Schizothorax plagiostomus* Heckel, 1838
5. *Triplophysa microps* (Steindachner, 1867)
6. *Triplophysa naziri* (Ahmad and Mirza, 1963)
7. *Triplophysa yasinensis* (Alcock, 1898)

These species are mostly restricted to the Indus above Attock Khurd. *Racoma labiata, Schizopyge esocinus* and *Schizothorax plagiostomus* are reportd upto Kalabagh. Among these species *Schizothorax plagiostomus* is predominant, while the other species are not common. During the winter, *Schizothorax plagiostomus* commonly known as "Mallah" is the only commercial species which is caught in hundreds every day and sold in the fish markets at Ghazi and Topi. The other species are rarely noticed. However, during the summer, several other species migrate from downstream to Ghazi and are seen in the markets. Even during the summer, *Schizothorax plagiostomus* is the commonest species. It dominates in the commercial catches between Tarbela and Attock Khurd. Only one West Asian species, viz., *Cyprinion watsoni* (Day, 1872) is known from this area. It is not found in the river Indus. It has, however, been collected from the rivers Soan, Haro and Kohat Toi. It has not been collected upstream of Attock Khurd.

All the remaining species are South Asian and are widely distributed in Pakistan, south of Tarbela. Upstream of Tarbela Dam, there are a few species belonging to genera *Aspidoparia, Chela, Salmostoma, Barilius, Crossocheilus, Puntius, Tor, Channa* and *Mastacembelus* recorded from the rivers Siran, Brandu and Unar about 1850 Km upstream from the mouth of the Indus. Upstream from here all the native fishes belong to schizothoracine genera (Family Cyprinidae) i.e. *Racoma, Schizopyge, Schizothorax, Ptychobarbus, Diptychus* and *Schizopygopsis*; noemacheiline genus *Triplophysa* and the sisorid catfish genus *Glyptosternum*. In addition, genus *Gymnocypris* has also been recorded from Ladakh. This fish fauna is supplemented by some species of exotic trouts. Of these, the brown trout (*Salmo trutta fario Linnaeus*) sometimes comes down to Tarbela and even through spillways to Ghazi (Mirza et al., 1995). Downstream of Tarbela, the South Asian species gradually increase in number. It is worth mentioning that following peripheral freshwater fishes have established in the river Indus.

1. *Gudusia chapra* (Hamilton, 1822)
2. *Xenentodon cancila* (Hamilton, 1822)
3. *Glossogobius giuris* (Hamilton, 1822)
4. *Chanda nama* Hamilton, 1822
5. *Parambassis baculis* (Hamilton, 1822)
6. *Parambassis ranga* (Hamilton, 1822)

These species, however, are not common in this area as only a few specimens have been collected. All the South Asian species are widely distributed in the Indus Plain and adjoining hilly areas.

The exotic species generally found in this part of the Indus and its tributaries are the following:

1. *Carassius auratus* (Linnaeus, 1758)
2. *Cyprinus carpio* Linnaeus, 1758
3. *Hypophthalmichthys molitrix* (Valenceinnes, 1844)
4. *Oreochromis aureus* (Steindachner, 1864)
5. *Oreochromis mossambicus* (Peters, 1852)
6. *Oreochromis niloticus* (Linnaeus, 1758)

A systematic list of all these species is presented in the Table I. It is also noteworthy that our major carps i.e. *Catla Giberion catla* (Hamilton, 1822), *Labeo rohita* (Hamilton, 1822) and *Cirrhinus mrigala* (Hamilton, 1822) are not naturally found in this area although introduced stock is present in the Rawal Lake. This region is the home of our common mahseer, *Tor putitora* (Hamilton, 1822). This fact along with the presence of the schizothoracine fishes led Mirza (1994) to recognize this part of the Indus and its tributaries as a separate ichthyogeographical division within the Mehran Province.

Table I: Biodiversity of fishes of the river Indus and its tributaries between Kalabagh and Tarbela

Taxa	I	II	III	IV	V
FAMILY CLUPEIDAE					
1 . *Gudusia chapra* .	+	+	-	-	+
FAMILY NOTOPTFRIDAE					
2. *Notopterus notopterus*	+	+	+	-	+
FX4MILY SALMONIDAE					
3. *Salmotrutta fario*	+	-	-	-	-

FAMILY CYPRINIDAE					
4. *Chela cachius*	+	+	-	-	+
5 . *Salmostoma bacaila*	+	+	-	+	+
6. *Salmostoma punjabensis*	+	+	-	-	-
7. *Securicula gora*	-	-	-	-	+
8. *Amblypharyngodon mola*	+	+	-	+	+
9. *Aspidoparia morar*	+	+	+	+	+
10. *Barilius modestus*	+	+	-	-	+
11. *Barilius naseeri*	-	-	-	-	+
12. *Barilius pakistanicus*	+	+	+	+	+
13. *Barilius vagra*	+	+	+	+	+
14. *Brachydanio rerio*	-	+	-	-	-
15. *Danio devario*	+	+	-	+	-
16. *Esomus danricus*	+	+	-	-	-
17. *Rasbora daniconius*	-	+	-	-	-
18. *Barbodes sarana*	+	+	-	+	-
19. *Cirrhinus mrigala*	+	-	-	-	+
20. *Cirrhinus reba*	+	+	-	-	-
21. *Cyprinion watsoni*	+	+	+	+	+
22. *Gibelion catla*	+	-	-	-	?
23. *Labeo calbasu*	+	-	-	-	+
24. *Labeo dero*	+	+	+	+	+
25. *Labeo dyocheilus pakistanicus*	+	-	+	-	+
26. *Labeo gonius*	+	-	-	-	-
27. *Labeo rohita*	-	-	-	-	+
28. *Naziritor zhobensis*	-	+	-	-	-
29. *Osteobrama cotio*	+	+	-	+	+
30. *Puntius chola*	+	+	-	+	+
31. *Puntius conchonius*	+	+	+	+	+
32. *Puntius punjabensis*	+	+	-	-	+
33. *Puntius sophore*	+	+	+	+	+
34. *Puntius terio*	-	-	-	-	+
35. *Puntius ticto*	+	+	+	+	+
36. *Puntius waageni*	-	-	-	-	+
37. *Tor putitora*	+	+	+	+	+
38. *Crossocheilus diplocheilus*	+	+	+	+	+
39. *Garra gotyla*	+	+	-	+	+
40. *Racoma labiata*	+	+	-	+	-
41. *Schizopyge curvifrons*	+	-	-	-	-
42. *Schizopyge esocinus*	+	-	-	-	-
43. *Schizothorax plagiostomus*	+	+	+	+	+
44. *Hypophthalmichthys molitrix*	-	-	-	-	+
45. *Carassius auratus*	+	+	-	-	+
46. *Cyprinus carpio*	+	+	+	+	+
FAMILY COBITIDAE					
47. *Botia birdi*	+	+	-	-	+
48. *Botia javedi*	-	+	-	-	-
49. *Lepidocephalus guntea*	+	+	-	-	-

FAMILY NOEMACHEILIDAE					
50. *Acanthocobitis botia*	+	+	+	+	+
51. *Noemacheilus corica*	+	+	+	+	+
52. *Schistura alepidota*	-	+	-	+	-
53. *Schistura fascimaculata*	-	-	+	+	-
54. *Schistura kohatensis*	-	-	+	-	-
55. *Schistura microlabra*	-	+	-	-	-
56. Schistura prashari	-	+	+	+	-
57. *Schistura punjabensis*	-	-	-	-	+
58. *Triplophysa microps*	+	-	-	-	-
59. *Triplophysa naziri* ad	-	+	-	-	-
60. *Triplophysa yasinensis*	+	-	-	-	-
FAMILY BAGRIDAE					
61. *Aorichthys aor sarwari*	+	+	-	+	+
62. *Batasio pakistanicus*	+	-	-	-	-
63. *Mystus bleekeri*	+	+	-	+	+
t4. *Mystus cavassius*	+	+	-	-	+
65. *Mystus horai*	+	-	-	-	-
66. *Mystus vittatus*	+	+	-	-	+
67. *Rita rita*	+	+	+	+	+
FAMILY SISORIDAE					
68. *Bagarius bagarius*	+	-	-	-	-
69. *Gagata cenia*	+	+	-	+	+
70. *Glyptothorax cavia*	+	+	-	-	-
71. *Glyptothorax naziri*	+	+	+	+	-
72. *Glyptothorar punjabensis*	+	+	-	+	+
73. *Glyptothorax stocki*	+	-	-	+	-
74. *Glyptothorax telchitta sufii*	+	-	-	-	-
75. *Nangra robusta*	+	-	-	-	-
76. *Sisor rabdophorus*	-	-	-	-	+
FAMILY SILURIDAE					
77. *Ompok pabda*	+	+	-	+	+
78. *Wallago attu*	+	+	+	-	+
FAMILY SCHILLEIDAE					
79. *Clupisoma garua*	+	-	-	-	-
80. *Clupisoma nazari*	+	+	-	+	+
81. *Eutropiichthys vacha*	+	+	-	-	-
82. *Pseudeutropius atherinoides*	+	+	-	-	+
FAMILY HETEROPNEUSTIDAE					
83. *Heteropneustes fossilis*	+	+	-	+	+
FAMILY BELONIDAE					
84. *Xenentodon cancila*	+	+	-	-	-
FAMILY CHANNIDAE					
85. *Channa gachua*	+	+	+	+	+

86. *Channa marulius*	+	+	-	-	-
87. *Channa punctatus*	+	+	-	+	+
88. *Channa striatus*	+	+	-	-	-
FAMILY MASTACEMBELIDAE					
89. *Macrognathus aral*	-	-	-	-	+
90. *Macrognathus pancalus*	-	-	-	+	-
91. *Mastacembelus armatus*	+	+	+	+	+
FAMILY CHANDIDAE					
92. *Chanda nama*	+	+	-	+	+
93. *Pararmbassis baculis*	-	-	-	+	-
94. *Parambassis ranga*	+	-	-	-	-
FAMILY BELONTTIDAE					
95. *Colisa fasciata*	+	+	-	-	+
96. *Colisa lalia*	+	+	-	-	+
FAMILY GOBIIDAE					
97. *Clossogobius giuris*	+	+	-	-	-
FAMILY NANDIDAE					
98. *Nandus nandus*					
FAMILY CICHLIDAE					
99. *Oreochromis aureus*	+	+	-	-	+
100. *Oreochromis mossambicus*	+	-	-	-	+
101. *Oreochromis niloticus*	-	-	-	-	+

I Indus; II Kabul; III Kohat Toi; IV Haro; V Soan

ACKNOWLEDGEMENTS

The author is indebted to Professor Dr. Azizullah, Head of the Department of Zoology, Government College Lahore for his kind suggesions. He is also greatful to Professor Dr. Masud-ul-Hasan Bokhari, Head of the Department of Geography, Government College Lahore and Ch. Ghulam Ahmad, Chief Engineer (C.D.O.) WAPDA, for their discussions about the source and length of the river Indus and other problems relating to the hydrography of Pakistan.

REFERENCES

ALI, S.R., AHMAD, M., MIRZA, M.R., ANSARI, M.A.S. AND AKHTAR, N., 1980. Hydrobiological studies of the Indus river and its tributaries above and below the Tarbela Dam. *Pakista J. Scient. Stud.*, **2**: 15-31.

BUTT, J.A., 1986. Fish and fisheries of North West Frontier Province (N.W.F.P.) Pakistan. *Biologia,* (spec. suppl.): 21-34.

BUTT, J.A., 1992. Some observations of the fishes of the North western Frontier Province, Pakistan. *Proc. Pakistan Congr. Zool.*, **12**: 25-29.

BUTT, J.A. AND MIRZA, M.R., 1981. Fishes of the vale of Peshawar, North-west Frontier Province, Pakistan. *Biologia,* **27**: 145-163.

FAIRLEY, J., 1993. *The Lion River: The Indus.* Brothers Publishers, Lahore.

MIRZA, M.R., 1973a. Aquatic fauna of Swat Valley, Pakistan, Part 1: Fishes of Swat and adjoining areas. *Biologia*, **19**: 119-144.

MARZA, M.R., 1973b. Fishes of Kohat and adjoining areas. *Pakistan J. Sci.*, **25:** 253-254.

MIRZA, M.R., 1975. Freshwater fishes and zoogeography of Pakistan. *Bijdr. Dierk.*, **45**: 143-180.

MIRZA, M.R., 1976. Fish and fisheries of the northern montane and submontane regions of Pakistan. *Biologia*, **22**: 107-120.

MIRZA, M.R., 1994. Geographical distribution of freshwater fishes in Pakistan: a review. *Punjab Univ. J. Zool.*, **9**: 93-108.

MIRZA, M.R., ADIL, S.F., GEORGE, W. AND CHOHAN, M.B., 1995. Systematic list of fishes of the river Indus at Ghazi and adjoining areas of NWFP, Pakistan, with a note on their parasites. *Proc. Parasitol.*, **19:** 69-74.

MIRZA, M.R. AND ALAM, M.K., 1994. A checklist of the freshwater fishes of Pakistan and Azad Kashmir. *Sci. Int.* (Lahore), **6**: 187-189.

MIRZA, M.R. AND AWAN, M.I., 1976. Fishes of the Sun-Sakesar Valley, Punjab, Pakistan, with the description of a new subspecies. *Biologia*, **22**: 27-49.

MIRZA, M.R. AND JAN, M.A., 1993. Fish fauna of Kalabagh, Pakistan. *Biologia*, **38**: 17-22.

MIRZA, M.R. AND KASHMIRI, M.K., 1973. Fishes of river Soan in Rawalpindi District. *Biologia*, **19**: 161-182.

MIRZA, M.R., NALBANT, T.T. AND BANARESCU, P., 1981. A review of the genus Schistura in Pakistan with description of new species and subspecies. *Bijdr dierk.*, **21**: 103-130.

MRIZA, M.R. AND OMER, T., 1974. A note on the fishes of the Haro river with the record of *Tor mosal* (Hamilton) from Pakistan. *Pak. J. Zool.*, **6** : 193-194.

OMER, T. AND MIRZA, M.R. 1975. A checklist of the fishes of Hazara District, Pakistan, with the description of a new subspecies. *Biologia*, **21**: 199-209.

QURESHI, N.A., RAFIQUE, M., AWAN, F.A. AND MIRZA, M.R., 1988. Fishes of the river Haro, Pakistan. *Biologia*, **34**: 179-191.

RAFIQ, M. AND JANJUA, M.H., 1983. Some fishes from the river Swat near Chakdara, North West Frontier Province, Pakistan. *Biologia*, **29**: 337-338.

RAZAQ, A. AND MIRZA, M.R., 1992. Some new records of fishes from the river Soan. *Proc. Pakistan Congr Zool.*, **12**: 291-293.

WAPDA REPORTS: Canadian International Agency Hydroelectric Inventory Ranking and Feasibility studies for Pakistan. Draft Inventory and Ranking study report. May, 1984 (CIDA Project No. 714/00603, Montreal Engineering Company Ltd.).

A CONTRIBUTION TO THE FISH AND FISHERIES OF AZAD KASHMIR

M. RAFIQUE AND M. YOUSAF QURESHI *

Pakistan Museum of Natural History, Islamabad
** Department of Fisheries and Wildlife , Azad Kashmir*

Abstract: Present study was undertaken to explore all possible habitats of the water bodies of Azad Kashmir. Sixty six fish species have been reported from the area and ranges of several species have been extended. Fish fauna of the three rivers of Azad Kashmir as well as of the Mangla reservoir have been compared with each other and based on water temperature and altitude, the water bodies have been ecologically categorized.

INTRODUCTION

The state of Azad Jammu and Kashmir is peculiar from ichthyogeographical point of view. It possesses both , the typical river fish fauna with cold water as well as warm water fishes and the reservoir fisheries. A long and narrow stretch of 14000 Km^2 of land is drained by three rivers, the Neelum, the Jhelum and the Poonch along with their tributaries. Water from all these three rivers has been collected downstream in Mangla reservoir having a catchment area of about 250 square kilometers.

The origin and drainage areas of these three rivers have different topographical and geographical features. This characteristic imparts specific physical factors to each river which in turn determine qualitative and quantitative aspects of their respective fish fauna. River Neelum originates from occupied Kashmir and enters Pakistan near Taubat at an altitude of more than 10,000 ft. It meets river Jhelum near Muzaffarabad at 2100 ft. Water temperature of this river is very low at higher altitude and comparatively high at lower altitude but remains below 20^0 C throughout the year. The river Jhelum enters Azad Kashmir at Chakothi area at an altitude of 3500 ft. Its water is warmer than that of river Neelum upstream of Muzaffarabad but as a result of its confluence with river Neelum at Muzaffarabad and then with river Kunhar further down, its temperature drops lower than its usual temperature upstream. At Chhattar plain area and below, it receives several side streams of warm water both from Pakistan and Azad Kashmir areas which raise its temperature again from here onwards till it ends in Mangla reservoir. The river Poonch enters Azad Kashmir at an altitude of 2300 ft. This is a small river and its water is warmer than the other two rivers in Azad Kashmir.

Fish fauna of Kashmir has been studied in perspective of the Jhelum river as well as the Indus river as the latter flows through Kashmir in its upper reaches. Heckle (1838) was the first ichthyologist who studied the fish fauna of Kashmir and recorded 16 species, which were all new to science. Later on, Steindachner (1866) described many new genera and species from the area. Since then many contributions have been made by Day (1889), Chaudhuri (1909), Hora (1936), Mukerjee (1936), Silas (1960), Talwar (1978), and Nath (1980). All this work however, has been carried out in the areas of Kashmir occupied by India. In Azad Kashmir some fragmentary reports (Mirza, 1975; Mirza and Waheed-ud-Din, 1976; Waheed-ud-Din, 1979; Mirza and Awan, 1979; Mirza and Janjua, 1984, and Mirza and Ejaz, 1992) about fish fauna of different areas have been published. Butt and Butt (1988), however, published a consolidated report on the fish fauna of Kashmir and reported 42 species from the area. They , however, remarked that records of many more species are expected from the region and much more is needed to be done on taxonomic side of fishes of Azad Kashmir. Inspite of their collections from various new localities in Azad Kashmir, more areas especially upper Neelum valley, Bhimber, Smahni, Upper Poonch river and the Mangla reservoir still needed exploration. Present study is thus undertaken to explore all possible habitats of the Mangla reservoir and the three main rivers and their tributaries within the geographical boundaries of Azad Kashmir.

MATERIALS AND METHODS

Fishes were collected from 1988 to 1995 on many different occasions which covered all seasons. Collection was made mostly by small-meshed cast nests and occasionally by drag nets and also by scoop nets. Large specimens were injected with 10 % formalin in the abdominal cavity while small specimens were put as such in the fluid. Relevant data like locality, altitude, and water temperature were recorded at the spot in the field book as well as on the tag that was put in the fluid along with the specimens. In the laboratory, the specimens were put in 70 % alcohol and identified following Mirza (1980) and Jayaram (1981).

Similarity between fish fauna of any two rivers was calculated following Swift et al. (1986), with the formula:

$$C = . R (N1 + N2) / 2 (N1 \times N2)$$

where 'C' is similarity coefficient; its value ranging from 0 (complete faunal dissimilarity) to 1 (entirely similar fauna), 'R' is number of species shared between two drainages, 'N1' is number of species in one drainage and, 'N2' is the number of species in the other drainage.

RESULTS

Total number of species found in water bodies of Azad Kashmir was 66, generally more species occurred in warm water areas with low altitude as compared to cold water areas with high altitude. More faunal similarity was observed among warm water bodies and among cold water bodies whereas warm water fauna was quite different from cold water fauna as shown in Table I. Family Cyprinidae was the most specious family and represented 54.4 % of the fish fauna of Azad Kashmir followed by the families Sisoridae and Noemacheilidae which were represented by 14.04 % and 10.5% respectively. The other families were represented by two or three species while six families were represented only by one species.

All the fish fauna found in Azad Kashmir belongs to Cohart Euteleostei, Infraclass Teleostei, subclass Actinopterygii and class Teleostomi. Results of the present studies along with some previous findings are given in Table II.

Table I: Similarity coefficient between different water bodies of Azad Kashmir

	Neelum river	Jhelum river	Poonch river	Mangla reservoir	Bhimber Nullah
Neelum river	-	0.86	0.16	0.17	0.05
Jhelum river	0.86	-	0.50	0.46	0.40
Poonch river	0.16	0.50	-	0.74	0.75
Mangla reservoir	0.17	0.46	0.74	-	0.64
Bhimber Nullah	0.05	0.40	0.75	0.64	-

Table II: Fish fauna of Azad Kashmir and its distribution in different water bodies

Families/ Species	Neelum river	Jhelum river	Poonch river	Mangla reservoir	Bhimber/ Smahni area
I- Clupeidae					
Gudusia chapra	-	-	-	+	-
II- Notopteridae					
Notopterus chitala	-	-	-	+	-
III- Salmonidae					
Salmo trutta fario	+	+	-	-	-
Oncorhynchus mykiss	+	+	-	-	-

IV- Cyprinidae					
Subfamily Cultrinae					
Salmostoma bacaila	-	-	+	+	+
Chela cachius	-	-	+	+	+
Subfamily Rasborinae					
Barilius vagra	-	-	+	+	+
Barilius pakistanicus	-	+	-	+	+
Aspidoparia morar	-	-	+	+	+
Esomus danricus	-	-	+	+	+
Subfamily Barbinae					
Labeo rohita	-	-	-	+	-
Labeo dyocheilus pakistanicus	-	+	+	+	+
Labeo boga	-	-	-	+	-
Labeo dero	+	+	+	+	+
Labeo calbasu	-	-	-	+	-
Cirrhinus reba	-	-	-	+	-
Cirrhinus mrigala	-	-	-	+	-
Tor putitora	-	+	+	+	+
Puntius sarana	-	+	+	+	+
Puntius sophor	-	-	+	+	+
Puntius chola	-	-	-	+	-
Puntius ticto	-	+	+	+	+
Puntius titius	-	-	+	+	-
Cyprinion watsoni	-	-	+	+	-
Osteobrama cotio	-	-	-	+	-
Catla catla	-	-	-	+	-
Subfamily Garrinae					
Crossocheilus diplochilus	-	+	+	+	+
Garra gotyla	-	+	+	+	+
Subfamily Schizothoracinae					
Schizothorax plagiostomus	+	+	+	+	-
Schizopyge esocinus	+	+	-	+	-
Schizopyge micropodon	+	+	-	-	-
Schizothorichthys longipinnis	-	+	-	-	-
Racoma labiatus	+	+	-	-	-
Subfamily Cyprininae					
Cyprinus carpio	-	+	+	+	-
Subfamily Hypophthalmichthyinae					
Hypophthalmichthys molitrix	-	-	-	+	-
V- Cobitidae					
Botia birdi	+	-	-	-	-
Botia lohachata	-	-	-	+	+
VI- Noemacheilidae					
Acanthocobitis botia	-	+	-	+	-
Schistura nalbanti	+	+	+	+	-
Schistura parashari	+	+	-	-	-
Schistura alepidota	-	+	+	+	-

Triplophysa kashmirensis	+	+	-	-	-
Triplophysa microps	+	-	-	-	-
VII- Bagridae					
Aorichthys seenghala	-	-	+	+	+
Mystus bleekri	-	-	-	+	-
VIII- Sisoridae					
Bagarius bagarius	-	-	-	+	-
Glyptothorax punjabensis	-	+	+	+	+
Glyptothorax kashmirensis	+	+	-	-	-
Glyptothorax stocki	-	+	+	+	+
Glyptothorax pectinopterus	-	-	+	+	+
Glyptosternum reticulatum	+	+	-	-	-
Gagata cenia	-	-	-	-	+
IX- Schilbeidae					
Clupisoma naziri	+	+	-	-	-
Clupisoma garua	-	-	-	+	-
X- Siluridae					
Wallago attu	-	-	-	+	-
Ompok pabda	-	-	-	+	+
Ompok bimaculatus	-	-	-	+	+
XI- Belonidae					
Xenentodon Cancila	-	-	-	+	-
XII- Channidae					
Channa punctatus	-	-	+	+	+
Channa orientalis	-	-	+	+	-
XIII- Chandidae					
Chanda nama	-	-	-	+	-
Chanda ranga	-	-	-	+	-
Chanda baculius	-	-	+	+	+
XIV- Mastacembelidae					
Mastacembelus armatus	-	-	-	+	-
XV- Gobiidae					
Glossogobius giuris	-	-	-	+	+
XVI- Nandidae					
Nandus nandus	-	-	-	+	-

DISCUSSION

River Neelum entres Pakistan in an area having an altitude of more than 10,000 ft. Due to this , mean annual temperature of the water of this river remains low throughout the year, i.e.,12° C. This river exclusively possesses the cold water fish fauna of Azad Kashmir and has three species viz., *Triplophysa kashmirensis*, *Triplophysa microps* and *Glyptothorax kashmirensis* as endemic. Temperature has a direct effect on the primary productivity of a water body (Huet, 1972), and due to cold water of this river its algal biomass is lower than any other river in Azad Kashmir. As a consequence of this, all the fishes found in this river are carnivorous, and to a lesser extent omnivorous but none is herbivorous.

River Neelum has a similarity coefficient of 0.86 with river Jhelum, 0.17 with the Mangla reservoir, 0.16 with the Poonch river and 0.05 with the Bhimber Nullah. This shows that its fauna is quite different than of river Poonch, Mangla reservoir and the Bhimber Nullah and only those species are common which have a wide range of temperature tolerance. The high value of similarity coefficient of this river with the river Jhelum is due to mixing of their waters at Muzaffarabad where some of the species found in river Jhelum migrate into river Neelum to a variable extent but they are never found beyond Kahori. This migration of fishes between two rivers is not reciprocal as none of the species typical to river Neelum is found in the river Jhelum.

Water temperature of the river Jhelum is comparatively high and variable at different points. It has a mixture of cold water and warm water fish fauna. This river not only shares its fauna with Neelum but has a similarity coefficient of 0.50 with Poonch, 0.46 with Mangla reservoir and 0.40 with the Bhimber Nullah. In its upper portion it shares its cold water fauna with river Neelum but in lower areas its maximum faunal similarity is with the other rivers having warm water fishes.

River Poonch enters Azad Kashmir in an area of low altitude. Its water is warmer than the other two rivers and has typical warm water fishes, all of which are also found in Mangla reservoir and further down in Pakistan. This river has a similarity coefficient of 0.74 and 0.75 with Mangla reservoir and Bhimber Nullah respectively. This indicates that it has maximum faunal similarity with these two water bodies as compared to the river Neelum and Jhelum.

Mangla resevoir has number of fishes, more than any other water body in Azad Kashmir. It has a similarity coefficient of 0.64 with Bhimber Nullah. Its highest faunal similarity with river Poonch and Bhimber Nullah shows that its fauna is not significantly influenced by the fish species found either in the river Jhelum or Neelum. It , however, shares some of its fishes with river Jhelum and a few with river Neelum. It has 21 species of fishes which are not found in any of the rivers of Azad Kashmir but all of its fish fauna is found in Pakistan downstream of the reservoir.

At least 18 out of 66 fish species found in Azad Kashmir gain a size at least a foot or more and are considered important as food fishes. Among them the Snow Trouts belonging to the subfamily Schizothoracinae and the Rainbow Trout and Brown Trout are mainly found in rivers Neelum and Jhelum. The other twelve species viz., *Labeo rohita*, *Labeo dero*, *Cirrhinus mrigala*, *Catla catla*, *Cyprinus carpio*, *Tor putitora*, *Hypophthalmichthys molitrix*, *Aorichthys seenghala*, *Wallago attu*, *Bagarius bagarius*, *Notopterus chitala*, and *Channa punctatus* are mainly found in Mangla reservoir. It also possesses the highest population of the fishes *Catla catla* (Thaila) and the *Tor putitora* (Mahaseer). These two big sized fishes only breed in nature and like the other Indian carps are not successfully reared in the hatcheries. Mangla reservoir is by far the major source for these fishes as compared to any other water body in Pakistan.

Mahaseer is not found in the river Jhelum upstream of the Nulla coming from Bagh and Hajeera areas. The reason is the ecological barrier made by the cold water of the rivers Kunhar and the Neelum joining river Jhelum upstream and also the profile of the river which is sort of a gorge in this area. Similarly, other major carps and the commercially important fishes found in Mangla reservoir are also not found in the main body of the river Jhelum. As these fishes do not breed in stagnant water, the main breeding ground for all the important fishes in Mangla reservoir is the river Poonch. This phenomena makes river Poonch very important inspite of its small size. This river and all the nullas and side streams feeding it deserve special attention, especially in terms of avoiding hunting during breeding season when these fishes migrate backwaters for breeding.

In general cold water fish fauna of Azad Kashmir is restricted to the river Neelum and in the upper parts of river Jhelum. From zoogeographical point of view this fauna is mainly High Asian in origin except a few species which, due to wide range of temperature tolerance, migrate from downstream areas. On the basis of difference in mean temperature of the water bodies of Azad Kashmir, river Neelum is completely included in the Rhithron zone established by Illies and Botosaneanu (1963). In this zone, water temperature of the river remains below 20^0 C, oxygen concentration is always high, flow is fast and bed is composed of rocks , stones or gravel. The Jhelum river below Chattar plain, Mangla reservoir, Poonch river and Bhimber Nullah are included in Potamon zone where temperature rises to over 20^0 C, the flow is slow and the bed is mainly sandy or muddy. This zone has warm water fishes and dominantly includes the South Asian fish fauna along with a few West Asian species like *Cyprinion watsoni.* The Rhithron zone in Azad Kashmir is suitable for cold water fish culture while the warm water fishes can be reared in Potamon zone such as is present in most of Pakistan.

REFERENCES

BUTT, J. A., AND BUTT, A. A., 1988. An addition to the Fishes of Azad Kashmir. *Scientific Khyber.*, **1**: 77-84.

CHAUDHURI, B. L. 1909. Description of new species of *Botia* (*B. birdi*) and Noemachilus (*N. mcmahoni*). *Rec. Ind. Mus.*, **3**: 339-42.

DAY, F., 1889. *The Fauna of British India including Ceylon and Burma, Fishes.* Vol. **1** & **II,** Tayler and Frances , Bombay.

HECKLE J. J., 1838. *Fische aus Cashmir,* pp., x+112. Wien, P. P. Mechitaristen.

HORA, S. L., 1936. Yale North India Expedition, Article XVII. Report on Fishes. Part I. Cobitidae. *Mem. Conn. Acad. Art. Sci.*, **10**: 299-321.

HUET, M., 1972. *Text Book of Fish Culture, Breeding and Cultivation of Fish.* Fishing News (Books) Ltd. England.

ILLIES, J. AND BOTOSANEANU, L.,1963. Problems et methodes de la classification de la zonation ecologique des eaux courantes, considerees surtout du point de vue faunistique. *Mitt. Int. Verein. Theor. Angew. Limnol.*, **12**: 1-57.

JAYRAM, K. C., 1981. The freshwater fishes of India, Pakistan, Bangladesh, Burma and Sri Lanka- A Handbook. *Zoological Survey of India*, Calcutta.

MIRZA, M. R., 1975. Freshwater fishes and Zoogeography of Pakistan. *Bijdr. Dierk.*, **45**: 143-180.

MIRZA, M. R., 1980. The systematics and zoogeography of the Freshwater Fishes of Pakistan and Azad Kashmir. *Proc. Ist. Pakistan Congr. Zool.*, **1**:. 1-41.

MIRZA, M. R. AND WAHEED-UD-DIN, 1976. A note on the Fishes of River Poonch in Azad Kashmir. *Pakistan J. Zool.*, **8**: 98-99.

MIRZA, M. R. AND AWAN, A. A., 1979. Fishes of the genus *Schizothorax* Heckel, 1838 (Pisces Cyprinidae) from Pakistan and Azad Kashmir. *Biologia*, **25**: 1-21.

MIRZA, M. R. AND JANJUA, H., 1984. Fishes of Muzaffarabad, Azad Kashmir. *Biologia,* **30**: 229-224.

MIRZA, M. R. AND EJAZ, A., 1992. Fishes of the river Neelum with record of *Botia birdi* and *Triplophysa micropes* from Azad Kashmir. *Pakistan J. Zool.*, **24**: 168-169.

MUKERJEE, D. D., 1936. Yale North India Expedition, XVIII. Report on Fishes. Part II. Sisoridae and Cyprinidae. *Mem. Conn. Acad. Arts. Sci.,* **10:** 323-359.

NATH, S., 1980. On the extension of range of two freshwater catfishes, *Glyptothorax conirostre* (Steind.) (Sisoridae) and *Clupisoma garua* (Ham.) (Schilbeidae), to Poonch Valley (Jammu and Kashmir), India. *J. Bombay Nat. Hist. Soc.*, **78**: 178-179.

SILAS, E. G., 1960. Fishes from Kashmir Valley. *J. Bombay Nat. Hist. Soc.*, **57**: 66-77.

STEINDACHNER, F., 1866. Ichthyologische Mitter-lungen (IX). VI- Zur Fisch-fauna Kaschmirs and der benachbarten landerstriche. *Verh. Zool. bot. Ges. Wien.*, **16**: 789-796.

SWIFT, C. C., GILBERT, C. R., BORTON, S. A., BURGESS, G. H. AND YERGER, R.W., 1986. Zoogeography of the Freshwater Fishes of the Southern United States: Savannah River to Lake Pontchartrain. In : C. H. Hocutt, and E. O. Wiley (eds.), *The zoogeography of North American Freshwater Fishes*, pp. 213-265. John Wiley & Sons, New York.

TALWAR, P. K., 1978. On the Fishes collected by the Ladakh Expedition. *J. Bombay Nat. Hist. Soc.*, **74:** 501-505.

WAHEED- UD-DIN, 1979. A report on the Fishes of the Poonch river, Azad Kashmir. *Pakistan J. scienst. Stud.*, **1**: 9-21.

FISH DIVERSITY OF RIVER CHENAB IN DISTRICT MULTAN, PAKISTAN

JAVED AKHTER MAHMOOD AND ABDUS SALAM*

Department of Fisheries, Govt. of Punjab. Fish Hatchery, Islamabad
**Institute of Pure and Applied Biology, Bahauddin Zakariya University, Multan*

Abstract: This paper deals with 36 species of fresh-water fish distributed over 27 genera, 13 families, 7 orders, 4 superorders and 3 cohorts belonging to the class Teleostomi sampled from various localities of river Chenab near District Multan-Pakistan. Fish diversity conservation has also been discussed.

INTRODUCTION

Fish exhibit enormous diversity in their morphology, in the habitats they occupy, and in their biology. Unlike other commonly recognized vertebrates, fish are a heterogeneous assemblage. Fishes constitute almost half of the total number of vertebrates. An estimated 21723 living species have been described. Other workers, for various reasons, have arrived at different estimates, most of which range between 17000 and 30000 (Nelson, 1984).

A great many of freshwater species occur in southeastern Asia. Estimated number of fish species in Asia is 1500 (Gilbert, 1976), or even greater (Nelson, 1984).

The river Chenab originates from Jummu & Kashmir. After receiving several tributaries, it enters Punjab near District Sialkot. In the Punjab, this river flows through Gujrat, Sargodha and Gujranwala Districts. It receives river Jhelum at Trimmu in District Jhang and river Ravi at Sidhnai in District Khanewal. It then flows through Districts of Multan and Muzaffargarh and joins river Sutluj in District Muzaffargarh in Punjab.

The fish diversity of this river in Punjab has been worked out in District Sialkot at Marala (Mirza & Khan, 1988) and in the Multan District Khan et al. 1991). They have described 33 species belonging to 27 genera. Although, Ahmed (1963) listed the fish fauna of Multan but without mentioning which of the species has been collected from river Chenab. The present study deals with 36 species of fish found in river Chenab near District Multan along with the threats they face and recommendations for management implications for their conservation and sustainable use.

MATERIALS AND METHODS

The fish were collected from October 1991 to December 1992 from river Chenab near Bosan, Lutfabad, Nawabpur, Surajmiani, Langrial, Murad abad and Thatti bakri, both from east and west banks of the river. Boats were used to cross the river. Fish were captured by using cast and drag nets of variable sizes. The fish were preserved in 8% Formalin solution. Large fishes were given the injection of 20% Formalin in the gut region following the method of Mirza and Ahmad (1987). Different body measurements were taken by using perspex measuring tray fitted with millimeter scale before putting the fish into jars. For rays and scale count, magnifying lens was used.

RESULTS

The following fish were found during the survey of the study area.

Family Clupeidae
Gadusia chapra (Hamilton)
Family Notopteridae
Notopterus notopterus (Pallas)
Notopterus chitala (Hamilton)
Family Cyprinidae
Securicula gora (Hamilton)
Aspidoparia morar (Hamilton)
Barilius vagra (Hamilton)
Labeo rohita (Hamilton)
Labeo calbasu (Hamilton)
Labeo dero (Hamilton)
Catla catla (Hamilton)
Cirrhinus mrigala (Hamilton)
Cirrhinus reba (Hamilton)
Puntius sophore (Hamilton)
Puntius ticto (Hamilton)
Osteobrama cotio (Hamilton)
Cyprinus carpio (Linnaeus)
Ctenopharyngodon idella (Valenciennes)
Noemacheilus corica (Hamilton)
Family Bagridae
Aorichthys aor sarwari (Hamilton)
Mystus cavasius (Hamilton)

Mystus vittatus (Bloch)
Family Sisoridae
Gagata cenia (Hamilton)
Family Siluridae
Wallago attu (Bloch & Schneider)
Ompok bimaculatus (Bloch)
Family Schilbeidae
Eutropiichthys vacha (Hamilton)
Ailia punctata (Day)
Pseudeutropius atherinoides (Bloch)
Family Channidae
Channa marulius (Hamilton)
Channa striata (Bloch)
Channa punctata(Bloch)
Family Chandidae
Chanda baculis (Hamilton)
Family Osphronemidae
Colisa fasciata (Bloch & Schneider)
Family Mugilidae
Sicamugil cascasia (Hamilton)
Family Mastacembelidae
Mastacembelus armatus (Lacepede)
Mastacembelus pancalus (Hamilton)
Macrognathus aculeatus (Bloch)

DISCUSSION

The present study gives more details and up-to-date information about the fish diversity of the area. Previously, (Khan et al. 1991) had listed 33 species without mentioning *Puntius ticto, Ompok bimaculatus, Pseudeutropius atherinoides, Channa marulius, Channa striata* and *Mastacembelus pancalus* which are found in the area. *Puntius ticto* and *Channa striata* have also been recorded from District Sialkot (Mirza and Khan, 1988). However, *Crossocheilus latius, Botia lohachatta* and *Rita rita* were not found in the study area. According to Mirza (1990), *Crossocheilus latius* and *Botia lohachatta* are found in waters of semi-hilly plains and hilly areas, which indicates that these fishes did not like warm waters like river Chenab in District Multan. *Rita rita* is common in river Indus near District Dera Ghazi Khan (Rafiq, 1992), but inspite of intensive netting, could not be found in the study area.

The fish namely *Gudusia chapra, Notopterus chitala, Labeo calbasu, Labeo rohita, Labeo dero, Cirrhinus mrigala, Cirrhinus reba, Aspidoparia morar, Puntius sophore, Osteobrama cotio, Securicula gora, Wallago attu, Eutropiichthys vacha, Aorichthys aor sarwari,*

Mystus cavasius, *Chanda baculis*, *Mastacembelus armatus* and *Channa punctata* are the common fish fauna of river Jhelum in District Jhelum (Islam and Siddiqi, 1971) and District Sargodha (Mirza and Ahmad, 1987); river Ravi in District Lahore (Mirza, 1982) and river Chenab in District Multan (Khan et. al 1991). and present study) showing their wide range of distribution in Punjab, and hence, adaptability to different habitats.

The fish fauna of Multan is almost the same as found in river Ravi in District Lahore with only one exception, that is, *Rita rita* is not found, though it had been recorded in 1990 from the same area (Khan et al., 1991).

The species found in District Multan but not recorded by Mirza and Khan (1988) in District Sialkot are *Gudusia chapra*, *Notopterus notopterus*, *Notopterus chitala*, *Labeo rohita*, *Catla catla*, *Cirrhinus mrigala*, *Cirrhinus reba*, *Crossocheilus latius*, *Eutropiichthys vacha*, *Ailia punctata*, *Mystus cavasius*, *Chanda baculis*, *Colisa fasciata*, *Macrognathus aculeatus* and *Sicamugil cascasia*. On the other hand, the species found in District Sialkot but not in District Multan are *Salmostoma punjabensis*, *Amblypharyngodon mola*, *Labeo dyocheilus pakistanicus*, *Puntius sarana*, *Puntius chola*, *Tor putitora*, *Schizothorax plagistomus*, *Schistura sp.* *Ompok pabda*, *Bagarius bagarius*, *Glyptothorax cavia*, *Glyptothorax punjabensis*, *Heteropneustes fossilis*, *Chanda ranga*, *Glossogobius giuris* and *Colisa lalia*.

The species *Puntius sarana*, *Esomus danricus*, *Salmostoma bacaila*, *Mystus bleekri*, *Rita rita*, *Mastacembelus pancalus* and *Glossogobius giuris* were found in District Sargodha (Mirza and Ahmad, 1987), but not in District Multan, whereas, *Notopterus notopterus*, *Barilius vagra*, *Catla catla*, *Puntius ticto*, *Cyprinus carpio*, *Ctenopharyngodon idella*, *Noemacheilus corica*, *Mystus vittatus*, *Gagata cenia*, *Ompok bimaculatus*, *Ailia punctata*, *Pseudeutropius atherinoides*, *Channa marulius*, *Channa striata*, *Colisa fasciata*, *Sicamugil cascasia*, and *Macrognathus aculeatus* have not been recorded from District Sargodha.

Three important Indian major carps namely *Catla catla*, *Labeo rohita*, and *Cirrhinus mrigala* are very commonly found in this area. These are most renowned and fast growing fishes and cultured commercially in India, Pakistan, Burma, Nepal, Bangladesh and Assam and many other countries. These are considered excellent eating, especially when the fish are of moderate size. Two carp species, *Cyprinus carpio* and *Ctenophrangodon idella*, commonly called as Common carp and Grass carp respectively, are also common in this area. The growth of the common carp is very rapid, particularly in favorable habitats. It's rapid growth, tasty flesh, and good reproductive ability have led to the carp's becoming the stable fish of warm water fisheries. It is cultured in ponds all over the world. In some of the countries, the main purpose of the introduction of Grass carp, in addition to culture, was biological aquatic weed control in natural waterways, lakes and man-made ponds. Due to its rapid growth and good flesh, it also has become very popular in freshwater fisheries. The important catfishes of the area are *Aorichthys aor sarwari* and *Wallago attu* and of considerable fishery value. They come on dead bait and provides a good sport. They are generally liked as their flesh contains only a few bones.

ACKNOWLEDGMENTS

Authors are thankful to the Fisheries Department, District Multan particularly Mr. Tufail Sajjad Qureshi, Assistant Director Fisheries; Mr. Riaz Hussain Abbasi, Assistant Warden Fisheries and Mr. Shehr Yar, Assistant Warden Fisheries for their consistent help throughout this work. Special thanks are due to Mr. Kashif M. Sheikh (QAU, Islamabad) for his support during the field work.

REFERENCES

AHMAD, N., 1963. Fishery Gazetteer of District Multan. Government Printing Press, Lahore.

GILBERT, C.R., 1976. Composition and derivation of the North American freshwater fish-fauna. *Florida Sci.,* 39: 104-111

ISLAM, A. & SIDDIQUI, M.N., 1971. Fishes of the Jhelum with some new records from the Punjab. *Biologia,* 17: 27-44

KHAN, M.I., IRSHAD, R. AND SAGA, F.H., 1991. Fishes of river Chenab in Multan District. *Biologia,* **37**: 23-25.

MIRZA, M.R. 1982., A contribution to the fishes of Lahore. 1st. ed. Polymer Publications, Lahore.

MIRZA, M.R. AND AHMAD, I., 1987. Fishes of the River Jhelum in Sargodha District. *Biologia,* **33**: 253-263.

MIRZA, M.R. AND KHAN, A.J., 1988. Fishes of Marala, Sialkot District, Pakistan. *Biologia,* **34:** 151-153.

MIRZA, M.R., 1990. Freshwater fishes in Pakistan. Urdu Science Board, Lahore.

NELSON, J.S., 1984. Fishes of the World. 2nd.ed. Jhon Willey. New York.

RAFIQ, M.R., 1992. Morphometry and body compositin of wild *Rita rita* from river Indus, Dera Ghazi Khan. M.Sc. Thesis, Government College, Multan.

GEOGRAPHIC VARIATION IN *BUFO STOMATICUS*, WITH REMARKS ON *BUFO OLIVACEUS*: BIOGEOGRAPHICAL AND SYSTEMATIC IMPLICATIONS

WALTER AUFFENBERG AND HAFIZUR REHMAN*

Florida Museum of Natural History, Gainesville, Fl. 32611, U.S.A.
** Zoological Survey Department, Karachi*

Abstract: Variation in the color and external morphology of *Bufo stomaticus* populations are analyzed and the resulting geographical clinal trends are discussed. There is no evidence that any of the morphoclines identified are related to those current environmental isophenes normally suspected as correlative factors, though the genetics responsible for color and pattern may indeed be selected on the basis of local environmental conditions. Variation in the morphological characters studied seems to be influenced more by genetic drift engendered by the presence of, and geologic changes in, major physiographic land forms.

Earlier judgements to the contrary, *Bufo olivaceus* is not conspecific with *B. stomaticus*. There is no evidence of morphological similarity in any of the nearly sympatric populations we have examined. We conclude that they are specifically distinct.

INTRODUCTION

The present paper is a continuation of our investigation of the geographic variation in common Pakistan reptiles and amphibians. These studies were (and those in the future are) largely stimulated by our commitment to produce a book-length compendium of the reptiles and amphibians of Pakistan (sponsored by the US Fish and Wildlife Service and the Zoological Survey Dept. of Pakistan). Additionally, these studies are intended to help form a sound basis for a future in-depth analysis of the zoogeography of Pakistan as illustrated by its extant herpetofauna.

The marbled toad, *Bufo stomaticus* Lütken was selected as the object of study for the simple reason that it is a common species, assuring statistically appropriate sample sizes. Its geographic range is from Bangladesh and Assam, India in the east to southeast Afghanistan and extreme eastern Iran in the west (We provisionally exclude the locality of Eiselt and Schmidler, 1973, from extreme northern Iran as highly unlikely on zoogeographical grounds). Except for this one record, *Bufo stomaticus* is a purely tropical to subtropical species. No in-depth study of the morphological variation of this widespread, nearly ubiquitous toad has ever been published; nor of any other amphibian species in the entire Indian subcontinent. Finally, the relationship of *B. stomaticus* to the poorly known *B. olivaceus* has always been in doubt - due mainly to the paucity of this once presumed rare species of toad living in a restricted range along the Mekkran Coast. Minton (1966) and Eiselt and Schmidler (1973) suggested that the two species were simply geographic races, though all agreed that a final determination depended on the availability of adult specimens from critical geographically intermediate localities. During our studies we obtained sufficient material to allow comparisons of both the skeletal and external anatomy of both species.

In certain previous instances, *Bufo stomaticus* was confused with other species. The most recent was by Khan, who in a series of papers (1965, 1968, 1969) published information purportedly dealing with this species in Pakistan, under the name *Bufo melanostictus.* He corrected his error in 1972.

Much more serious nomenclatorial confusion occurred earlier in the history of this species. *Bufo stomaticus* was first decribed by Lütken in 1863 on the basis of material collected in "eastern India", presumably Assam, according to Boulenger (1891a,b). In 1883, Boulenger described *Bufo andersonii* on the basis of material from an unknown locality in India. Adding to this confusion, Murray (1884) described a *Bufo· andersoni* (non-Boulenger 1883) from Pakistan (type locality Thatta and Jungshahi, Sindh Province). Later, Parker (1938), believing *andersonii* to be a species distinct from *stomaticus*, selected Ajmere, Rajputana (Rajisthan), India as the type locality of Boulnger's species. Boulenger's characters to distinguish the two species are now known due to modifications caused by preservation. Compounding matters, the type series of Boulenger's *andersonii* was found to include still another species of toad. Additionally, *B. stomaticus* was discovered to be more variable throughout its range than previously thought (Annandale, 1909). As a result, *B. andersonii* Boulenger was declared a synonym of *B. stomaticus* Lütken. The latter (as *andersonii*) was believed to extend as far west as Muscat, but this population was later found to represent a closely related, but distinct species, *B. dhufarensis* (Parker, 1938). The probable subspecific relationship between *B. stomaticus* and the Mekkran toad, *B. olivaceus* Blanford has already been alluded to. The *stomaticus* populations from southern peninsular India have been described by Rao (1920) as a separate subspecies - *Bufo stomaticus peninsularis* (type locality Mysore, India). However, this taxon is considered a synonym of *Bufo stomaticus* (see Daniel, 1963) by most herpetologists.

Bufo stomaticus is one of the most common toads in the northern part of the Indian subcontinent. The soil types on which it is found are extremely variable, including almost all those characteristically found in arid, tropical to subtropical areas (chernozemic, latosolic, grumusolic, desertic and alluvial). Within this broad soil spectrum, *B. stomaticus* can be found from humid, warm, mixed forests to almost perennially arid, stony places supporting only scattered shrubs and grasses. However, it is usually most dense in regions experiencing alternately wet and dry seasons dominated by a monsoonal climate. Within this broad characterization, the plant associations in which it is most common are dry tropical lowland forests of deciduous and thorny types (and their degraded forms). Though a species of drier habitats, it avoids true desert and even in arid regions it is more or less restricted to oases, watercourses, and the vicinity of temporary ponds. It is particularly common in sites modified by man, including the largest cities of the Indus and Ganges River Plains, where it regularly forages below porch and street lights. The geographic range extends from eastern Iran (Eiselt and Schmidler 1973) and from southeastern Afghanistan (Anderson and Leviton 1969) southeast through the entire Indo-Gangetic Plain to West Bengal, India; the vertical range is from sea level to 1,650 m (in the Himalaya Mountains, Annandale 1908; possibly to 2,330 m Annandale 1907).

Bufo stomaticus is a monsoon breeder throughout its range. That the species is adapted to dry environments is proven by the speed with which the eggs hatch (ca 24 hrs, Minton 1966), the habit of soaking in moist spots before and after foraging at night (McCann 1938, Khan 1979), that individuals regularly dig themselves completely under the earth or large stones every night (Daniel 1963, Sarkar 1984), and that they are physiologically capable of aestivation during the dry season. Young toads reach sexual maturity in about one year (Minton 1966).

In spite of its abundance, very few studies have been conducted on it. In addition to the workers and their publications mentioned above, Bhati has described some of the musculature (1955, as *Bufo andersoni*), Hashmi (1965) the larval development, Khan (1979) its life history and predators, and Sarkar (1984) food and general habits.

In addition to our studies on *Bufo stomaticus* morphology and color and its geographic variation, we take this opportunity to compare this species with *Bufo olivaceus* Blanford, from the extreme southwestern

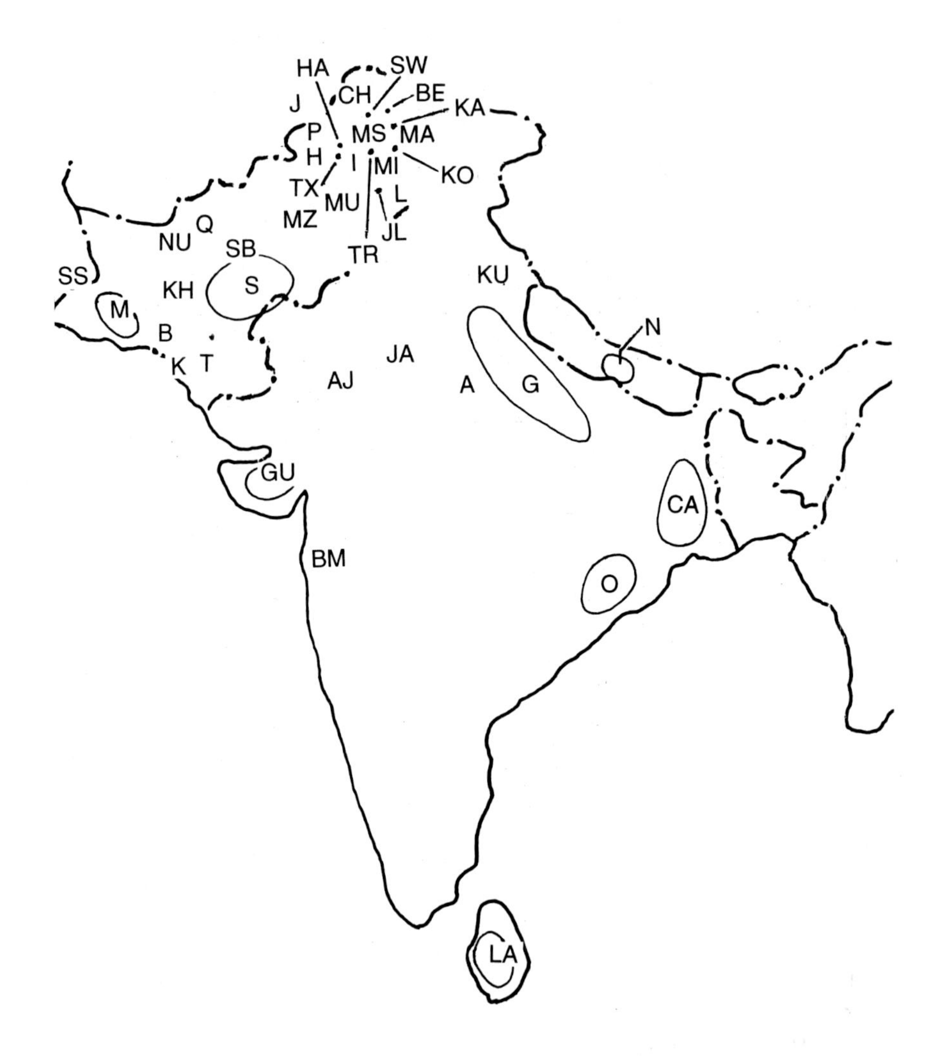

Fig. 1. Locations and general geographic area of sample subsets used in this study.

Abbreviations: numbers in parentheses refer to number of individuals in subset: A, Agra (25); AB, Abbotabad (35); AJ, Ajmer; B, Bannu (19); BL, Las Bela (17); BE, Besham (8); BM, Bombay (15); CS Central Sindh (17); CA, Calcutta (11); CH, Chitral (5); DG, Dera Ghazi Khan (15); F, Faisalabad (32); G, Ganges (19); GU, Gujarat (13); H, Hungu (8); HA, Haripur (14); I, Islamabad (27); J, Jalalabad (7); JA, Jaipur (12); JL, Jhellum (25); K, Karachi (24); KA, Kagan (12); KH, Khuzdar (11); KO, Kotli (6); KU, Kumaon (5); L, Lahore (11); LA, Sri Lanka (6); M, Mekkran (1); MA, Muzaffarabad (17); MI, Mirpur (7); MS, Manshera (11); MU, Multan (13); MZ, Muzafaragar (32); N, Nepal (17); NU, Nushki (18); O, Orissa (7); P,Peshawar (26); PG, Paghman (2); Q, Quetta (13); S, Sindh (12); SB, Sibi (4); SS, Seistan (3); SW, Swat (12); T, Tarnah (12); TX, Taxilla (24); TH, Thatta (13); TR, Tret (2).

part of the Pakistan Mekkran coast and adjacent coastal areas of Iran. Both Minton (1966) and Eiselt and Schmidler (1973) suggest that *Bufo olivaceus* Blanford may eventually be considered a geographic race of *Bufo stomaticus*, though the former admits that the westernmost populations of the latter are no more like those of *olivaceus* than are specimens from eastern Pakistan. At least Minton seems to not have an opportunity to examine specimens of *B. olivaceus*, which are represented by only a very few specimens in museum collections. Fortunately, we have been able to collect good series (University of Florida/Florida Museum of Natural History collection) from several localities which form the basis of our study and conclusions. We were also able to examine the preserved types of both species.

MATERIALS AND METHODS

The study is based on 524 specimens from the following institutions: American Museum of Natural History, New York; British Natural History Museum, London; Bombay Natural History Society, Bombay; California Academy of Sciences, San Francisco; Museum of Natural History, Vienna; Field Museum of Natural History, Chicago; Museum of Comparative Zoology, Harvard University, Cambridge; State Zoological Collection, Munich; Pakistan Museum of Natural History, Islamabad; Senckenberg Museum of Natural History, Frankfurt; Florida Museum of Natural History, Gainesville; University of Michigan Museum of Zoology, Ann Arbor; United States National Museum, Washington; Zoological Survey Department of Pakistan, Karachi; Zoological Survey of India, Calcutta; Alexander Koenig Museum, Bonn.

Figure 1 shows the location and extent of the areas encompassed by each of the sample subsets upon which morphological comparisons are based.

Twenty characters were analyzed on each specimen. The conventional ones included sex and snout-vent length (SVL hereafter, measured to closest mm). Additionally, the following characters and their definitions as they apply to this study were used. All measurements were made to the closest 0.5 mm.

HL, Head length: from tip of snout to most anterior part of tympanum.
HW, Head width: across widest part, almost always just anterior to anterior edge of tympanum.
EYED, Diameter of eye: greatest horizontal diameter.

IW, Interorbital width: least transverse distance at crease located where tissue covering orbit contacts that of frontal part of skull.

EW, Upper eyelid width: greatest transverse distance of eye lid.

TYMP, Diameter of tympanum: greatest vertical diameter.

PAL, Parotid gland length: greatest antero-posterior distance of parotid gland on basis of observed swollen glandular tissue.

PAW, Parotid width: greatest transverse distance of parotid glandular tissue.

TL, Tibia length: distance between most posterior outer surface of tarso-metatarsal area ("heel") to most anterior part of articulation between tibia and femur ("knee").

WS, Wart size: greatest diameter of largest and most medial of warts located paravertebrally at about level of anterior insertion of hind limb.

Additionally the color of the following anatomical features were noted: general dorsal head and body color and pattern; whether a stripe darker than the adjacent ground color exists from the posterior part of the parotid gland to near the insertion of the hind limb, whether the darker dorsal markings were restricted to dorsal warts, or extended beyond them onto the dorsal surface, the dorsal tibial surface, the swollen part of the finger tips, any dark markings on the sternal area.

All illustrations are.the work of the senior author.

RESULTS

General Remarks.-- Table I shows the length and ratio statistics for all characters used in this study.

Table I: Statistics for all length and ratio parameters used in this study; sexes and all localities combined.

	Mean	STD	O.R.	SEM
SVL	52.9	8.9	28.2-87.0	0.5
TIBIA	19.9	3.1	22.4-28.5	0.2
PARL	12.9	2.8	6.4-27.0	0.1
PARW	6.4	2.2	6.0-12.4	0.1
TYMP	3.9	0.7	2.4--6.0	0.0
SVL/TIBIA	2.6	0.4	1.9--2.1	0.0
SVL/PAL	4.1	0.9	2.0--8.0	0.0
PAL/PAW	2.0	0.5	1.0--4.2	0.0
SVL/TYM	13.6	0.9	8.4--9.3	0.0
SVL/PAW	6.7	3.0	4.5--9.9	0.2

To assess the relationships between each of the parameters against size, linear regressions of all parameters (dependent variable) were run against SVL (independent variable, 52.6 ± 10.6 mm) (Table II).

Table II: Linear regression of statistical parameters used in the study. Independent variable SVL.

	Mean Y	**Corr. Coef.**	**R2**	**p**
PARL	12.7 ± 3.6	0.6	0.37	< 0.000
PARW	6.4 ± 2.1	0.6	0.33	< 0.000
SVL/TIBIA	2.6 ± 0.4	0.6	0.37	< 0.000
SVL/PAL	4.1 ± 0.9	0.4	0.15	< 0.000
SVL/TYMP	0.1 ± 0.1	-0.1	0.01	0.039
SVL/PAW	6.7 ± 3.0	0.2	0.03	0.003
PAL/PAW	1.9 ± 0.5	0.2	0.04	< 0.000
TYMP	3.9 ± 0.9	0.7	0.49	< 0.000

The results show that all linear regressions except SVL/TYMP are highly significant, with very low probable errors, the anatomical parameters increased in more or less direct proportion to SVL. Thus they are appropriate parameters for statistical analyses. The linear relationship between SVL and TYMP is complicated by the fact that the growth relationship between them is not the same as the others, due partly to the variability within the total sample caused by sexually dependent significantly different means of TYMP in males and females, and the fact that TYMP tended to become proportionately smaller with increasing SVL.

Highly significant multiple coefficients (p 0.000 - 0.015) between various sets of three parameters each show that the tibia contributes the most variation (T test 21.3, df 295, p 0.000).

The correlation coefficients were obtained from a pair-wise comparison of all of the measured parameters used in this study (Table III).

Table III: Comparison of correlation coefficients of parameters measured in *Bufo stomaticus.*

PARL				
0.59	SVL			
0.43	0.73	TYMP		
0.62	0.85	0.70	TIBIA	
0.59	0.60	0.50	0.63	PARL

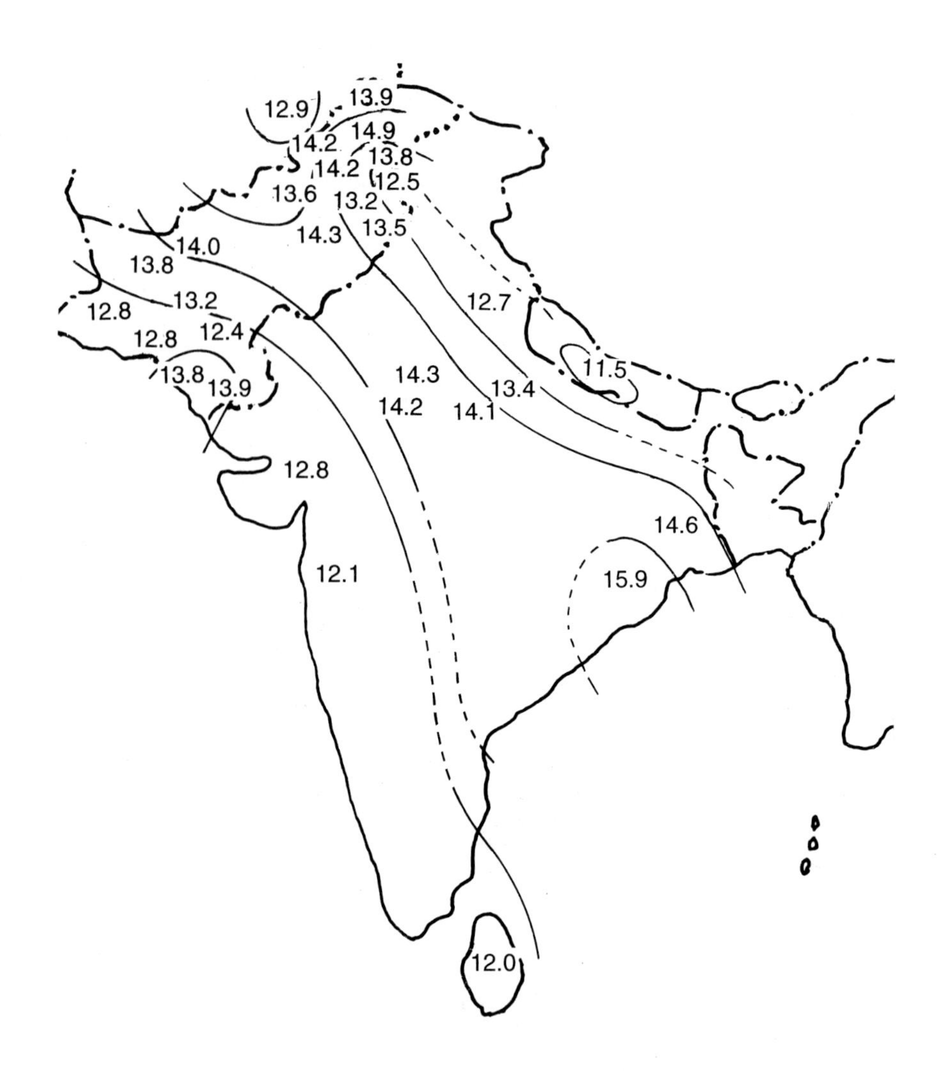

Fig. 2. Geographic variation in the statistic SVL/TYMP of *Bufo stomaticus* throughout its geographic distribution.

For reasons given above, the TYMP measurements for the total sample are divided into adult male and adult female subsets. When this is done, a statistically significant correlation can be demonstrated. The entire data set suggests that the best correlation among parameter pairs occurs with SVL-TIBIA, SVL-TYMP, and TYMP-Tibia. Therefore these pairs are given special attention in the following discussions.

Sexual Variation.-- As is common in all bufonid species that have been studied, females of *B. stomaticus* are statistically larger than males (mean adult female SVL 55.5 ± 9.7mm, mean adult males 49.8 ± 6.5 mm). The proportional length of several characters in males and females are shown in Table IV; none of the proportionate values are significantly different between adults of either sex.

Table IV. Summary of morphological statistics for adult male and female *Bufo stomaticus* examined.

	Mean	STD	O.R.	SEM
Females (N 245)				
SVL	55.5	9.7	28.2-87.0	0.7
TIBIA	20.4	3.1	11.4-28.5	0.2
PARL	13.4	2.8	7.8-27.0	0.2
PARW	06.9	1.8	0.6-12.0	0.1
TYMP	04.0	0.7	2.5--6.0	0.0
Males (N 255)				
SVL	49.8	6.5	33.4-70.6	0.5
TIBIA	19.3	2.9	2.4-26.3	0.2
PARL	12.3	2.7	6.4-26.7	0.2
PARW	06.3	1.5	3.6-12.4	0.1
TYMP	03.8	0.7	2.4--6.0	0.0

The skin of adult females is very noticeably more roughened than that of adult males. This roughness is largely produced by a single black-colored asperity, of which there are fewer in males, giving them a smoother skin texture. However, during the breeding season, the skin of the chest of adult males becomes slightly roughened, due to the growth of similar, but smaller asperities in that area. Furthermore, during this time of year fingers 1 and 2 of adult become partially covered with patches of blackish, cornified tissue, missing the remainder of the year.

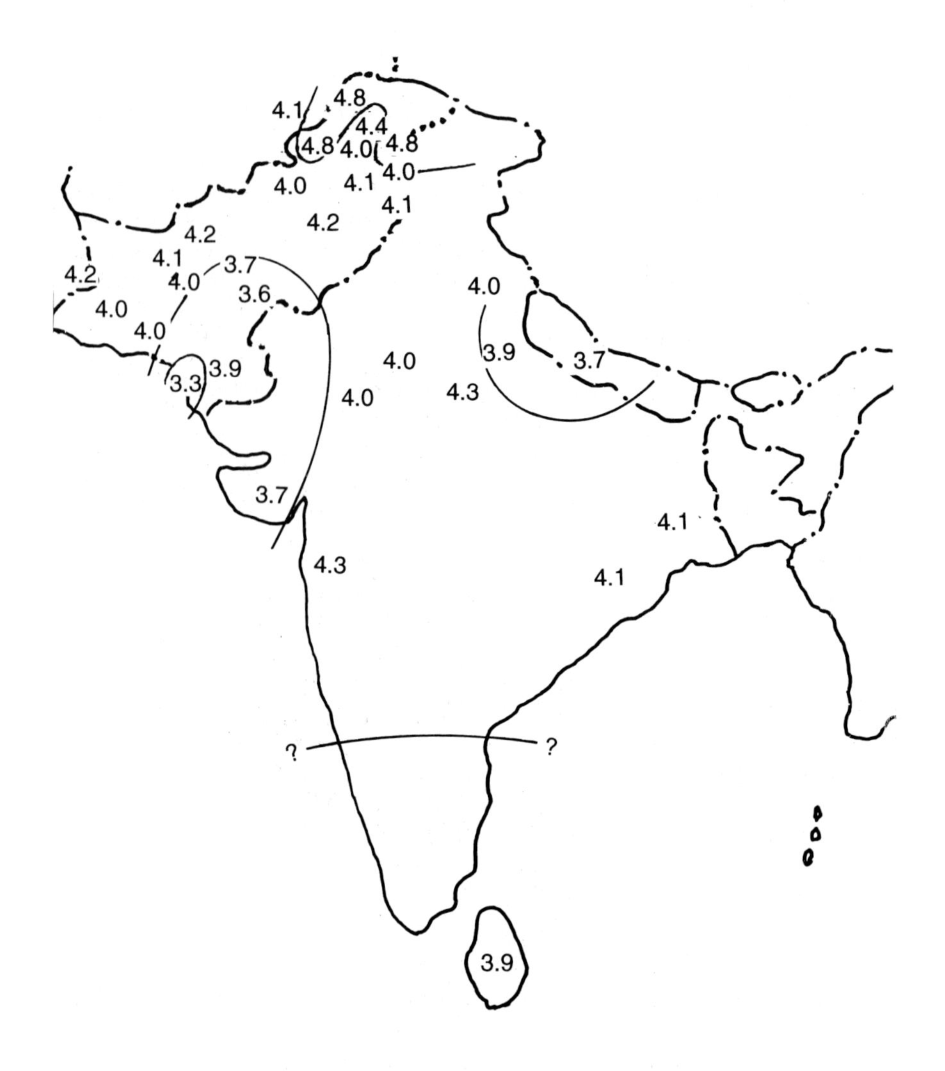

Fig. 3. Geographic variation in the statistic SVL/PAL of *Bufo stomaticus* throughout its geographic distribution.

Geographic Variation.-- The following section describes the geographic variation in the morphological parameters studied.

SVL/TYMP: Figure 2 illustrates the pattern of geographic variation in the vertical diameter of the tympanum standardized against snout vent length (SVL/TYMP). The average mean value for the total sample is 13.60. This value is found throughout much of the central part of the geographic range of the species and on this basis we consider these populations as representing the "average" condition for proportionate tympanum size. These populations are the standard against with the other geographic subsets are compared. The highest values (15.91, = smallest TYMP) are in Orissa, and the lowest (12.69, = largest TYMP) occur in Nepal.

That tympanum size is not related to elevation or even to general habitat is shown by a comparison of this statistic from east to west among only the more northern subsets -- all in mountainous regions, yet containing from the lowest to some of the highest values calculated. Nor is moisture a clearly correlative factor, as demonstrated when comparing the Nepal and Kumaon samples (11.54 and 12.69 respectively) from mesic, more or less forested (mountainous) areas with similar low values in desertic subsets at low elevations (i.e., Sindh, 12.91). Finally, compare the values for the Jalalabad and Peshawar subsets, one being among the highest values in the total sample and the other among the lowest, yet both are at about the same elevation, in the same general environment.

When viewed in broad perspective, the general geographic pattern of variation is one in which values over most of the range hover around the average, though several geographic morphoclines are apparent. In one of these the values decrease (tympanum larger) southward through the Indian peninsula and along the Mekkran Coast. Similarly, values become greater eastward, including Calcutta and Orissa, though this cline is steeper than the one passing southward. Still another morphocline of gradually increasing values (tympanum smaller) runs northwest, culminating in the high values of the Peshawar, Swat and Haripur subsets. However, the mean value of the Peshawar subset is significantly different from the values of the Jalalabad subset (t = 4.01), being separated from one another by the north-south Lundi Hills.

Another cline, in which values become significantly smaller, runs northwest from the Gangetic "average" populations, extending through Kumaon into Nepal.

SVL/PAL: Figure 3 illustrates the geographic variation in parotid gland, standardized against snout-vent length (SVL/PAL). The values for the various sample subsets vary from 3.30 (Karachi, proportionately

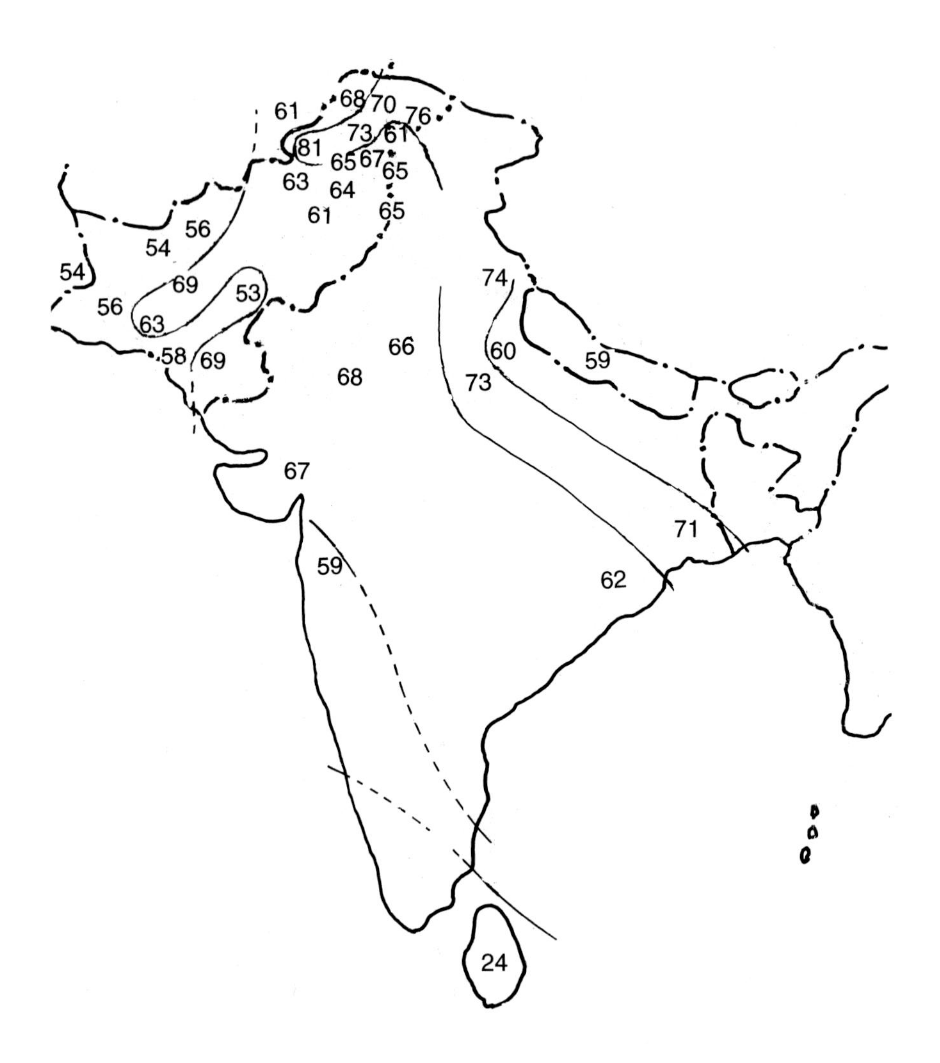

Fig. 4. Geographic variation in the statistic SVL/TIBIA of *Bufo stomaticus* throughout its geographic distribution.

large parotid gland) to 4.77 (Chitral, proportionately smaller parotid gland). The average for all the subsets is 4.17, with values at or near this level found mainly in the central Indogangetic-Indus River Plains. Thus these centrally located populations are considered by us as representing the "average" condition in respect to the character PAL. In all directions from this area, the proportionate parotid gland length in toads from geographically surrounding sampled populations tends to decrease (=large parotid) (the only statistically significant exception being Bombay, with 4.30). However, this trend is very notably reversed in populations along the entire northern edge of the species range from Peshawar eastward to Muzaffarabad and Kotli. The highest values in this reversal zone are in Chitral and Peshawar. The Jalalabad population along the Kabul River west of the Lundi Hills is again significantly different from that of the adjacent sample area of Peshawar, as they are in tympanum diameter (above).

The Nepal subset is also again significantly different, showing a substantially different trend than in the western Himalayan foothill populations -- in this case being more similar to the Indian peninsular populations than to the western Himalayan populations.

If *Bufo dhufarensis* is more similar to the ancestral stock of this species group in this character, then the coastal populations *of B. stomaticus* are closest to it, and the parotid gland becomes shorter towards and in the mountains (except Nepal).

SVL/TIBIA: Figure 4 illustrates the geographic variation in tibial length, standardized against snout-vent length (SVL/TIBIA). The mean value of the total sample is 2.71. Populations having the same or nearly the same values are found throughout most of the entire species distribution. However, subsets of this toad from the southern part of peninsular India, Quetta, and the Mekkran Coast are all significantly lower (below 2.60). The Sindh subset is distinct, having a significantly lower value than all surrounding populations. The Peshawar subset is again significantly different from that of Jalalabad. The highest mean value of the total sample is found at Tarnah (3.94), being significantly higher than that for all adjacent subsets. The lowest value in the total sample occurs in Sri Lanka (2.24). Nepal is again significantly different from populations in the Himalayan foothills to the west.

WARTS: Minton (1966) states that individuals from northeastern Pakistan have more, and larger warts than those from the rest of the country. However, he had only four individuals from this area and we suspect they may have been females, for our work shows that adults of this sex are often more warty on the dorsal body surfaces than adult males.

In some specimens there is an ill defined row of enlarged warts along the lateral body stripe. This was noted as present or not, and if so, whether large, medium, or small. Our analysis shows that strongly developed lateral body warts are found in 100% of the Kumaon subset, in the foothills of the Himalaya Mountains north of the Ganges Plain in India and that they occur in from 25 to 65 % of several populations in these foothills both east and west of this area. They are not, however, found in any individuals in our Nepalese, or Muzaffarabad, Pakistan subsets. Thus the character cannot be described as always well developed in populations from mountainous areas. However, it is clear that there is a morphocline in which these warts become smaller southward in both Pakistan and India, and are smallest and usually absent at the southern range limits. Thus such warts never occur on individuals in Pakistan subsets of Thatta and Sindh, or in that from Sri Lanka. Geographically intermediate populations in the lower and upper coastal plains tend to be represented by intermediate values. We found no correlation between the development or presence of these warts and sex, though they become most obvious when the individuals become adults.

COLOR: Color pattern variables investigated in this study are extent of dark speckling on the chest and the prominence of a lateral stripe (both noted in each specimen as strong, weak or none), whether each had light interorbital bars or not, presence or absence of a light median dorsal body stripe, size of dorsal body spots (as large, larger, or smaller, than largest dorsal body warts) presence or absence of a dark, posteriorly directed V-shaped mark in the shoulder area, general dorsal ground color (grayish or brownish, light, medium to dark).

In life, the entire ventral surface is normally dirty white. Minton (1966) states that individuals from northeastern Pakistan have dark mottling on the chest and throat. Though this conclusion was based on only four adults, it is partly largely substantiated by the present study. Here some populations have the chin and/or chest flecked with dark brown or black, sometimes forming a thin, median stripe. The flecking is equally found in both sexes. It is completely absent in all individuals examined from all subsets in the lower coastal plains of both India and Pakistan, and is rare (1-3 %) in a few subsets from the upper coastal plain. While found in mountainous parts of both Nepal, India, and Pakistan and the adjacent most southern foothills (2 - 6 %.), it occurs in 25 - 30% of the individuals in those subsets where the Indus River is leaving the mountainous zone (Haripur and Manshera), becoming 100% in Besham subset upriver along this valley at the northern limits of the species range. The presence of a dark lateral body stripe shows almost the same pattern, being absent in the coastal areas, and only common (to 30 % of the subset) in the same part of the Indus Valley in Pakistan. Light-colored median body stripes are relatively rare throughout the range; light-colored interocular bars are more common, but neither show any clear geographic pattern. Dorsal ground color is, however, clearly correlated with substrate color, being darkest in mesic zones with darker soils, and light in arid zones with light-colored soils. Though there is considerable individual variation throughout the range, individuals with larger and darker dorsal spots tend to be found in more mesic, upland areas.

Though there is no indication that the darkness of the ground color or of the dorsal spots are sex related, during the breeding period in at least Pakistan, the dorsal body color of adult males changes to a lighter, sometimes decidedly yellow-grey, or even (rarely) to a lemon-yellow. We did not notice any color changes in females during the breeding period. Juveniles have a color pattern similar to the adults, except that the interorbital area on top of the head and the middorsal scapular region are often provided with orange speckling and the sooty-colored dorsal mottling often has pinkish centers (latter especially in subadults).

Bufo olivaceus Blanford

In his herpetology of Pakistan, Minton (1966) fails to describe or provide notes on this species, except to suggest that it may be a subspecies of *Bufo stomaticus*, based on general similarity and non-sympatry of the then available material. Perhaps because of the rarity of this species in museum collections, he apparently did not examine any, but lists their differentiating characteristics largely on the basis of available literature (parotid gland shape, dorsal skin roughness and webbing on toe IV. Fortunately, we have been able to secure small series of *Bufo olivaceus* from several Pakistan localities along the Mekkran Coast, one of which was prepared as a skeleton.

On the basis of this new material we are able to demonstrate many morphological differences between the two species. The crania differ as follows (based on *olivaceus*, adult, Univ. Florida/Florida Mus. Nat. Hist. 82724; *stomaticus*, adult, Univ. Florida/Florida Mus. Nat. Hist. 74673).

In that of *stomaticus*, the head is proportionately narrower, and higher, exemplified by narrower frontoparietals and narrower and more triangular-shaped nasals. Both species have a prootic of similar shape from above, but that of *stomaticus* is proportionately shorter. A probably more important difference is that the dorsal

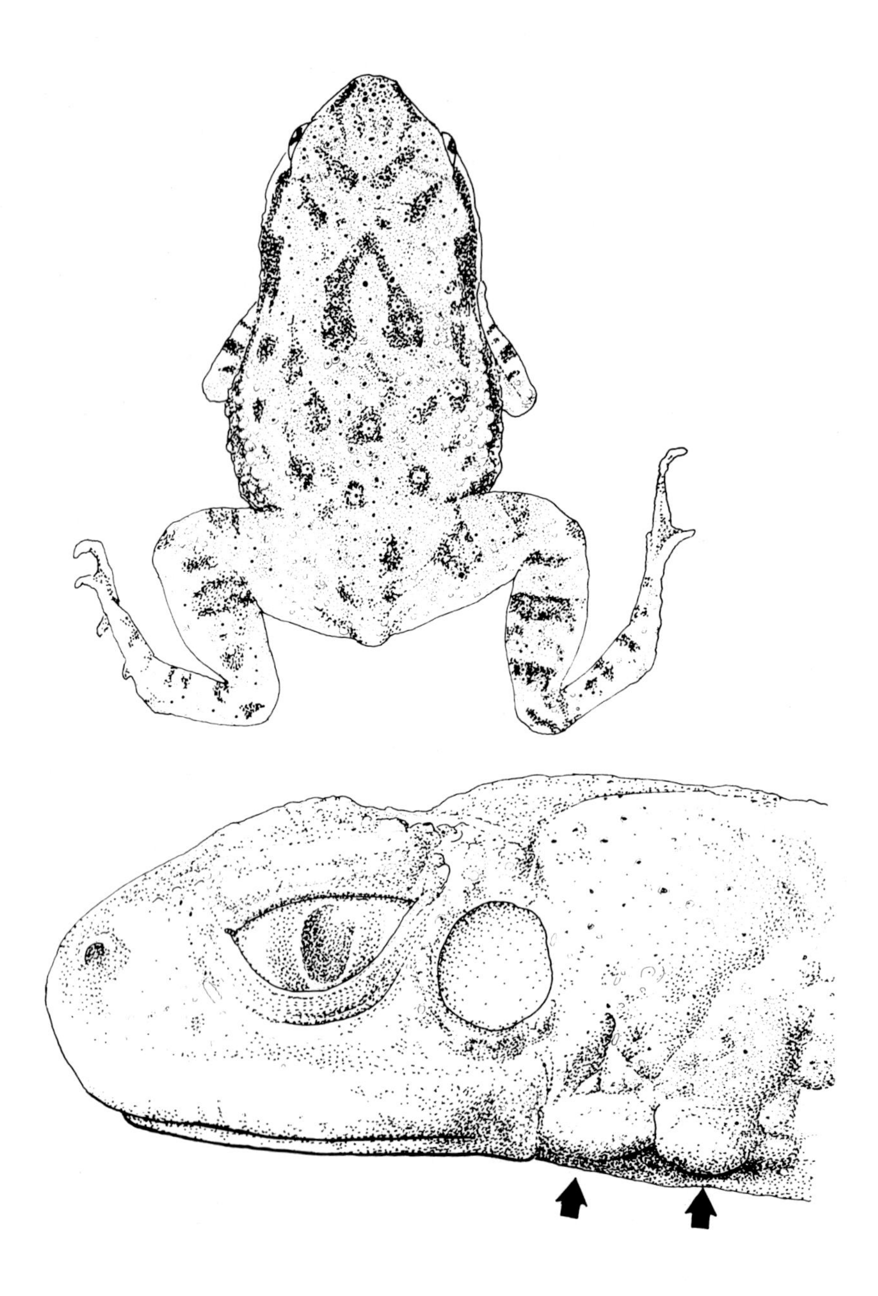

Fig. 5. Near extremes of pattern in adult *Bufo stomaticus*. Left, adult male, Makli, Sindh Prov., Pakistan. W. A. photo No. 436; Right, Bombay, Maharashtra State, India, Brit. Mus. Nat. Hist. 83.11.36.105.

surface of this bone and the adjacent temporal plate each has a large concavity, located mainly near the posterior edge; missing in *olivaceus*. In *stomaticus*, the epiotic eminence is not well developed, possessing a small depression lateral to it and between it and the prootic depression. Each of these depressions have a narrow ridge between them.The epiotic eminence continues ventrally as a ridge on the posterior surface of the cranium.

While the columella of *olivaceus* is rather exposed when compared to many other bufonid species, that of *stomaticus* is even more so. The parasphenoid stem and its lateral wings are proportionately wider in *stomaticus*, the former being un-notched posteriorly, whereas it is notched in *olivaceus*. In the former species, the squamosal stem is proportionately narrower and longer than in the latter, and the "ocular foramen" is much larger than in *olivaceus*. Because there is none available in the literature, we provide a description of the salient features of the cranium of the adult *olivaceus*. The premaxillaries are about as high as they are wide; nasals large, flat dorsally, medially in contact throughout their length. Anterior contact of the frontoparietal with the nasals is somewhat transversely straight, in intimate contact with the nasals along the entire articulation. The sagital suture between the frontoparietals is separated in a small oval area anteriorly, filled by a longitudinal strip of the ethmoid and marked by a shallow, unraised junctional line in the middle and posterior sections. There are no cranial ridges. The frontoparietal is unfused, flattened both longitudinally and transversely. Contact with the unfused nasals is broad, almost straight edged transversely. Lateral perpendicular lamina narrow, without a dorsal seam, but completely fused to true frontoparietal part; antero-dorsal portion plate-like, resting on the dorsal edge of the sphenethmoid (ethmoid). Occipital grooves (for occipital artery) between posterior part of frontoparietal and the otoparietal plate. The exoccipital is well developed, ending posteriorly in a subtriangular area of spongy bone where these areas contact the atlas. Otoparietal plate not well differentiated, but largely fused to the prootic. On the dorsal surface its transverse part is small, subtriangular in shape, and restricted to an area at the anterolateral portion of the occipital groove. The longitudinal portion is not visible dorsally, but forms the outer wall of the occipital groove. The occipital groove is well-developed, open along its entire length. The temporal portion is vertical with greater lateral than ventral length; a large foramen is present in the external ventrolateral corner. The prootic is transversely elongate, with a flat dorsal surface, but with well-developed, subpyramidal-shaped epiotic eminence, rising above the level of the postorbital shelf. The lateral part of the postorbital shelf is formed of a bony plate fused to the squamosal externally and

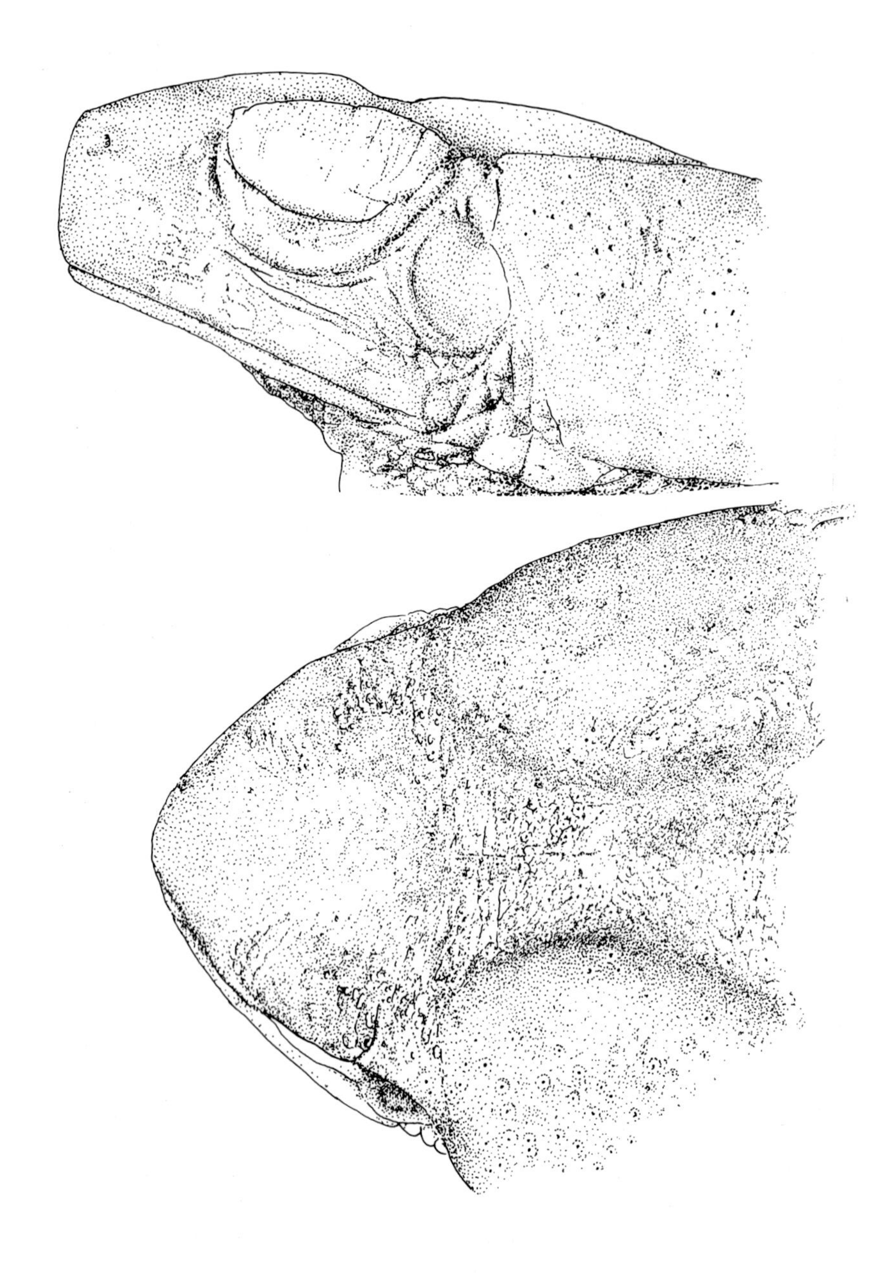

Fig. 6. Dorsal anterior body and head surfaces of *Bufo olivaceus* showing large, oval, parotid glands and smooth skin. Lateral view of head showing distinctive bi-lobed parotid glands.

articulated with the prootic medially. The columella is free, durably constructed, and not encased in either bone or cartilage.The sphenethmoid is ossified. The parasphenoid stem is longer than its cross piece. On the external ventral surface the anterior stem end has a central furrow; the posterior end is notched.

In color there are also important differences between these two species. Dorsally the body of *B. stomaticus* is usually variously flecked and reticulated with medium brown or gray through black, on a ground color of gray, cream, or light brownish to greenish-gray. Laterally a dark stripe often starts from behind the eye and runs to the groin. A lighter parallel band is often seen above it. One to two dark-colored, V-shaped marks often occur on the top of the head, between and/or just posterior to the eyes (Fig. 5). The dorsal color of *B. olivaceus* is almost always more or less uniform olive to olive brown. There are no markings laterally or cranially (Fig. 6).

When handled, *Bufo stomaticus* often produce a strong odor similar to crushed ailanthus leaves. This odor was never detected in *B. olivaceus.*

Bufo olivaceus is much larger than *B. stomaticus*. The parotid gland of *olivaceus* (Fig. 6) is proportionately much larger than that in *stomaticus*, and frequently much more raised off the body surface (Fig. 5). Though Minton states that its shape is wedge-shaped in *olivaceus*, and reniform in *stomaticus*, we have found that this is variable in both species. Indeed, *B. stomaticus* from Pakistan usually have reniform parotid glands, though those in the eastern part of the range (Ganges River Plain to Calcuta and Orissa) often a sub-triangular shape; the parotid gland of most adult *olivaceus* are reniform to oval in shape (Fig. 6), though sub-triangular shaped ones do occur, and are the rule in subadults.

The skin texture of these species is also very different. That of *olivaceus* is relatively smooth, rather shiny (even on the parotid glands),. without well differentiated warts on the dorsal body surface. That of *stomaticus* is much more warty dorsally, with each of the larger warts usually provided with one to two short, hair-like spines (asperites). There are also more pores on the entire dorsal surface.

The extent of the webbing on the hind foot is noticeably greater in *olivaceus*, where on toe IV it is 25 to 30 percent of the toe length, and in *stomaticus* the web is 30 to 45 percent (as correctly stated by Minton 1966).

CONCLUSIONS

On the basis of our morphometric analyses we suggest the following points are potentially important from both systematic and zoogeographic stand points:

Zoogeographical Implications

1). The Indus River Valley is probably a more important genetic barrier than any of potential barriers within peninsular India (ex.: TYMP).

2.) The Lundi Hills (forming the Khyber Pass) form an important genetic barrier (ex.: TYMP, PAL, TIBIA) between lowland Afghanistan and adjacent lowland Pakistan populations. The importance of the Lundi Hills as a genetic barrier is further illustrated in the fact that in each of the characters in which significant differences can be demonstrated on either side of the Lundi uplands, the east-west morphoclinal trends running through lowland northwestern Pakistan populations are all reversed northwest of the isolating Lundi Hills. All of this is rather unusual in view of the fact that there are much higher mountain ranges nearby (i.e., Suliman and Brahui Ranges) that seem much less important zoogeographically.

3). The *Bufo stomaticus* populations of the western parts of the Himalaya mountain complex (Pakistan) are morphologically significantly different from those of the eastern part (Nepal) of the same upland complex. The upper reaches of the Ganges River, within the Himalayan foothills, may represent the most important isolating feature, at least during parts of the Pleistocene, when the river was presumably much larger.

4). The Sindh Desert populations are morphologically distinct at a statistically significant level from all surrounding populations (ex.: TIBIA).

5). The middle section of the Indus River from Besham south to where it leaves the foothills represents a distinctly different genetic entity (e.: VENTRAL COLOR). This is also reflected in the local endemism of other amphibian species (ex. *Rana hazarensis*) in the same area.

Systematic Implications. The results outlined above suggest that about five geographic populations of *Bufo stomaticus* are recognizable on the basis of significantly different morphometry. However, the rather low level of differences exhibited between these genetic subsets suggest to us that we do not recognize any of them as worthy of subspecies rank on the basis of our results.

If the trend lines we have demonstrated in the morphoclines studied throughout this species geographic range ultimately emanate from the closely related western species of the complex (*B. dhufarensis*), as we believe they do, then the "primitive" condition is one of a larger tympanum in respect to SVL and it follows that almost all trend lines within the "central" part of the species range run from the coastal area of the Arabian Sea (SW Pakistan to W India) northward.

ACKNOWLEDGMENTS

Thanks are extended to those individuals in charge of the museum collections listed above for their generosity in making important material in their care available for our study. We are also highly appreciative of the funding furnished us by several agencies for visits to many museums (United States Fish and Wildlife Service, Deutscher Akademischer Austauschdienst, and Office of Sponsored Research, University of Florida). The work is an outgrowth a major project concerning the herpetology of Pakistan, funded through the Office of International Affairs, United States Fish and Wildlife Service. Finally we wish to acknowledge the support offered by our respective institutions.

REFERENCES

ANDERSON, S. C AND A. E. LEVITON. 1969. Amphibians and reptiles collected by the Street Expedition to Afghanistan. *Proceedings California Academy Science*, 4th ser. **37**(2):25-56.

ANNANDALE, N. 1907. The distribution of *Bufo andersoni. Records Indian Museum.* **1**:171-2.

ANNANDALE, N. 1908. Notes on some Batrachia recently added to the collection of the Indian Museum. *Records Indian Museum.* **2**:304-5.

ANNANDALE, N. 1909. Notes on Indian Batrachia. *Records Indian Museum.* **3:**282-286.

BHATI, D. P. S. 1955. The pectoral musculature of *Rana tigrina* Daudin and B. andersoni Boulenger. *Acad. Zool.* **1**(2):13-17.

BOULENGER, G. A. 1891a. Descriptions of new oriental reptilia and batrachians. *Annals Magazine Natural History.* ser 6, **7**:279-283.

BOULENGER, G. A. 1891b. Notes on *Lycodon atropureus* Cantor and *Bufo stomaticus* Lütkin. *Annals Magazine Natural History*, ser 6, **7**:7:462-463.

BOULENGER, G. A. 1893. Description of new species of reptiles and batrachians in the British Museum. *Annals Magazine Natural History*, ser. 5, **12:**161-167.

DANIEL, J. C. 1963 Field guide to the amphibians of western India. *Journal Bombay Natural History Society.* **60**:415-438.

EISELT, J. AND J. F. SCHMIDLER. 1973. Froschlurche aus dem Iranenter Berucksichtung auseriranischer Populations gruppen. *Annals Natural. History Museum, Vienna.* **77**:181-243.

HASHMI, T. H. 1955. *The amphibians of Lahore with descriptions of the early development of Bufo melanostictus Schneider.* M. Sc. Thesis, University of Punjab, Lahore.

KHAN, M. S. 1968. Morphogenesis of digestive tract of *Bufo melanostictus* Schneider. *Pakistan Journal Scientific Research* **20:**93-106.

KHAN, M. S. 1972. The 'commonest toad' of West Pakistan and a note on *Bufo melanostictus* Schneider. *Biologia* **18**:131-3.

KHAN, M. S. 1979. On a collection of amphibians from northern Punjab and Azad Kashmir, with ecological notes. Biologia **25**:37-50.

MCCANN, C. 1938. The reptiles and amphibians of Cutch State. *Journal Bombay Natural History Society.* **40**:425-427.

MINTON, S. A., JR. 1966. A contribution to the herpetology of West Pakistan. *Bulletin American Museum Natural History*, **134**:29-184.

MURRAY, J. A. 1884. Additions to the present knowledge of the vertebrate zoology of Persia. *Annals Magazine Natural History*, ser 5, **14**:101-106

PARKER, H. W. 1938. Reptiles and amphibians from southern Hejaz. *Annals Magazine Natural History* (London) ser 11, **1**:481-492.

SARKAR, A. K. 1984. Taxonomic and ecological studies on the amphibians of Calcutta and its environs. *Records Zoological Survey India.* **81**(3-4):215-36.

DISTRIBUTION OF *LAUDAKIA* (SAURIA:AGAMIDAE) AND ITS ORIGIN

KHALID J. BAIG

Pakistan Museum of Natural History, Garden Avenue, Shakarparian, Islamabad

Abstract: Different species of *Laudakia* presently deposited in various natural history museums of the world have been thoroughly studied and all relevant literature reviewed. The studies address different issues related to *Laudakia* including its centre of origin, age, distribution and dispersal pattern.

INTRODUCTION

Laudakia Gray has recently been resurrected by partitioning *Stellio* (sensu Moody, 1980) into *Laudakia* Gray and *Acanthocercus* Fitzinger on the basis of biochemical, karyotypic, anatomical, zoogeographical and morphological evidence (Baig, 1992; Baig and Böhme, in press). The main objective of the present study is to outline the distribution of this genus and propose the age and centre of origin of *Laudakia.* It includes 16 species viz. *himalayana, bochariensis, badakshana, agrorensis, pakistanica, nuristanica, tuberculata, dayana, stoliczkana, lehmanni, erythrogastra, caucasia, microlepis, nupta, melanura* and *stellio.* Of these five species are further subdivided and altogether make 25 taxa in this genus.

More than 600 specimens have been studied and all related literature reviewed. The studies suggest that *Laudakia* has a very wide distribution (Map I), from Mongolia in the North to Makran Coast of Pakistan in the South; Lhasa (Tibet) in the East to the Mediterranean Islands, Europe and NE Africa in the West. According to the zoogeographic distribution of Sclater (1858), Wallace (1876) and Darlington (1957) it is Palaearctic. The peripheral limits of this genus are marked by the presence of single species in each direction but the rest of the taxa are mainly distributed in the areas around Pamir Knot. Keeping in view the present distribution of the species and fossil records of the ancestors of this group, centre of origin and age of *Laudakia* is being proposed.

CENTRE OF ORIGIN

The area around "Pamir Knot" is being proposed as centre of origin for this group of agamids, using some selected criteria adopted by

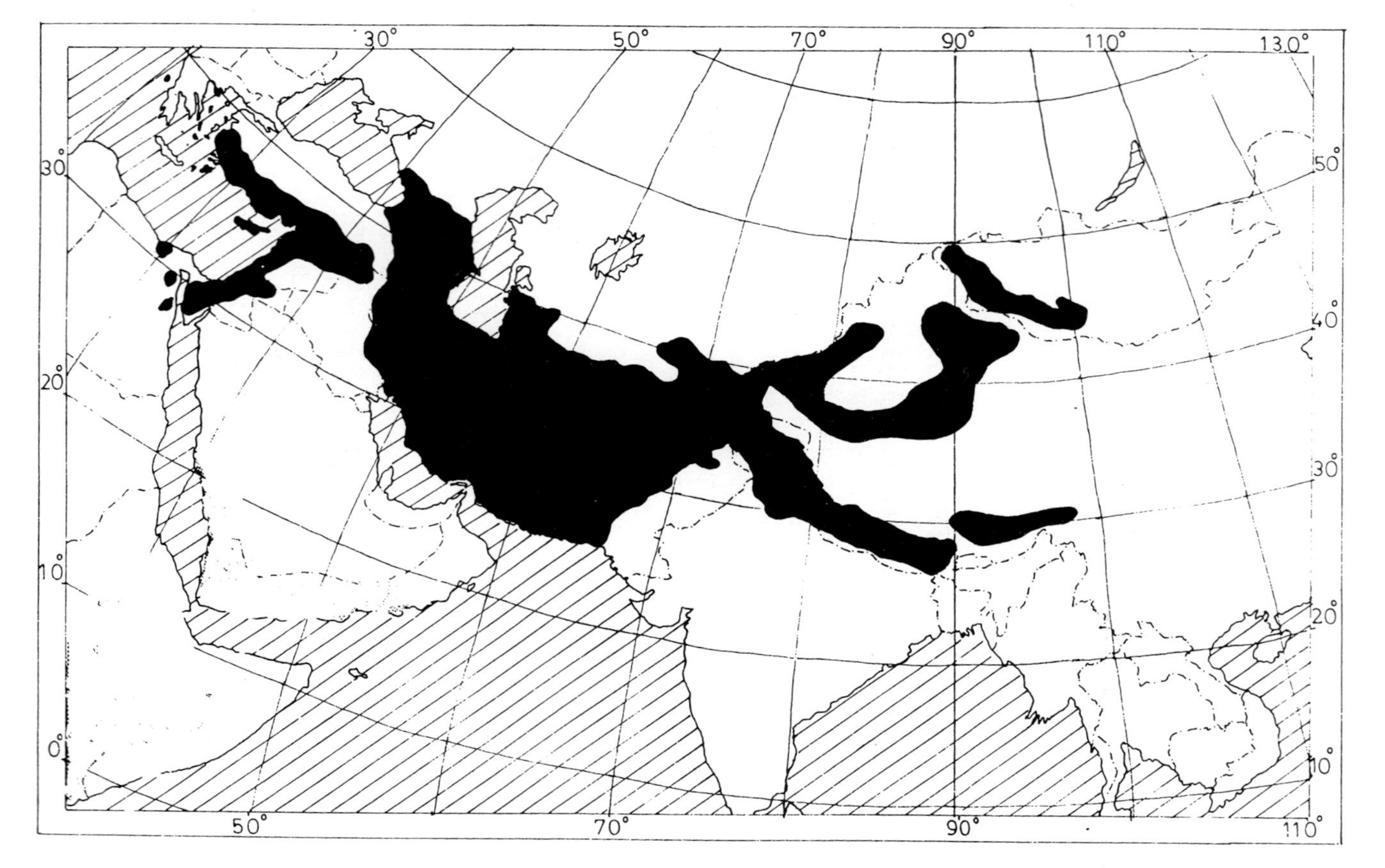

MAP. I *LAUDAKIA* (●)

Moody (1980) for the family Agamidae.

1- Higher Number of Taxa: The area with high number of species is indicative of the origin of that group. This assumes that rate of speciation exceeds the rate of extinction.

There are about 12 taxa viz. *stoliczkana, himalayan, bochariensis, badakshana, lehmanni, pakistanica, nuristanica, tuberculata, caucasia, nupta,* which are represented here. If the radius would be extended a little more, then *agrorensis, melanura, microlepis, dayana* and *erythrogastra* can also be included. This practically leaves only peripheral taxa i.e. *stellio, tuberculata sacra* and *stoliczkana altaica.*

2- Diversity of Endemism: The region where there is a marked degree of phenotypic differentiation between species indicates the centre of origin.

The proposed centre of origin represents the members showing wide morphological range. It includes small *Laudakia* forms i.e. *himalayana, bochariensis, badakshana* and also large ones, they are *nupta, pakistanica, stoliczkana, tuberculata;* species of biwhorled tail segment (*caucasia*), triwhorled (*pakistanica, lehmanni, himalayana,* and tetrawhorled (*tuberculata, stoliczkana*); strongly carinated (*pakistanica, lehmanni*) and less carinated (*stoliczkana, tuberculata*); regarding colour, jet black *pakistanica* (nominate form) and light coloured *stoliczkana* (n.f.) and *caucasia*, all are represented in this area.

3- Centre of Distribution Range: Assuming that expansion occurrs in all directions, then the place of origin would be approximately the middle of the distribution range.

Vertical and longitudinal axes of the present distribution of the group approximately cross each other at the proposed centre of origin, showing uniform distribution in all directions.

4- Continuity of the Area: The geographical regions which show continuity of the habitat required for the dispersal would indicate long period of occupation. The centre of origin would by hypothesized where different dispersal routes join each other.

Laudakia is basically a mountainous group which lives upto an elevation of 4000 m. Except the area occupied by *stellio*, all remaining distribution areas are more or less continuous. The major mountain ranges, Himalaya, Hindu Kush, Karakoram, and Pamir meet each other at Pamir Knot. The other mountain ranges are continuous at the distal end of these ranges. The disjunct areas are occupied by distinct taxa e.g. *tuberculata sacra, stoliczkana altaica.*

This again supports the present point of view regarding the centre of origin.

5- Distribution of sister groups: Hypothesized origin of any sister group, phylogenetically derived from a common ancestor would also indicate centre of origin for the other group.

Joger (1991) in molecular phylogenetic studies indicated *Phrynocephalus* and *Stellio* close to each other in Wagner Tree. Moreover, Joger and Arano (1987) showed *Trapelus* very close to *Stellio* (sensu Moody, 1980) in phyletic branching order. Studies made by Macey et al. (unpublished) on *Phrynocephalus* presented some facts which again support the present point of view about *Laudakia.*

i) To account for the pre-Pamir diversification of the genus they proposed Afghanistan more feasible place for initial speciation.

ii) They indicated *Trapeleus* and *Stellio* (now *Laudakia*) as sister groups in the Wagner Tree based on weighted and unweighted analysis.

iii) Except the presence of two endemic species, one in Thar Desert of India and other in Arabian Peninsula all other distribution of *Phrynocephalus* was parallel to *Laudakia.*

These facts again prove common phylogenetic ancestry and habitat oriented parallel distribution.

6- Distribution of Primitive Species: The primitive taxa should be distributed near the centre of origin (Hennig, 1966). Any vertical or longitudinal distance from the centre of origin would encourage derived characters suitable for the new habitat. In case of disjunct habitat it may give rise to distinct populations which could be subspecies. The primitive species would remain close to the centre of origin unless the habitat itself changes.

Among all studied morphological characteristics, caudal segment has been found the most stable characteristic at the level of species or in some cases subspecies. Large size, higher scale count and greater number of whorls in tail segment have been considered as primitive.

Laudakia tuberculata, nuristanica and *stoliczkana stoliczkana* qualify all criteria laid down for primitive species within this group. They all have a distribution to the proposed centre of origin and with the exception of *nuristanica* other two are distributed over a wide area in two separate directions. Keeping in view all these facts it is concluded that *tuberculata, stoliczkana* and *nuristanica* are probably the first evolved *Laudakia* species. *Laudakia nupta* which is also unique in certain characteristics, might have independently evolved in early times and followed the route along Hindu Kush range. Its wide distribution also suggests its long history.

7- Fossils Record: Fossils, assumed to represent the ancestors of the group, should be located near the centre of origin. They indicate approximate age of the group and their distribution accurately represent the total distribution of that taxon.

The oldest fossil record which shows very primitive characteristics of Agamidae is *Mimeosaurus crassus* Gillmore, 1943 which was discovered from upper Cretaceous in Mongolia. The other records are *Tinosaurus* Marsh, 1872 from the Eocene of north America and Mongolia; *Agama galliae* Filhol, 1877 from the Eocene in France; *Agama sinensis* Hou Lian-Hai, 1975 from Paleocene in Anhui province of China.

Looking on the distribution of these fossil records it is quite clear that the distribution of the ancestors of *Laudakia* can not be justified until the inclusion of presently proposed centre of origin.

These traditional methods listed by Moody (1980) are equally useful for analysing the biogeography of the presently studied group, i.e. *Laudakia*. In some cases they are rather more marked and clearly visible. The species within a genus are usually product of habitat adaptation which enable them to live there successfully. Any discontinuity in the habitat beyond the reachable limit of the group or unfavourable climatic condition would mark the terminal end.

Keeping in view the aforementioned criteria the area around Pamir Knot qualifies all standards laid down for the centre of origin.

AGE OF *LAUDAKIA* AND ITS DISPERSAL

Depending on the fossil records and geological history of the area the approximate age of *Laudakia* is being proposed and simultaneously correlated to the present distribution pattern.

It is assumed that until Cretaceous when there was very little undulation on the surface of earth, *Mimeosaurus* evolved in Mongolia or any nearby area. By the end of Cretaceous the breaking off of Gondwana Land and microcontinental plate accretion had already been on its way. One event took place in the vicinity of Mongolia where Tarim Plate collided with Siberian and Kazakhstan Blocks of Eurasia and several other cimmerian blocks completed suturing on the southern border of Eurasia approximately by the onset of Cretaceous. These geological events evoked evolution of new forms, best suited to the new environmental condition. *Tinosaurus* was probably the result of it. By the end of Eocene it was widely distributed in the northern hemisphere, including both Asia and America. Moody (1980) and Estes (1983) proposed splitting of *Tinosaurus* into two separate genera because of some dissimilarities among them. It might be possible that any one form of it represents some link between Agamidae and Iguanidae. Apart from this debate *Tinosaurus* or its relatives would have reached upto Pamir or even more towards east, by the end of Eocene. *Agama galliae* which is the next subsequent record from upper Eocene or lower Oligocene from France, was originally identified as a member of present *Agama* and was confirmed by Moody (1980) as closely related species of *Agama* or *Stellio* (sensu Moody). Estes (1983) showed some doubts about its generic name with reference to unpublished information of Moody. The present studies disapprove the presence of *Agama* in whole Palaearctic Region and has replaced *Stellio* with *Laudakia* in this region. Estes (1983) described a new species of *Agama* from upper Miocene in France. This is not clear whether it resembles with *Agama* (s.str.) or any other form of *Agama* (sensu Wermuth, 1967). *Agama sinensis* has been reported from Paleocene in Anhui, China. Moody (1980) doubted about it and proposed it more close to *Mimeosaurus.*

The fossil records show that the ancestors of this group were widely distributed until Miocene, in the west upto France and in the east upto eastern side of China. It is supposed that they remained on the surface of earth until Pliocene. During glacial age drastic climatic changes took place on this planet. Many parts of the world including northern Europe came under snow. Being cold blooded animals they could not survive and disappeared from that part of the world.

During Tertiary, major surface changes occurred and mountain ranges formed. The collision of Indian Plate with Asia (ca. 50 MYBP) acted as wedge and extruded blocks in Afghanistan to the southwest and blocks in Tibet to east. The Pamir, Karakoram, Himalaya, Hindu Kush, and Tien-Shan all underwent substantial uplifting. Until Miocene the mountain ranges were fairly high and gave birth to different types of habitats i.e. mountainous, semi-mountainous or flat, and deserts. These habitats had their own peculiarities which initiated evolutionary changes in three different directions. It is assumed that *Laudakia, Trapelus* and *Phrynocephalus* are product of this habitat oriented evolution.

During the process of mountain and desert formation one group developed characteristics suitable for mountain (it is *Laudakia*), the other adopted itself for deserts (it is *Phrynocephalus*). The intermediary habitat was occupied by *Trapeleus.*

The sequence of events which were started in early Tertiary transformed the earth into high mountains and true deserts by late

Miocene. The Pamir Knot where Pamir, Karakoram, Himalaya, and Hindu Kush join each other, provides an ideal place for the dispersal of species in all directions. The earliest presence of *Phrynocephalus* has already been proposed as in the late Miocene (Macey et al., unpublished). The same period is being proposed for the birth of *Laudakia* or may be a little earlier, mid or early Miocene, supposing *Laudakia* as primary product of surface changes during Tertiary and *Phrynocephalus* as secondary.

DISPERSION OF *LAUDAKIA*

After being evolved around Pamir Knot *Laudakia stoliczkana* went into northeastern side along the Tien-Shan Range. It occupied all favourable habitats on its way upto Mongolia. It seems that the population which reached in Altai area was separated from the rest of *stoliczkana* by topographic or some climatic factors in the past. With the passage of time it developed some characteristics which categorized it as a distinct population of *stoliczkana. Laudakia tuberculata* followed Himalayan Range and occupied almost entire range upto Tibet. Macey et al. (unpublished) mentioned that the Tibetan Plateau lost flat plain topography by late Miocene (10 MYBP) and until Pliocene (5 MYBP) it get elevated another 2000 m. It is conceivable that this event caused the separation of Tibetan Plateau and lowland Chinese clades and most of the speciation on the Tibetan Plateau is probably an outcome of the third phase of uplifting that took place since the early Pliocene in the past 5 MYBP (Mancey et al., unpublished). I suppose that the population of *tuberculata* in Tibet was separated from the rest of the range at that time and since then developed some distinguishable characters which categorized it as a separate subspecies of *tuberculata. Laudakia agrorensis* and *dayana* could be a product of vertical evolution of *tuberculata.*

Laudakia nupta was probably independently evolved during the formation of Hindu Kush range and it followed the same range towards south. After reaching Makran Coast it invaded the southern mountain ranges of Iran and reached upto west Iraq. *Laudakia melanura* was apparently evolved from *nupta* and occupied some adjacent low mountain ranges mostly on eastern side of Hindu Kush with or without the presence of *nupta*.

Laudakia pakistanica and *L. lehmanni* which are strongly carinated and triwhorled species, were either independently evolved from a common ancestor or from any four whorled species but their distribution suggests that *lehmanni* followed the northwestern side along Pamir from centre of origin whereas *pakistanica* remained restricted to Karakoram Range.

The eastern distribution of *Laudakia* is rather more complicated. Generally it constitutes a group of biwhorled *Laudakia* with the exception of newly described *triannulata* (Ananjeva, 1984). Whether they evolved independently from the present form of *caucasia/ microlepis* or derived from any four-whorled *Laudakia* is yet to be proved, may be through some biochemical studies. Secondly, whether all of the members of this group are the product of a single invasion or multiple invasions is also needed to be proven.

In case of *Laudakia stellio* it is supposed that more than one invasions are likely because some of its subspecies, e.g. *s. cypriaca, s. picea* are very distinct and restricted to very small isolated areas. During glacial ages of Pleistocene when different land connections were made and broken and different habitats were also largely affected, this species invaded and disappeared leaving only some isolated populations. These populations were either disconnected through sea waters e.g. *cypriaca* or developed some special characteristics only suitable for that isolated habitat e.g. *picea.*

Laudakia himalayana-group is a product of vertical distribution. I suppose that this group evolved from already existing laudakias when the mountain ranges in this area attained enormously high elevations. Whether its members *himalayana, badakshana, bochariensis* evolved independent of each other or they occupied the present distribution range after being evolved around Pamir Knot, is yet to be proved, may be through some biochemical analysis.

REFERENCES

ANANJEVA, N.B. & ATAEV, C.H., 1984. *Stellio caucasus triannulatus* ssp. nov: A new species of Caucasian Agama from south-western Turkmenia. *Proc. Zool. Inst., Academy of Sciences, USSR*: 4-11. (in Russian).

BAIG, K.J., 1992. *Systematic studies of the Stellio-group of Agama (Sauria: Agamidae).* Ph.D. Thesis, Q. A. Univ., Pakistan.

BAIG, K.J. AND BÖHME, W., 1997. Partition of the *Stellio*-group of *Agama.* Into two distinct genera : *Acanthocercus* Fitzinger, 1843, and *Laudakia* Gray 1845 (Sauria: Agamidae).- pp 21-25, in: Böhme, W., Bischoff, W. & T. Ziegler: *Herpetologia Bonnensis, Bonn, Germany* (SEH), 416 pp.

DARLINGTON, P.J. JR., 1957. Zoogeography: *The geographical distribution of the Animals.* John Wiley, N.Y.

ESTES, R., 1983. *Handbuch der Palaeoherpetologie/ Encyclopedia of Palaeoherpetology,* Teil 10 A/Part 10 A Gustav Fischer Verlag, Stuttgart/New York.

HENNIG, W., 1966. The Diptera fauna of New Zealand as a problem in the systematics and zoogeography. *Pac. Insects Monograph*, **9**: 1-81 (Translated by P. Wygodzinsky).

JOGER, U., 1991. A molecular phylogeny of agamid lizards. *Copeia,* **3:** 616-622.

JOGER, U., AND ARANO, B., 1987. Biochemical phylogeny of the *Agama* genus group. *Proc. Ordinary General Meeting of the Societas Europaea Herpetologica,* **4**: 215-218

MACEY, J.R., ANANJEVE, N.B., ZHAO, E. AND PAPENFUSS, T.J., (in press). An allozyme-based phylogenetic hypothesis for *Phrynocephalus* (Agamidae) and its implication for the historical biogeography of Arid Asia.

MOODY, S.M., 1980. *Phylogenetic and historical biogeographical relationship of the genera in the family Agamidae (Reptilia Lacertilia).* Ph.D. dissertation (unpub.).

SCLATER, P.L., 1858. On the general geographical distribution of the members of class Aves. *Proc. Linn. Soc. (London),* **2:** 130-145.

WALLACE, A.R., 1876. *Die geographische Verbreitung der Tiere.* R.V. Zahn Verlag, Dresden.

WERMUTH, H., 1967. Liste der rezenten Amphibien und Reptilien: Agamidae. *Das Tierreich* **86**: 1-127.

BIODIVERSITY OF GECKKONID FAUNA OF PAKISTAN

MOHAMMAD SHARIF KHAN

Herp Laboratory, 15/6 Darul Saddar North, Rabwah, Pakistan

Abstract: Thirty six species of geckkos are recorded and described from diversified habitats of Pakistan. Morpho-ecological groups of Pakistani geckkos are proposed and their relationships are discussed.

INTRODUCTION

Our knowledge about the biodiversity of amphibians and reptiles of Pakistan dates way back (Boulenger, 1890). Later, ranges of already known species were expanded and several new species were addedd (Smith, 1931, 1935, 1943; Minton, 1962, 1966; Mertens, 1969, 1971; Khan, 1980a, 1980b, 1987, 1988, 1989, 1992, 1993 a, 1993b). However, lots of work is still needed to be done in many areas of Pakistan , especially in N.W.F.P, Baltistan, southern and nothern Balochistan, alpine Punjab and southern Sindh.

In this paper biodiversity of geckkonid fauna of Pakistan has been discussed. Geckkos have always been a favourite group of animals among herpetologists to work with. geckkos are one of the most prolific and widely distributed elements of the Pakistani reptilian fauna. They inhabit all basic ecological natural habitats and readily invade day-to-day man generated modifications in natural habitat. Despite their frail and soft body they are one of the most readily colonizing Pakistani reptiles.

Geckkos are included in almost all fauna analysis reports (to mention a few, Boulenger, 1890; Smith, 1935; Annandale, 1913; Constable, 1949; Minton , 1962, 1966; Mertens, 1969; Khan, 1972, 1989; Khan and Mirza, 1977; Khan and Baig, 1988, 1992). Monographic works on African (Loveridge, 1947) and Asian geckkos (Szcerbak and Golubev, 1986) have already appeared.

Ecology and zoogeography of geckkonid fauna of Pakistan

Since the publication of list of amphibians and reptiles of Pakistan (Khan, 1980a), several new geckkos have recently been added (Khan, 1980b, 1988, 1991, 1992, 1993a, 1993b; Khan and Baig, 1992; Khan and Tasnim, 1990). Moreover, several geckkos, which were included as Pakistani geckkos, appear to be extra limital thus not belonging to Pakistani fauna (Khan, 1993c).

More than 34 geckkonid taxa falling in 12 genera have so far been ascertained from diverse habitats throughout Pakistan. Despite fragile morphology and simple habits, geckkos range from southern hot sands of Chagai and Thar Deserts to Himalayan heights among perpetual snow-fields in the north (Khan and Baig, 1992; Khan, 1993c).

Table I and Table II give the checklist of the geckkos so far reported from Pakistan and summarize their preferred habitat and distribution in Pakistan and neighbouring Azad Kashmir.

Table I: Checklist, habitat and distribution of geckkonid lizards of Pakistan.

(1=North Western Frontier Province; 2=Punjab; 3=Sindh: 4=Balochistan; 5= Kashmir; 6=Gilgit Agency; Distribution patternin provinces: C=Central; E= East; G=Widely distributed; N=North; S=South; W=West; Hab= Habitat types: i =ground; ii = sandstone; iii=rock; iv=mesic vegetation;v=desert vegetation; vi=arboreal; vii=ice fields;** = anthropogenic; + =extends; - = does not extend)

Taxa	**	**Hab**	**1**	**2**	**3**	**4**	**5**	**6**
Agamura femoralis	-	v	-	-	-	SW	-	-
A. persica	-	iii+v	S	-	S	G	-	-
Alsophylax tuberculatus	+	v	-	-	S	CS		
Crossobamon lumsdenii	-	v	-	-	-	W	-	-
C. maynardi	-	iii	-	-	-	W	-	-
C. orientalis	-	v	-	S	G	-	-	-
Cyrtodactylus battalensis	+	iii	SE	-	-	-	-	-
C. dattanensis	+	iii	SE	-	-	-	SW	-
C. mintoni	+	iii	NW	-	-	-	-	-
Cyrtopodion kachhensis	+	i	S	W	S	W	-	-
C. scaber	+	i	S	NW	W	E	-	-
C. watsoni	+	i	S	W	W	E	-	-
C. kohsulaimanai	+	iii	-	W	-	-	-	-
C. montiumsalsorum	+	i	S	N	-	-	-	-
Eublepharis macularius	+	i	G	N	S	G	S	-
Hemidactylus brooki	+	iv	G	G	G	G	S	-
H. flaviviridis	+	vi	G	G	G	S	G	-
H. frenatus	+	iv	-	S	S	-	-	-
H. leschenaultii	+	iv	-	-	S	-	-	-
H. persicus	+	iii	S	N	S	G	-	-
H. triedrus	+	iv	-	-	S	-	-	-
H. turcicus	+	iv	-	-	S	-	-	-
Ptyodactylus homolepis	-	iii	-	-	W	-	-	-
Tenuidactylus baturensis	+	vii	-	-	-	-	-	E
T. fortmunroi	+	ii	-	W	-	-	-	-
T. indusoani	+	ii	S	W	-	-	-	-
T. rohtasfortai	+	ii	-	N	-	-	SW	-
T. walli	+	iii	W	-	-	-	-	-
Teratolepis fasciata	-	i	-	-	-	WS	-	-
Teratoscincus microlepis	-	v	-	-	-	W	-	-
T. scincus	-	v	-	-	-	W	-	-
T. persicus euphorbicola	-	v	-	-	S	S	-	-

Table II. Zoogeographical analysis of geckkonid fauna of Pakistan.

(A=total number of species reported; B=number of species shared; C= endemic species

Province	A	B	C	i	ii	iii	iv	v	vi	vii
NWFP	11	8	3	2	2	6	1	1	1	-
Punjab	15	11	3	5	3	3	-	2	1	-
Sindh	17	12	5	5	-	3	2	3	1	-
Balochistan	15	9	6	4	-	3	-	4	-	-
Azad Kashmir	2	2	0	1	1	1	1	-	1	-
Gilgit Agency	1	0	1	-	-	-	-	-	-	1

Pakistani geckkonid genera occupy following specific habitat types:

Agamura: long legged peculiar Balochistani geckkos are represented by two species, confined to cliff and rocky terraces with intevening sand fields, with little xeric vegetation. Their genus is represented by two species i.e. *femoralis* which is confined to Southwestern Balochistan, while wide ranging *persica* extends northerly into southern NWFP and southernly to coastal Sindh.

Alsophylax: peculiar tuberculated geckko , inhabits open sandfields with sparse bushes, invading edificeal structures, boulders and other stony formations. Represented by a single species *tuberculatus* common in deserticolous central to southern Balochistan extending into southern Thar desert.

Crossobamon: small geckkos of open sandfields with sparse scrubby vegetation. In Pakistan the genus is represented by three species, *lumsdenii* and *maynardi* which are confined to Balochistan, while *orientalis* is widely distributed in deserticole Punjab and Sindh with no records from Balochistan.

Cyrtodactylus: plum bodied Tibeto-Himalayan geckkos inhabit humid mountainous areas, with lush grass and bush vegetation, readily extend into anthropogenic habitat. Three species are known from Pakistan, *battalensis* and *dattanensis* are reported from eastern N.W.F.P, while *mintoni* is known from Swat.

Cyrtopodin: small geckkos which inhabit arid barren submountain badland areas, descending in dry mud flats in the plains. They are primarily ground geckkos, readily extend into buildings and other edifices. Five species are know from Pakistan: *kachhensis* from lower Sindh, while *scaber* and *watsoni* are known badland geckkos recorded from Salt Range, Balochistan and N.W.F.P: They extend into plains of Punjab and Sind. While *kohsulaimanai* inhabits mudflats along the foot of Sulaiman Range in the west and *montiumsalsorum* along Salt Range in north.

Eublepharis: large ground or crevice geckkos, frequent rocky deserts with sparse vegetation and intervening clay soil, avoiding sand. Single species *macularius* inhabits sub mountainous and tableland areas in all four provinces of Pakistan, not extending into plains. Its relic population in the plain rocky outcrops --the Karana Hills is seriously injured and under constant threat of extermination due to recently increased human intervention. The last pair of these geckkos in hills around Rabwah city was probably killed back in 1965.

Hemidactylus: The dilated digited geckkos are met with in every possible habitat in mesic to subxeric conditions, avoiding as a rule high mountains and open deserts. They are mostly anthrophilic. Several species have been recorded from Pakistan; *brooki* is almost universal Pakistani barn geckko, while *flaviviridis* is universal Pakistani house geckko, it avoids mounainous areas so it is rarer in Balochistan and NWFP, while *frenatus*, *leschnaulti*, *triedrus* and *turcicus* are typical coastal forms and are recorded along Sindh and Mekran coast. On the other hand, *persicus* is mostly Balochistani house geckko, it recently has been reported from Rohtas Fort, in Salt Range, Punjab. Perhaps it was transported from Balochistan to Rohtas Fort along rock blocks used in the building of the fort (Khan and Tasnim, 1990).

Ptyodactylus: Single species *homcies homolpis* is known from Pakistan which inhabits rocky areas with little vegetation along Kirthar Range, north-western Sindh.

Tenuidactylus: group of sandstone long limbed geckkos which inhabit crevices in sandstone rocks with little scrubby vegetation around. It avoids mudflats and sandy areas, readily invading edificial structures. Five species are known from Pakistan: two highland stone geckkos, *baturensis* which inhabits rocky outcrops in snowfields in Baltistan, Gilgit Agency, while *walli* is reported from Chitral, westren NWFP. The lowland sandstone geckkos are: *indusoani* recorded from western Salt Range, while *rohtasfortai* from eastern Salt Range extending eastwards in to Azad Kashmir. The third species *fortmunroi* is a western Indus form, reported from eastern hilly tracts of Koh Sulaiman along western border of Indus, Punjab.

Teratopepos: small sand geckkos frequent patches of high ground rising from flat flooded delta with little vegetation in lower Sindh. Only single species *fasciata* is know from southern Sindh Pakistan.

*Teratoscincus:*These geckkos are characteristic of open fine sand dunes with sparse bushy vegetation. Western wide ranging *scincus* is widely distributed in Chagai Desert, Balochistan, while second species *microlepis* is more western extending into neighboring Iran.

Tropiocolotes: small geckkos frequent flat country with stony outcrops and are generally associated with common desert spiny plant *Euphorbia caducifoia.* Single species*T. persicus euphorbicola* has been recoded in Thar Desert in lower Sindh and Balochistan.

As it is apparent from the above account of the distribution, the geckkos are the most diversified group of our vertebrate fauna. They rapidly colonize and differentiate into geographical races on adjacent hilly ranges (genus *Cyrtodactylus*), badland areas (genus *Cyrtopodion*), desert (genus *Crossobamon* and *Teratoscincus*) and sandstone rock (*Tenuidactylus*). Toads present similar cofusing distribution pattern (Eiselt and Schmidtler, 1973). Lot of work remains to be done in these natural habitats to sort out distribution patterns and relationship among these congeneric forms.

REFERENCES

ANNANDALE, N. 1913. The Indian geckkos of the genus *Gymnodactylus. Rec. Indian Mus.,* **9**: 309-326.

BOULENGER, G. A. 1890. *Fauna of British India, including Ceylon and Burma. Reptilia and Batrachia.* London.

CONSTABLE, J. D. 1949. Reptiles from Indian peninsula in the Museum of Comparative Zoology. *Bull. Mus. Zool. Harvard,* **103** ; 59-160.

EISEL, V. J. AND SCHMIDTLER, J.F. 1973. Froschlurche aus dem Iran under Berucksichtigung auBeriranischer Population sgruppen. *Ann. Naturhistor. Mus. Wien.,* **77**: 181-243.

KHAN, M. S. 1972. Checklist and key to the lizards of Jhang district, West Pakistan. *Herpetologica,* **28**: 94-98.

KHAN, M. S. 1980a. A new species of geckko from northern Pakistan. *Pakistan . J. Zool,* **12**:11-16.

KHAN, M.S. 1980b. Affinities and Zoogeography of herpetiles of Pakistan. *Biologia,* **26**:113-171.

KHAN, M. S. 1987. Checklist, distribution and zoogeographical affinities of amphibians and reptiles of Balochistan . *Proc. Pak. Cong. Zool.,* **7**:105-112.

KHAN, M. H. 1988. A new Cyrtodactylid geckko from northwestern Punjab, Pakistan. *J. Herpetol.,* **22**: 241-243.

KHAN, M. S. 1989. Rediscovery and redescription of *Tenuidactylus motiumsalsorum* (Annandale,1913).*Herpetologica.*, **45**:46-54.

KHAN, M. S. 1991. A new *Tenuidactylus* geckko from Sulaiman Range, Punjab, Pakistan. *J. Herpetol.*, **25**:199-204.

KHAN, M. S. 1992. Validity of mounain geckko *Tenuidactylus walli* (Ingoldby, 1922). *Birish J. Herpetol.*, **2**:106-109.

KHAN, M. S. 1993a. A new angular-toed geckko from Pakistan , with remarks on the taxonomy and a key to the species belonging to genus *Cyrtodactylus* (Reptilia:Sauria: geckkonidae). *Pakistan. J. Zool.*, **25**: 67-73.

KHAN, M. S. 1993b. A new sandstone geckko from Fort Munro, Dera Ghazi Khan District, Punjab, Pakistan. *Pakistan J. Zool.,* **25**: 217-221.

KHAN, M. S. 1993c. A checklist and key to the geckkeonid lizards of Pakistan. *Hamadryad.,* **18**: 35-41.

KHAN, M. S., AND BAIG, K. J., 1988. Checklist of the amphibians and reptiles of Distrist Jhelum, Punjab, Pakistan. *The Snake,* **20**:156-161.

KHAN, M. S. AND BAIG, K. J. 1992. A new tenuidactylid geckko from northeastern Gilgit Agency, North Pakistan. *Pakistan J. Zool.,* **24**: 273-277.

KHAN, M. S. AND MIRZA, M. R. 1977. An annotated checklist and key to the reptiles of Pakistan, Part II:Sauria (Lacertalia). *Biologia,* **23:** 41-64.

KHAN, M. S. AND TASNIM, R. 1990. A new geckko of the genus *Tenuidactylus* from northwestern Punjab, Pakistan and southwestern Azad Kashmir. *Herpetologica,* **46**:142-148.

LOVERIDGE, A. 1947. Revision of the lizards of the family Geckkonidae. *Bull. Mus. Zool .Harvard,* Cambridge, USA.

MERTENS, R. 1969. Die Amphibien und Reptilien West-Pakistans. *stittg. Beitr. Naturk.*, **197:**1-96.

MERTENS, R. 1971. Die Amphibien und Reptilien West-Pakistan. *Senckenb. biol.,* **55**:35-38.

MINTONS, S. A. JR. 1962. An annotated key to the amphibians and reptiles of Sind and Las Bela, West Pakistan. *Am. Mus. Novit.,* **2081**:1-21.

MINTON, S. A. JR. 1966. A contribution to the herpetology of West Pakistan. *Bull. Mus. Nat. Hist.,* **134**:1-184.

SMITH, M. A. 1931. *The fauna of Bitish India, including Ceylon and Burma. Reptilia and amphibia.* 1: Loricata, Testudines. Taylor and Francis Ltd. London.

SMITH, M. A. 1935. *The Fauna of British India, including Ceylon and burma. Reptilia and amphibia.* II: Sauria. Taylor and Francis Ltd. London.

SMITH, M. A 1943. *The Fauna of British India, including Ceylon and Burma. Reptilia and amphibia.* III: Serpentes. Taylor and Francis Ltd. London.

SZCERBAK N. N..AND GOLUBEV, M.L. 1986. geckkos of the USSR and adjoining conutries. *Sci. Acad. Uckr SSR Zool. Inst.* **1986**:1-232, (in Russian).

THE BIRDS OF PALAS VALLEY (DISTRICT KOHISTAN, NWFP)

N:A. RAJA AND G.R.J. DUKE

Himalayan Jungle Project, Islamabad

Abstract: The temperate forests of the Western Himalayas have been identified as an 'Endemic Bird Area' (EBA), a centre of global importance for biodiversity. The Palas Valley, District Kohistan, NWFP, probably contains the most outstanding remaining tract of such forests in Pakistan, and is certain to qualify as an Asian 'Important Bird Area' (IBA). In particular, Palas contains the largest known population of the threatened Western Tragopan *Tragopan melanocephalus*, and breeding populations of seven of the eight restricted range species characteristic of the EBA. The results of surveys of the birds of Palas between 1988 and 1995 are presented, including an annotated checklist and notes on records of particular interest. Major bird habitats in Palas are briefly described and threats to birds and bird habitats outlined. Bird conservation priorities in Palas are identified and conservation activities to date under the Himalayan Jungle Project outlined.

INTRODUCTION

Geographical location

The Palas Valley lies in Pattan Tehsil, District Kohistan, NWFP Pakistan, between 34°52'E to 35°16'E and 72°52'N to 73°35'N. The Palas Valley lies east of the River Indus with the valley of Jalkot to the north, Kaghan to the east and Allai to the south. The main river, the Musha'ga, is c.75 km long and joins the River Indus at 73°05'E, 35°09'N about 15 km north of the town of Pattan on the Karakorum Highway. The Palas Valley covers an area of c.1300 sq km. Altitudes ranges between c.700 m (River Indus) to c.5550 m (Bahader Ser peak).

Biogeographical location

Palas lies at the north-western extremity of the Great Himalaya, close to the meeting point with the Hindu Kush and Karakorum ranges. This great geological meeting point is also a meeting point for two zoogeographical realms - the Palaearctic and the Oriental (or Indo-Malayan). The avifauna of Palas can therefore be expected to include elements of these two realms. Palas also shows the influences of four phytogeographical regions, the Irano-Turanian, Sudano-Zambezian,

Indian and Eastern Asiatic. A rich floral diversity has been recorded (Rafiq, 1994) and this will also tend to support a rich avian diversity.

MATERIALS AND METHODS

No specific census technique was used, owing to the difficulties of field work in the Himalaya. Records were made using visual and/or aural observations, and with reference to a number of field guides (Ali and Ripley 1983, Ali and Ripley 1987, Roberts 1991, 1992). A variety of makes of binoculars were used for visual observations. Some audio recordings were made of bird vocalisations.

The annotated checklist is based predominantly on observations made by experienced ornithological teams between 1988 and 1995. The most comprehensive observations were made in the course of detailed survey work on the pheasants of Palas as follows: Paul Walton and Guy Duke, May-June 1987 (Duke & Walton 1988); Guy Duke, May-June 1988 (Duke 1989); Jonathan Eames and Guy Duke, May-June 1989 (Duke 1990); Nigel Bean, Philip Benstead, Dave Showler, Philip Whittington and Naeem Ashraf Raja, May-June 1994; Derwyn Liley, Dave Gandy, Guy Thompson, Ahmed Khan and Abdul Ghafoor, February 1995. Additional records are provided by Richard Grimmett in June 1993 (pers comm.) and between 1991 and 1995 by Guy Duke and Naeem Ashraf Raja in the course of numerous field visits. There has been a strong bias towards surveys in the early summer breeding season (May to July). Thus the checklist is probably almost complete for summer breeding species, but less so for species occurring on passage or in winter. Species recorded in Palas in the breeding season may generally be assumed to be breeding; direct evidence of this is given wherever recorded.

RESULTS

The bird species observed to date number 150, of which 54 are resident, 77 are summer breeding, 6 are winter visitors and 4 are vagrants. The residential status of remaining 9 species is not known due to single or very few sightings.

The western Himalaya (in particular, the temperate forests) has been identified as an Endemic Bird Area, based on the occurrence of nine 'restricted range species' or 'west Himalayan endemics'. Of these nine species, no fewer than eight are found in Palas, namely: Western Tragopan *Tragopan melanocephalus*, Tytler's Leaf Warbler *Phylloscopus tytleri*, Brook's Leaf Warbler *Phylloscopus subviridis*, White-cheeked Longtailed Tit *Aegithalos leucogenys*, White-throated Longtailed Tit *Aegithalos niveogularis*, Kashmir Nuthatch *Sitta cashmirensis*, Red-browed Finch *Callacanthis burtoni* and Orange Bullfinch *Pyrrhula aurantiaca*.

Palas holds the world's largest known population of Western Tragopan with the current population estimated at 325 breeding pairs. The species is resident in Palas. Red-browed Finch is also resident and fairly common. Both Tytler's and Brook's Leaf Warbler are summer breeding and locally common in Palas. White-cheeked Longtailed Tit, White-throated Longtailed Tit, Kashmir Nuthatch and Orange Bullfinch are resident but quite uncommon.

ANNOTATED CHECKLIST OF THE BIRDS OF PALAS

Common Teal *Anas crecca*. One female seen flying downstream over Musha'ga between Paro and Shuki Ser, 4 September 1992. Two birds seen near Pichbela flying downstream over Nila Nullah, 24 September 1994. M. Buzerg observed 11 birds flying upstream over Musha'ga near Gadar, 25 September 1994.

Shikra *Accipiter badius*. Recorded regularly during the breeding season, most often below 2000 meters, presumably breeding. Two, Paro, probably nesting, 11 July 1992. One recorded Gaber (1900 m), 11 February 1995 and one male between Kundal and Shared, 22 February 1995.

Eurasian Sparrow Hawk *Accipiter nisus melaschistos*. Usually recorded above 2000 m suggesting an altitudinal separation from the previous species.

Long-legged Buzzard *Buteo rufinus*. One, Tiko Sar, 19-20 June 1989. One seen above Mukchaki, taking a lizard *Agama* sp. and a medium sized bird in a period of fifteen minutes, was almost certainly feeding young.

Booted Eagle *Hieraaetus pennatus*. Regularly recorded throughout Palas. All records are of dark phase individuals except one pale phase adult seen displaying above Gadar, 31 May 1994.

Golden Eagle *Aquila chrysaetos*. Fairly common between 1200-3200 m, probably breeding. Single immatures observed, Kana Nullah and upper Ledi, September 1991. One sub-adult, Ishaq Bek, June 1994. A pair noted calling, one of the birds performing a territorial display of undulating dives and upward swoops, 24 February 1995.

Eurasian Griffon Vulture *Gyps fulvus*. Single sighting of three birds soaring low over mixed forest near Gaber (1900 m), 9 February 1995.

Himalayan Griffon Vulture *Gyps himalayensis*. Common and widespread in Palas, recorded almost daily in all surveys. 30 feeding on dead cow, Ilo Bek, 18 June 1989. 40+, upper Ledi near Kaghan

Pass, 25 September 1991. Recorded regularly in lower parts of Bar Palas during February 1995; seven circling with a Golden Eagle in Kuz Palas, 8 February 1995.

Lammergeier *Gypaetus barbatus*. Probably breeding. One feeding at a dead cow, Ilo Bek, 18 June 1989. One, near Kana-Kunari Pass, 20 September 1991. One immature, near Ishaq Bek, 25 September 1991. One, Gidar, 12 September 1992. Two, between Bar Paro and Gana, 16 May 1993. One, between Bar Paro and Kuz Ser, 17 May 1993. One immature, Mukchaki, 16 June 1994. One adult, Ishaq Bek, 22-23 June 1994. Six sightings of adult birds, Bar Palas, 8-25 February 1995.

Eurasian Kestrel *Falco tinnunculus*. Fairly common and widespread throughout Palas. Several individuals and a few pairs recorded from Musha'ga (1500 m) to high alpine pastures (4000 m), September 1992.

Himalayan Snowcock *Tetraogallus himalayensis*. Two above Tiko Sar, 20 June 1989. One pair, Kana-Kunari Pass (c.4200 m), 20 September 1991. Three heard calling, Malik Siri Gali Pass, 23 June 1994.

Chukar ***Alectoris chukar***. Probably breeding, recorded mostly above 2500 m. Frequently encountered in lower parts of Palas. Maximum 5 recorded in one day, May-June 1994. Three birds flushed from scree slopes and scrub between Kundal and Sartoe (2200 m), 21 February 1995.

Western Tragopan *Tragopan melanocephalus*. Resident. Palas contains the largest known population of this globally threatened species, with an estimated 660 birds. During the breeding season, these are concentrated in temperate forests of tributary valleys south of the Musha'ga from Moru east to Pichbela, though a few occur north of the Musha'ga from Gorkhar east to Pichbela. Population estimates are based on dawn call counts, which recorded 42-49 calling birds 8-22 June 1989 (Moru, Khajil, Kabkot and Unser nullahs, 2650-3350 m), 11-13 calling birds 12-16 June 1991 (Diwan nullah, 2650-3000 m), 22-27 calling birds, May-June 1994 (Gorkhar, Shared, Gadar, Pharrogah, Kundal, Shaman-Dumbela and Dumbela-Pichbela nullahs, 2400-3350 m). In winter, birds have been flushed (using dogs) at lower altitudes (2000-2350 m) and nearer to the River Indus: one bird Gaber (1800 m), 9 February 1995 (lowest altitude recorded for the species in winter); one bird Kabkot, 16 February 1995; two females and a male, Shared Nullah, 23 February 1995; one male above Shared, 24 February 1995.

Himalayan Monal *Lophophorus impejanus*. Fairly common and widespread in suitable habitat throughout Palas. A total of of 35 males and 16 females, 2050-2500 m, typically in single-sex flocks, the largest containing nine males (above Shoman), with a maximum day-count of 21 birds, 23 February 1995.

Koklass Pheasant *Pucrasia macrolopha biddulphi.* Very common and widespread in suitable habitat. Six birds, Kuz Palas (lowest record, 1600 m), 12 February 1995; a total of 22 birds, Bar Palas (1700-2450 m), 16-24 February 1995. Roberts (1991) gives lowest altitudinal record as 2100 m but 12 records were made below this altitude in February 1995.

Black-winged Stilt *Himantopus himantopus.* One, flying downstream over Musha'ga, near Shuki Ser, 10 September 1992.

Common Sandpiper *Actitis hypoleucos.* Probably breeding, odd pairs seen between Dumbela and Wulbela, and at Pichbela and Pulbela, June 1994.

Eurasian Woodcock *Scolopax rusticola.* One, near Sar Bek, 14 June 1989. One, Tiko Sar, 19-20 June 1989. One, upper Diwan, birch/yew and fir/kail forest, 14 June 1991. One heard, between Moru-Gidar, 27 June 1992. One, above Rajaser (Kundal nullah), 26 May 1994. One heard, Takhto, 27 May 1994. Two birds roding in nullah above Mukchaki, 19 June 1994.

Snow Pigeon *Columba leuconota.* One at edge of moist deciduous forest, Diwan, 15 June 1991. Up to four birds, Ishaq Bek, 21-22 June 1994. One, near Malik Siri Gali Pass, 23 June 1994. Five between Ishaq Bek and Malik Siri Gali Pass (c. 4060 m), 30 June 1994.

Speckled Wood Pigeon *Columba hodgsonii.* Two, Sar Bek, 14 June 1989. Two to three birds seen at each of Belgi, Chakala, Wulbela, Dumbela, Pichbela, and above Mukchaki, May-June 1994.

Oriental Turtle Dove *Streptopelia orientalis.* Common and widespread throughout Palas.

Spotted Dove *Streptopelia chinensis.* Several, between Bar Paro and Kuz Ser, 17 May 1993. One, near Karat, 9 June 1993.

Slaty-headed Parakeet *Psittacula himalayana.* Regularly recorded especially in sub-tropical broadleaf forest. 10+ between Sherakot and Moru, 26 June 1992. 15+ between Shuki Ser, Ilo Bek and Gidar (Kabkot nullah), 11 September 1992. 7, Pochmoru, 20 May 1994.

Eurasian Cuckoo *Cuculus canorus.* Common throughout Palas.

Oriental Cuckoo *Cuculus saturatus.* Common around Kabkot June 1989. Odd birds June 1991, 1992, 1993, but none May-June 1994.

Little Cuckoo *Cuculus poliocephalus.* Common but very shy and seldom seen. Often calls at night.

Scop's Owl *Otus scops.* One flushed from roost in *Quercus baloot* and observed for twenty minutes being mobbed by two Black-throated Jays, Pochmoru, 20 May 1994.

Collared Pygmy Owlet *Glaucidium brodiei.* One heard, Sar Bek, 15 June 1989. One seen being mobbed by Western Crowned Warbler and Crested Black Tit, below Shalkho, 1 June 1994.

Tawny Owl *Strix aluco.* Fairly common in both sub-tropical broadleaf and evergreen temperate forests.

White-throated Needle-tail Swift *Hirundapus caudacutus.* 5+ Gidar, 27 June 1992. One flying north, Bush (above Massi/Gorkhar), 18 May 1994. 6+ near Malik Siri Gali Pass, 30 June 1994.

Alpine Swift *Apus melba.* Fairly common mostly around Musha'ga and around bigger nullahs. 50+ between Pattan-Sherakot, 25 June 1992. Several, between Paro to Pattan, 30 June 1992. 50+ near Khorghi, 11 July 1992. Several between Taghi and Paro, 14 May 1993. Several, Bar Paro, 10 June 1993. 30+ between Karat and Paro, 15 May 1994.

Common Swift *Apus apus.* Fairly common and probably breeding at suitable cliff sites throughout Palas.

Indian House Swift *Apus affinis.* Many, Karat, 23 June 1989. 3-4, Shambela, 8 June 1991. Several between Paro and Pattan, 30 June 1992. Several, between Pattan and Sherakot over *Quercus baloot* forest, 10 July 1992. One between Kuz Paro and Bar Paro, 15 May 1993. Ten between Paro and Gorkhar, 16 May 1994. 2+, Mohran, 12 June 1994. 'Few', Pichbela, 13 June 1994.

Kingfisher *Alcedo atthis.* One, between Karat-Paro, 15 May 1994.

European Roller *Coracias garrulus.* One pair near Indus between Pattan and Sherakot, 6 June 1991. Two between Pattan and Sherakot, 25 June 1992. Four between Paro and Pattan, 30 June 1992. One between Paro and Pattan, 20 July 1992.

Hoopoe *Upupa epops.* Occasionally recorded in ones or twos, probably breeding.

Eurasian Wryneck *Jynx torquilla.* One, between Sherakot and Moru, 26 June 1992. One seen and heard calling, Bush, 19 May 1994. One, Sartoe, 25 May 1994. Two heard high above Rajaser, 27 May 1994. One in Tamarisk scrub by Musha'ga, Dumbela, 19 June 1994.

Speckled Piculet *Picumnus innominatus.* Two agitated adults observed by a nest hole in a *Quercus baloot* tree, near Musha'ga between Shared and Bangaha, 22 May 1994. A juvenile may have been calling from the nest.

Scaly-bellied Green Woodpecker *Picus squamatus.* One, Chuk Bek, 8 June 1989. One, between Sharial and Magri, 21 August 1993. Six birds, observed generally in oak forest at lower altitude, May-June 1994. Single bird in *Quercus baloot* trees, Khalyar nullah, 22 September 1994. Flushed single bird from *Q. baloot* forest, between Khalyar Nullah and Kuz Paro, 25 September 1994. One male, seen at 2200 m, in open blue pine forest, Shoman, 21 February 1995. One male, Shared, 23 February 1995. One flushed out of *Quercus* scrub, Kuz Paro, 24 February 1995.

Himalayan Pied Woodpecker *Dendrocops himalayensis.* Widespread and fairly common throughout forests. Commonly recorded February 1995, Kuz Palas; only three birds recorded in dense coniferous forest from Bar Palas, the highest altitudinal record being 2400 m, Kabkot, 18 February 1995.

Brown-fronted Woodpecker *Dendrocops auriceps.* Two observed foraging in *Quercus baloot*, above Pochmoru, 19 May 1994. One male seen in coniferous forest at 1850 m, near Gabir, 9 February 1995.

Oriental Skylark *Alauda gulgula.* At least three singing birds above pastures, Gadar (Chor Nullah), 18 June 1994.

Collared Sand Martin *Riparia riparia.* 5+ near Gale Bek, Kana Nullah, 21 September 1991. One, between Bar Paro and Gana, 16 May 1993. Six, Ishaq Bek, 3200 m, 22 June 1994.

Crag Martin *Ptyonoprogne rupestris.* Odd pairs recorded throughout the valley. One colony of 30+ birds near Gorkhar, 16 May 1994.

Common Swallow *Hirundo rustica.* One, Bar Ser, 17/18 July 1992.

Red-rumped Swallow *Hirundo daurica.* Frequently recorded, generally in the vicinity of habitations. Several, near Badakot, 6 June 1991. 18 between Kuz Paro and Gorkhar, 16 May 1994.

European House Martin *Delichon urbica.* Recorded frequently throughout the valley up to 4300 m. Breeding. Several flocks (5-100+) in Bar Palas sometimes with *D. dasypus.* Several, between Pattan and Sherakot, over *Quercus baloot* forest, 10 July 1992. A nesting colony of 10+, above Karoser, 21 May 1994. 30+, Breathbek (Kundul Nullah), 23 May 1994.

Kashmir House Martin *Delichon dasypus cashmiriensis.* 30, Galli Bek, 11 June 1989. Large flock (c. 100) near Musha'ga with several *D.*

urbica, between Moru and Gidar, 27 June 1992. 20+, near Shuki Ser, 4 September 1992. 30+, between Ilo Bek and Tiko Sar, mixed with European House Martin, 7 September 1992. 20+, between Kuz Ser-Sertay, 18 May 1993.

Rufous-backed Shrike *Lanius schach*. Several, between Pattan and Sherakot, 6 June 1991. One, between Pattan and Sherakot, 25 June 1992. Several, between Sherakot-Moru, 26 June 1992. One, near Sherakot, 10 July 1992. One, between Paro-Pattan, 20 July 1992. One, between Paro and Shuki Ser, 4 September 1992. Singles recorded, between Karat and Paro, 15 May 1994 and between Paro and Gorkhar, 16 May 1994.

Golden Oriole *Oriolus oriolus*. Fairly common recorded usually from lower altitudes, generally near habitations.

Black Drongo *Dicrurus macrocercus*. 'Few', Kuz Paro, 8 June 1991. One, between Paro and Pattan, 20 July 1992. One, between Shuki Ser, Ilo Bek and Gidar, 11 September 1992. 12+, between Paro and Bar Paro, 15 May 1993. Two, Shared, 12 June 1993.

Ashy Drongo *Dicrurus leucophaeus*.Commonly recorded 800-3000 m.

Brahminy Starling *Sturnus pagodarum*. One, near Badakot, 6 June 1991.

Black-throated Jay *Garrulus lanceolatus*. Fairly common and widespread in broadleaf forest, *Quercus baloot* in particular. Freshly fledged juveniles recorded above Seri and Khalyar, 22 June 1989. Recorded as fairly common in *Quercus* scrub, February 1995.

Yellow-billed Blue Magpie *Urocissa flavirostris*. Two, between Pattan and Sharial, 25 June 1992. Two, between Sharial and Magri, 21 August 1993. Three in mixed forest, 1800 m, near Gaber, 9 February 1995. One bird heard in mixed *Quercus* and coniferous forest at 1900 m, above Shared, 23 February 1995.

Nutcracker *Nucifraga caryocatactes*. Regularly recorded in suitable habitat. A single bird heard calling from *Quercus* scrub above Kundal (2500 m), 20 February 1995. A flock of six birds observed in Deodar forest, Shoman (2300 m), February 1995.

Alpine Chough *Pyrrhocorax graculus*. 20+ near Kana-Kunari Pass, 20 September 1991. 20, Gadar (Chor Nullah), 17 June 1994. 32, Pulbela, 21 June 1994. 70+, Ishaq Bek, 22 June. 20+, Malik Siri Gali Pass, 23 June 1994. Two birds seen circling high above crags on the ridge between Shared and Gadar, 24 February 1995.

Red-billed Chough *Pyrrhocorax pyrrhocorax*. Several, near Quh Ledi, 23 September 1991. Two, Ishaq Bek, 22 June 1994.

Jungle Crow *Corvus macrorhynchos.* Common and widespread throughout Palas. Breeding. Nest containing two pulli located in *Abies pindrow* 20 m above the ground, Ilo Bek, 18 June 1989.

Tibetan Raven *Corvus corax.* Regularly recorded throughout Palas.

Long-tailed Minivet *Pericrocotus ethologus.* Jonathan Eames in May-June 1989 considered it to be uncommon in Kabkot area but subsequent surveys reveal that it is quite common in parts of Palas in the breeding season. Guy Duke observed family parties with young, between Bar Ser and Shuki Ser, 18 July 1992. May-June 1994 survey found it to be common and widespread.

White-cheeked Bulbul *Pycnonotus leucogenys.* Several, Karat, 23 June 1989. Several below Badakot, 6 June 1991. One, Karat, 20 July 1992. Singles around Karat, 15 May 1994 and 3 June 1994.

Black Bulbul *Hypsipetes madagascariensis.* Fairly common and widespread especially around sub-tropical broadleaf forest in lower parts of Palas. Several pairs, near Musha'ga, between Kundal and Pulbela, 22-23 September 1991. Several, Bar Paro, 10 June 1993. Several, Shared, 12 June 1993. Nine, between Wulbela-Pichbela, 16 June 1994. 15 seen, Kuz Paro, 14 February 1995. 15 birds seen, Kuz Paro, 14 February 1995.

Variegated Laughing Thrush *Garrulax variegatus.* Fairly common above 2300 m and probably breeding. Two noted stripping bark from *Betula utilis*, Sar Bek, 13 June 1989. One of a pair seen carrying a small piece of bark at Belgi, 31 May 1994. February 1995 survey recorded it as common in broadleaf and mixed forest around Gabir (2000 m), Kuz Palas, February 1995, with a maximum count of 25 birds on 9 February 1995, but did not record it in Bar Palas.

Streaked Laughing Thrush *Garrulax lineatus.* Often recorded in scrub around cultivation between 1600-2900 m, probably breeding. One, seen carrying food, Pichmoru, 20 May 1994. Common in *Quercus* scrub and around villages to 2600 m in lower parts of Bar Palas, February 1995, with maximum count of 30 birds between road-head and Kuz Paro, 14 February 1995.

Green Shrike Babbler *Pteruthius xanthochlorus.* One seen at close range for about 1 minute in canopy of mature *Quercus* forest, eastern side of Shared Nullah (1900 m), 23 February 1995. This is the western-most sighting of the species (westernmost record in Roberts is Murree Hills). It is considered rare with only two previous sightings from Pakistan this century Roberts (1992).

Spotted Flycatcher *Muscicapa striata.* One, near Gorkhar, 16 May 1994.

Sooty Flycatcher *Muscicapa sibirica*. Abundant, usually at higher altitudes 2850-3000 m. Adult with one young, Ilo Bek, 6 September 1992.

Rufous-tailed Flycatcher *Muscicapa ruficauda*. Widespread throughout Palas both in subtropical broadleaf forest and temperate forest. One young in oak scrub between Khorghi and Sharial Nullah, 11 July 1992. 3+, with young in oak forest between Bar Ser and Shuki Ser, 18 July 1992.

Ultramarine Flycatcher *Ficedula superciliaris*. Common, probably breeding. One young, between Ilo Bek-Tiko Sar, 7 September 1992.

Slaty-blue Flycatcher *Ficedula tricolor*. Fairly common mostly above 2200 m.

Rufous-bellied Niltava *Niltava sundara*. One male, between Ilo Bek and Paro, 29 June 1992; this may be the most westerly record for the species.

Verditer Flycatcher *Muscicapa thalassina*. One, between Sherakot and Moru, 26 June 1992. One, between Bar Paro and Gana, 16 May 1993.

Grey-headed Canary Flycatcher *Culicicapa ceylonensis*. One, Chuk Bek, 8 June 1989.· Several heard, Sar Bek, 14 June 1989. Common, Ilo Bek, 17 June 1989. Three in oak forest at Pochmoru, 20 May 1994. Single at Sartoe, 25 May 1994.

Asian Paradise Flycatcher *Terpsiphone paradisi*. Mostly recorded around habitations, usually below 2000 m. Several, near Khalyar, 22 June 1989. Two, Karat, 23 June 1989. Two males, one female, between Pattan and Sherakot, 6 June 1991. One female, Kuz Paro, 8 June 1991. One male, one female, near Paro, 8 June 1991. One, between Kuz Paro and Bar Paro, 15 May 1993. Several males and females, Shared, 12 June 1993. Three, around Kuz Paro, 16 May 1994 and 12 June 1994. Three, Pochmoru, 20 May 1994. Three, Shared, 31 May 1994. Four, Gadar, 2 June 1994.

Brownish-flanked Bush Warbler *Cettia fortipes*. Common, Khorghi, 6 June 1989. Several heard, Ilo Bek, 17 June 1989. One, Shuki Ser near lower fringe of Deodar forest, 9 June 1991. One, between Bar Ser and Ilo Bek, 10 June 1991. Many singing, between Sherakot and Moru, 26 June 1992. One, between Gidar and Moru, 12 September 1992. One heard, between Bar Paro and Gana, 16 May 1993. One heard, between Bar Paro and Kuz Ser, 17 May 1993. May-June 1994 survey found it to be widespread from Gadar eastwards.

Lesser Whitethroat *Sylvia curruca*. Singles recorded, around Kuz Ser and Ilo Bek in June 1991 but May-June 1994 survey found it to be fairly common throughout Palas.

Chiffchaff *Phylloscopus collybita tristis*. One heard, between Paro and Bar Ser, 12 July 1992.

Tytler's Leaf Warbler *Phylloscopus tytleri*. Locally common in Palas. Found among *Betula utilis* and *Abies pindrow* forest near the tree-line.

Tickell's Leaf warbler *Phylloscopus affinis*. Found among low bushes and boulder scree above the tree-line. Three, Moru, 7 June 1989. One in *Populus/Salix* scrub near Pichbela, 22 September 1991. Two, below Breathbek, 24 May 1994. One, above Kundal, 27 May 1994. One in Juniper *Juniperus* scrub above Ishaq Bek, 22 June 1994.

Yellow-browed Leaf Warbler *Phylloscopus inornatus*. One seen and many heard, between Moru and Gidar, 27 June 1992. May-June 1994 survey found it to be fairly common between Breathbek and Sartoe (Kundal Nullah).

Brooks' Leaf Warbler *Phylloscopus subviridis*. Locally common in temperate forest. Rarely confiding and difficult to observe, usually remaining near the treetops.

Pallas's Leaf Warbler *Phylloscopus proregulus*. Widespread in Palas, mostly above 2400 m, probably breeding. Jonathan Eames found it very common in lower parts of Palas. Many heard, between Moru and Gidar, 27 June 1992. Many, Gidar, 27 June 1992. 10, between Breathbek-Sartoe, 24 May 1994. Fairly common above Kundal, 27 May 1994, above Belgi 31 May 1994 and at Shalkho, 2 June 1994. A pair alarm calling and carrying food above Pulbela, 21 June 1994. One observed feeding in scattered *Quercus* trees near road-head of Bar Palas, 14 February 1995 - the only *Phylloscopus* record in February 1995.

Large-billed Leaf Warbler *Phylloscopus magnirostris*. Typically found in forest along nullahs. Common, Ilo Bek, 16-19 June 1989. One calling, Diwan, 12 June 1991. Several heard, between Gidar and Ilo Bek, 28 June 1992. One, between Bar Ser-Ilo Bek, 16/17 July 1992. One, between Ilo Bek and Tiko Sar, in *Viburnum* bushes, 7 September 1992. One heard, between Shuki Ser, Ilo Bek and Gidar, 11 September 1992. May-June 1994 survey only recorded the species from Belgi where five were present.

Greenish Leaf Warbler *Phylloscopus trochiloides*. Jonathan Eames found it common in lower parts of Palas in May-June 1989. Guy Duke observed several, particularly in shrubby gulleys, in September 1992.

One, Kabkot forest, 6 September 1992. One singing at Bush, 19 May 1994. Two worn birds lacking wingbars, above Dumbela, 18 June 1994.

Western Crowned Leaf Warbler *Phylloscopus occipitalis.* Abundant, the most common leaf warbler in Palas, presumably breeding. May-June 1994 survey recorded many birds carrying food.

Grey-headed Flycatcher Warbler *Seicercus xanthoschistos.* Locally common in lower parts of Palas, usually below 1600 m. Recorded frequently around Paro. Probably breeding. One alarm calling with food, near Paro, 12 June 1994. Recorded commonly in small flocks with White-cheeked Tits in *Quercus* scrub along Musha'ga in lower parts of Bar Palas, February 1995; maximum day-count of 20, between Kuz Paro and Shuki Ser, 15 February 1995.

Goldcrest *Regulus regulus.* Jonathan Eames found it common in lower parts of Palas, especially Kabkot Nullah, May-June 1989. Probably breeding. One seen at Bush carrying nesting material, 18 May 1994. Two observed feeding in a tit flock above Kundal (2600 m), 20 February 1995.

Himalayan Rubythroat *Luscinia pectoralis.* Two singing males, Moru 7 June 1989. One male, between Sherakot and Moru, 26 June 1992. A male and a female observed in dwarf juniper *Juniperus* scrub, above Ishaq Bek 22 June 1994.

Indian Blue Robin *Luscinia brunnea.* Common and widespread, 1600-3000 meters, probably breeding. Jonathan Eames found it common in lower parts of Palas and around Kabkot Nullah. Many, recorded singing, between Sherakot and Moru, 26 June 1992. Maximum count in May-June 1994 of 10 above Karoser including a female carrying moss, 21 May 1994. A male was seen carrying food above Dumbela, 19 June 1994.

Orange-flanked Bush Robin *Tarsiger cyanurus.* Jonathan Eames found it common around Kabkot Nullah (he considered it replacing the previous species at higher altitudes, but May-June 1994 survey found a zone of overlap). Many heard calling, between Bar Ser and Ilo Bek, 10 June 1991. Several heard, Diwan eastern aspect, c. 2850 m, 12 June 1991. Common, upper Diwan, 14 June 1991. One immature, between Sharial and Magri, 21 August 1993. Recorded from six localities all above 2600 m in May-June 1994, with a maximum count of nine at Bush, 18 May 1994. An immature male observed singing, Bush, 18 May 1994.

Blue-capped Redstart *Phoenicurus caeruleocephalus.* Six recorded on two dates, Galli Bek and Sar Bek, June 1989. One male between Moru and Gidar, and two at Gidar, 27 June 1992. May-June 1994 survey found it to be fairly common above 2400 m with maximum count of five pairs at Bush, 18 May 1994 (including two females carrying food).

Blue-fronted Redstart *Phoenicurus frontalis*. Two, Galli Bek, 11 June 1989. Recorded from three localities all above 2750 m in May-June 1994; single male, Bush, 18 May; a pair, above Rajaser, 27 May; three pairs at Ishaq Bek, 22-23 June.

Plumbeous Redstart *Rhyacornis fuliginosus*. Fairly common, breeding. Found along Musha'ga and nullahs throughout Palas to c. 2750 m. Several, with young, Kabkot Nullah, 6 September 1992. Fledged juveniles observed throughout Palas, May-June 1994 survey. Found fairly common along lower reaches of Mushaga River in Bar Palas, February 1995; maximum day-count of 10, between road-head and Kuz Paro, 14 February 1995.

Hodgson's Shortwing *Hodgsonius phoenicuroides*. 1-2 males and one female, between Sherakot and Moru, 26 June 1992. One male, between Moru-Gidar, 27 June 1992. Observed running in the open, never far from *Viburnum* shrubs, and singing from the top of a *Viburnum* clump.

Little Forktail *Enicurus scouleri*. Uncommon, usually found along fast flowing nullahs. Two, Ilo Bek, 17 June 1989. Two, Paro, 23 June 1989. One pair, on Sharial Nullah near Musha'ga, 9 June 1993. One on nullah from Bar Paro to Kuz Paro, 10 June 1993. One, between Shared-Bangaha, 22 May 1994. Three, Pharroga Nullah, 28-29 May 1994. One, Shared, 29 May 1994. One, Belgi, 30-31 May 1994. One, Sharial Nullah, 26 September 1994. One near Sharial bridge, Sharial Maidan, 24 October 1994. Recorded from 5 sites along Musha'ga, February 1995, with two seen together between Kundal and Shared, 22 February 1995, and at Gadar, 24 February 1995.

Spotted Forktail *Enicurus maculatus*. One, Paro, 23 June 1989. One, between Badakot and Sherakot, 21 December 1994.

Stonechat *Saxicola torquata*. Common around villages and cultivation. Breeding. A nest found with female incubating four eggs at Karin, 31 May 1994. Recently fledged young observed at Nawn, 14 June 1994, and a male carrying a faecal sac at Dumbela, 18 June 1994.

Pied Stonechat *Saxicola caprata*. One, Shuki Ser, near lower fringe of Deodar forest, 9 June 1991. Two, between Pattan and Sherakot, 25 June 1992. Common in Deodar/Kail forest between Sherakot and Khorghi, 11 July 1992. One, between Paro and Bar Ser, 12 July 1992.

Dark-grey Bush-chat *Saxicola ferrea*. Mostly recorded above 2000 m often in the vicinity of cultivation. Two, Khorghi, 6 June 1989. One, Chuk Bek, 9 June 1989. One, Kundal, 23 September 1991. Four, between Sherakot and Moru, 26 June 1992. One, between Paro and Pattan, 20 July 1992. Several, between Bar Paro and Gana, 16 May 1993. One male, Dhar in Kundal Nullah, 26 May 1994 and another at

Belgi, 31 May 1994. 6+ birds, between Karin-Shalkho, 1-2 June 1994. One, below Kuz Ser, 13 June 1994.

White-capped Redstart *Chaimarrornis leucocephalus*. Common and widespread especially along large nullahs to 3200 m. Breeding. Nest containing four eggs found, Galli Bek, 11 June 1989. Two adults feeding four juveniles, near Nawn, 12 June 1994. Recorded as common along lower reaches of Mushaga River, February 1995; maximum day-count of 10 between road-head and Kuz Paro, 14 February 1995.

Blue-headed Rock Thrush *Monticola cinclorhyncha*. Two, Chuk Bek, 8 June 1989. One, Khalyar, 22 June 1989. One, Diwan, 12 June 1991. One, between Shuki Ser and Ilo Bek, 5 September 1992. 2+ between Paro-Bar Paro, 15 May 1993. Several heard, near Bar Paro, 9 June 1993. May-June 1994 survey found it fairly common mostly in broadleaf forest throughout the valley. Probably breeding. A female observed collecting nesting material (moss and lichen) from an oak tree and from the ground, Pochmoru, 20 May 1994.

Blue Rock Thrush *Monticola solitarius*. One, Kana-Kunari Pass, c. 3800 m, 22 September 1991. One, between Sharial Nullah and Paro, 11 July 1992. One, between Bar Paro-Kuz Ser, 17 May 1993. One, between Sharial-Magri, 21 August 1993. One male flushed along Mushaga, below Karoser, 21 May 1994. One female, between Pharroga and Shared, 29 May 1994.

Blue Whistling Thrush *Myiophoneus caeruleus*. Very common along nullahs throughout the valley to 3000 m. Breeding. Common, between Pattan and Sherakot, 6 June 1991. Several, between Taghi and Paro, 14 May 1993. Birds recorded collecting nesting material and carrying food in May-June 1994 survey. Single birds regularly recorded in suitable habitat in Kuz and Bar Palas, February 1995, the highest altitudinal record being Kundal, 2300 m, 20 February.

White's Mountain Thrush *Zoothera dauma*. Two, Sar Bek, 14 June 1989. One seen briefly, between Ilo Bek and Paro, 29 June 1992. 1+, Belgi, 31 May 1994.

Eurasian Blackbird *Turdus merula*. Usually found in alpine pastures. Probably breeding. Singles in Sar Bek and Tiko Sar, June 1989. One, upper Diwan, 14 June 1991. One with one juvenile in moist deciduous forest, Diwan, 15 June 1991. One, near Kunari Bek, 20 September 1991. Single male, above Gadar (Chor Nullah), 18-19 June 1994. 4+, Ishaq Bek, 22 June 1994.

Grey-headed Thrush *Turdus rubrocanus*. Fairly widespread in forest above 2400 m. Probably breeding. One noted carrying worms Sar Bek, 14 June 1989. A female carrying nesting material, above Rajaser,

27 May 1994. Three flocks totalling about 30 birds, observed in dense broad-leaf forest near Gabir (1900 m), 9 February 1995.

Black-throated Thrush *Turdus ruficollis atrogularis.* 10+, Pichbela, 25 October 1993. Singles seen near Gabir at 1900 m and 1700 m on 9 and 12 February 1995 respectively.

Mistle Thrush *Turdus viscivorus.* Two, Galli Bek, 11 June 1989. Several, Sar Bek, 14 June 1989. Two, Sar Bek, 15 June 1989. One, Tiko Sar, 20 June 1989. One heard, Magri, 18 September 1991. Two, between Sherakot-Moru, 26 June 1992. Two, between Moru-Gidar, 27 June 1992. 1+, between Sharial-Magri, 21 August 1993. May-June 1994 survey recorded it from five localities all above 2750 m: three pairs, Bush, 18 May and Shalkho 2 June; three at Dhar (Kundal Nullah), 26 May; a pair above Mukchaki, 16 June; and a pair at Ishaq Bek, 22 June.

Wren *Troglodytes troglodytes.* Ten, Galli Bek, 11 June 1989. Several, Tiko Sar, 20 June 1989. One, upper Diwan, 14 June 1991. One, near Kunari Bek, 19 September 1991. One, near hut Gidar, 27 June 1992. One, Pichbela, 25 October 1993. May-June 1994 survey recorded it from three localities all above 2750 meters: three Bush, 18 May; singles above Rajaser, 26 May and at Shalkho, 1-2 June. February 1995 survey found it to be common in Kuz and Bar Palas, most frequently observing it on scrėe slopes and around villages, the highest altitudinal record being Kundal, 2300 m, 20-21 February 1995.

Brown Dipper *Cinclus pallasii.* Common along main river and nullahs, breeding. Fledged juveniles recorded at six localities in May-June 1994 survey. February 1995 survey found it to be very common in Bar Palas, the highest altitudinal record being Kundal Nullah, 2400 m, 20 February 1995, and highest day-count 20, between road-head and Kuz Paro, 14 February 1995. Two, observed repeatedly entering a nest carrying food, below Gadar, 24 February 1995. The nest was constructed in a mossy rock crevice on a cliff, c. 4 m above the river surface.

Alpine Accentor *Prunella collaris.* First time recorded by February 1995 survey. Two birds observed in rock scree along Musha'ga, just east of Kundal (2000 m), 21 February 1995.

Rufous-breasted Accentor *Prunella strophiata.* Two, Chuk Bek, 9 June 1989. Two, Galli Bek, 11 June 1989. Two, Sar Bek, 14 June 1989. Two, Tiko Sar, 20 June 1989. One, Kana-Kunari Pass, 22 September 1991. Several, between Sherakot and Moru, 26 June 1992. Many, Gidar, 27 June 1992. May-June 1994 survey recorded it from three localities all above 2850 meters: six, Bush, 17-19 May; two, Shalkho, 1 June and one, at Ishaq Bek, 21 June.

Great Tit *Parus major.* Fairly common, recorded frequently from lower parts of Palas especially in sub-tropical broadleaf forest. Probably breeding. Several, near Paro, 23 June 1989. Several pairs, in *Quercus baloot* forest, between Pattan and Sherakot, 6 June 1991. Abundant, between Khorghi and Kuz Sharial, 7 June 1991. Common, Kuz Paro, 8 June 1991. Several, on path between Khalyar to Ser, 9 June 1991. One seen carrying food, near Kuz Paro, 2 June 1994. February 1995 survey frequently recorded it in mixed tit flocks at 1500-2400 m in Kuz and Bar Palas, most commonly in scrub and mixed forest at lower altitudes, with a maximum day-count of 10 between Bar Ser and Kundal, 18 February 1995.

Green-backed Tit *Parus monticolus.* Two, Ilo Bek, 16 June 1989. One, between Ilo Bek and Paro, 29 June 1992. Heard between Bar Ser-Ilo Bek, 16/17 July 1992. One, between Bar Ser and Ilo Bek, 5 September 1992. One in *Viburnum* scrub, Ilo Bek, 6 September 1992. One mixed with leaf warblers, Pichbela, 25 October 1993. One, Sartoe, 25 May 1994. Two, Belgi, 30 May 1994. Four, above Mukchaki, 15 June 1994. One, above Dumbela, 19 June 1994. February 1995 survey frequently recorded it, in mixed tit flocks, both in Kuz and Bar Palas, the highest altitudinal record being Kabkot, 2400 m, 18 February 1995 and maximum day-count 15 in broad-leaf forest, Gabir, 9 February 1995.

Crested Black Tit *Parus melanolophus.* Common and widespread generally above 2000 m in both coniferous and broadleaf forest. Breeding. One seen nest-building, Bush, about 6-7 m above ground in a fir tree, 19 May 1994. Pairs observed courtship feeding in oak forest, above Karoser, 21 May 1994 and at Sartoe, 23 May 1994. February 1995 survey recorded it to be the most common tit species, seen in large flocks to 2600 m; maximum day-count 75, including a single feeding flock of 50, Kabkot Nullah, 18 February 1995.

Simla Tit *Parus rufonuchalis.* Common and sympatric with previous species in coniferous forest. Probably breeding. A pair was seen at Bush carrying fine grass or hair presumably as nest lining, 18 May 1994. February 1995 survey found it to be less common than Crested Black Tit, with only the following two records; 3, Gabir, 9 February 1995; 2, Kabkot Nullah, 16 February 1995.

Fire-capped Tit *Cephalopyrus flammiceps.* Two males and two pairs, one courtship feeding, in oak forest above Karoser, 21 May 1994. Two males feeding in a seeding *Salix* tree at Sartoe, 25 May 1994. One seen, Dhar, in a seeding *Salix* tree, 26 May 1994, and three males and a female seen at the same spot, 27 May 1994.

Red-headed Longtailed Tit *Aegithalos concinnus.* First recorded February 1995. Monospecific feeding flocks of 30 birds, 12 February

1995 and 15 birds, 18 February, in *Quercus* forest below Gabir, 1700 m. Lilley et al. record that each of the three *Aegithalos* species in Palas occupies a distinct habitat without overlap.

White-cheeked Longtailed Tit *Aegithalos leucogenys*. Uncommon. Recorded in oak forest to 2200 m in summer. Probably breeding. Six, near Paro in *Quercus baloot* forest, 23 June 1989. A feeding flock of 10+, Pochmoru, 20 May 1994. A pair seen carrying nesting material, Pochmoru, 21 May 1994. Two, above Karoser, 21 May 1994. Two, Karin, 31 May 1994. 3+, in the main river valley opposite Gadar, 12 June 1994. February 1995 survey recorded it to be fairly common in secondary *Quercus* scrub in the lower reaches of Mushaga, highest day-count 35, between Kuz Paro and Shuki Ser, 15 February 1995. A flock of five, in low scrub above Kundal, 2600 m, 20 February 1995.

White-throated Longtailed Tit *Aegithalos niveogularis*. Uncommon. Recorded from areas of open coniferous forests, 2400-3000 m. Probably breeding. Many, Tiko Sar, 21 June 1989. 20+, comprising at least two separate family parties, Gidar, 27 June 1992. 6+, between Bar Paro-Kuz Ser, 17 May 1993. Two, Bush, 17 May 1994. One seen collecting nesting material (chicken feathers), Sartoe, 25 May 1994. Two, Dhar, 27 May 1994. A flock of 15+ on a slope high above Nawn, 15 June 1994.

Kashmir Nuthatch *Sitta cashmirensis*. Uncommon. Breeding. One, between Sherakot and Moru, 26 June 1992. One, between Shuki Ser, Ilo Bek and Gidar, 11 September 1992. A pair observed at a nest hole 5 m above ground in a *Cedrus deodara* tree on a *Quercus* covered ridge above Pochmoru, the male taking food into the nest, 19 May 1994. Another pair seen in similar habitat above Karoser, 21 May 1994. Three, between Breathbek and Sartoe, 24 May 1994. Two, Pharroga Nullah, 27 May 1994. February 1995 survey found it occasionally in coniferous forest up to 2400 m, maximum day-count of 4 near Gabir, 9 February 1995, and again in Kabkot Nullah, 16 February 1995.

White-cheeked Nuthatch *Sitta leucopsis*. Jonathan Eames found it common in May-June 1989 in Kabkot Forest. Several in higher forests at Moru and Ledi, September 1991. One, between Sherakot and Moru, 26 June 1992. Several, Gidar, 27 June 1992. One, between Bar Ser and Ilo Bek, 5 September 1992. One heard between Ilo Bek and Tiko Sar, 7 September 1992. One heard between Ilo Bek and Bar Ser, 8 September 1992. One heard between Sharial and Magri, 21 August 1993. May-June 1994 survey recorded it from most areas of coniferous forest above 2400 meters. February 1995 survey recorded it to be less common than the above species; sighted at only four locations in Bar

Palas, all from coniferous forests above 2100 m, the highest altitudinal record being 2500 m, Kabkot Nullah, 19 February 1995.

Common Tree Creeper *Certhia familiaris*. Found sympatric with *C. himalayana* in coniferous forests above 2600 m. Several, Gidar, 27 June 1992. Two, Bush, 19 May 1994. One, Breathbek, 24 May 1994. A pair courtship feeding, Shalkho, 1 June 1994. One, Mukchaki, 16 June 1994.

Bar-tailed Tree Creeper *Certhia himalayana*. Common and breeding. Appears to be more widespread than the previous species. Three Chuk Bek 8 June 1989. A pair feeding young, Chuk Bek, 9 June 1989. Nest containing two pulli Sar Bek, 13 June 1989. One, fir/spruce forest Gidar, 12 September 1992. Three pairs with nests at Bush, 18 May 1994. One carrying food, near Sartoe, 24 May 1994. February 1995 survey found it occasionally in mixed species flocks. Four, Kuz Palas, 9 February 1995. A total of four, on four dates, Bar Palas, the lowest altitude being 1600 m, near road-head, 14 February 1995.

Tree Pipit *Anthus trivialis*. One seen, in dwarf juniper *Juniperus* sp. at 3200 m, Gadar (Chor Nullah), 18 June 1994.

Rosy Pipit *Anthus roseatus*. Usually found in alpine pastures. Breeding. One, Galli Bek, 11 June 1989. 30, Tiko Sar, 20 June 1989. Several small groups, upper Ledi, 24 September 1991. One, between Paro-Shuki Ser, 4 September 1992. 10+, feeding, Bush 18 May 1994. Six, above Dhar, 25 May 1994. A common breeding species in grassland around Ishaq Bek, one did a distraction display when flushed from the nest, 21 June 1994. The nest consisted of a grassy cup situated on the ground in a wet flush (with *Caltha palustris*) and contained three brown eggs covered with dark brown blotches.

Upland Pipit *Anthus sylvanus*. Several between Ilo Bek and Bar Ser, 8 September 1992. One, between Shuki Ser, Ilo Bek and Gidar, 11 September 1992.

Yellow-headed Wagtail *Motacilla citreola*. One, Galli Bek, 11 June 1989. One, Tiko Sar, 19 June 1989. A pair and 2+ males, Pichbela, 13-21 June 1994. ·Two pairs and one juvenile at Gadar (Chor Nullah), 18 June 1994. 2+ males by river, Ishaq Bek, 21-23 June 1994, including a leucistic bird with back of the neck, mantle, rump and tail white instead of black, chased by the normal coloured male.

Grey Wagtail *Motacilla cinerea*. Very common along nullahs and habitations to at least 3200 m. Breeding. Males seen song-flighting at several localities May-June 1994. A pair watched copulating and displaying at Dumbela, 20 June 1994. Two fledged juveniles, Mukchaki, 20 June 1994.

White Wagtail *Motacilla alba*. Fairly common but rarely seen away from the main river and not as widespread as the previous species. Breeding. A fledged juvenile was observed being fed by adults, Wulbela, 16 June 1994. A pair alarming with food, near Gadar, 12 June 1994.

Large Pied Wagtail *Motacilla maderaspatensis*. Several, between Taghi and Paro, 14 May 1993. Several, between Kuz Ser and Sertay, 18 May 1993.

Oriental White-eye *Zosterops palpebrosa*. Scarce. One, Kuz Paro, 16 May 1994. Two, Pochmoru, 20 May 1994.

House Sparrow *Passer domesticus*. Two, Kuz Paro, 8 June 1991. Two, between Karat and Paro, 15 May 1994.

Cinnamon Tree Sparrow *Passer rutilans*. Recorded around habitation to c. 2400 m. One flock, between Pattan and Sherakot, 6 June 1991. One, between Sherakot and Sharial, 7 June 1991. One, between Bar Ser and Ilo Bek, 10 June 1991. Common around Sherakot, 26 June 1992. Many, Bar Ser, 17/18 July 1992. Many, Shuki Ser, 10 September 1992. One, between Kuz Ser and Sertay, 18 May 1993. Several, Shared, 12 June 1993. 10+, Pochmoru 20 May 1994.

Red Munia *Estrilda amandava*. Small flock in ripe maize fields near Sharial, 18 September 1991.

Spotted Munia *Lonchura punctulata*. Two, between Shuki Ser and Bar Ser, 5 September 1992.

Chaffinch *Fringilla coelebs*. First recorded by February 1995 survey; a flock of 12 birds seen feeding in an open grassy area, on Palas side of Indus River near Pattan, 8 February 1995.

Black and Yellow Grosbeak *Mycerobas icterioides*. Frequently recorded usually above 2200 m. A flock of approximately 15 observed in Deodar forest, Kundal Nullah, 2500 m, 20 February 1995. Single male above Shoman, 2400 m, 21 February 1995. One heard calling, mixed forest above Shared, 2000 m, 23 February 1995.

White-winged Grosbeak *Mycerobas carnipes*. Single male, Bush, 19 May 1994.

Goldfinch *Carduelis carduelis*. Two seen feeding, in low scrub above terraced maize field near 'hospital', Pichbela 17 June and 20 June 1994.

Himalayan Greenfinch *Carduelis spinoides*. Five, Ser, 22 June 1989. Several, Gale Bek and lower Ledi, 22 September 1991, between Sherakot and Moru, 26 June 1992. Several, Shuki Ser, 10 September 1992. One, between Gidar and Moru, 12 September 1992. One, between Sharial and Magri, 21 August 1993. Two, Karin, 1 June 1994. 10+, Wulbela, 18 June 1994.

Red-fronted Serin *Serinus pusillus*. First recorded February 1995; one pair accompanied by an immature seen feeding among riverside scree (almost certainly the same individuals sighted twice), Kundal, 2000 m, 21-22 February 1995.

Plain Mountain Finch *Leucosticte nemoricola*. 60, Galli Bek, 11 June 1989. One, Sar Bek, 15 June 1989. 100, Tiko Sar, 20 June 1989. Two small flocks; Sherakot, 18 September 1991. Large flocks common over 3000 m in Moru, Kunari, Ledi, September 1991. May-June 1994 survey recorded it from four localities all above 2600 m: 70+, Bush, 18 May; four, Breathbek, 23 May; single, Dhar, 27 May; common on the grassland above Ishaq Bek, 21-23 June. February 1995 survey found it common around villages, in open slopes and in cultivated areas. Largest flock c. 100+, Gabir, 12 February 1995. Smaller numbers seen, above Gadar, to 2500 m, 24 February 1995.

Red-browed Finch *Callacanthis burtoni*. Well distributed and fairly common in coniferous forest usually near tree-line. Breeding. Six, Moru, 7 June 1989. Three, Galli Bek, 11 June 1989. Three, Sar Bek, 14 June 1989. Several small groups at upper edge (3000-3300 m) of forests, Moru and Ledi, September 1991. One, Gidar, 27 June 1992. One, between Shuki Ser-Bar Ser, 5 September 1992. Heard often between Ilo Bek and Tiko Sar, 7 September 1992. 12+, Moru, 12 September 1992. Small flock near Magri, 21 August 1993. Three pairs, Bush, 18 May 1994; one pair carrying nesting material and another accompanied by a begging juvenile. 15+, including an immature, Shalkho, 1 June 1994. Six, Gadar (Chor Nullah), 19 June 1994. A male seen feeding two young, above Pulbela, 21 June 1994. February 1995 survey sighted it on three dates in Bar Palas, all sightings in Deodar forest to 2500 m: 20, Kabkot Nullah, 16 February 1995; 41, Kabkot Nullah, 18 February 1995; three, Kundal, 21 February 1995.

Common Rosefinch *Carpodacus erythrinus*. A widespread species occurring from Musha'ga to 3000 m. 10+, Gidar, 27 June 1992. 10+, near Shuki Ser, 5 September 1992. 20+, in three flocks between Paro and Gorkhar, 16 May 1994.

Pink-browed Rosefinch *Carpodacus rhodochrous*. One, Chuk Bek, 8 June 1989. Six, Galli Bek, 11 June 1989. Two, Tiko Sar, 20 June 1989. One male, between Moru-Gidar, 27 June 1992. One male and one female, Gidar, 27 June 1992. Heard between Bar Ser and Ilo Bek, 16/17 July 1992. Seven, above Rajaser 2750 m, 27 May 1994. One female, Belgi, 2650 m, 30 May 1994. One male, Karin, 2200 m, 2 June 1994.

Orange Bullfinch *Pyrrhula aurantiaca*. Two, Chuk Bek, 8 June 1989. Two, Jhil Bek, 12 June 1989. Two, Sar Bek, 14 June 1989. May-June 1994 survey recorded it from five localities between 2400-3000 m: 15+, feeding on *Salix* seed, Sartoe 26 May; four males and two females, Dhar, 26 May; single pairs recorded at Bush, 18 May; Belgi, 30-31 May; and Mukchaki, 20 June. One seen in flight over mixed scrub, just below Gabir, 1650 m, 12 February 1995.

White-capped Bunting *Emberiza stewarti* Common around habitation and cultivation. Breeding. Several, Ser, 22 June 1989. Several, between Paro and Bar Paro, 15 May 1993. Several, between Bar Paro and Gana, 16 May 1993. Common, around Bar Paro, 9 June 1993. One female with nest material, Pochmoru, 20 May 1994. One male with an immature, Sartoe, 25 May 1994.

Rock Bunting *Emberiza cia*. Replaces the previous species altitudinally, especially common around the tree-line. Breeding. Nest containing three eggs found, Sar Bek, 12 June 1989. A male carrying food above Mukchaki, 2900 m, 20 June 1994. February 1995 survey found it common around Gabir, seen daily, maximum day-count 30, 9 February 1995. In Bar Palas recorded from two locations; four, Kundal, 22 February 1995; 15, Gadar, 24 February 1995.

DISCUSSION

Biodiversity Importance of the Palas Valley

The West Himalayan temperate forests of Pakistan and India have been identified an Endemic Bird Area (EBA), a centre of global importance for conservation of biodiversity (ICBP 1992).

Palas contains the largest known population of the Western Tragopan *Tragopan melanocephalus*, a rare pheasant species vulnerable to extinction (Collar *et al.* 1994), and probably contains the most outstanding pristine remaining tract of such forests in Pakistan. Palas will undoubtedly qualify as an Important Bird Area (IBA) in a forthcoming analysis of the IBAs of Asia. Recent studies - on mammals and higher plants - confirm the importance of Palas not only for birds but also for other taxa, supporting the hypothesis that avian diversity may often be a good indicator of high overall biological diversity. Palas also is an important centre of floral endemism, in addition to nearly 100 western Himalayan endemics, Palas also contains a good population of endangered Western Himalayan Elm *Ulmus wallichiana*.

Habitat Diversity in Palas

Subtropical in latitude, lying in a transitional climatic belt (Palas marks the approximate northern limit of the influence of the summer monsoon), and covering an altitududinal range of almost five vertical kilometres, Palas contains a remarkable variety of avian habitats. These include riverine and lacustrine habitats, a dry subtropical zone, broadleaf and evergreen temperate forests, subalpine birch forests, alpine scrub and meadows, and high mountain peaks with large areas of permanent snow. Transition zones (ecoclines) between these major habitat types are also of importance. Some areas of Palas remain in a near-primary condition, but many areas have been modified by the activity of man and his domestic animals. This has created new habitats such as agricultural land, grazed pastures and ponds.

Threats to Birds in Palas

a) Commercial Logging

Unsustainable commercial logging poses the most serious threat to birds and their habitats in Palas. Logging plans prepared by the NWFP Forest Department are thought to over-estimate the growing stock and annual yield, and in practice extraction often exceeds the prescriptions. Antiquated technology and inadequate controls lead to damaging selection and felling practices. Over-logging is linked to local poverty, which drives the forest owners to sell their rights to outside contractors with little vested interest in sustainable logging. Prior to 1992, logging was mostly restricted to the more accessible parts of Kuz Palas, and all logging is currently suspended under the ongoing moratorium. However, accelerated logging and forest destruction is anticipated if and when the moratorium is lifted, unless an improved forest management plan and improved harvesting systems are first put in place.

b) Lopping in Oak Forest for Fuel and Fodder

Oak forest is mostly present in the lower reaches of Palas along the River Indus and Musha'ga. It is traditionally lopped for local bona fide use as firewood and fodder. However, construction of the Pattan-Bar Palas road is opening these forests to exploitation for fuelwood to supply the demands of Pattan and other bazaars on the KKH. To date, this exploitation has been to some extent restricted by the organised opposition of the owning tribes. Birds directly affected by degradation of these dry subtropical woodlands would include the rare Speckled Piculet and Slaty-headed Parakeet.

c) Hunting

Hunting does not seem to pose a serious threat to most birds in Palas. However, hunting of Monal may be a problem - the skin and crest feathers fetch Rs 300-400 in Pattan. Monal is usually hunted in winter when the birds move to lower altitudes in search of food. Rarely, skins of Western Tragopan are observed for sale.

Conservation Measures

Threats to birds in Palas are being addressed under the Himalayan Jungle Project. HJP has been operating since 1991, and a five-year follow-on project, the Palas Conservation and Development Project, is proposed. HJP/PCDP aims to safeguard the biodiversity of the Palas Valley by enabling local communities to tackle the linked causes of poverty and incipient natural resource degradation. Project components include: community organisation and participation; natural resource management; development of basic infrastructure; and health, nutrition and sanitation.

HJP/PCDP is a joint venture of BirdLife International, the NWFP Wildlife Department, National Council for Conservation of Wildlife, WWF-Pakistan and WPA-Pakistan.

REFERENCES

ALI, S. AND RIPLEY, S.D., 1983. *A Pictorial Guide to the Birds of the Indian Subcontinent.* Oxford University Press, Delhi.

ALI, S. AND RIPLEY, S.D., 1987. *Compact Handbook of the Birds of India and Pakistan.* Oxford University Press, Delhi.

COLLAR, N.J., CROSBY, M.J. AND STATTERSFIELD, A., 1994. *Birds to Watch 2: The World List of Threatened Birds.* Bird Life International (Bird Life Conservation Series No. 4), Cambridge, UK.

DUKE, G. AND WALTON, P., 1988. *Indus Kohistan: Refuge of the Western Tragopan.* Unpublished report to ICBP, Cambridge, UK.

DUKE, G., 1988. *Survey of the Western Tragopan, Palas Valley,Indus Kohistan.* Operation Raleigh Pakistan Expedition 14E Science Report. Unpublished report, Operation Raleigh, London.

DUKE, G., 1989. *Survey of the Western Tragopan Tragopan melanocephalus and its montane forest habitat, Palas and Kandia Valleys, Indus Kohistan, Pakistan.* Unpublished report to ICBP, Cambridge, UK.

DUKE, G., 1990. Using call counts to compare western tragopan populations in Pakistan's Himalayas. In: Hill, D.A., Garson, P.J. and D. Jenkins (Eds). *Pheasants in Asia* 1989, WPA, Reading, UK.

ICBP, 1992. *Putting Biodiversity on the Map: Priority Areas for Global Conservation.* ICBP, Cambridge, UK.

EAMES, J., 1989. *Birds Recorded During the Expedition.* 15 May to 23 June 1989. Unpublished.

MUSHTAQ, M., 1988. *Revised Working Plan for Palas Forests of Kohistan Forest Division, NWFP.* Forestry Pre-investment Centre, Peshawar.

RAFIQ, R., 1994. *A Preliminary Botanical Inventory and Ethnobotanical Checklist, Palas Valley, District Kohistan, NWFP.* Unpublished report to HJP, Islamabad.

RIPLEY, S.D., 1982. *A Synopsis of the Birds of India and Pakistan,* Bombay Natural History Society, Oxford University Press, Bombay.

ROBERTS, T.J., 1991. *The Birds of Pakistan,* **1.** *Non-passeriformes.* Oxford University Press, Karachi.

ROBERTS, T.J., 1992. *The Birds of Pakistan,* **2.** *Passeriformes.* Oxford University Press, Karachi.

CRANES: AN INTEGRAL COMPONENT OF BIODIVERSITY

MEENAKSHI NAGENDRAN

U.S. Fish and Wildlife Services, Arlington, Virginia, U.S.A.

Abstract: Biodiversity across our planet is seriously threatened. Both habitats and species are headed more toward extinction now than they were before, primarily because of the human population explosion. Large human populations have a direct impact on natural habitats. Biodiversity faces a greater threat in the tropics. Pakistan and India (the Indian Sub-continent) were rich in biodiversity during the last century, but in this century several species have been decimated to extinction, for example megafauna such as the Cheetah and the Siberian Crane. Many species are critically endangered and numbers of most species are being further reduced annually. Three species of cranes migrate through Pakistan, and the Sarus Crane used to breed in Pakistan. Among the 3 migratory species of cranes, the Siberian Crane is close to extinction, and the numbers of Demoiselle and Eurasian Cranes appear to be decreasing in numbers. Cranes are excellent indicator species of the health of wetlands. Although there are factors other than habitat loss that have also contributed to the loss of cranes, what has resulted in the decimation of cranes is cause for concern for many other species. Current international crane research and conservation efforts include species reintroduction efforts, documenting migratory routes, identifying important staging and stopover areas, conserving critical habitats, assessing environmental hazards, and education at all levels.

INTRODUCTION

Four species of cranes occur in Pakistan. Three species, the Eurasian Crane *Grus grus*, the Demoiselle Crane *Anthropoides virgo* and the Siberian Crane *Grus leucogeranus*, are migratory stopping over in areas such as Lakki and Bannu (Malik, pers. comm.; Ferguson, pers. comm.). The Sarus Crane *Grus antigone* used to breed in Pakistan. There were recent sightings of Sarus Cranes (with young) in Sindh (Ahmad, pers. comm.), but it was not confirmed that these cranes nested in Pakistan; it was believed that the Sarus Cranes that were observed perhaps bred in India not far from the Indo-Pakistan border.

The western population of Siberian Cranes wintering in India after declining steadily over the last 30 years (Sauey, 1985), reached an all-time low of 5 birds in 1992-93, and did not return to winter at Keoladeo National Park (KNP) in India for 2 years thereafter. During the winter of 1995-96, 3 adults and one juvenile Siberian Crane arrived at KNP. The juvenile was banded in the Kunovat River Basin in 1995. This group is on the verge of extinction. The population wintering in Iran in 1995-96 had 9 birds, their numbers here remaining relatively unchanged for the last decade. The larger eastern population winters in China on mudflats beside a lake, Poyang Lake, that is threatened by the construction of a dam across the Yangtze River, and oil exploration nearby.

Hunting along their long migration corridors, and loss of wetlands along their migration routes and on their wintering grounds, have resulted in Siberian Cranes becoming one of earth's most endangered species of birds. Furthermore, the breeding biology of the species has not helped the species over- proliferate either. Cranes are long lived and reach sexual maturity around 4-7 years of age. Cranes typically lay 2 eggs and frequently are only able to raise one chick. Siberian Crane chicks display considerably more sibling rivalry/aggression than other crane species' chicks. It appears (Nagendran 1995), and this aggression frequently results in the death of one chick (usually the younger and weaker chick) in the wild. To add to all this the Siberian Crane perhaps has the longest migration route, and migration can be quite hazardous to especially juveniles. All of a sudden, indiscriminate hunting will result in extinction. Crane hunting in Pakistan is an age old tradition (Landfried, pers. comm.; Malik, pers. comm.), and the hunters have acknowledged that the numbers of cranes migrating through, i.e., Demoiselle and Eurasian Cranes, appear to be reduced considerably. In order for this activity to be sustainable, it is essential to have season limits and bag limits. It is unclear at this time what the level of impact is due to hunting in the former Soviet Union.

Wetland habitat is also at a premium, both in terms of quantity and quality. This is a direct consequence of an exploding human population. In the Indian Subcontinent (as in other parts of the world) any land not under the plow or developed otherwise, is considered to be wasteland. Wetlands are constantly being drained for development.

Most crane species are on top of the wetland food chain, and Siberian Cranes in particular are an obligate wetland species. Wetland species are good indicators of the health of the land, as wetlands are good sinks of agricultural chemicals (such as herbicides and pesticides). It would be naive to surmise that all of this has no bearing on the health and future of humans. While there is regulation of pesticide/herbicide usage in the western hemisphere, there appears to be no such regulation in the Indian Sub-continent (and perhaps several other parts of the world). Several dead Sarus Cranes were examined for pesticides/herbicides in a study conducted in India in the 1980's and high levels of residues were found in the brain tissues of these dead birds (Vijayan 1992).

It appears that the primary cause for the disappearance of the Sarus Crane from Pakistan was habitat loss. In time this species may return to nest in Pakistan.

CONSERVATION EFFORTS

Breeding grounds to wintering areas:

A group of breeding Siberian Cranes was discovered in the Kunovat River Basin near the Ob river in the early 1980's and efforts to assist this group/population began shortly thereafter. An attempt was made at cross-fostering by putting a viable Siberian Crane egg into a Eurasian Crane nest in Kunovat. While this method of rearing has been shown to produce cranes with sexual imprinting problems with cross-fostered (under Sandhill Cranes *G. canadensis*) Whooping Crane *G. americana* chicks at Grays Lake in North America, this is being employed on a small scale in western Siberia under the justification that the young cross-fostered Siberian Crane chicks are still being "exposed" to normal wild Siberian Cranes. Therefore these cross-fostered Siberian Cranes, it is believed, "will not" experience a sexual imprinting problem. This justification perhaps lacks validity because it appears that filial and sexual imprinting may be on a continuum. Notwithstanding, cross-fostering Siberian Cranes in western Siberia might be worthwhile to the degree that once the current wild population is extirpated, there may be at least other "white" cranes available in the flyway, that could perhaps be used as guide birds for released cranes.

One egg, from nests with 2 eggs, was removed to add to the captive species bank (primarily from the eastern population, but there are a few individuals in captivity that are from the western population as well), during the 1980's. A nature reserve was also established to ensure the safety of the western breeding population. Established captive crane breeding centers, particularly for Siberian Cranes, are in North America (the International crane Foundation), Germany (Vogelpark-Walsrode), and Russia (Oka State Nature Reserve Crane Breeding Center).

Beginning in 1990, with the satellite tracking of Eurasian Cranes by Ellis and Markin (1991), and continued since 1991, Eurasian Cranes and Siberian Cranes have been raised in captivity using costume-rearing techniques (Horwich 1989, Archibald and Archibald 1992, Horwich et al. 1992, Urbanek and Bookhout 1992), native peoples are subsistence hunters, fishermen, and caribou herders, and pose no significant threat to the endangered Siberian Cranes and other wildlife species that occur in Siberia. What might happen in the current epidemic of pursuit of capitalism in collaboration with western oil companies and other industry remains to be seen; habitat degradation is always the first concern. One can only hope that better sense will prevail.

Once the Siberian Cranes leave their refugia in northwestern Siberia they are more likely to encounter decimating factors and stresses in the way of human population, advanced agricultural practices resulting in loss of wetland habitat, possible hunting and other disturbances. While these appear logical our knowledge of the migration route of Siberian Cranes, their stopover areas and the obstacles they encounter is limited.

On the crane scale of time wetlands were more prominent on the Gangetic Plain. This grain belt is heavily populated now and has seen the disappearance of most wetlands. The wetland ecosystem of KNP is the only authentically recorded winter resort of the Siberian Crane in India (Ali and Vijayan 1986). It is a 29 sq. km (one of the smallest national parks in India) relatively flat basin, lying in the Gangetic Plain of northern India, 2 km southeast of Bharatpur city.

Originally a natural depression, KNP was developed into a (BNHS) with support from the USFWS-Office of International Affairs conducted research on the winter ecology of Siberian Cranes. All the efforts by individuals and agencies resulted in establishing Keoladeo National Park, a park that is home to approximately 370 species the most famous of them being the wintering Siberian Crane which may now be extinct.

In 1992-93 and 1993-94 costume-reared and parent-reared Siberian Cranes were released at KNP with the hope that they would migrate with wild Siberian Cranes (1992-93) or wild Eurasian Cranes (1992-93 and 1993-94). While wintering ground releases may not result in migration (Nagendran 1995), this effort in India was a last ditch effort. The 1992-93 effort had the added element of symbolism--it was the first project undertaken by the Russians and Indians under the Migratory Bird Treaty signed between the 2 nations at perhaps the highest ministerial levels in the 1980's. The releases accomplished in 1993 and 1994 also indicated that the more time these cranes spent in captivity, the less inclined they were to display any migratory behavior.

Since the wintering population of Siberian Cranes in India was on the verge of extinction, more needed to be learned about the Eurasian Cranes that wintered in and around KNP in order to re-establish a wintering population of Siberian Cranes at KNP. Using Eurasian Cranes as "guide" birds for released Siberian Cranes would be the only option available in the near future.

In 1993 and 1994 .wild Eurasian Cranes were captured at KNP and fitted with satellite transmitters (PTTs). Two of the migrating Eurasian Cranes provided exciting information in 1993. For the first time the exact migration route, critical stopover areas, and breeding grounds of some cranes wintering in India were discovered (Higuchi et al. 1994). The 2 Eurasian Cranes migrated from KNP to their breeding grounds in southern Siberia, a little east of Omsk, not far from the headwaters of the Ob river.

Once the Siberian Cranes' migration route and critical stopover areas were confirmed, conservation programs could be initiated to lessen the impact of hunting and habitat loss. The same holds true for the more abundant Eurasian Cranes, that might end up as guide birds for the released, young Siberian Cranes. Unfortunately, we were denied permission to deploy PTTs on Siberian Cranes, and lost a rare opportunity to document with a level of certainty their migration route, critical stopover areas and their summering grounds. This opportunity was offered us in 1995-96 but at the very last minute, just a few days prior to the migration of the small group of 4. This make it very difficult to capture the only unpaired adult in the group; the group had got back together, from the time permission to deploy a satellite transmitter was granted until they left in migration. Once again another crucial opportunity was lost.

The research efforts in 1993 and 1994 were a collaborative effort among several agencies, governmental and non-governmental, and numerous individuals including international scientists, conservationists and local people at KNP.

A survey of Demoiselle Crane wintering sites conducted in Gujarat in November 1995 (Nagendran, unpubl. data) revealed the degree and nature of stresses faced by these long distance migrants on their wintering sites, in addition to the stresses encountered during migration. Several thousand cranes (Demoiselle and Eurasian Cranes) were dependent on very small wetlands which were scattered, and heavily used by humans and domestic livestock. Furthermore, almost all natural foraging sites had been converted to agricultural fields or developed otherwise. This has also resulted in cranes adapting and becoming dependent on planted cereal grains and other crops (such as peanuts) for forage.

The capture, marking and satellite tracking of Eurasian and Demoiselle Cranes should continue for several more years to have a better understanding of the migratory routes, stopover areas, and breeding grounds of these migrant cranes. Cranes should be marked on as many wintering areas as possible in India. Besides satellite telemetry, radio transmitters should also be fitted each winter on a few cranes to study their localized movements. Without the knowledge that we acquire from radio and satellite telemetry it will only become more difficult to implement any conservation measures, whether it is habitat conservation or species conservation. Satellite telemetry has progressed along very well, both in terms of technology and knowledge of the cranes instrumented, since the first time that a satellite transmitter was used on a crane (Nagendran 1989). The Wild Bird Society of Japan has conducted extensive research in this area. Many important areas used by cranes during migration continue to be established into nature reserves, the knowledge of these areas coming to light

CONCLUSION

Cranes that visit Pakistan are very long distance migrants, stopping in Pakistan for a much needed sojourn, rest and refurbishment. Most crane species are heavily dependent upon wetlands during many stages of their life, during the summer (e.g., breeding), at the time of migration and on their wintering areas. Many crane species are obligate wetland species. Furthermore, many crane species are on top of the wetland food chain.

Maintaining crane habitats in essence maintains wetland habitats for several other wetland species. Cranes require fairly large hectares of habitat. Cranes, as a group of birds, have existed on our planet for more than 38 million years.

These cranes that know no political boundaries, will hopefully bring more nations together in efforts to conserve cranes and their habitats. Satellite telemetry will continue to provide a wealth of knowledge on migration routes, stopover areas and breeding grounds. This telemetry information combined with Landsat analysis of important rest sites will provide information that is critical for habitat conservation and to manage crane hunting. The irreversible decimation of cranes from any of their current ranges would be great cause for concern, and a further serious threat to biodiversity.

REFERENCES

ALI, S., AND VIJAYAN, V.S., 1986. Keoladeo National Park Ecology Study Summary Report 1980-1985. *Bombay Natural History Society Report* 180p.

ARCHIBALD, K., AND ARCHIBALD, G. W., 1992. Releasing puppet-reared Sandhill Cranes into the wild: a progress report, p. 251-254. In D. A. Wood [ed.], *Proceedings 1988 Crane Workshop.*

ELLIS, D. H., AND MARKIN, Y., 1991. Satellite monitors cranes migrating from Siberia. *ICF Bugl.,* **17**: 4-5.

HIGUCHI, H., OZAKI, K., FUJITA, G., SOMA, M., KANMURI, N., AND UETA, M., 1992. Satellite tracking of the migration routes of cranes from southern Japan. *Strix* **11**: 1-20.

HIGUCHI, H., NAGENDRAN, M., AND SOROKIN, A. G., 1994. Satellite tracking of Common Cranes *Grus grus* migrating north from Keoladeo National Park, India, p. 26-31. In H. Higuchi and J. Minton [eds.], *Proceedings of the International Symposium and Workshop on "The Future of Cranes and Wetlands", Tokyo, Japan*, 1993.

HORWICH, R. H. 1989. Use of surrogate parental models and age periods in a successful release of hand-reared Sandhill Cranes. *Zoo. Biology* **8**:379-390.

HORWICH, R. H., WOOD, J., AND ANDERSON, R., 1992. Successful release of Sandhill Crane chicks, hand-reared with artificial stimuli, 255-262. In D. A. Wood [ed.], *Proceedings 1988 North American Crane Workshop. Kissimmee, FL.*

NAGENDRAN, M. 1989. Satellite tracking of a Greater Sandhill Crane. *The Unison Call. Summer.*

NAGENDRAN, M. 1992. Winter release of isolation-reared Greater Sandhill Cranes in south Texas. *Proceedings North American Crane Workshop,* **6**:131-134.

NAGENDRAN, M. 1995. *Behavioral ontogeny and release of costume-reared Siberian and Sandhill Crane chicks.* Ph.D. dissertation, North Dakota State University, Fargo, ND.

NAGENDRAN, M., HIGUCHI, H., AND SOROKIN, A. G., 1994. A harnessing technique to deploy transmitters on cranes, p. 57-60. In H. Higuchi and J. Minton [eds.], *Proceedings of the*

International Symposium and Workshop on "The Future of Cranes and Wetlands". Tokyo, Japan. *1993.*

SAUEY, R. T., 1985. *The range, status, and winter ecology of the Siberian Crane (Grus leucogeranus).* Ph.D. dissertation, Cornell University, Ithaca, NY.

VIJAYAN, V. S., 1992. Keoladeo National Park Ecology Study 1980-1990. Executive Report. Bombay Natural History Society. 253p.

URBANEK, R. P., 1990a. Reintroduction studies: a summer release. *ICF Bugle* **16**: 4-5.

URBANEK, R. P., 1990b. Behavior and survival of captive-reared juvenile Sandhill Cranes introduced by gentle release into a migratory flock of Sandhill Cranes. *Final Report., Ohio Cooperative Fish and Wildlife Research Unit, Columbus. 98p.*

URBANEK, R. P. AND BOOKHOUT, T. A., 1992. Development of an isolation-rearing/gentle release procedure for reintroducing migratory cranes. *Proceedings North American Crane Workshop,* **6**:120-130.

AVIFAUNA OF THE MANGROVES OF BALOCHISTAN COAST

S. ALI GHALIB AND SYED HASNAIN

Zoological Survey Department, Pakistan Secretariat, Karachi

Abstract: The Mangrove sites on Balochistan Coast were surveyed during 1992-95 to study the distribution and status of the Avifaua. During this period, 82 species of birds belonging to 7 orders and 22 families were recorded from the area.

INTRODUCTION

On the coast of Balochistan in Pakistan, the mangrove swamps are located at Miani Hor near Somiani/Dam in Lasbella district and Kalmat/Chundi near Pasni and Jiwani Hor/Gwader Bay near Jiwani in Gwadar district. They provide ideal habitat for many of the bird species for their food, shelter and breeding. Many birds visit this area for feeding and nesting. They feed upon fishes, small snakes, molluscs, crustaceans and insects etc., which are abundant in the mangroves and the exposed mudflat located nearby.

Karim (1988) recorded 48 species of birds belonging to 15 families and 6 orders from the mangrove swamps of Sindh. Groombridge (1989) in his report on Marine Turtles of Balochistan has mentioned briefly about the bird fauna of that coast. Ahmed et. al. (1992) recorded 91 species of aquatic birds from Makran Coast but they did not specifically mention the mangrove birds.

MATERIALS AND METHODS

The surveys were generally undertaken at low tide at the edge of the mangrove swamps and the intertidal mudflats using spottingscope. The mangrove vegetation itself was also searched for bird nests. When mangroves were inaccessible by foot, surveys were made by using a motorboat. The birds were identified and counted from the boat by getting close to the swamp edge and by looking through binoculars at the field marks. Identification of the birds was made by using the publication of Heinzel *et al.* (1972).

RESULTS AND DISCUSSION

During our brief surveys of the mangroves of Balochistan coast, we have recorded 82 species of birds belonging to 7 orders and 22 families. The present paper provides a checklist of the birds seen with information about their status (Table I).

Table I: Bird species of Mangrove forests of Balochistan (WV-winter visitors, R-resident, OS-over summering, LM-local migrant)

Family / Species	Common Name	Satus
Pelecanidae		
Pelecanus onocrotalus	White Pelican	WV
Pelecanus philippensis	Dalmatian Pelican	Rare, WV
Phalacrocoracidae		
Phalacrocorax carbo sinensis	Large Cormorant	WV
Phalacrocorax fuscicollis	Indian Shag	R
Halietor niger	Little Cormorant	R
Ardeidae		
Ardea cinerea	Eastern Grey Heron	R/WV
Ardea purpurea manilensis	Eastern Purple Heron	R
Butorides striatus chloriceps	Little Green Heron	Rare, R
Ardeola grayii grayii	Indian Pond Heron	R
Egretta alba alba	Great Egret	WV
Egretta intermedia intermedia	Intermediate Egret	R
Egretta garzetta garzetta	Little Egret	R
Egretta gularis asha	Indian Reef Heron	R
Ciconiidae		
Mycteria leucocephala	Painted Stork	LM
Ephippiorhynchus asiaticus	Black necked Stork	WV
Threskiornithidae		
Platalea leucorodia major	White Spoonbill	LM
Phoenicopteridae		
Phoenicopte ruber roseus	Greater Flamingo	LM
Phoeniconaias minor	Lessor Flamingo	LM
Pandionidae		
Pandion haliaetus haliaetus	Osprey	WV
Accipitridae		
Elanus caeruleus caeruleus	Black-shouldered Kite	R
Pernis ptilorhynchus ruficollis	Oriental Honey Buzzard	WV
Milvus migrans migrans	Black Kite	R
Haliastur indus indus	Brahminy Kite	R
Buteo buteo vulpinus	Common Buzzard	WV

Aquila rapax nipalensis	Eastern Steppe Eagle	MV
Aquila heliasa	Imperial Eagle	WV
Neophron percnopterus percnopterus	Egyptian Vulture	R
Circus aeruginosus aeruginosus	Marsh Harrier	WV
Gruidae		
Athropoides virgo	Demoiselle Crane	WV
Haematopodidae		
Haematopus ostralegus ostralegus	Oyster catcher	WV/OS
Burhinidae		
Esacus recurvirostis	Great Stone Plover	Rare, R
Charadriidae		
Vanellus indicus indicus	Redwattled Lapwing	R
Pluvialis squatarola	Grey Plover	WV
Charadrius leschenaultii	Great Sand Plover	WV
Charadrius hiaticula tundrae	Eastern Ringed Plover	WV
Charadrius dubius curonicus	European Little Ringed Plover	WV
Charadrius dubius jerdoni	Indian Little Ringed Plover	R
Charadrius alexandrinus alexandrinus	Kentish Plover	R
Charadrius mongolus atrifrons	Mongolian Plover	WV/OS
Scolopacidae		
Numenius phaeopus phaeopus	Whimbrel	WV/OS
Numenius arquata orientalis	Eastern Curlew	WV/OS
Limosa limosa	Black tailed Godwit	WV
Limosa lapponica lapponica	Bartailed Godwit	WV
Tringa totanus eurhinus	Redshank	WV/OS
Tringa stagnatilis	Marsh Sandpiper	WV
Tringa nebularia	Greenshank	WV
Xenus cinereus	Terek Sandpiper	WV/OS
Actitis hypoleucos	Common Sandpiper	WV
Calidris alba	Sanderling	WV
Calidris minuta	Little Stint	WV/OS
Calidris alpina alpina	Dunlin	WV/OS
Calidris ferruginea	Curlew Sandpiper	WV
Limicola falcinellus falcinellus	Broadbilled Sandpiper	WV
Arenaria interpres interpres	Trunstone	WV
Recurvirostridae		
Himantopus himantopus himantopus	Blackwinged Stilt	R
Recurvirostris avosetta	Avocet	LM
Larus argentatus heuglini	Yellow-legged Herring Gull	WV
Larus fuascus fuascus	Lesser Blackbacked Gull	WV

Larus ichthyaetus	Great Blackheaded Gull	WV
Larus brunnicephalus	Indian Brown-headed Gull	WV
Larus ridibundus	Black-headed Gull	WV/OS
Larus genei	Slonder-billed Gull	WV/OS
Gelochelidon nilotica nilotica	Gull-billed Tern	WV/OS
Hydroprogne caspia caspia	Caspian Tern	WV
Sterna hirundo hirundo	European Common Tern	WV
Sterna albifrons albifrons	Little Tern	R
Thalasseus bergii velox	Large Crested Tern	WV
Thalasseus bengalensis bengalensis	Indian Lesser Crested tern	WV
Thalasseus sandvicensis sandvicensis	Sandwich Tern	WV
Alcedinidae		
Ceryle rudis	Indian Pied Kingfisher	R
Halcyon smyrnensis	White breasted Kingfisher	R
Alaudidae		
Eremopterix grisea	Ashy-crowned Finch Lark	R
Alaemon alaudipes doriae	South-Asian Hoopoe Lark	R
Galerida cristata	Crested Lark	R
Hirundinidae		
Riparia paludicola chinensis	African Sand Martin	R
Hirundo rustica rustica	Eurasian Barn Swallow	WV
Motacillidae		
Anthus campestris campestris	Eurasian Tawny Pipit	WV
Muscicapidae		
Saxicoloides fulicata cambiensis	Black-backed Robin	R
Oenanthe deserti	Desert Wheater	WR
Oenanthe monacha	Hooded Wheatear	R
Emberizidae		
Emberiza melanocephala	Black-headed Bunting	R
Corvidae		
Corvus ruficollis ruficollis	Brown necked Raven	R

The evergreen forests of mangroves provided ideal habitat for the avifauna. It was found that the winter visitors were numerous while the residents were few. The charadriids (plovers) and the scolopacids (sandpipers, stints, curlew and godwits etc) visited the area in large numbers in winter, some staying in summer also. Waders like oystercatcher, red-shank, curlew, whimbrel, dunlin, little stint, mongolian plover, greater sand plover and little ringed plover were mainly winter visitors and a small population of these birds were observed in summers in the mangrove areas. Grey plover was also observed in summer at Jiwani while terek sandpiper was recorded in summer from Jiwani and Miani Hor. The cormorants and ardeids (egrets and herons) roosted at night on the mangroves and in day time searched for food. The birds of prey were seen perching on mangrove trees and resting on the exposed mudflats, usually in the middle of day and also searching for mudskippers and crabs. Flamingos were occasionally seen resting and searching for food. The larids (gulls and terns) were common in the area; black headed gull was seen in summer too. The demoiselle cranes were observed in the adjacent exposed areas of mangroves at Kalmat in January of 1994 and 1995.

Among the particular localities, Miani Hor and Jiwani Hor were found to be the important sites in respect of bird populations. Jiwani Hor was a less disturbed area and level of exploitation of either the mangroves or of the bird fauna was not very high. However, the shorebirds were trapped by locals using nets at Chundi and Jiwani Hor and used them as food. The extent of bird exploitation in Balochistan is not alarming at present because of fewer human settlements and because of its being an isolated area.

Miani Hor is comparatively more disturbed and hunting of birds, cutting of mangroves and other disturbances were observed in the area, still the population of birds in the area was considerable.

ACKNOWLEDGEMENT

The authors are deeply grateful to the Pakistan Agricultural Research Council for financing the project viz. "Vertebrate Fauna of Mangrove Swamps of Balochistan Coast" under which this work was undertaken.

REFERENCES

AHMED, M.F., GHALIB, S.A. & HASNAIN, S.A. 1992. The Waterfowl of Makran Coast. *Proceedings of National Conference on Problems and Resources of Makran Coast and Plan of Action for its development.* 113-123. Pakistan Council for Science and Technology, Islamabad.

GROOMBRIDGE, B. 1989. Marine Turtles in Balochistan: Report on an aerial survey 9-11 Sept. 1988 with notes on wetland sites and a proposed marine turtle conservation project. World Conservation Monitoring Centre, Cambridge.

HEINZEL, H., FITTER, R. & PARSLOW, J. 1972. *The Birds of Britain and Europe with North Africa and the Middle East.* Collins, London.

KARIM, S.I. 1988. Avifauna of Sindh Mangroves. *Proceedings of the International Conference - Marine Science of the Arabian Sea.* American Institute of Biological Sciences, Washington, D.C.

BIODIVERSITY OF AVIFAUNA OF LAL SUHANRA NATIONAL PARK, BAHAWALPUR, PAKISTAN

WASEEM A. KHAN

Federal Directorate of Education, Islamabad.

Abstract: An extensive survey of Avifauna of Lal Suhanra National Park has been conducted with special reference to population density of waterfowls.Two major categories of birds i.e Migratory & Resident, have been recognized. A decline in the density of migratory birds especially waterfowls, was noticed when compared with the data obtained in the past.

INTRODUCTION

Lal Suhanra National Park is located in the southern region of Punjab, within the administration boundaries of Bahawalpur Division (Punjab-Pakistan). It lies between 29^0-23' Northern latitude & 71^0-39' Eastern longitude on the world map, approximately 350-375 ft. above sea level. The total ground area of the Park is 127480 acres. Out of 127480 acres area of the Park, 20974 acres are forest plantation, 4780 acres are lake area and the rest 101726 acres are desert.

While the preservation of game animals for hunting has been the main trend during the past ages, the idea of large scale conservation of wildlife for aesthetic, scientific, bioecological and economic purposes is relatively new. The movement for the protection of nature had its beginning barely 60 years ago. The spread of agriculture and industry that threatened the destruction of indigenous fauna, was countered by establishing National Parks. Game reserves and Sanctuaries, which not only provided protection to wild animals, but also offered the people an added attraction because of their scenic beauty and their historical, geographical or archaeological interest.

In Pakistan there are a few reports on the Avifauna of different areas. Baker (1930), Ali (1945), Ripley (1961) and Roberts (1991, 1992) have given important information about the avifauna of Pakistan. Lal Suhanra National Park is one of the few areas in Pakistan which has desert, wetland and forest plantation. The varied habitat supports a great variety of fauna and flora. A diverse fauna from the three habitats could be anticipated but not much scientific data is available about this important Park of Punjab. The present study was conducted to find out the latest status of the Avifauna of the Park and to compare presently observed density of migratory birds with similar data obtained in the recent past.

MATERIALS AND METHODS

The study of the Avifauna of the Park was based upon the visual observation with the help of binoculars and spotting scope. For the description of external features and identification of various bird species, Baker (1923), Ali and Ripley (1961) and Khanum et. al. (1980) were followed. Status of the species was indicated by R for Resident bird, M/Wv for Migratory Winter visitor birds, V for Vagrant and sb for summer breeders.

RESULTS

The birds form the largest segment of the fauna of Lal Suhanra National Park (LSNP). All the birds observed in the LSNP belong to 16 orders, 43 families, 81 genera and 117 species. The density of resident birds was high as compared to migratory birds. Most of the migratory birds were observed during winter season. The birds observed in the in the Park during different seasons are listed below:

ORDER PODICIPEDIFORMES	STATUS
Family Podicipedidae	
Tachybaptus ruficollis (Little grebe or Dabchick)	R
ORDER PELECANIFORMES	
Family Phalacrocoridae	
Phalacrocorax niger (Little or Javanese cormorant)	R
P. carbo sinensis (Common or Large cormorant)	R
P. fuscicollis (Indian shag)	R
Family Anhingidae	
Anhinga melanogaster (Indian darter or Snake bird	R
ORDER CICONIIFORMES	
Family Ardeidae	
Ardea cinerea cinerea (European grey heron)	R
A. purpurea menilensis (Eastern purple heron)	R
Ardeola grayii grayii (Indian pond heron)	R
Bubulcus ibis coromandus (Cattle egret)	R
Egretta alba alba (Great white heron)	R
E. garzetta garzetta (Little egret)	R
E. intermedia intermedia (Median egret)	R
Nycticorax nycticorax nycticorax (Night heron)	R

ORDER ANSERIFORMES

Family Anatidae

Anas acuta acuta (Pintail)	M/Wv
A. crecca crecca (Green winged teal)	M/Wv
A. platyrhynchos (Mallard)	M/Wv
A. strepera strepera (Gadwall)	M/Wv
A. penelope (European wigeon)	M/Wv
A. clypeata (Common Shoveller)	M/Wv
Aythya ferina (European Pochard)	M/Wv
A. fuligula (Tufted Duck)	M/Wv
Netta rufina (Red crested Pochard)	M/Wv

ORDER ACCIPITRIFORMES

Family Pandionidae

Pandion haliaetus haliaetus (Osprey)	M/Wv

Family Accipitridae

Elanus caeruleus caeruleus (Black-shouldered kite)	R
Neophron percnopterus gingincanus (Egyptian vulture)	R
Gyps bengalensis (Indian White backed vulture)	R
G. idicus indicus (Indian griffon)	R
G. fulvus fulvescens (Griffon Vulture)	R
Circaetus gallicus gallicus (Short-toed eagle)	R
Circus aeruginosus aeruginosus (Marsh harrier)	M/Wv
C. macrourus (Pallid harrier)	M/Wv
Accipiter nisus meloschistoes (Indian Sparrow hawk	M/Wv
A. badius dussuminck (Indian Shikra)	R
Butastur teesa (White-eyed buzzard eagle)	R
Aquila clanga (Greater spotted eagle)	R
A. rapax vindhiana (Tawny eagle)	R
Hieraaetus fasciatus fasciatus (Bonelli's eagle)	M/Wv

ORDER FALCONIFORMES

Family Falconidae

Falco tinnunculus tinnunculus (Comm. Eurp. Kestrel)	R
F. biarmicus (Lanner falcon)	M/Wv
F. cherrug cherrug (Saker falcon)	M/Wv
F. peregrinus peregrinator (Peregrine shaheen falcon)	M/Wv

ORDER GALLIFORMES

Family Phasianidae

Francolinus francolinus henria (Pak.black partridge)	R
Francolinus pondicerianus pondicerianus (Indian grey partridge)	R
Coturnix coturnix coturnix (Common quail)	R

ORDER GRUIFORMES

Family Rallidae

Porphyrio porphyrio seistanicus (Purple swamphen)	M/Wv

Fulica atra atra (Common Coot)	M/Wv
Family Otididae	
Chlamydotis undulata macqueenii (Houbara bustard)	R
Family Gruidae	
Grus antigone antigone (Indian sarus crane)	M/Wv
ORDER CHARADRIIFORMES	
Family Jacanidae	
Hydrophsianus chirurgus (Pheasant-tailed jacana)	R
Metopidius indicus (Bronze winged jacana)	R
Family Recurvirostridae	
Himantopus himantopus himantopus (Black winged stilt)	M/Wv
Family Charadridae	
Hoplopterus indicus indicus (Red wattled lapwing)	R
Charadrius dubius jerdoni (Indian little ringed plover)	R
Family Scolopacidae	
Tringa totanus totanus (Common red shank)	M/Wv
T. nebularia (Green shank)	M/Wv
T. ochropus (Green sand piper)	M/Wv
T. glareola (Wood or spotted sand piper)	M/Wv
T. hypoleucos (Common sand piper)	R
Gallinago gallinago gallinago (Common snipe)	M/Wv
Family Laridae	
Sterna aurantia (Indian river tern)	R
ORDER PTEROCLIDIFORMES	
Family Pteroclididae	
Pteroclesexustus hindustan (Indian chestnut bellied sandgrouse)	R
ORDER COLUMBIFORMES	
Family Columbidae	
Columba livia neglecta (Blue rock pigeon)	R
Streptopelia decaocta decaocta (Indian collard dove)	R
S. tranquebarica tranquebarica (Indian, red-collard dove)	R
S. senegalensis cambayensis (Indian laughing dove)	R
ORDER APODIFORMES	
Family Apodidae	
Apus affinis galilejensis (Southern house swift)	R
ORDER PSITTACIFORMES	
Faily psittacidae	
Psittacula eupatria nipalensis (Large Indian alexandrine parakeet)	M/Wv
P. krameri borealis (Northern rose-ringed parakeet)	R

ORDER CUCULIFORMES

Family Cuculidae

Eudynamys scolopacea scolopacea (Indian Koel)	R
Centropus sinensis sinensis (Indochinese common crowpheasant)	R

ORDER STRIGIFORMES

Family Strigidae

Bubo bubo bengalensis (Indian eagle owl)	R
Athene brama indica (Indian spotted little owl)	R

ORDER CORACIIFORMES

Family Alcedinidae

Ceryle rudis lecuomelanura (Indian lesser pied kingfisher)	R
Alcedo atthis bengalensis (Indian common blue kingfisher)	R
Halcyon symrnensis symrnensis (Southern white breasted Kingfisher)	R

Family Meropidae

Herops apiaster (European bee-eater)	R
Merops orientalis beludschicus (Si ndh green bee-eater)	R

ORDER CORACIFORMES

Family Coraciidae

Coracias benghalensis (Indo-arabian roller)	R

Family Upupidae

Upupa epops epops (Hooproe)	R

ORDER PICIFORMES

Family Picidae

Dinopium benghalensis dilutum (Balochistan lesser golden backed woodpecker)	R
Picoides mahrattensis mahrattensis (Northern yellow crowned woodpecker)	R

Family Capitonidae

Megalaima haemacaphala indica (Southern crimson breasted barbet)	R

ORDER PASSERIFORMES

Family Alaudidae

Mirafa erythroptera sindiana (Sindh red-winged bushlark)	R
Galerida cristata chandoola (Indian crested lark)	R
Alauda gulgula inconspicua (Afqhan-oriental skylark)	M/Wv

Family Hirundinidae

Hirundo rustica rustica (European barn swallow)	M/Wv
H. smithii filifera (South Asian wiretailed swallow)	M/Sb

Family laniidae

Species	Status
Lanius excubitor lathora (Indian great grey shrike)	R
L.schah erythronotus (Indian black headed shrike)	R
L. vittatus vittatus (Indian baybacked shrike)	R

Family Dicruridae

Species	Status
Dicrurus macrocercus albirictus (North Indian black drongo)	R

Family Sturnidae

Species	Status
Acridotheres tristis tristis (Indian common mynah)	R
A. ginginianus (Bank mynah)	R

Family Corvidae

Species	Status
Corvus spelendens zugmayeri (Sindh house crow)	R
Dendrocitta vagabunda pallida (North Indian tree pie)	R

Family Paridae

Species	Status
Parus major ziaratensis (Balochistan grey tit)	R

Family Pycnonotidae

Species	Status
Pycnonotus cafer intermedius (Punjab redvented bulbul)	R
P. leucogenys leucogenys (Himalayan white-cheeked bulbul)	R

Family Muscicapidae

Species	Status
Turdoides caudatus caudatus (Indian common babbler)	R
T. striatus sindianus (Sindh white-headed jungle babbler)	R
Prinia subflava terricolor (Northwesteren tawny-flanked prinia)	R
Copsychus saularis saularis (Indian magpie robin)	R
Saxicola caprata bicolor (Northern pied bushchat)	M/sb

Family Motacillidae

Species	Status
Anthus nokaereelandiae richaridi (Pakistan Richard's pipit)	M/Wv
Motacilla flava melanogrisea (Black-headed yellow wagtail)	M/Wv
M. f. beema (Blue headed yellow wagtail)	M/Wv
M. alba elukhunensis (Indian pied wagtail)	M/Wv
M. maderaspatensis (Large pied wagtail)	M/Wv

Family Estrildidae

Species	Status
Amandava amandava amandava (Red munia)	R
Lonchura melabarica (Indian silver bill)	R

Family Ploceidae

Species	Status
Passer domesticus indicus (Indian House sparrow)	R
Ploceus philippinus philippinus (Indian baya weaver)	R
P. manyar flaviceps (Indian streated weaver)	R

Family Emberizidae

Species	Status
Emberiza bruniceps (Red headed bunting)	M/Wv
E. melanocephala (Black headed bunting)	M/Wv
Tephrodornis pondicerianus pallidus (Sindh wood shrike)	R

ORDER CAPRIMULGIFORMES

Family Caprimulgidae

Caprimulgus asiaticus asiaticus (Indian night jar) R

DISCUSSION

The study of different species of birds showed that the most common species of the Park were Indian common mynah. (*Acridotheres tristis tristis*), Indian house sparrow. (*Passer domesticus indicus*), Indian collered dove. (*Streptopelia decaocta decaocta*) Pakistan black partridge, (*Francolinus francolinus henria*), Indian grey partridge, (*Francolinus pondicerianus pondicerianus*), Little egret, (*Egretta garzetta garzetta*), Black kite, (*Milvus migrans migrans*), Northern rose-ringed parakeet, (*Psittacula krameri borealis*), Sindh house crow, (*Corvus spelendens zugmayeri*) and Indian red collered dove, (*Streptopelia tranquebatica tranquebarica*). Observations also showed that nearly 32 species of birds actually breed in the Park. Approximately 40 species of birds were migratory and seen in the Park only during the winter months.The most abundant family of the birds was Accipitridae whose 16 species were observed while 15 families of birds namely Podicipitidae, Anhingidae, Otididae, Gruidae, Recurvirostridae, Laridae, Pteroclididae, Apodidae, Coraciidae, Upupidae, Capitonidae, Dicruridae, Paridae, Campephagridae and Caprimulgidae are represented only by one species.

In the Park 79 species were resident and seen throughout the year. A continuous decline in the migratory birds was recorded, both in population and species since the last 8 years. In 1990, 934 birds of 40 species belonging to 15 families were recorded. In 1992 ,773 birds of 30 species belonging to 12 families were observed. In 1994, 538 birds of 31 species belonging to 12 families were recorded. This decrease in number was due to many reasons i.e. hunting in the route to the Park, fluctuation in the quantity of water in the lake, fishing in the lake, deforestation for new plants and human interference in the landing and roosting sites.

REFERENCES

ALI, S., 1945. The birds of Bahawalpur, Punjab. *JBNHS,* **43**: 703-747.

ALI, S. AND RIPLEY, S.D., 1968-1974. *Handbook of the birds of India and Pakistan.* 10 volumes, Oxford University Press, Bombay.

BAKER, E.C.S., 1922-1930. *Fauna of British India. Birds.* 8 vols. Taylor & Francis, London.

KHANUM, AHMED, Z. M. AND AHMED, M. F., 1980. A checklist of birds of Pakistan with illustrated keys to their identification. *Rec. Zool. Sur.*, **9**: 1-138

MOUNTFORT, G., 1969. *The Vanishing Jungle.* Collins, London, 1-286

RIPLEY, S.D., 1961. *A synopsis of the birds of India and Pakistan together with those of Nepal, Sikkim, Bhutan and Ceylon.* Bombay Natural History Society

ROBERTS, T.J., 1991. *The birds of Pakistan.* Vol. 1 Non-passeriformes. Oxford University Press, Karachi.

ROBERTS, T.J., 1992. *The birds of Pakistan.* Vol 2 Passeriformes. Oxford University Press, Karachi.

SIDDIQI, M.S., 1969. *Fauna of Pakistan*, Agricultural Research Council, Government of Pakistan, Karachi.

BIODIVERSITY OF SMALL MAMMALS IN THE MOUNTAINS OF PAKISTAN

(high or low?)

CHARLES A. WOODS & C. WILLIAM KILPATRICK*

Florida Museum of Natural History, Gainesville, FL , U.S.A.
**Department of Biology, University of Vermont, Burlington, VT U.S.A.*

Abstract: The biodiversity of small mammals in the mountains of Pakistan includes 56 species. In few locations, however, there are more than five or six species occurring together, giving the impression that most communities of small mammals in the mountains of Pakistan are depauperate. However, a close comparison with comparable mountainous regions of the world, especially the Rocky Mountain massif of North America, indicates that the small mammal communities of Pakistan are nearly comparable. The highest levels of biodiversity in Pakistan are along the western mountain ranges. Here several closely related species can occur sympatrically, and there is a rich mixture of Arabian, Hindu Kush, Palearctic and Indian Subcontinent species. The lowest levels of biodiversity are found in northeastern Pakistan, especially in the eastern Karakoram massif. This zone appears to be the end of a great intermountain cul-de-sac, and many species have not dispersed there.

INTRODUCTION

The diversity of small mammals in montane niches of Pakistan appears to be lower than in other comparable mountainous regions of the world, such as the high deserts and intermountain valleys of the American southwest. Habitats that are teaming with small mammals in New Mexico and Colorado are almost devoid of small mammals in northern Pakistan. Examples of these nearly depauperate habitats are the rolling sand hills in the upper Indus Valley near Skardu, and the sandy deserts in the Indus Valley near Chilas. How diverse are the small mammal communities of Pakistan when compared with other regions of the world? How much biodiversity of small mammals is there in the mountains of Pakistan? How has this biodiversity been established and maintained? And finally, at what taxonomic levels is the biodiversity highest?

Small mammals are defined in taxonomic terms rather than strictly on the basis of body size. Most species under consideration are less than 3,000 grams, and most are less than 200 grams. Taxonomically this survey includes all insectivores (in this case shrews), bats, lagomorphs (rabbits and pikas), rodents (including marmots, porcupines and giant flying squirrels which can exceed 3 kg in mass), and mustelid carnivores. The survey does not include the remaining carnivores.

Much of the information discussed in this analysis is based on a survey of the small mammals of Pakistan carried out as a joint project of the Florida Museum of Natural History (FLMNH) in collaboration with the Pakistan Museum of Natural History (PMNH) and the Zoological Survey Department of Pakistan (ZSD). This ongoing project began in 1991. As part of this survey investigators visited sites throughout northern and western Pakistan at least two times, and where possible at different times of years. Collections of small mammals were made using a variety of traps, and trapping techniques. Historically, some sites were also collected at earlier dates by investigators from the British Museum (Natural History) and the Bombay Natural History Survey as part of the original survey of mammals of the Indian Subcontinent. Specimens from these collections are at the British Museum in London, the Zoological Survey Department in Karachi, the Indian Museum in Calcutta and the Bombay Natural History Society. Between 1964 and 1966 many areas of Pakistan were surveyed for small mammals as part of a study of parasites and disease vectors by the University of Maryland (USA). Small mammals collected as part of this survey are mostly cataloged in the collections of the United States National Museum of Natural History (Smithsonian) in Washington. Collections of small mammals also exist at the Vertebrate Pest Control Centre in Karachi. A collection was made of the Swat Valley area by investigators from the Field Museum of Natural History and the Pakistan Museum of Natural History (specimens now at both institutions).

The analysis presented here draws on our own recent collections and surveys as well as specimens, field notes and published material gathered as part of the above surveys. It also makes use of published information in Siddiqi (1961), Mirza (1969), and Roberts, (1977). Many changes in nomenclature and the classification of small mammals have occurred in the past decade. We have reviewed as many of these as possible, and synthesized the classification followed here from Corbet and Hill (1992), Wilson and Reeder (1993) and Hoffmann (1996).

GEOGRAPHY

The geographical area under consideration in this survey includes all of the major mountain ranges of northern and western Pakistan. These ranges are the Murree Hills and adjacent areas in Azad Kashmir along the Indian frontier, as well as the mountains and valleys associated with the Chitral, Swat, Upper Indus, Kaghan and Neelum Valleys. Farther to the north it includes the Western Himalaya as far west as Nanga Parbat, the Deosai Mountains and Deosai Plateau, the Karakoram Mountains, and the Hindu Kush. In Balochistan it includes the Toba Kakar, Sulaiman, Central Brahui, and Central Makran Ranges. In geographic terms the area is located between 26 and 31 degrees latitude in Balochistan, and 34 and 37 degrees latitude in northern Pakistan. It is located between 62 and 77.5 degrees longitude. In elevation the area is mostly above 1,000 meters on the valley floors (up to 2,600 meters at Khaplu), while the upper ridges and nullahs surveyed include areas to the edge of the permanent snowfields at about 5,000 meters.

The mountains of northern Pakistan (excluding Balochistan) occupy an area of approximately 130,000 km^2. In comparative terms, it is an area equivalent to the western half of the State of Colorado in the region of the southern Rocky Mountains of the United States (maximum elevation 14,494 feet or 4,421 m), or the combined area of the northern New England states of Vermont, New Hampshire and Maine (maximum elevation 5,268' or 1,607 m).It is also approximately the combined area of Austria, Switzerland and the northern Alps of Italy (maximum elevation 15,771' or 4,810 m), or of South Island in New Zealand (maximum elevation 12,349' or 3767 m). The mountains of Balochistan form a great S-shaped curve from near the Iranian border to just east of Zhob, a distance of over 1,000 km, and occupy a total area of approximately 180, 000 km^2. This long narrow series of ranges is approximately the same size as the peninsula of Baja California, and located at approximately the same latitude.

The small mammals of northern Pakistan are mostly Palearctic in affinity, and are usually associated with montane forest or mountain steppe. This region is located in the NWFP from Dir northwards, and includes all of the Northern Areas. Central and southern Pakistan, which is the area south of 32 °, include almost no small mammals from the Palearctic Region. Most small mammals from these areas have Ethiopian and Indo-Malaysian affinities. The overall area includes the foothills of the northern mountains in the Safed Koh Range, the Kala Chitta Hills, Margalla Hills and the Salt Range, as well as the Sulaiman and Kirthar Ranges to the south. Also included are the Makran Coast Range along the Arabian Sea, the various isolated ranges of Balochistan including the dry, sandy Chagai Hills on the southern border with Afghanistan, and the grasslands and deserts of the Punjab, Sindh and Balochistan.

There are many other contrasts between the small mammals of southern and northern Pakistan. For example, dry sandy habitats in central and southern Pakistan have numerous small mammals specialized for xeric habitats (with mainly Arabian affinities), while similar habitats in the valleys of the intermountain north such as the Indus Valley near Chilas or Skardu are completely devoid of comparable small mammals in these same niches. The colonization of these areas by small mammals from Balochistan and the Punjab is blocked by the high passes leading into the valleys, and by the extreme environment of the Indus Valley. No comparable Palearctic species can disperse into these lowland habitats over the high passes and permanent snow fields of the Karakorams and Hindu Kush. So, the small mammal communities of the intermountain north and the isolated ranges and xeric lowlands of central and southern Pakistan have very different characteristics.

The mountains of northern Pakistan are actually a series of interconnected mountain ranges. The area in northern Pakistan where they come together (and in some cases overlap), is called "the Pamir Knot", since it appears to be rooted in the Pamir Mountains just north of Pakistan. This area is sometimes referred to as the “roof of the world”. The mountains of northern Pakistan are centered on Nanga Parbat, and radiate outward as a series of well defined mountain ranges, as well as a jumble of high ridges and isolated valleys in Hunza and Kohistan. The area is important not only because it is in the center of so many radiating massifs, but also because it is on the edge of so many biological, climatilogical and geological zones. For example, the high mountains of northern Pakistan are located on the edge of:

1. Central Asia and Siberia;
2. South Asia and the India Subcontinent;
3. Asia Minor and the Persian Highlands;
4. The Taiga (Great Boreal Forest) stretching southward from Siberia;
5. The seasonal monsoon extending northwest from India and the Arabian Sea;

   The moist westerly winds from the area of the Black and Caspian Seas.

The area, therefore, is of great importance as a biogeographical laboratory. Few areas on earth have as many life zones, faunal elements, and community types so closely associated with one another, or as many endemic species of small mammals. Northern and western Pakistan also is located at the confluence of seven major biogeographical regions of Udvardy (1975, 1985):

Biogeographical Realm	Approximate area in Pakistan
Anatolian-Iranian Desert	Makran Range & Mts. of central Balochistan
Iranian Desert	Mountains of northern and northwestern Balochistan
Thar Desert	Mountains of eastern Balochistan
Hindu Kush Highlands	Waziristan and western NWFP
Pamir-Karakoram Highlands	Karakorams, Hunza and Baltistan
Himalayan Highlands	Nanga Parbat and east, plus part of Deosai Plateau
Sal Forest	Margala Hills and part of Salt Range

The concentration of so many of Udvardy's biogeographical realms in Pakistan, as well as similar analyses indicating unusually high levels of biogeographical diversity in the northwestern area of the Indian Subcontinent by Ali and Qaiser (1986), Mani (1974), Troll (1969), Swan (1961, 1963), von Wissmann (1961), and Schweinfurth (1957) all suggest that levels of diversity of small mammals should also be unusually high.

Northern Pakistan- The major mountain ranges of northern Pakistan are:

1. **Western Himalayas.-** Ending in the mega-mountain Nanga Parbat (8,126 m, 26,660 feet), the last great mountain in the Himalayan Range, the Western Himalayas lie south of the Indus River. The Indus River, with its deep, rugged, xeric gorge flows southward after passing around the north flank of Nanga Parbat. From the top of Nanga Parbat (8,126 m, 26,660 feet) to the valley of the Indus River near Chilas (1,200 m or 4,000 feet) there is a nearly vertical wall of 22,660 feet, the greatest vertical drop of any terrestrial mountain on earth.

2. **Karakoram Ranges.-** All of these ranges are north of the Indus River, and including the second highest mountain in the world (K-2, 8,611 m, 28,250'), as well as other giants such as Masherbrum (25,660'), Gasherbrum (26,470'), Rakaposhi (25,550'), and the

Siachen Glacier and several other glaciers that are the largest glaciers in the world outside of Antarctica. The major ranges include: a) the Rakaposhi-Haramosh Mountains near Skardu; b) the South Ghujerab Mountains in the Shimshal area; c) the North Ghujerab Mountains in the region of Khunjerab National Park; d) the Pamir-i-Wakhan Mountains west of Khunjerab National Park; e) the Hispar Mountains; the Masherbrum Mountains; and several other ranges in the region of the Shyok River.

3. **Hindu Kush Range.-** This range extends westward from the Karakorams west of Khunjerab Pass, and arching southward forms the border between Pakistan and Afghanistan, including Tirich Mir (7,690 m, 25,230').

4. **Deosai Highlands.-** South and east of the Indus River and north of the Astor Valley are a series of high peaks and rolling plains known as the Deosai Plateau and the Deosai Mountains. This unique region is the western extreme of a complex of ranges that include elements of the Zanszkar mountains, the Ladakh Mountains and even parts of the Western Himalayas. We separate this region from the Western Himalaya because it represents a transition zone between Himalaya and Karakoram faunas.

5. **Hunza Kingdoms.-** This area of high mountains and steep valleys lies east of the Ishkuman River and west of the Hunza River. It includes the Batura Muztagh, the Naltar Mountains, but possibly not the Gilgit Mountains(west of the Indus and south of the Gilgit Rivers). The fauna of the region appears to differ from adjacent ranges and highlands, which is the reason that we designate it as "kingdoms" rather than ranges or highlands.

Intermountain Highlands- South of the above major mountain ranges in Pakistan are a series of north-south orientated highland areas between the major valleys of the Indus and Jhelum watersheds.

6. **Hindu Raj Mountains.-** This well defined mountain range forms the east side of the Chitral Valley, and separates this valley from the Swat Valley. It also includes the uplands west of the Ishkuman River. The Chitral Valley contains the Kunar River which flows southwest into Afghanistan where it joins the Kabul River, a major tributary of the Indus.

7. **Swat Kohistan Highlands.-** The Swat Valley is separated from the massive Indus Valley by another north-south orientated

highland that forms Swat Kohistan. The Swat Valley contains the Swat River, which flows southward to join the Kabul River in Pakistan near Charsadda.

8. **Indus Kohistan Highlands.-** To the east of the Indus Valley is another massive upland area known as Indus Khohistan, which includes such pristine intermountain valleys as the Palas Valley. As defined here, these highlands are east of the Indus, west of the Kaghan Valley and south of Babusar Pass (near Chilas).

9. **Kaghan-Neelum Highlands.-** East of the Indus are the north-south orientated Kaghan (with the Kunhar River) and Neelum Valleys (with the Neelum River), both tributaries of the Jhelum River. The Jhelum River flows west from the highlands surrounding the Vale of Kashmir (an immense, beautiful valley). The Kunhar and Neelum Rivers join the Jhelum near Muzafarabad. South of this the Mahl River drains the high Pir Panjal Mountains (with heights of 4,314 m just south of the Vale of Kashmir), and flows west to join the Jhelum.

10. **Murree Hills.-** The well forested Murree Hills form an interesting alpine transition zone between the dry plains of the Punjab to the south, and the towering mountains of the north. They are at the edge of the seasonal monsoon, and receive abundant rainfall (over 60"). They are covered with a rich forest of spruce, silver fir and blue pine, and usually have deep soils.

11. **Margalla/Salt Ranges.-** The last out-riders of the Himalayas are foothill ranges such as the Margalla Hills and the Salt Range. These ranges are dry, rocky, and the first uplifts of the Himalaya. They are collectively known as the Shiwaliks.

A series of isolated, well forested upland areas which resemble "sky islands are found in this area of seven rivers, seven valleys and ten high mountain ranges (excluding the Margalla/Salt Range).Uplands in these ranges receive water from seasonal rains and snowfall, or from melting glaciers and snow fields of the surrounding peaks, and mountain valleys above 3000 meters are characterized by luxurious forests and alpine meadows. Valley floors are dry, sandy or rocky, and are often desert-like. The highest ridges above the valleys (or "nullas") are often far above tree line, and permanently covered with snow, glaciers or bare rock. These dry lowlands and frozen mountain peaks are a barrier to most small mammals. The patches of forests and alpine meadows that are often found in the uppers valleys (nullahs) are the sky islands mentioned above. These mesic patches usually are located between 3000 and 4200 meters, and are geographically isolated by desert-like habitats below and snow, ice and bare rock above.

Balochistan Highlands

South of the above intermountain areas of northern Pakistan are the highlands of Balochistan. The mountains of Balochistan are not nearly as high in elevation. The highest peaks occur in the Quetta area, with the highest being Zarghun (3637 m) along a towering rocky wall of peaks southeast of Quetta. There are also high peaks in the Ziarat area (Khalifat at 3485 m), and at the north end of the Suleiman Range near Zhob (Takht-i-Suleiman at 3375 m). All of these ranges and peaks receive snowfall in the winter, and have areas of forest habitats, especially in the Ziarat area, and in the northern Suleiman Range near Zhob. The mountains of Balochistan are interesting to compare with the mountains of northern Pakistan because they are close to Iran and Afghanistan, where numerous small mammals occur (Lay, 1967; Hassinger, 1973). They are also connected with the mountains of northern Pakistan via the northern extension of the Suleiman Range into Waziristan, with peaks up to 3513 m, and the Safed Koh Mountains with several high snowy peaks in the Parachinar area. There is a mountainous corridor between the Hindu Kush, Safed Koh, Sulaiman, and Toba Kakar ranges. This corridor would have allowed small mammals from the Palearctic to disperse to the mountains of central Balochistan, such as the high peaks surrounding Ziarat and Quetta. The coastal mountains of southern Pakistan and Iran are known as the Makran Ranges. These ranges curve northward to include and connect with the Siahan Range, the Central Brahui and Ras Koh Ranges. They all interconnect together in the Quetta area as a knot of ridges, peaks and ranges which is analogous to the Pamir Knot of northern Pakistan. We call this area the Balochi Knot. The definable ranges of Balochistan are:

12). **Suleiman Range.-** This rugged range forms the northeastern boundary of Balochistan, and parallels the Indus Valley.The mountains are mostly dry, and are transected by river valleys that drain the Balochistan uplands. The peaks and ridges are low and weathered in the south, but can exceed 3000 m in the north near the border between Balochistan and the NWFP

13).**Toba Kakar Range.-** This range curves northeast from near the border with Afghanistan south of Kandahar towards the high ridges of the Suleiman Range. The two ranges merge together to form the highlands of Waziristan in the NWFP They are rugged, dry mountains with broad intermountain valleys. Examples of sub-units of this overall range include the Spingar and Torgar Mountains north of Qila Saifullah.

14). **Ziarat Highlands.-** Near the mountain village of Ziarat (2460 m) are a series of higher peaks that are formed by the same massif. Higher ridges near Ziarat include Khalifat (3485 m). The area is well forested with juniper trees, which in some valleys form a mature and dense juniper forest. The overall area is more mesic than most surrounding mountains.

15). **Central Brauhi Range.-** This north-south range extends from near Khuzdar to southeast of Quetta, and is transected by the Bolan Pass. It has a number of high ridges above 3000 m, and has a sparse cover of juniper trees. In the north it approaches the Zarghun Range, with peaks up to 3575 m (the highest in Balochistan). South of Quetta it includes Koh-i-Maran, as well as the Harboi Hills.

16).**Central Makrań Range.-** Beginning near the Iranian border at Zamuran Pass, this range first parallels the coast before curving north towards the town of Besima. East of Besima it forms the rugged Kalghali Hills. The mountains are mostly dry, with numerous inter-valley lowlands and nullahs. Few peaks are over 1500 m, but in the north high ridges can exceed 2200 m. Most of the tree-like vegetation is old pistachio trees.

These high ranges are biogeographical laboratories. They form the transition zone between South Asia and Central Asia. The point of contact between the Indian Plate and the Asian Plate passes through the center of the intermountain region of Kohistan. The point of contact is preserved in the rocks, and can be seen in the Indus Valley near the town of Jujial just south of Pattan (in central Kohistan). This occurs at 35° North Latitude.

As described in Mani (1974:707) "...the Himalayan uplift dominates practically the entire range of events culminating in the shaping of the climate and composition of the flora and fauna of the whole of India. The Himalaya presides over the ecology and biogeography of India". This is even more true of Pakistan, which lies closer to the Himalaya, and is a much smaller more mountainous country. The history of the Himalaya is manifested everywhere within the boundaries of Pakistan.

The western borderlands of the Himalaya touch on Asia in Afghanistan, and were the last to make contact with Asia as the Indian Plate collided with Asia. Contact started in the east, and slowly compressed the Tethys Sea westward. According to Mani (1974:684), the western borderlands of the Himalaya remain dynamic and active today, with the faunal elements mainly drawn from the Turkmenian Subregion of the Palaerctic, as well as Mediteranian and Ethiopian elements. The eastern boundary of this region has been shifting gradually eastward (even within historical times).

As outlined above, these mountains are not a uniform series of high mountain ridges, but rather divided into many distinct biogeographical zones. One of the best documented zones lies to the east of Pakistan. As reported by Mani (1974:673) "Ecologically and biogeographically the Himalayan forests, west of the great defile of the R. Sutlej, are strikingly different from those in the east." The flora and fauna of the steppes of the Turkmenian element dominate here. This is an important observation that adds importance to the high ranges of Pakistan as a biogeographical laboratory. Mani (1968) has shown that the Northwest Himalaya has a very different history from the rest of the Himalaya, and that this region is really a component of the Pamir Knot. The Northwest Himalaya, along with the Pamir Ranges of Tajikistan, constitute an independent biogeographical center. The major trends that he reported (mainly for insects) were:

1) The modification of a relatively few species into high altitude specialists (mainly during the Pliocene);

2) A pronounced tendency towards an increase in the number of taxa by a rapid process of sub-speciation and isolation on single high massifs.

The trends are also valid for the small mammals.There is a remarkable pattern of subspecies, and even species level diversity, in the isolated "sky islands" of the Northwestern Himalaya.

<u>The latitudinal characteristics are summarized below</u>:

Panjgur (Makran Range)	27°
Besima (near Kalghali Hills)	28°
Zarghun (near Quetta)	30°
Ziarat	30°
Torgar Range	31°
Zhob	32°
Margalla Range	32°
Murree Hills	33-34°

Kaghan Valley	35°
Nanga Parbat	35.5°
Gilgit	36°
Karakorams	36°
Pakistan/Tajikistan	38°
Pamirs of Kyrgystan	40-42°
So. Russia/Siberia	50°
Lake Baikal	52°

To the south are grasslands and deserts of the Punjab, Sindh, Rajasthan and the Deccan Plateau (to 20° latitude in central India). To the north is Siberia (to 70° latitude)

Sky Islands and Fragmented Forests

The plant and animal species associated with the once broadly interconnected inter-montane ecosystems of the region became fragmented into patterns of disjunct distributions. For some species of small mammals these isolated areas of suitable habitat were close enough to similar habitats in adjacent valleys or ranges that dispersal was possible, and corridors of genetic continuity persisted. This was especially true for larger mammals such as the wild goats and sheep, and for larger carnivores. It is also true of forest dwelling small mammals such as flying squirrels (*Eoglaucomys* [=*Hylopetes*] *fimbriatus* and *Petaurista petaurista*) and perhaps for forest species such as the wood mouse (*Apodemus wardi*) or more xeric tolerant forms such as the gray hamster (*Cricetulus migratorius*). It might also be true for the forest burrowing voles of the *Hyperacrius wynnei* complex. These are species of the fragmented forest, and patterns of sub-speciation or polymorphism are observable.

However, some species have become confined to higher alpine zones where frequently they are very isolated from similar forms living in other suitable habitats separated by miles of unsuitable habitat (dry valleys, desert-like xeric soils, raging rivers). Species associated with the isolated alpine pockets of alpine meadows and northern forest are living in ecological islands. Some are maintained by high mountain phenomena such as melting glaciers, streams supplied by high peaks, precipitation combed from blowing clouds through small needled conifers, and protection from desiccation by shadows and north facing slopes. These high, isolated ecosystems have been called "sky islands". In the far north of the area, deep in the intermountain ranges of Baltistan, Hunza, Ishkoman and Chitral, the rain shadow effect from the monsoon is nearly total and very little precipitation falls. An excellent example of the entire range of a species being confined to a series of sky islands is illustrated by the distribution of the woolly flying squirrel, *Eupetaurus cinereus* (Zahler, 1996; Zahler and Woods, this volume).

In some regions, however, forest (especially poplars) and grasslands can spread along major rivers, such as in the Shigar Valley north of Skardu. We call these areas of suitable habitat "riparian islands".

The opposite effect is also possible. In some regions, unique ecological conditions can create habitats that should be rich in certain species of small mammals. The desert-like sand dunes along the upper Indus River Valley in the broad Skardu Valley are a good example. Conditions here strongly resemble the sandy deserts of Balochistan and Iran, which are filled with small mammals such as jerboas, gerbils and jirds. The sandy hills and dunes surrounding the Indus are isolated from the lower Indus Valley, and areas of the Punjab and Balochistan with almost exactly the same ecological conditions.

Dispersal of small mammals from the deserts of southern and central Pakistan and India is blocked by the extreme conditions of the narrow, rocky, geologically active Indus Valley, and the high peaks of Himalaya, many of which are permanently covered with glaciers, snow and rocks without any soil. Gerbils, jirds and even hamsters are not able to disperse through these habitats. The areas are also devoid of other organisms that once would expect to find in such habitats, such as lizards (Auffenberg, pers com). These areas are "eco-vacuums".

Faunal List

The following list of small mammals of the mountains of Pakistan is based on our field work between 1990-1995. The nomenclature for the small mammals of the Indo-Malaysian Sub-region has changed dramatically during the past decade. We are following the current nomenclature in Wilson and Reeder (1992) and Corbet and Hill (1992). The identifications of several of the species are still questionable, however, and we are currently studying the systematics of all of the small mammals of northern Pakistan using biochemical as well as standard morphometric techniques. There are several new taxa. However, for the purpose of this paper we are using the following list of small mammals of northern and western Pakistan as our standard for analysis and comparison. Appendix I lists all of the nomenclatorial changes of the past decade.

Small Mammals of the Mountains and Intermountain Valleys of Pakistan

Sorex thibetanus planiceps	Tibet Red-toothed Shrew
Crocidura pullata	Western Himalayan White-toothed Shrew
Crocidura pergrisea	Pale Gray Shrew
Crocidura zarudnyi (sensu latu)	Persian Desert Shrew
Crocidura gmelini (sensu latu)	Balochistan Short-tailed Shrew
Rhinolophus ferrumequinum	Greater Horseshoe Bat
Rhinolophus hipposideros	Horseshoe Bat
Myotis mystacinus	Whiskered Bat
Eptesicus isabellinus	Isabelline Serotine
Eptesicus serotinus	Common Serotine
Pipistrellus pipistrellus	Common Pipistrelle
Pipistrellus coromandra (marginal)	Indian Pipistrelle
Pipistrellus mimus (marginal)	Pygmy Pipistrelle
Barbastella leucomelas	Asiatic Wide-eared Bat or Asian Barbastelle
Otonycteris hemprichi	Hemprich's Long-eared Bat
Plecotus austriacus	Gray Long-eared Bat
Murina huttoni	Tube-nosed Bat
Ochotona roylei	Himalayan Pika (Himalayas,
Ochotona macrotis	Karakoram Pika (esp. Baltistan & Karakorams, possibly Hindu Kush)
Lepus capensis	
Apodemus wardi	Himalayan Wood Mouse
Apodemus (pale color)	Baltistan Wood Mouse
Rattus turkestanicus	Turkestan Rat
Rattus rattus	Black Rat
Mus musculus	House Mouse
Nesokia indica	Short-tailed Mole-rat or Short-tailed Bandicoot
Bandicota bengalensis	Indian Mole-rat or Bandicoot
Millardia meltada	Soft-furred Field Rat or Metad
Golunda ellioti	Bush Rat
Alticola argentatus	Himalayan High Mountain Vole
Alticola glacialis	Baltistan High Mountain Vole
Hyperacrius wynnei (sensu lato)	Murree (or Forest Burrowing) Vole

Hyperacrius fertilis (sensu lato)	Mountain Burrowing Vole
Ellobius fuscocapillus	Afghan Mole Vole
Sicista concolor	Chinese Birch Mouse
Cricetulus migratorius	Migratory (or Gray) Hamster
Calomyscus baluchi	Balochistan Mouse-like Hamster
Calomyscus hotsoni	Makran Mouse-like Hamster
Dryomys nitedula	Forest Dormouse
Dryomys niethammeri	Balochistan Dormouse
Gerbillus nanus	Balochistan Gerbil
Meriones persicus	Persian Jird
Meriones libycus	Libyan Jird
Tatara indica	Indian Gerbil
Marmota caudata	Long-tailed Marmot
Marmota himalayana	Karakoram Marmot
Eoglaucomys fimbriatus	Small Kashmir Flying Squirrel
Petaurista petaurista	Giant Red Flying Squirrel
Eupetaurus cinereus	Woolly Flying Squirrel
Hystrix indica (marginal)	Indian Porcupine
Mustela altaica	Alpine Weasel (Wilson & Reeder 322 = *M. kathiah)*
Mustela erminea	Ermine
Martes foina	Stone Marten
Martes flavigula	Yellow-throated Marten
Vormela peregusna (marginal)	Marbled Pole-cat ("Sarmantier")
Paguma larvata	Masked Palm Civit

Table I: Comparisons of Small Mammal Communities

N. Vermont	W. Montana	NW. Pakistan
Characteristics		
45 degrees North	45 degrees North	35 degrees North
Well Forested	Moderate Forested	Isolated Nullah Forests
Year-round Precip.	Winter Rain/Snow	Edge of Monsoon
Deep Winter Snow	Deep Winter Snow	Deep Winter Snow
Conifer/N. Hardwoods	Pines, Some Hardwoods Grasslands	Spruce-Fir,Pine, Hardwoods w/ very dry surroundings
Semi-uniform	Somewhat Patchy	Sky Islands
Insectivores		
Sorex hoyi	*Sorex hoyi*	*Sorex thibetanus*
S. cinereus	*S. cinereus*	None
S. fumeus	*S. merriami*	*Crocidura pullata*

S. dispar	*S. monticolus*	None
S. palustris	*S. palustris*	*Crocidura pergrisea*
Blarina brevicauda	None	None
Parascalops breweri	None	None
Condylura cristata	None	None
Rodents		
Tamias striatus	*Tamias amoenus*	*Dryomys nitedula*
None	*T. minimus*	*Calomyscus baluchi*
None	*Spermophilis umbrinus*	None
None	*T. armatus*	None
None	*Spermophilis lateralis*	None
Marmota monax	*Marmota flaviventris*	*Marmota caudata*
Sciurus carolinensis	None	None
Tamiasciurus hudsonicus	*T. hudsonicus*	None
Glaucomys sabrinus	*G. sabrinus*	*Eoglaucomys fimbriatus*
None	None	*Petaurista petaurista*
None	None	*Eupetaurus cinereus*
None	*Cynomys leucurus*	None
None	*Thomomys talpoides*	*Nesokia indica*
Castor canadensis	*Castor canadensis*	None
Peromyscus maniculatus	*P. maniculatus*	*Apodemus wardi*
Peromyscus leucopus	None	*Mus musculus*
None	*Neotoma cinerea*	*Rattus turkestanicus*
Clethrionomys gapperi	*C. gapperi*	*Hyperacrius wynnei*
Microtus pennsylvanicus	*M. pennsylvanicus*	None
Microtus pinetorum	None	*Hyperacrius fertilis*
Microtus chrotorhinus	*M. montanus*	*Alticola argentatus*
None	*M. richardsoni*	None
None	*M. longicaudus*	None
None	*Lemmiscus curtatus*	*Cricetulus migratorius*
Ondatra zibethicus	*O. zibethicus*	None
Synaptomys cooperi	*Phenacomys intermedius*	None
Zapus husonicus	*Zapus princeps*	*Sicista concolor*
Napaeozapus insignis	None	None
Erethizon dorsatum	*E. dorsatum*	*Hystrix indica* (marginal)
Lagomorphs		
Lepus americanus	*Lepus americanus*	None
None	*Lepus townsendii*	*Lepus capensis*
Sylvilagus floridanus	*Sylvilagus nuttali*	None
Sylvilagus transitionalis	*S. audubonii*	None
None	*Ochotona princeps*	*Ochotona roylei*
Small Carnivores		
Mustela erminea	*Mustela erminea*	*Mustela erminea*
Mustela frenata	*Mustela frenata*	*Mustela altaica*
Mustela vison	*Mustela vison*	None
Martes americana	*Martes americana*	*Martes foina*
Martes pennanti	None	*Martes flavigula*
None	None	*Paguma larvata*
Mephitis mephitus	*Mephitis mephitus*	*Vormela peregusna*
Lutra canadensis	*Lutra canadensis*	*Lutra lutra*

Table II: Biodiversity of Small Mammals in the Mountains of Pakistan (excluding bats)

Location	No. Species	Affinity	Elevation
Ziarat Valley (Balochistan) (**Hot Spot**)	11 (3 endemic)	Palearctic,Arabian,Indian	2,480 m
Sundalai (Bumburet)	10	Palearctic, Indian	2,600 m
Zarghun (Balochistan)	8	Palearctic, Arabian, Indian	2,300 m
Changa Manga	7	Indian	Low
Palas Valley	7	Palearctic, Indian	2,400 m
Kaghan Valley	7	Palearctic, Indian	2,400 m
Babusar Village	6	Palearctic	2,800 m
Tarshing (Nanga Parbat)	6	Palearctic	2,933 m
Khunjerab Top	5	Palearctic	4,520 m
Hushe area, Baltistan	4	Palearctic	3,200 m

Comparisons with other Intermountain Ecosystems

In making comparisons with other parts of the world, it is hard to find geographical regions that are appropriate because there are few areas with comparable massifs. For example:

Asia Proper: The mountains are lower and wetter, and too far south.

Africa: The mountains are less extensive, and sit on the equator (Kilimanjaro is 5,899 m)

South America: The mountains are along an extended long narrow strip, and many are near the equator. The larger southern ones (Aconcagua of Argentina (6,964 m) are high enough, but there is not enough land mass to the east (i.e. South America tapers to a point with too little land mass). There is no direct contact with the Holarctic since it is in the southern hemisphere, and the groups of small mammals are different. So, even though there many peaks over 6,100 m the comparison with the Himalayas breaks down.

Europe: The mountains are generally high, but not high enough (Mt. Blanc is 4,810 m).The latitude is too high (about 48° N). There is no extensive dry zone to the south (the Mediterranean blocks appropriate areas in north Africa).

The best comparison in the world is the Rocky Mountains of western North America. The mountains at first glance appear too low:

McKinley	20,320' or 6198 m	(Alaska)
Whitney	14,494' or 4421 m	(California)
Elbert	14,433' or 4402 m	(Rockies)

However, there are many important reasons for selecting the Rockies as the best massif for comparison with the mountains of Pakistan.

1). They are located at approximately the same latitudes:

31.5°	Southern New Mexico	Lahore
34°	Central New Mexico/central Arizona	Islamabad/Margalla Hills
35.5°	Southern Colorado Rockies (Santa Fe)	Nanga Parbat
37°	San Juan Mountains of Colorado	Central Hunza
41°	Rocky Mountain National Park	Pamir Mountains
43°	Wind River Range in Wyoming	Tian Shan Range
45°	Yellowstone Ecosystem of northern Rockies	Altai Mountains
50°	Southern Alberta	Southern Siberia
52°	Central Alberta (Banff/Lake Louise)	Lake Baikal

2). The eastern and southern slopes extend outwards in grasslands and deserts, just as in Pakistan (to 20° degrees N in Mexico north of Mexico City).

3). The northern reaches of the Rockies extend into northern Canada and Alaska to about 70° North Latitude)

4). There are an extensive series of intermountain ridges (i.e. Front Range, Central Rockies, Sangre de Cristos, San Juans, Wasatch Mts. Etc.). Just as in Pakistan, there are many rivers, intermountain valleys and plateaus (some orientated east-west, and others north south).

5). The taxonomic mix of small mammals is almost exactly the same at the sub-family and in many cases even at the genus level (i.e. arvicolines, murines, soricines, mustelids).

Note: Among larger mammals there are even wolves and grizzly bears, and the ecological equivalent (=ecomorph) of the puma (the snow leopard).

6). Both areas are at the southern limit of several important indicator species.

	Pakistan	Rockies
Voles	*Hyperacrius fertilis* (32°)	*Clethrionomys gapperi* (34°)
	Alticola argentatus (34°)	*Microtus montanus* (34°)
Shrews	*Sorex thibetanus* (35°)	*Sorex cinereus* (34°)
	Crocidura pergrisea (35°)	*S. palustris* (34°)
	Crocidura zarudnyi (26°)	*Notiosorex crawfordi* (30°)
	Crocidura gmelini (30°)	*Sorex nanus* (30°)
		Sorex merriami (35°)
Martes	*Martes flavigula* (35°) (Yellow-throated Marten)	*Martes americana* (35°)
	Martes foina (25°) (Beech or Stone Marten)	
Mustela	*Mustela erminea* (35°)	*Mustela erminea* (35°)
Otters	*Lutra lutra* (38°)	*Lutra canadensis* (32°)

7). Some species have dispersed into the intermountain region from the west in both ecosystems. For example, in Pakistan several species have dispersed into the mountains via Iran and Afghanistan, such as *Calomyscus baluchi*, *Calomyscus hotsoni*, *Dryomys nitedula*, *Dryomys niethammeri*, *Cricetulus migratorius*, and *Ellobius fuscocapillus.* The distribution of all of these species is only in far western Pakistan. In the Rocky mountains similar distributions occur in several chipmunks of the genus *Tamias* (ecomorphs of *Dryomys*), pocket gophers of the genus *Thomomys* (ecomorphs of *Ellobius*), and mice of the genus *Peromyscus* (ecomorphs of *Calomyscus*).

8). Both regions are continental in position, and have huge plains to the southeast (the plains of the Punjab and the Deccan Plateau in Pakistan mountains; the Great Plains in the Rocky Mountains).

The best point of comparison is the Southern Rockies in northern New Mexico and southern Colorado (i.e. the Sangre de Cristo Mts. & San Juan Mts.) at approximately 35° latitude.

Faunal Descriptions

Now that we have established the overall similarity between the Himalaya and Rocky Mountain geological, climatilogical and faunal features, a common set of terms are needed if biogeographical analyses and comparisons are to be successful. A foundation for such comparisons was formulated by Armstrong (1972), and is carefully utilized in the last chapter of his book. The chapter is titled "Zoogeography of Mammals of Colorado" and in it he describes both the ecological characteristics and faunal elements of Colorado. This approach was expanded by Jones et al. (1983) for the habitats and mammals of the northern Great Plains, and by Smith (1993) for Alberta.

The small mammals of southern Colorado fall into two major biogeographic assemblages (Table III).

Table III: Small Mammals of Southern Colorado

SOUTHERN COLORADO		**PAKISTAN HIMALAYA**
Cordilleran Faunal	Boreal-Cordilleran Faunal	Pakistan Cordilleran
Sorex vagrans	*Sorex cinereus*	*Sorex thibetanus*
Sorex nanus	*Sorex palustris*	*Crocidura pullata*
	Sorex hoyi	*Suncus etruscus*
		Crocidura pergrisea
Ochotona princeps	*Lepus americana*	*Ochotona roylei*
		Ochotona macrotus
Tamias umbrinus	*Tamias minimus*	*Dryomys nitedula*
	Tamiasciurus hudsonicus	
Marmota flaviventris		*Marmota caudata*
		Marmota himalayana
Spermophilis richardsonii		*Cricetulus migratorius*
Spermophilis lateralis		
Thomomys talpoides		*Ellobius fuscocapillus*
Neotoma cinereus		*Rattus turkestanicus*
		**Apodemus wardi*
Microtus montanus	*Clethrionomys gapperi*	*Alticola argentatus*
Microtus longicaudus	*Microtus pennsylvanicus*	*Hyperacrius fertilis*
	Phenacomys intermedius	*Hyperacrius wynnei*
Zapus princeps		*Sicista concolor*
	Martes americana	*Martes flavigula*
		Martes foina
	Mustela erminea	*Mustela erminea*
		Mustela kathiah
	Gulo gulo	
	Lynx canadensis	*Lynx lynx*
		Calomyscus baluchi

* *Peromyscus* sp is an ecological equivalent, but it occurs in dryer habitats in Colorado.

At first review (Table I) the small mammal communities of northern Pakistan appear to be very depauperate. Many habitats filled with small mammals in North America are vacant in Pakistan, and some have only 3 to 5 small terrestrial mammals living in them where comparable areas in Colorado have 6-10 species. However, if specific faunal elements are compared (i.e. compared with the data in Armstrong, as in Table III)), the small mammal communities in the mountains of northern Pakistan Himalayas are not nearly as depauperate.

Table IV: Small Mammals of the Southern Rocky Mountains (Colorado)

Taxon	Elevational Range (feet)	Mass (grams)	Pakistan Equivalent
Sorex cinereus	to 10,400	5-6	*Sorex thibetanus*
Sorex vagrans	5,300-11,400	5	*Crocidura pullata*
Sorex nanus	6,200-9,800	3	*Crocidura gmelini (sensu latu*
Notiosorex crawfordi	5,300-6,800	4	*Crocidura zarudnyi (sensu latu)*
Sorex palustris	6,000-10,000	15	*Crocidura pergrisea*
Sorex hoyi	10,000	3	*Suncus etruscus*
None			*Pipistrellus mimus*
Myotis lucifugus	6,000-11,000	7	*Myotis mystacinus*
Myotis evotis	6,000-8,400	5	*Pipistrellus pipistrellus*
Myotis thysanodes	7,500	7	*Pipistrellus coromandra*
Myotis volans	7,000-11,000	9	*Eptesicus isabellinus*
Lasionycteris noctivagans	9,000	9	*Murina huttoni*
Eptesicus fuscus	5,000-8,500	12	*Eptesicus serotinus*
Lasiurus cinereus	5,000	28	*Barbastella leucomelas*
Plecotus townsendii	5,000-8,000	11	*Plecotus austriacus*
None			*Otonycteris hemprichi*
None			*Rhinolophus ferrumequinum*
None			*Rhinolophus hipposideros*
None			*Ochotona rufescens*
None			*Ochotona macrotis*
Ochotona princeps	above 8,500	200	*Ochotoan roylei*
Sylvilagus nuttalii	6,000-11,500	750	None
Sylvilagus audubonii	below 7,000	1000	None
Lepus americanus	8,000-11,000	1500	None
Lepus townsendii	5,000-11,000	3000	*Lepus capensis*
Tamias minimus	8,500-10,000	50	None
Tamias quadrivittatus	4,200-10,500	30-50	*Dryomys nitedula*
Tamias umbrinus	6,500-12,000	30-50	*Dryomys niethammeri*
Marmota flaviventris	5,400-13,760	4000	*Marmota caudata*
None			*Marmota himalayana*
Spermophilus richardsonii	5,000-11,000	300	None
Spermophilus tridecelineatus	to 9,000 in grass	120	*Cricetulus migratorius*

Spermophilus variegatus	to 8,500	300	None
Spermophilus lateralis	6,500-13,000	300	None
Cynomys leucurus	to 8,500	1000	None
Cynomys gunnisoni	6,000-12,000	900	None
Sciurus aberti	5,300-10,200	680	None
Tamiasciurus hudsonicus	6,000-12,000	200	None
None			*Eoglaucomys fimbriatus*
None			*Petaurista petaurist*
None			*Eupetaurus cinereus*
Thomomys bottae	in valleys in south	190	*Nesokia indica*
Thomomys talpoides	5,000-10,000	150	*Ellobius fuscocapillus*
Castor canadensis	5,000-11,000		None
Peromyscus maniculatus	6,500-11,000	25	*Apodemus wardi*
Peromyscus difficilis	5,000-8,500	29	*Calomyscus baluchi*
None			*Calomyscus hotsoni*
Neotoma mexicana	4,300-8,300	150	None
Neotoma cinerea	4,500-8,000	260	*Rattus turkestanicus*
Clethrionomys gapperi	8,000-11,000	20	*Hyperacrius wynnei*
Phenacomys intermedius	7,000-12,000	30	None
Microtus pennsylvaticus	valleys	60	*Hyperacrius fertilis*
Microtus montanus	6,000-12,000 (dry)	75	*Alticola sp* (pale color)
Microtus longicaudus	5,000 (moist)	50	*Alticola argentatus*
Microtus mexicanus	7,000-8,000	30	None
Ondatra zibethicus	to 11,000	1200	None
Zapus princeps	6,000-11,500	22	*Sicista concolor*
Erethizon dorsatum	5,000-10,000	10000	*Hystrix indica*
Martes americana	above 9,000	1000	*Martes flavigula*
Martes frenata	5,000-13,000	250	*Martes foina*
Martes erminea	5,300-13,000	120	*Mustela erminea*
Mustela vison	5,000-10,000	1200	*Mustela altaica*
Mephitis mephitis	5,000-10,000	2600	*Vormela peregusna*
Lutra canadensis	rare	10000	*Lutra lutra*

Table V: Summary of Small Mammal Groups

	Southern Rockies	No. & Western Pakistan
Shrews	5	5
Bats	8	11
Lagamorphs	5	3
Squirrels (niche)	12	1
Flying Squirrels	None	3
Pocket Gophers (niche)	2	2
Beavers	1	None
Hamsters	None	2
Mice & Rats	4	3
Voles	6	5
Jumping Mice	1	1
Porcupines	1	1
Mustelids	6	6
Totals	**51**	**43**

Table VI: Distribution of small Mammals in Pakistan Mountains

		Mountain Range															
		1 Him	2 KKr	3 Hku	4 Deo	5 Hnz	6 HRj	7 Swa	8 Koh	9 Kag	10 Mur	11 Mr	12 Sul	13 Tor	14 Zia	15 CBr	16 Mak
Taxon	**OTU**																
Sorex thibetanus planiceps	1	1	1	0	1	0	0	0	0	1	0	0	0	0	0	0	0
Crocidura pullata	2	1	1	1	1	1	1	1	1	1	1	0	0	0	1	0	0
Crocidura pergrisea	3	0	1	0	1	0	0	0	0	0	0	0	0	0	0	0	0
Crocidura zarundyi (sensu latu)	4	0	0	0	0	0	0	0	0	0	0	0	1	1	1	1	1
Crocidura gmelini (sensu latu)	5	0	0	0	0	0	0	0	0	0	0	0	0	0	1	0	0
Ochotona roylei	6	1	0	0	0	0	1	1	1	1	0	0	0	0	0	0	0
Ochotona macrotis	7	0	1	1	1	1	0	0	0	0	0	0	0	0	0	0	0
Ochotona rufescens	8	0	0	0	0	0	0	0	0	0	0	0	1	1	1	1	0
Marmota caudata	9	1	1	1	1	0	0	0	1	1	0	0	0	0	0	0	0
Marmota himalayana	10	0	1	0	0	0	0	0	0	0	0	0	0	0	0	0	0
Eoglaucomys fimbriatus	11	1	0	1	0	1	1	1	1	1	1	0	0	0	0	0	0
Pletaurista petaurista	12	1	0	1	0	0	1	1	1	1	1	0	0	0	0	0	0
Eupetaurus cinereus	13	1	0	0	0	1	0	0	1	1	0	0	0	0	0	0	0
Apodemus wardi	14	1	0	1	1	1	1	1	1	1	1	0	0	0	1	0	0
Apodemus sp (pale color)	15	0	1	0	0	0	0	0	0	0	0	0	0	0	0	0	0
Ratus turkestanicus	16	1	1	1	1	1	1	1	1	1	1	0	0	0	0	0	0
Bandicota bengalensis	17	0	0	0	0	0	0	0	0	0	0	1	0	0	0	0	0
Acomys cahirinus	18	0	0	0	0	0	0	0	0	0	0	0	0	0	0	0	1
Golunda ellioti	19	0	0	0	0	0	0	0	0	0	0	1	0	0	0	0	0
Nesokia indica	20	0	0	1	0	0	0	0	0	0	0	1	1	0	1	1	1
Alticola argentatus	21	1	0	1	1	0	1	1	1	1	0	0	0	0	0	0	0
Alticola glacialis	22	0	1	0	0	0	0	0	0	0	0	0	0	0	0	0	0

Hyperacrius wynnei (SL)	23	0	0	0	0	0	0	1	1	1	1	0	0	0	0	0	0
Hyperacrius fertilis (SL)	24	1	1	1	1	0	0	1	1	1	0	0	0	0	0	0	0
Ellobius fuscocapillus	25	0	0	0	0	0	0	0	0	0	0	0	0	1	1	1	0
Sicista concolor	26	1	0	0	1	1	0	0	0	1	0	0	0	0	0	0	0
Cricetulus migratorius	27	0	1	1	0	1	1	1	0	0	0	0	1	1	1	1	0
Calomyscus baluchi	28	0	0	1	0	0	1	1	0	0	0	0	1	1	1	1	0
Calomyscus hotsoni	29	0	0	0	0	0	0	0	0	0	0	0	0	0	0	0	1
Dryomys nitedula	30	1	0	1	0	0	1	1	0	0	0	0	0	0	0	0	0
Dryomys niethammeri	31	0	0	0	0	0	0	0	0	0	0	0	0	0	1	1	0
Gerbillus nanus	32	0	0	0	0	0	0	0	0	0	0	0	0	0	0	0	1
Meriones persicus	33	0	0	0	0	0	0	0	0	0	0	0	1	1	1	1	1
Meriones libycus	34	0	0	0	0	0	0	0	0	0	0	0	0	1	0	1	1
Tatara indica	35	0	0	0	0	0	0	0	0	0	0	1	1	0	0	0	1
Hystrix indica	36	0	0	1	0	0	1	1	1	1	1	1	1	1	1	1	1
Mustela altaica (*M. kathiah*?)	37	1	1	0	1	1	0	0	0	0	0	0	0	0	0	0	0
Mustela erminea	38	1	1	1	1	1	0	0	0	0	0	0	0	0	0	0	0
Martes foina	39	1	1	1	1	1	1	1	1	1	0	0	1	1	1	1	0
Martes flavigula	40	1	0	?	0	?	1	1	1	1	1	1	0	0	0	0	0
Vormela peregusna	41	0	0	0	0	0	0	0	0	0	0	0	0	1	1	1	1

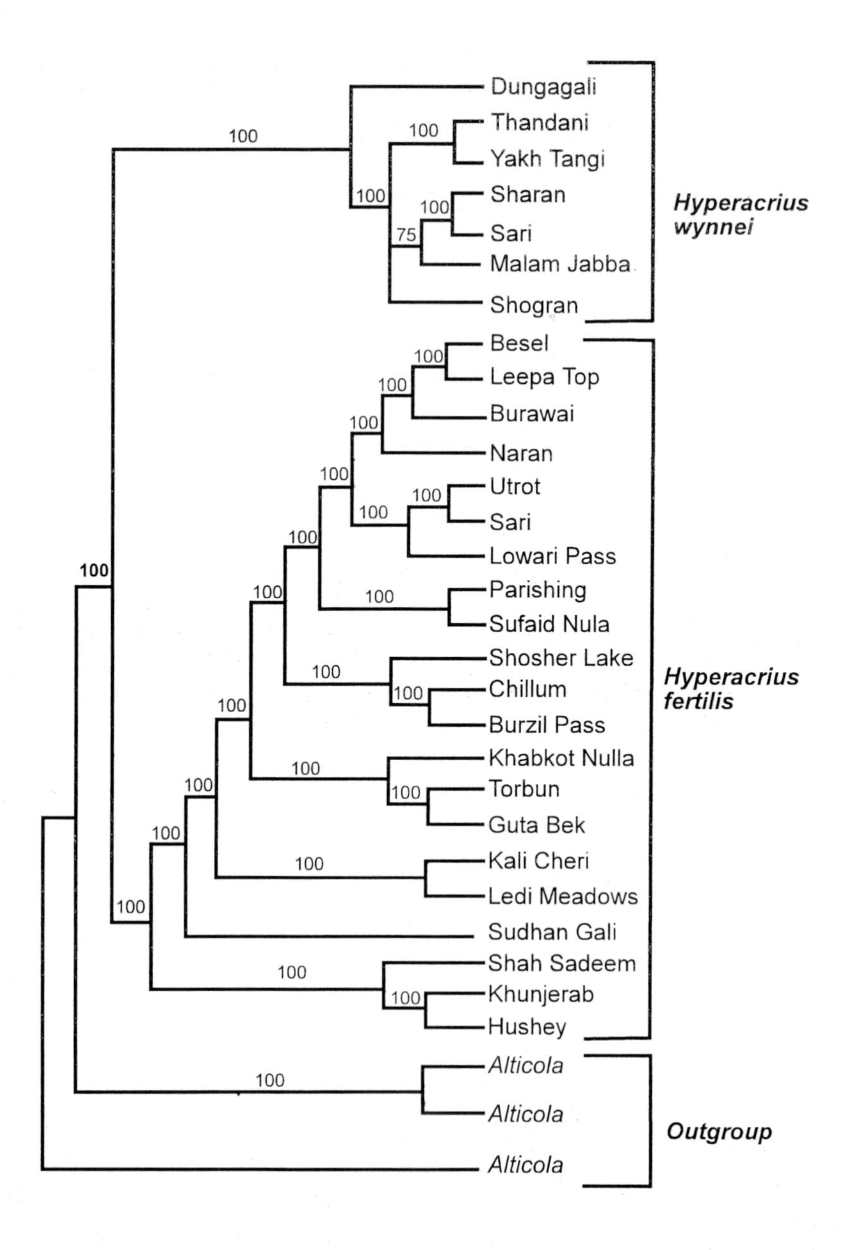

Fig. 1 : Phylogenetic analysis of populations of *Hyperacrius* based on analysis of cytochrome b sequence data.

CONCLUSIONS

There are fewer terrestrial small mammals in the mountains of Pakistan than would be predicted on the basis of suitable habitats. A comparison of small mammal biodiversity in mountainous areas of the southern rocky mountains indicates that Pakistan has approximately 20 % fewer small mammals than would be expected. Some regions of Pakistan have very low levels of diversity of small mammals, such as the upper Indus Valley and certain areas of Hunza. Habitats in these areas can have as few as two species, while comparable areas farther to the west may have as many as 10 species of small terrestrial mammals. Areas of especially high diversity of small mammals are found in the valley of western Chitral, such as the Bumburet Valley (Kafir Kalash valleys), and the Ziarat area of Balochistan. These areas are more xeric, and have less diverse habitats, so it is a surprise that they are so rich in small mammal biodiversity. However, they are along the western boundary of the Indian Subcontinent where small mammals from both the Arabian and Nearctic regions are part of the fauna. These include jirds (*Meriones persicus*), shrews [*Crocidura zarudnyi* (sensu latu) and *Crocidura gmelini* (sensu latu)] long-tailed hamsters (*Calomyscus baluchi* and *C. hotsoni*), mole rats (*Ellobius fuscocapillus*) and oak mice (*Dryomys nitedula and D. niethammeri*), all from the Hindu Kush highlands, Anatolian-Iranian Desert and Iranian Desert biogeographical provinces.The ranges of these species do not extend far east in Pakistan, and are largely confined to the western highlands of Balochistan and NWFP. None of these small mammals is found in Baltistan (Skardu and Ghanshe), Azad Kashmir or the Neelum and Jhelum valleys.

The biodiversity of small mammals of western Pakistan is as diverse as any comparable region in the world, even though conditions in some of these high, cold mountain ranges are extreme. Ziarat and the Bumburet Valley are "hot spots" of small mammal biodiversity in Pakistan. Other areas, such as the Palas Valley, also have high levels of biodiversity However, the ranges of some species do not extend this far east, and the overall diversity in the Palas Valley is about 30% less than it would be if the ecosystem was located west of the Swat Valley.

Comparable habitats farther to the east in Pakistan, such as the Kaghan Valley, and valleys surrounding the last major peak of the Himalayas (Nanga Parbat) are less diverse in small mammals. These valleys are out of the range of Hindu Kush and Iranian species, but still suitable for such species as flying squirrels and burrowing voles. Farther to the north and east in the Karakorams and interior

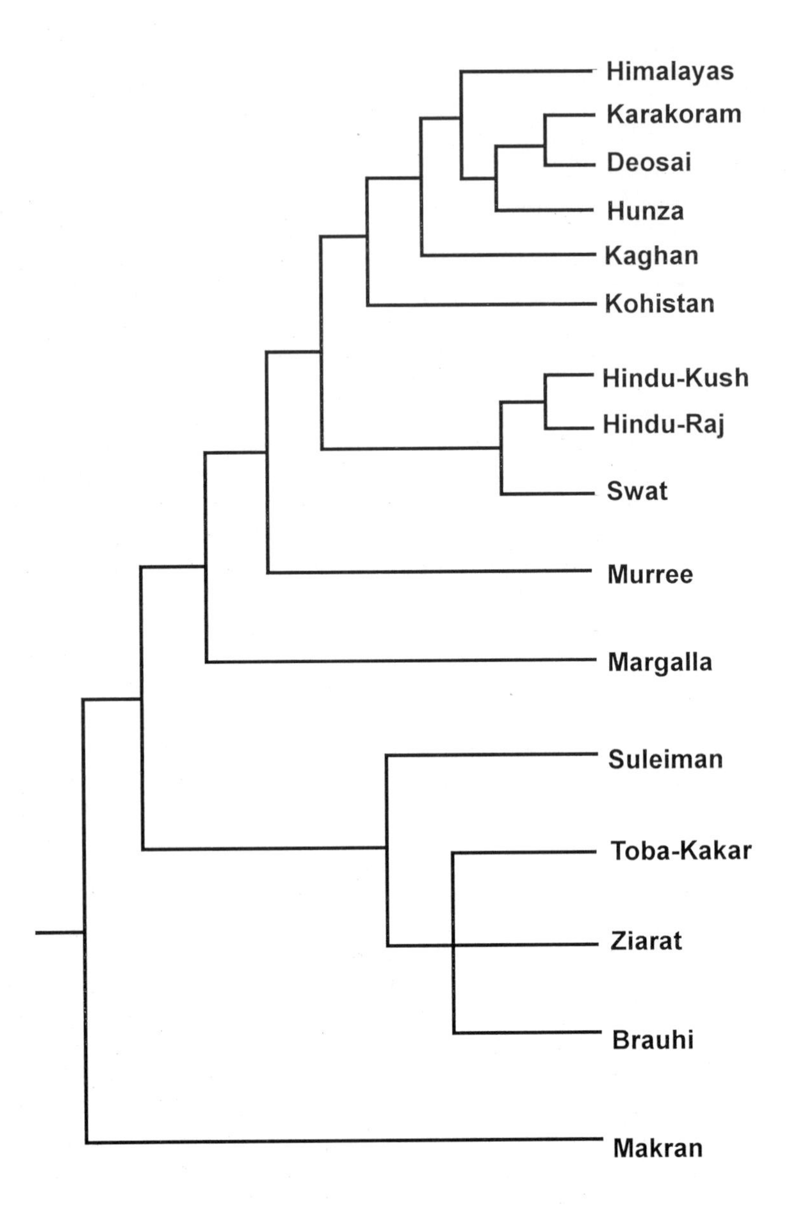

Fig. 2 : Relationship of small mammal fauna inhabiting various mountain ranges based on cladistic analysis of shared taxa of small mamals.

Baltistan (Skardu and Ghanche), the dispersal of small mammals from the south and west is blocked by the dry, rocky Indus Valley. Flying squirrels and burrowing voles drop out. The towering peaks of the Karakorams form a barrier to the north across which Palearctic shrews and voles cannot disperse. As a result, these areas of far northwestern Pakistan are depauperate.The dominant small mammals include species that have dispersed into the region from the northeast (Ladakh and Tibet) such as the Tibet red-tooth shrew (*Sorex thibetanus*) and the Chinese birch mouse (*Sicista concolor*). Small mammal diversity in some apparently ideal settings, such as the Hushe Valley or Khaplu area, consist of only 2 or 3 species (*Apodemus* sp, pale form; *Alticola* sp. pale form; *Ochotona macrotis*).

The diversity of small mammals is highest in the Ziarat Valley (Table II) of Balochistan, which not only has a large number of small mammals, but also a high level of endemicism. We recommend that Ziarat be designated a biodiversity "hot spot" and that a management plan to protect the biodiversity of the area be initiated. The reasons that Ziarat has such a high level of biodiversity and endemicism are not totally clear to us at this time, but in part relate to its location in western Pakistan near the Iranian and Palaearctic faunal sources, as well as to its role as a refugium for species from the Hindu Kush Highlands to the north. The diversity of small mammals in many areas of northern Pakistan, especially in the Murree Hills, the Kaghan Valley and the Deosai Plateau, appears to be richer than expected on the basis of geography alone because of the presence of "keystone species" (see Brown and Heske, 1990) of the genus *Hyperacrius*. These burrowing voles create elaborate underground burrow systems that are utilized by invertebrates as well as other small mammals. Where *Hyperacrius* occur, the small mammal community is more diverse and robust. Conversely, in areas of northeastern Pakistan such as the Skardu and Ghanshe where *Hyperacrius* do not occur, biodiversity is extremely limited. Because these voles depend on deep soils and mesic conditions, the distribution of *Hyperacrius* in northern and northwestern Pakistan is very disjunct. Our analyses indicate that there are many more than the currently recognized number of species and subspecies in the genus *Hyperacrius*. The large number of isolated populations of *Hyperacrius* in northern Pakistan (Fig. 1) is similar to the pattern of distribution of land snails (see K. Auffenburg, this volume). Auffenberg designated such a disjunct distribution in mesic high mountain habitats with deep soils as his " Alpine sub-province". We have illustrated the relationships of the small mammal communities in Pakistan by creating a matrix (Table VI) of mammal species plotted against the major mountain ranges of Pakistan. This data base is then used in a cladistic analysis (using PAUP) to illustrate the relationships of the ranges to each other (Fig. 2). The cladogram is rooted by the Makran Range, and demonstrates that the ranges and valleys in northern Pakistan are the most derived (and most depauperate).

When the pattern of speciation and distribution of new species and subspecies is fully documented, we expect the levels of biodiversity of small mammals in Pakistan to increase. This will increase the diversity of small mammals, especially in some communities in northern Pakistan such as the Palas Valley, and increase the number of endemic species documented in these locations as well.

ACKNOWLEDGEMENTS

This study would not have been possible without the support and encouragement of David Ferguson and the Office of International Affairs of the U. S. Fish and Wildlife Service. We also acknowledge with thanks the support of Mohammad Ahmad Farooq, the Director of the Zoological Zurvey Department, and the following members of his staff: Shamin Fakhri, Chaudrey Shafique, Aleem Khan and Hafizur Rahman. The project has also been carried out in close association with our counterpart institution the Pakistan Museum of Natural History, and we thank Dr. Shahzad Mufti (Director General) and Dr. S. Azhar Hasan (Director, Zoology) for their very important contributions and support. We also thank our PMNH colleagues Mohammad Rafique and Khalid Baig as well as Ghulam Mustafa, M. Fidah and Fiaz. Waseem Khan was a tireless and important contributor. For work in Balochistan we thank Naseer Tareen and Paind Khan of the Society for Torghar, Dr. Farooq Hameed and Tariq Hammed Dumar for work in Zarghoon, and WWF-Pakistan in Quetta (Mr. Tahir Rahshid and Ameer Mohammad Khan). Guy Duke of the Himalayan Jungle Project made our work in the Palas Valley possible.

REFERENCES

ALI, S. I. AND QAISER, M., 1986. A phytogeogeographical analysis of the phanerogams of Pakistan and Kashmir. *Proceedings Royal Society Edinburgh,* **89B**:89-101.

ARMSTRONG, D.M., 1972. Distribution of Mammals in Colorado. *Monographs Museum Natural History, University of Kansas*, **3**:1-415.

ARMSTRONG, D. M., 1987. *Rocky Mountain Mammals: a handbook of mammals of Rocky Mountain National Park and vicinity.* Colorado Associated University Press. 233 pp.

ARSHAD, M., 1991. *Ecology of Murree Vole, Hyperscrius wynnei Blandford in Galliat, Hazara, Pakistan.* Unpublished Ph.D. Dissertation, University of Peshawar, 182 pp.

BEG, A. R., 1975. Wildlife habitats of Pakistan. *Bulletin Pakistan forest Institute* (Botany Branch), **5**:56 pp.

BEG, M. A. AND HUSSAIN, I., 1990. Important rodent pests of Pakistan. Ecology and distribution. pp. 27-38 in: J. E. Brooks, E. Ahmed, I. Hussain, S. Munir and A. A. Khan (eds.). *Vertebrate Pest Management.*Pangraphics, Islamabad.

BROWN, J. H. AND HESKE, E. J., 1990. Control of a desert-grassland transition by a keystone rodent guild. *Science,* **250:**1705-1707.

CORBET, G. B. & HILL, J. E., 1992. *The Mammals of the Indomalayan Region.* Oxford. 488 pp.

ELLERMAN, J. R., 1961. *The Fauna of India including Pakistan, Burma and Ceylon*, Vol. **3,** Rodentia (in two parts). Part I, 482 pp.

HALL, E. R., 1981.*The Mammals of North America.* 2 vols. 1181 pp.

HASSINGER, J. D., 1973. A survey of the mammals of Afghanistan resulting from the 1965 Street Expedition (excluding bats). *Fieldiana, Zoology,* **60**:1-195.

HOLDEN, M. E., 1996. Description of a new species of *Dryomys* (Rodentia, Myoxidae) from Balochistan, Pakistan, including morphological comparisons with *Dryomys laniger* Felton & Storch, 1968, and *D. nitedula* (Pallas, 1778). *Bonner Zoologische Beiträge,* **46**:111-131.

HOFFMANN, R. S., 1996. *Noteworthy shrews and voles from the Xizang-Qinghai Plateau. pp 155-168 in Contributions in Mammalogy*: A memorial volume honoring Dr. J. Knox Jones Jr. Museum of Texas Tech University.

HOFFMANN, R. S., 1992. The Tibetan Plateau fauna. A high altitude desert associated with the Sahara-Gobi. pp. 285-297 in *Mammals in the Palaearctic desert: Status and trends in the Sahara-Gobian region.*(J. A. McNeely and V. Veronov, eds), Russian Academy of Sciences, Moscow, 289 pp.

HOFFMANN, R. S. AND TABER, R. D., 1968. Origin and history of Holarctic tundra ecosystems with special reference to their vertebrate faunas. pp. 143-170 in *Arctic and Alpine Environments* (H. E. Wright and W. H. Osburn, eds.), Indiana University Press, 308 pp.

JONES, J.K., ARMSTRONG, D., HOFFMANN, R. AND JONES, C., 1983. *Mammals of the Northern Great Plains.* Univ. Nebraska Press. 379 pp.

LAY, D. M., 1967. A study of the mammals of Iran resulting from the Street expedition of 1962-63. *Fieldiana, Zoology*, **54**:1-282.

MANI, M. S., 1968. *Ecology and Biogeography of High Altitude Insects.* Junk, The Hague. 527 pp.

MANI, M. S., 1974. *Ecology and Biogeography in India.* Junk, The Hague, 773 pp.

MIRZA, Z. B., 1969. *The small mammals of West Pakistan. Volume 1: Rodentia, Chiroptera, Insectivora, Lagomorpha, Primates and Pholidota.* Central Urdu Board, Lahore. 145 pp. (in Urdu).

MIRZA, Z. B., (no date). *Pakistan Mammals.* World Wildlife Fund Pakistan. Lahore. 58 pp. (in Urdu).

NUMATA, M., 1966. Vegetation and conservation in Eastern Nepal. *Journal of the College of Arts and Sciences, Chiba University*, **4**:559-69

PHILLIPS, C. J., 1969. A review of Central Asian voles of the genus *Hyperacrius*, with comments on zoogeography, ecology and ectoparasites. *Journal of Mammalogy,* **60**:457-474.

RASUL, G., (no date). *Inhabitants of the Jungle.* World Wildlife Fund Pakistan, Lahore, 62 pp. (in Urdu).

ROBERTS, T. J., 1977. *The Mammals of Pakistan.* Ernest Benn. London, 361 pp.

ROSSOLIMO, O. L., 1989. [Revision of Royal's high mountain vole *Alticola (A.) argentatus* (Mammalia: Cricedidae)]. *Zoologicheskii Zhurnal*, **68**:104-114 (in Russian).

SCHWEINFURTH, U., 1957. Die horizontale und vertikale Verbreitung der Vegetation im Himalaya. *Bonner Geograph. Abhandlungen.*, **20**:1-373.

SIDDIQI, M. S., 1961. Checklist of mammals of Pakistan with particular reference to the mammalian collection in the British Museum (Natural History)-London, *Biologia*, **7**(1-2): 93-225.

SMITH, H. C., 1993. *Alberta Mammals.* An Atlas and Guide. Provincial Museum of Alberta, Edmonton. 239 pp.

STREUBEL, D., 1989. *Small Mammals of the Yellowstone Ecosystem.* Roberts Reinhart, Boulder, CO. 152 pp.

SWAN, L. W., 1961. The ecology of the High Himalaya. *Scientific American,* **205** (4):68-78.

SWAN, L. W., 1963. *Ecology of the heights.* Natural History, 23-29.

TROLL, C., 1969. Die Klimatische und Vegetationsgeographische Gliederung des Himalaya Systems. *Khumbu Himal.,* **1**:353-88

UDVARDY, M. D. F., 1975. *A classification of the biogeographical provinces of the world.* IUCN Occasional Paper, 18:1-49.

UDVARDY, M. D. F., 1985. *Biogeographical Provinces of the Indomalayan Realm, Western Part and Eastern Part.* Revised maps for a classification of the Biogeographical Provinces of the World (UNESCO Man and the Biosphere Program, Project no. 8.

VON WISSMANN, H., 1961. Stufen und Gürtel der Vegetation und des Klimas in Hochasien und seinen Randgebieten. *Erdkunde, Arch. f. Wiss. Geogr.*, **15**:19-44.

WILSON, D.E. & REEDER, D. M., 1992. *Mammal Species of the World. Smithsonian*, Washington, 1206 pp.

ZAHLER, P., 1996. Rediscovery of the Woolly Flying Squirrel (*Eupetaurus cinereus*). *Journal of Mammalogy*, **77**(1):54-57.

ZAHLER, P. AND WOODS, C A., (this volume). The status of the woolly flying squirrel (*Eupetaurus cinereus*) in northern Pakistan. (In association with Naeem Dar, Akhtar Karim, Waseem Ahmad Khan, C.W. Kilpatrick & Mohammad Rafique).

STATUS OF SULEIMAN MARKHOR AND AFGHAN URIAL POPULATIONS IN THE TORGHAR HILLS, BALOCHISTAN PROVINCE. PAKISTAN

KURT A. JOHNSON

15250 S.E. 43 Court, Apt. H-202, Bellevue, WA 98006-2573, U.S.A.

The Torghar Hills of Balochistan Province, Pakistan, part of the Pushtun (or Pathan as they are called in the West) tribal areas located on the border with Afghanistan, have long been known for their abundant wildlife, most notably their populations of Suleiman markhor (*Capra falconeri jerdoni*; also called straight-horned markhor, *C. f. megaceros*) and Afghan urial (*Ovis vicfnei cycloceros* or *O. orientalis cycloceros*) (District Gazetteers of Balochistan 1906). The Torghar Hills are inhabited by the Jalalzai branch of the Sanzar Khel tribe of Kakar Pathans (Naseer A. Tareen, pers. comm.). The Pathans are the world's largest tribal society, with their own language (Pushto), culture, and complex tribal hierarchy (Caroe, 1958; Spain 1972; Quddus, 1987). The local ruling family of Pathans, the Jogezais, exercises authority over the tribesmen of Torghar.

In the early 1980s the local Pathan leader, the late Nawab Taimur Shah Jogezai, and one of his relatives, Sardar Naseer A. Tareen, became alarmed at what they perceived to be a dramatic decline in Suleiman markhor and Afghan urial populations in Torghar (Tareen no date, 1990a, 1990b) . They attributed this decline to an increase in poaching brought about by a dramatic influx of weapons and, especially, cheap ammunition into the area during the Afghan war (the 14-year war that started in 1978 between Muslim guerrillas and a series of Soviet-backed governments). The Nawab asked the Balochistan Forest Department (BFD) to station wildlife officers in Torghar to solve the poaching problem. The BFD did not respond to the request, so the Nawab and Tareen decided they would have to solve the problem themselves. In 1984, they sought assistance from professional wildlife biologists in the USA, with whom they developed the idea for the Torghar conservation Project (TCP).

As a private, "grassroots" initiative, the TCP's goals were to conserve local populations of the Suleiman markhor and Afghan urial, and to improve the economic condition of local Pathan tribesmen in the Torghar Hills. The Project's design was simple. Local Pathan tribesmen under the authority of Nawab Jogezai would be requested to refrain from hunting in exchange for being hired as salaried game guards to prevent poachers from entering the Torghar Hills to hunt. Game guard salaries and other costs of the TCP would be defrayed entirely by trophy fees paid by foreign hunters for the privilege of hunting a small, strictly controlled number of Suleiman markhor and Afghan urial. The TCP was instituted in 1986 and run informally by Tareen and members of the Jogezai family until April 1994. At that time, they formed an officially-registered non-governmental organization, the Society for Torghar Environmental Protection (STEP), to administer the TCP.

As of November 1994, the TCP employed 33 local Pathan game guards who were protecting an area of approximately 1,000 sq. km. in the Torghar Hills and the adjacent Tora Range. Game guards are concentrated in the "core protected area", an area of approximately 300 sq. km. in the heart of Torghar. The Project has achieved a virtually-complete cessation of poaching of markhor and urial by both locals and outsiders, especially within the core protected area. As a result, the Suleiman markhor and Afghan urial populations of Torghar, which were virtually extirpated by 1983-84, have grown steadily since 1985-86. The TCP has been largely self-sufficient since its inception, having relied primarily on the income generated through the trophy harvest of only 14 markhor and 20 urial over its 10-year history.

Although the TCP has succeeded in curtailing poaching, and this has lead to recovery of the markhor and urial populations in Torghar, STEP has received little recognition for this success, in large measure because the Project has lacked quantitative data on the size and growth of markhor and urial populations. To address this need, STEP decided to conduct a systematic field survey of the Suleiman markhor and Afghan urial populations of Torghar in November 1994. This paper presents the results of the survey, addresses some problems encountered during the survey and discusses management implications of the results.

STUDY AREA

The TCP is located in the Torghar Hills (Tor Ghar means "Black Mountain" in Pushto), a chain of ruggedly-upturned, predominantly sedimentary ridges approximately 90 km. long and 15-30 km. wide in the NE-SW trending Toba Kakar Mountains of Qila Saifullah District, Zhob Division, Balochistan Province. The approximately 300 sq. km. "core protected area" of Torghar consists of three parallel ridges separating two NE-running stream drainages. The southernmost ridge has a north-facing slope which rises gradually to an elevation of almost 2,800 meters, and is dissected by a number of deeply-incised drainages. From the ridge, the southfacing slopes drops precipitously in a step-like series of cliffs to the Khaisore Valley. The northern ridges consist of rock layers that are so steeply upturned that they resemble a series of parallel, jagged-toothed combs. The climate of Torghar is seasonal and arid. The vegetation is semi-desert or shrub-steppe dominated by perennial bunchgrasses, shrubs and scattered trees--primarily junipers (*juniperus macropoda*) and wild pistachios (*Pistacia khaniak*). Domestic livestock--primarily sheep and goats but also some donkeys and camels--are grazed on the valley bottoms and lower slopes.

SPECIES

The TCP is primarily concerned with conservation of the Suleiman markhor and Afghan urial. Both species are listed on the Third Schedule of The Balochistan Wildlife Protection Act 1974; these are "protected animals, i.e., animals which shall not be hunted, killed or captured" except as permitted under specific circumstances (Government of Balochistan Agriculture Department, 1977). The Suleiman subspecies of markhor has a limited distribution that includes the rugged mountains of western Pakistan (e.g., Suleiman, Takatu and Toba Kakar Ranges) and some of Afghanistan (Roberts 1977). It has been extirpated from much of its former range, and now occurs in very low numbers (Roberts 1977). It is listed as "Endangered" under the U.S. Endangered Species Act (ESA) (Fish and Wildlife Service 1994) and is included in Appendix I of the Convention on International Trade in Endangered Species of Wild Fauna and Flora (CITES) (Fish and Wildlife Service 1992). The Afghan urial is more widespread and common than the markhor, although it is not abundant (Roberts 1977). It is not listed on either ESA or CITES.

The local Pathan tribesmen of Torghar have a wealth of firsthand knowledge about the natural history of both the markhor and urial. They report that markhor prefer extremely rugged terrain, with bare rock surfaces and precipitous cliffs intermingled with small patches of vegetation (trees, shrubs, and/or bunchgrasses). Urial prefer more gradual, open slopes with a more continuous vegetative cover of bunchgrasses and shrubs. Habitat usage, however, is not mutually exclusive. Both species will utilize many of the same areas.

MATERIALS AND METHODS

Field Surveys

Population surveys of desert ungulates such as markhor and urial are notoriously difficult because of the species' high mobility, cryptic-coloration, and shy nature (i.e., they can detect and will flee from humans at a great distance). The situation at Torghar is even more difficult because the extremely rugged terrain makes aerial surveys from fixed-wing aircraft all but impossible, helicopters are not readily available, and the politically sensitive location of Torghar near the Afghan border precludes all flights except by the military. Ground surveys are the only viable option. Although systematic transacts would be the best approach from the standpoint of quantifying results, the rugged terrain at Torghar precludes this choice. The only viable option at Torghar is fixed-point counts of areas which can be readily identified on maps. And, while fixed points rarely afford a full view of the survey area (e.g., because canyon bottoms may be hidden) , utilizing several observers and observing throughout the morning and afternoon hours (active periods for the markhor and urial) can increase the probability that animals will be detected.

The nine-day survey was conducted from 3-13 November 1994. November is the rut season for both Suleiman markhor and Afghan urial. This time period was selected because markhor and urial form large groups during the rut, and larger groups have a higher probability of detection by human observers. Five major survey areas within the TCP in core protected area were pre-selected: Tanishpa, Khand, Shin Narai, Tor Aghbarg, and Kundar/Uria. These spatially-separated areas were selected in an attempt to minimize the possibility of double counting.

Each day during the survey, from one to three survey teams (consisting of two to five game guards and STEP's Manager for Natural Resources or the author) went into the field at sunrise and selected a point from which to conduct the day's survey. A prominent, readily-identifiable point was selected to maximize the amount of terrain that could be surveyed, and to ensure that the point could be relocated for use in future surveys.

The survey team(s) spent morning and afternoon hours continuously surveying the area through binoculars (10 x 35) and spotting scope (15-60 . variable power). The game guards are particularly adept at locating distant ungulates as the animals rest against a background of shadows, boulders, cliffs, and vegetation. The tribesmen see with binoculars what most nonnative people can see only through a spotting scope set at 30x. Survey points were occasionally changed in the middle of the day if few animals were observed in the morning.

For each group of markhor or urial located, data were recorded on total number in group and number in each sex/age category: lambs (5 years old) , yearling females (1.5 years old), yearling males (1.5 years old), adult females (2.5 years or older), adult males I (2.5 to 5.5 years of age) , and adult males II (trophy animals of 6.5 years or more). It was not always possible to accurately classify animals observed at a distance, and the author suspects that some yearling males of both species may have been incorrectly classified as adult females. Conversely, the tribesmen are very good at classifying adult males at a distance.

Population Size Calculation

Estimated population sizes for markhor and urial in the core protected area and for the entire TCP area were calculated as follows. First, the size of each of the five major survey areas was estimated. Using an old (surveyed in 1910-11) topographic map of Torghar (1:50, 000-scale) , the approximate boundaries of each major survey area were drawn (exact boundaries were impossible to determine because of the terrain's ruggedness, the lack of Global Positioning System (GPS) receivers in the field which precluded determination of precise locations, and the inadequate quality and scale of the map). The boundaries of each survey area were then enlarged in order to produce a conservative estimate of markhor/urial density, and altered into geometric shapes (i.e., straight lines and right angle corners) in order to utilize the map's 1,000-yard by 1,000-yard grids to calculate area. The size of each major survey area was then calculated by counting complete and partial grid squares contained within the drawn boundaries.

Second, a population density for each species was calculated for each major survey area by dividing the number of individuals counted in the survey area by the size of the survey area. This calculation was based on the assumption that the number of markhor or urial counted in each major survey area was the total number of markhor or urial actually in that area (i.e., there was no "visibility bias" and no double counting took place)—a questionable assumption that will be discussed later in the paper.

Third, a population density for markhor in high-quality habitat was calculated by averaging population densities in the three major survey areas that had the most markhor (Tanishpa, Khand, and Kundar/Uria). Population density for markhor in low-quality habitat was calculated from the remaining two major survey areas (Shin

Narai and Tor Aghbarg) . A population density for urial in high- and low-quality habitat was calculated in the same manner. Population densities for three major survey areas (Shin Narai, Tanishpa, Tor Aghbarg) were averaged to obtain a population density for urial in high-quality habitat. Population densities for the remaining two survey areas (Khand, Kundar/Uria) were averaged to obtain low-quality habitat population density.

Fourth, the amount of high-quality markhor habitat and high-quality urial habitat in Torghar were crudely estimated. Habitat quality types (i.e., high and low) were based on tribesmen's, descriptions of preferred habitat for both species and actual numbers counted in the field. High-quality markhor habitat is extremely rugged terrain, with bare rock surfaces and precipitous cliffs intermingled with small patches of vegetation (trees, shrubs, and/or bunchgrasses). High-quality urial habitat is more gradual, open slopes with a more continuous vegetative cover of bunchgrasses and shrubs. The areal extent of each habitat type was roughly estimated from the map.

An estimated population size for markhor in the core protected area was then calculated by: (1) multiplying high-quality population density by the total size of the core protected area (300 sq. km.) by the percent of high-quality habitat in the core protected area; (2) multiplying low-quality population density by 300 sq. km. by the percent of low-quality habitat in the core protected area; and (3) adding the results of (1) and (2) . Results were then extrapolated to the remainder of the TCP project area.

RESULTS AND DISCUSSION

Markhor

During the nine-day survey, 135 Suleiman markhor were counted (Table I) , including 24 lambs, 6 yearling females, 2 yearling males, 75 adult females, 12 adult males under six, and 16 adult males older than six. It is suspected that a few yearling males may have been incorrectly classified as adult females.

The population density for markhor in high-quality habitat was calculated to be 3.22 markhor per sq. km.; the population density in low-quality habitat was 0.47 markhor per sq. km. High-quality markhor habitat was estimated to cover 30 percent of the core protected area, leaving the remaining 70 percent as low-quality habitat. An estimated population size for markhor in the core protected area was then calculated by: (1) multiplying high-quality population density (3.22 markhor per sq. km.) by the size of the core protected area (300 sq. km.) by the percent of high-quality habitat in the core protected area (30 percent), yielding an estimate of 290 markhor in high-quality habitat; and (2) multiplying low-quality population density (0.47 per sq. km.) by 300 sq. km. by 70 percent low-quality habitat, yielding an estimate of 99 markhor in low-quality habitat. Adding the two, the estimated markhor population size in the core protected area is 389 animals.

Although this simple calculation may not be subject to statistical treatment (i.e., no confidence limits can be derived), it is a conservative approach because the assumptions and approximations used to calculate population size and density were conservative. For instance, the size of each major survey area was overestimated, resulting in an underestimate of markhor population density in each survey area. It is also likely that there was considerable "visibility bias" (sensu Pollock and Kendall 1987), i.e., the probability of detection of an individual markhor or urial within each major survey area was considerably less than the assumed 100 percent. It has been repeatedly demonstrated that visibility bias affects aerial survey estimates of big game species, including bighorn sheep (*Ovis canadensis*) (Bodie et al. 1995, Neal et al. 1993) and Dall sheep (*Ovis dalli*) (McDonald et al. 1990). Bodie et al. (1995) estimated that they observed, on average, 67.1 % of the bighorn sheep population during helicopter surveys of sagebrush-steppe canyon habitats in southwestern Idaho. Visibility bias for ground surveys would be expected to be even greater than for aerial surveys. Any visibility bias at all would result in an underestimate of markhor (and urial) populations.

To calculate an estimate of the markhor population in the remainder of the TCP area, it was assumed that the remainder of the TCP area covers 650 sq. km. (a conservative approximation), and that all markhor habitat in that area is of low quality (another conservative approximation) . Thus, multiplying 650 sq. km. by 0.47 markhor per sq. km. yields a population estimate of 306 markhor in the remainder of the TCP area. The two population estimates can then be combined for an estimated total population of 695 Suleiman markhor in the TCP area.

Assuming that a few yearling males were incorrectly classified as adult females and that lambs had a lower probability of detection than other age categories, then survey results suggest that the Suleiman markhor population in Torghar is reproducing well and that survival of adult males into older age categories is good. This further suggests that markhor habitat in Torghar is in good condition (because it is not limiting reproduction) and that the population is not being over-hunted (an over-hunted population would be expected to have fewer males in the older age categories).

Most markhor were observed in Tanishpa (Malao, Art, Garai), Khand, and Kundar/Uria (Uria, Murdara, Zerzha, Salawat, Surkham). These are among the most rugged parts of Torghar, with exposed rocks and precipitous cliffs making up most of the habitat. These results thus corroborate the tribesmen's first-hand knowledge of markhor habitat preferences.

Urial

During the nine-day survey, 189 Afghan urial were counted (Table I), including 12 lambs, 10 yearling females, 0 yearling males, 122 adult females, 25 adult males under six, and 20 adult males older than six. The author suspects that, as with markhor, a few yearling male urial may have been incorrectly classified as adult females.

Estimated urial population sizes for the core protected area and for the entire TCP area were calculated in the same manner as the markhor estimates. An average population density for high-quality urial habitat was calculated to be 4.45 urial per sq. km. average population density for low-quality urial habitat was calculated to be 0.77 urial per sq. km. High-quality urial habitat was estimated to cover 40 percent of the core protected area, while low-quality habitat covered the remaining 60 percent. In comparison, Edge and Olson-Edge (1987) documented urial population densities in the range of 1.7-2.5 urial per sq. km. in their study area in Kirthar National Park, Sindh Province, Pakistan, although they concluded that the study area could sustain a higher density of urial

An estimated population size for urial in the core protected area was calculated by: (1) multiplying 4.45 urial per sq. km. by 300 sq. km. by 40 percent, yielding an estimate of 534 urial in high quality habitat; and (2) multiplying 0.77 urial per sq. km. by 300 sq. km. by 60 percent, yielding an estimate of 138 urial in low-quality habitat. Thus, the estimated population size of urial in the core protected area is 672.

An estimate of the urial population in the remainder of the TCP area was calculated in the same manner as the markhor estimate. It was assumed that the remainder of the TCP area covers 650 sq. km., and that all urial habitat in the area is low-quality. This yields an estimate of 501 urial (650 sq. km. x 0.77 urial per sq. km.). These two figures can then be combined for an estimated total population size of 1,173 urial in the entire TCP area. Most urial were counted in Shin Narai, Tor Aghbarg (Saliwata, Whuchhwokai, and Bazalai) - some of Torghar's less rugged and more gently sloping habitats. But a substantial number of urial were also seen in areas where markhor were also abundant, most notably Khand and Tanishpa. These results, therefore, also corroborate the tribesmen's' first-hand knowledge of urial habitat preferences.

Table I. Numbers of Suleiman markhor and Afghan urial counted in survey areas in the Torghar Hills, Balochistan Province, Pakistan, November 1994. (Tot = total number counted, LB = lamb, YF = yearling female, YM = yearling male, AF = adult female, AMI = adult male less than 6 years old, AMII adult male greater than 6 years old)

Location(Date)	Markhor							Urial						
	Tot	LB	YF	YM	AF	AMI	AMII	Tot	LB	YF	YM	AF	AMI	AMII
Tanishya														
Malao(11/3)	17	2	0	0	13	0	2	16	0	0	0	12	4	0
Garai(11/3)	24	3	2	1	12	3	3	12	2	1	0	5	1	3
Art (11/ 3)	19	5	0	0	9	3	2	0	0	0	0	0	0	0
Shin Narai														
Thakarai(11/4)	0	0	0	0	0	0	0	25	7	0	13	0	2	3
Thakarai(11/4)	0	0	0	0	0	0	0	0	0	0	0	0	0	0
Khand														
Khand(11/5)	21	2	3	0	12	1	3	6	0	2	0	4	0	0
Khand(11/5)	0	0	0	0	0	0	0	8	0	0	0	6	1	1
Tor Aghbara														
Walla(11/7)	0	0	0	0	0	0	0	0	0	0	0	0	0	0
Saliwata(11/8)	0	0	0	0	0	0	0	35	0	0	0	22	2(7)	4
Whuchhokai(11/9)	4	0	1	0	1	1	1	59	2	7	0	40	5	5
Bazalai(11/10)	0	0	0	0	0	0	0	17	0	0	0	12	2	3
Saiduchina(II/II)	7	0	1	1	4	1	0	7	0	0	0	6	0	1
Kundar/Uria														
Uria(II/12)	0	0	0	0	0	0	0	0	0	0	0	0	0	0
Murdara(11/12)	15	3	0	0	9	3	0	0	0	0	0	0	0	0
Zerzha(11/13)	15	5	0	0	8	0	2	4	1	0	0	2	1	0
Salawata(11/13)	0	0	0	0	0	0	0	0	0	0	0	0	0	0
Surkham(11/13)	13	3	0	0	7	0	3	0	0	0	0	0	0	0
TOTALS	**135**	**24**	**6**	**2**	**75**	**12**	**16**	**189**	**12**	**10**	**0**	**122**	**25**	**20**

Markhor and Urial Population Changes

Numerous first-person accounts from game guards (pers. comm. with Sagzai, Khoshalay, Noordad and others), managers of the TCP (pers. comm. with Naseer A. Tareen and Mirwais Jogezai), and visitors to Torghar (pers comm. with Reza Abbas) as well as observations made by a wildlife biologist from Fall 1985 to Spring 1988 (Mitchell 1989) and the literature (Roberts 1977) strongly suggest that the Suleiman markhor and Afghan urial populations of the Torghar Hills were at very low levels in the early 1980s before the TCP was instituted (Johnson 1994). After the TCP began in 1986, both populations appear to have begun growing steadily, a conclusion supported by results of the November 1994 survey.

The most likely cause of this population growth is the substantial reduction in human-caused mortality that occurred when the TCP went into effect and uncontrolled hunting was stopped. A large source of mortality (uncontrolled hunting) was replaced by a much smaller source (controlled, limited trophy hunt). Other than stopping uncontrolled hunting, the TCP has not instituted any management practices (e.g., elimination of competition from domestic livestock, creation of water holes) that could have contributed to this degree of population growth.

Populations of markhor and urial in Torghar are likely to continue growing as long as hunting is controlled. Reducing the size of domestic herds of sheep and goats should also help both species, but especially the urial, by reducing competition for forage and reducing the possibility of disease transfer from domestic to wild animals. Carrying capacity should be reached, and the Torghar Hills may become a source of emigrants for other mountain ranges in the area (intermountain movement of markhor and urial is probably already taking place (see Bleich et al. 1990).

Population Viability for Sustainable Harvesting

Today, the Suleiman markhor and Afghan urial populations of Torghar are of adequate size and condition (in terms of sex and age ratios and reproduction) to be considered "viable" for both population and genetic processes (e.g., Soule 1987, Hebert 1991). Concern about population isolation and fragmentation appear to be unfounded, especially if intermountain movement of urial and markhor is taking place (Bleich et al. 1990).

Such populations should be able to sustain an annual "trophy harvest of males, in numbers equivalent to 1-2 percent of the total population size," without negative consequences to the population according to Harris' (1993) review of the literature on similar species. This is true, in part, because markhor and urial have a polygynous mating system, and the populations' overall reproductive rates would be little affected by the loss of a small number of males (Caughley 1977, Schaller 1977).

Assuming a total markhor population of 300 animals in the core protected area, a sustainable annual trophy harvest in the core area should be 3-6 markhor. And, assuming a total urial population of 675 animals in the core protected area, a sustainable annual trophy harvest in the core area should be 7-13 urial. Since no more than 3 adult male markhor and 4 adult male urial have been harvested in any given year at Torghar, it is apparent that harvest levels have been conservative in accordance with the "precautionary principle" (Freese 1994). The simple fact that both populations have continued to grow steadily while subject to a strictly-controlled trophy hunt is ample evidence that harvest levels have been conservative.

CONCLUSION AND RECOMMENDATIONS

Results of the November 1994 survey of Suleiman markhor and Afghan urial, perhaps the first formal survey ever conducted on these two species, indicate that the TCP area may have a population of around 700 markhor and 1,200 urial. Although most mountain ranges in Balochistan have not been formally surveyed, these results suggest that Torghar may be one of the last remaining strongholds for both species. Surveys results also support the conclusions that: (1) these populations have increased, most likely substantially, from their levels in the mid-1980s; and (2) these populations are of adequate size and condition to support a limited trophy hunt. Taken together, these conclusions support the further conclusion that the Torghar Conservation Project has been a successful tool for conserving the markhor and urial of the Torghar Hills (Johnson 1994). If applied elsewhere in Balochistan (e.g., Takatu, Tora, Balol Nikah), the TCP model could help recover other markhor and urial populations which now verge on extirpation.

The TCP should begin conducting markhor and urial population surveys each year (during the rutting season) . Only data collected over a number of years will adequately document population changes that have resulted from the TCP's management program. Spring surveys would be useful for obtaining an index to each population's annual reproductive performance. Annual survey data would allow greater 'fine-tuning" of harvest limits in accord with the principles of adaptive management (Walters 1986).

The population size estimates presented in this paper are only as accurate as the assumptions and approximations used in the calculations. Additional effort must be expended to further refine and improve the accuracy of these assumptions/approximations. First, major survey areas must be delineated more precisely on an accurate, large-scale topographic map. Such delineation will improve the precision of estimates of the total area surveyed, thereby improving the precision of population estimates. Use of GPS technology would allow survey points and major survey area boundaries to be accurately located on topographic maps.

Second, more major survey areas must be identified and surveyed, especially outside the core protected area, in order to improve the precision of density estimates in each habitat type. Third, the actual areal extent of each habitat type needs to calculated more accurately, thus allowing further refinement of density estimates. Analysis of satellite imagery in a Geographic Information System (GIS) framework, followed by ground truthing, will provide more accurate estimates of the quantity and distribution of habitat types.

The field methodology used in the November 1994 survey could benefit from a number of improvements. First, all major survey areas should be surveyed on the same day. This would reduce the opportunity for double counting animals that move long distances from day to day. Second, all game guards participating in the survey should have modern, high-powered binoculars. In November 1994, many guards used old, poor-resolution binoculars made in the former Soviet Union. This undoubtedly contributed to visibility bias (i.e., increased the number of animals missed). Third, combining GPS with accurate rangefinding equipment would allow observed groups of ungulates to be pinpointed accurately on maps.Novel approaches to population sampling may prove useful in Torghar.

Ultra-light aircraft or a hot air balloon may be sufficiently quiet and safe to allow for more-accurate aerial surveys to be conducted. In addition, it may be possible to use capture-resight methodology (e.g., Minta and Mangel 1989) to improve ground survey estimates of markhor and urial populations. Although it would be very difficult to capture and mark markhor and urial by traditional means (e.g., dart gun or remotely fired net-gun (Edge et al. 1989)), it may be possible simply to mark them with splotches of dye fired from a powerful paint ball gun. Re-sightings made during ground surveys could then be analyzed to obtain a statistically-valid population estimate. These approaches should be given further consideration.

Additional ecological studies of markhor and urial need to be undertaken in Torghar. Studies of habitat use/preference and breeding biology could be undertaken for relatively little cost or need for sophisticated equipment.

ACKNOWLEDGMENTS

Financial and logistical support for this survey was provided by the office of International Affairs of the United States Fish Service (OIA, FWS) and the Society for Torghar and Wildlife Service (OIA, FWS) and the Society for Torghar Environmental Protection (STEP). Special thanks in this regard are due to David Ferguson (OIA, FWS) and Naseer A. Tareen (STEP) . Additional thanks go to TCP managers (Maboob Jogezai, Mirwais Jogezai, Paind Khan), workers (Sharafhuddin), and game guards (especially Khoshalay, Abdullah, Sagzai, Noordad, Safferkhan, Khodaidad, Baqidad, Mohammad Afzal, Abdul Raziq, Janan, Piao, Mohammad Khan) without whose tireless assistance the fieldwork would not have succeeded. Drafts of the manuscript benefited from a critical review by Dave Ferguson (OIA, USFWS).

REFERENCES

BALOCHISTAN GOVERNMENT AGRICULTURE DEPARTMENT. 1977. The Balochistan Wildlife Protection Act 1974. And Rules Notified Thereunder with (Urdu Translation) (As amended up to December 13, 1977). 26 pp. + Urdu translation.

BLEICH, V.C., WEHAUSEN, J .D. AND HOLL, S.A., 1990. Desert-dwelling sheep: Conservation implications of a naturally fragmented distribution. *Conservation Biology*, **4**: 383-390.

BODIE, W. L., GARTON, E .O., TAYLOR, E .R. AND MCCOY, M., 1995. A sightability model for bighorn sheep in canyon habitats. *Journal Wildlife Management,* **459**: 832-840.

CAROE, O., 1958. The Pathans 550 B.C - A.D. 957. MacMillan and Company Ltd . , London, U.K. 521 pp. + 1 map

CAUGHLEY, G. 1977. *Analysis of vertebrate populations*. John Wiley & Sons, New York. 234 pp.

DISTRICT GAZETTEERS OF BALOCHISTAN, 1906. *The gazetteer of Balochistan (Zhob*). Government of Balochistan, Quetta, Balochistan. 351 pp. + folding map.

EDGE, W. D. AND OLSON-EDGE, S. L., 1987. Ecology of wild goats and urial in Kirthar National Park: A Final Report. Montana Cooperative Wildlife Research Unit, University of Montana, Missoula. 34 pp.

EDGE, W .D., OLSON-EDGE, S. L. AND O'GARA, B. W., 1989. Capturing wild goats and urial with a remotely fired net-gun. Aust. *Wildlife Research,* **16**: 313-315.

FISH AND WILDLIFE SERVICE, 1992. Appendices I, II and III to the Convention on International Trade in Endangered Species of Wild Fauna and Flora, September 30, 1992. U.S. Department of the Interior, Washington, D.C. 22 pp.

FISH AND WILDLIFE SERVICE, 1994. Endangered and threatened wildlife and plants, 50 CFR 17.11 & 17.12, August 20, 1994. U. S. Department of the Interior, Washington, D.C. 42 pp.

FREESE, C. (compiler), 1994. The commercial, consumptive use of wild species: Implications for biodiversity conservation. A WWF International Interim Report. October 1994. 60 pp.

HARRIS, R.B., 1993. Wildlife conservation in Yeniugou, Qinghai China: Executive summary. Unpublished Ph.D. Dissertation. University of Montana, Missoula, MT. 10 pp.

HEBERT, D., 1991. Appendix 4: Theoretical considerations in determining proper ESA listing for argali. Pages 120-121 in M. Lillywhite, ed. *Status review, critical evaluation, and recommendations on proposed threatened status for argali Ovis ammon)* prepared by Domestic Technology International, Evergreen, CO for Safari Club International. 123 pp.

JOHNSON, K. A., 1994. Torghar Conservation Project, Balochistan Province, Pakistan. Technical report to WWF-International, Gland, Switzerland. 52 pp. + appendices.

MCDONALD, L .L., HENRY, H.B. MAUER, F. J. AND BRACKNEY, A. B., 1990. Design of aerial surveys f or Dall sheep in the Arctic National Wildlife Refuge Alaska. *Bi-annual Symposium Northern Wild Sheep and Goat Council.* **7**: 176-193.

MINTA, S. AND MANGEL, M., 1989. A simple population estimate based on simulation f or capture-recapture and capture-resight data. *Ecology,* **70**: 1738-1751.

MITCHELL, R., 1989. Status of large mammals in the Torghar Hills, Balochistan. Unpublished report to U.S. Fish and Wildlife Service, Office of International Affairs. 17 pp. + 2 figures and 2 tables.

NEAL, A.K., WHITE, G.C., GILL, R.B. REED,D.F. AND OLTERMAN, J. H. 1993. Evaluation of mark-resight model assumptions for estimating mountain sheep numbers. *Journal Wildlife Management,* **57**: 436-450.

POLLOCK, K. H. AND KENDALL, W. L., 1987. Visibility bias in aerial surveys: A review of estimation procedures. *Journal Wildlife Management,* **51**: 502-510.

QUDDUS, S. A., 1987. *The Pathans.* Ferozsons (Pvt.) Ltd., Lahore. 342 .

ROBERTS T. J., 1977. *The Mammals of Pakistan.* Ernest Benn Ltd., London, UK. 361 pp.

SCHALLER, G., 1977. *Mountain monarchs: Wild sheep and goats of the Himalaya.* Wildlife Behavior and Ecology Series, University of Chicago Press. Chicago, IL. pp.

SOULE, M., 1987. *Viable populations for conservation.* Cambridge University Press, Cambridge, U.K. 189 pp.

SPAIN, J.W., 1972. *The way of the Pathans,* 2nd edition. Oxford University Press, Karachi. 190 pp.

TAREEN, N.A., no date. Sworn declaration regarding the Torghar Conservation Project. 11 pp.

TAREEN, N.A., 1990a. Torghar Conservation Project. Unpublished manuscript. 13 pp.

TAREEN, N.A., 1990b. Torghar: The Black Mountain of hope. *Natura* (World Wide Fund for Nature-Pakistan), July 1990 issue, pp. 18-20.

WALTERS, C.J., 1986. *Adaptive management of renewable resources.* McGraw-Hill, New York.

DISTRIBUTION, POPULATION STATUS AND CONSERVATION OF BARKING DEER IN THE MARGALLA HILLS NATIONAL PARK

MAQSOOD ANWAR

Rangeland Research Institute, NARC, Park Road, Islamabad

Abstract: A study of Barking deer *(Muntiacus muntjak)* was conducted in the Margalla Hills National Park (MHNP) to determine its distribution range, population & habitat status and conservation problems. In the MHNP, barking deer is associated with dense scrub forest in the northern and central part of the Park between 500 and 1200m elevation. Presently, about 7000 ha (55 % of total Park area) is occupied by barking deer. Another 3000 ha (23 % of total Park area) is potential for becoming barking deer habitat but, currently, no animal is found there. This might be due to low vegetation cover and disturbance caused by extensive activities of human and livestock populations in these areas. Barking deer population in the MHNP has been estimated between 100 and 120 individuals. They are found in forest areas with more than 70% hiding cover. Such plant species as *Carissa opaca, Dodonaea viscosa, Adhatoda vasica, Acacia modesta, Maytenus royleana, Oleaferruginea, Zizyphus nummularia, Woodfordia fruticosa, Myrsine africana* etc. make up more than 60 % of vegetation cover in barking deer habitat. Habitat disturbance caused by grazing livestock populations and hay & fuelwood cutting by human populations are major problems faced by barking deer in the MHNP. Distribution range and population of barking deer can be increased through controlling livestock and human activities by reducing their number inside the Park and restricting their activities outside the barking deer habitat.

INTRODUCTION

Barking deer is a smaller ungulate as compared to most other Cervids. A large buck is reported to weigh 22 kg and female 15 kg (Roberts, 1977). Males develop two long bony pedicles covered with skin. Antlers are supported by pedicles which are longer (75-lOOmm) than in any other deer. Two bony ridges extend down from the base of these pedicles and front of the skull. These prominent ridges are the basis for the name "rib-faced deer" often given to this animal. Females also develop pedicles but much shorter than males. Males

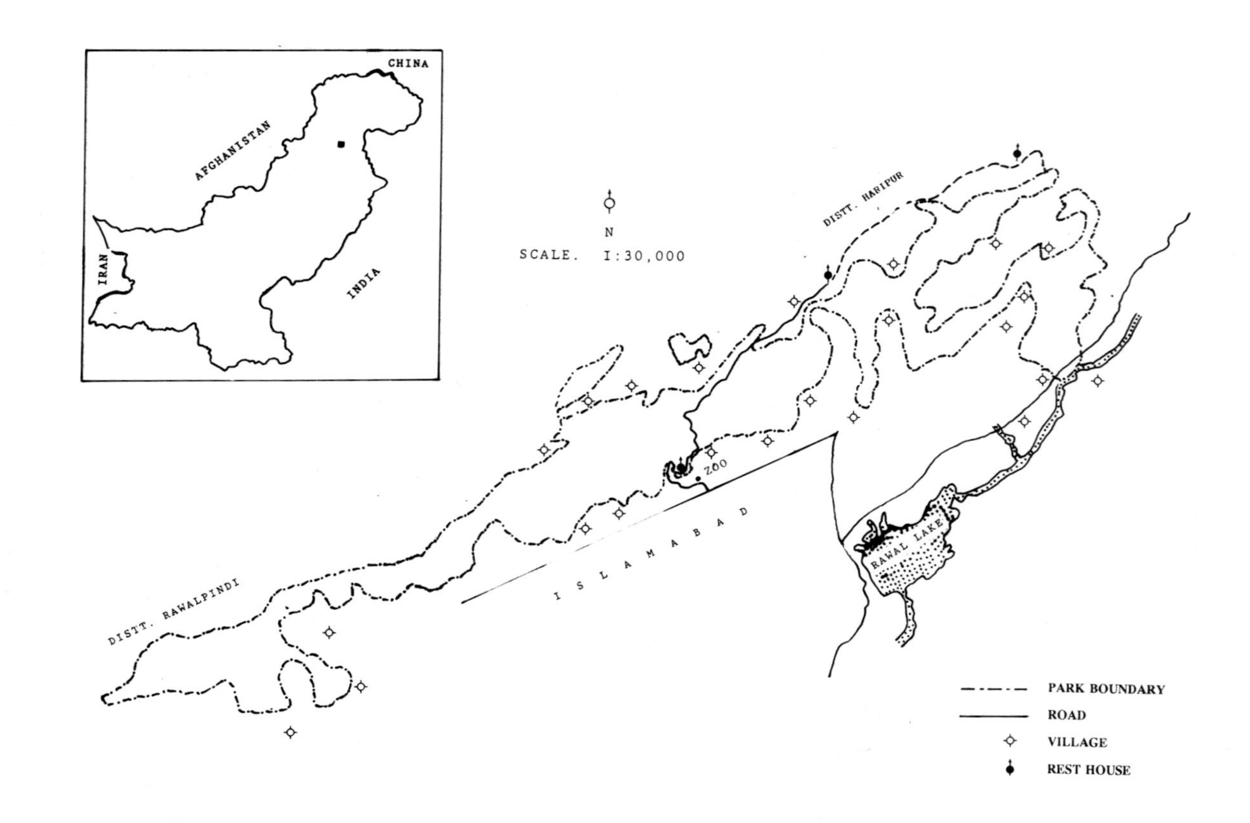

Figure 1. Map of the study area in the Margalla Hills National Park, Islamabad.

have very sharp elongated upper canines which curve downward and outward. Barking deer is restricted in range in Pakistan and is confined to Himalayan foothill zone where there are remains of tropical dry deciduous forest. It is associated with dense, low thorn scrub of phulai *(Acacia modesta),* wild olive *(Olea ferruginea)* with understory of *Zizyphus nummularia.* It does not ascend above 1200 m and has no association with tropical pine forest (Roberts, 1977).

They lie up in the densest thorny thicket available. Some individuals spend whole life in the same territory but are seldom seen. Their movements are limited and restricted to an area less than 1 sq.km. for most of the year (Dinerstein, 1979). Densities of barking deer have been reported in the range of 0.5-7.0 animals/km^2 in Sri Lanka (Barrette, 1987) and 1.7-6.7 animals/km^2 in Nepal (Dinerstein, 1980, Sidensticker, 1976).

In Pakistan, Margalla Hills National Park (MHNP), Islamabad and some adjacent areas is the only remaining habitat of barking deer found in the country. Population of barking deer was last estimated in 1972 and it was found that 20 to 30 animals still existed in the Margalla Hills. Since then, no studies have been conducted on this species in Pakistan and its present status in the country is unknown. But it has been declared endangered in Pakistan. The objective of present study was to collect basic data on distribution range, population status and habitat requirements of the barking deer in the Margalla Hills National Park and use this information for the management and conservation of this endangered species.

The Margalla Hills National Park was established in 1980 with the total area of 15,883 hectares and includes Margalla Hills range, Rawal lake and Shakar Parian. The Margalla Hills range (12,605 ha), immediately north of Islamabad city constitutes a remarkable diversity of ecological, cultural, and recreational environments (Fig. 1). Prior to 1960, Margalla Hills forest was managed for extraction of fire wood, hay and grazing. Hunting and poaching was common in the area. Excessive exploitation depleted the vegetation, degraded wildlife habitat and adversely affected the wild fauna. Management of the forest changed from Forest Department to the Capital Development Authority (CDA) in 1960. The latter banned the extraction of fuelwood, domestic livestock grazing, quarrying and all other activities which affected the landscape. Afforestation and Soil Conservation works were started in addition to protection of fauna and flora. These measures and operations had desirable impact and gradual restoration of habitat started.

The general aspect of the Margalla Hills is southern but due to hill configuration other aspects are also met with. Topography is rugged and elevation ranges from 450m to 1580m. The terrain is interspersed with small and sizeable valleys towards its northern boundary. The slopes are generally steep and even precipitous at places with localized patches of gentle and moderate slopes.

Climatically, the area falls in the far end of Monsoon zone and there are five distinct seasons of the year that is Winter (December - February), Spring (March-April), Summer (May-June), Monsoon (July - September) and Autumn (October). The average maximum and minimum temperature are 33.3°C and 19.5°C respectively. The hottest months are May and June when the temperature may rise upto 42°C and the coldest months are December and January when the temperature falls below Zero. Annual average precipitation is 1018 mm, most of which is received during the months of July and August. The mean relative humidity during the year ranges from 35 % to 76 %.

Flora can be classified into two distinct and major vegetation types viz. Sub-tropical dry semi-evergreen and sub-tropical Pine forests. The principal tree species of the former are *Acacia modesta* and *Olea ferruginea* occurring as admixture in varying proportions depending upon the edaphic factors. The second type is found on higher elevation that is above 1000 m and above the first type. *Pinus roxburghi* is the principal tree species with the under-growth of *Myrsine africana, Woodfordia fruticosa, Berberis lycium* and *Carissa opaca. Pinus rowburghi* occurs in groups and is better stocked on cooler aspect. Fires are common during summer which suppress *Quercus incana* (climax) in upper reaches. Other effects on vegetation and fauna have not yet been studied.

Based on the physiography and vegetation following five habitat types can be recognized in Margalla Hills National Park (Amin et al., 1982; Ridley and Islam, 1982; PARC, 1984).

i) <u>Open thorn scrub:</u> This habitat type occupies the eastern and southern aspects of the National Park in lower reaches. Typical plant species met with in varying proportions and densities are *Acacia modesta, Olea ferruginea, Dodonaea viscosa, Maytenus royleana* etc.

ii)<u>Sub-tropical dry thorn forest:</u> This type occurs in the inner valleys and on northern slopes in lower reaches and is densely vegetated with *Carissa opaca, Acacia modesta, Buxus sempervirens* etc.

iii) <u>Cliff ledge scrub:</u> This habitat is characterized by the presence of steep broken cliffs with scattered scrub in middle altitudes above type (i) and (ii). The plant species which dominate this type are *Olea ferruginea, Carissa opaca* and *Dodonaea viscosa.*

iv) Transition zone scrub: This habitat type occurs in higher and cooler altitudes. Typical plant species are *Myrsine africana, Carissa opaca, Bwsus sempervirens, Pistacia integerrima, Mallotus philippensis, Dodonaea viscosa* and scattered pine. The vegetation is thick in ravines and open on slopes.

v) Sub-tropical pine: This type is met with on higher elevations. Vegetation is open with scattered patches of pine trees. The dominant plant species are *Pinus roxburghi, Quercus incana* and *Myrsine africana.*

MATERIALS AND METHODS

Thorough foot surveys were conducted in potential barking deer habitat areas in the Park to confirm its presence by observing foot prints, droppings or actually seeing the animals. It was supplemented by the information from local residents and employees of the CDA. Being secretive, it is very difficult to observe and count the animals. Hence, an indirect method of population estimation has to be used. Barking deer tracks at water points were used as an index to estimate the population. The indices method has been used elsewhere for population estimation of various animals (Lindzey et al. 1977, McCaffery 1976, Sarrazin & Bider 1973). Within the determined range of barking deer in the Park, sample sites were randomly selected and deer tracks on these sites were counted for three consecutive days. Based on the observations at these sites, total population in the park was calculated. Habitat quality and extent of disturbance in and around these areas were also considered while estimating the population.

Vegetation survey of currently occupied barking deer habitat was conducted to determine its habitat requirements. Thirty quadrates of 5x 5 m each were laid within the barking deer habitat area in the park on different locations at various aspects and elevations. Plant species falling inside the quadrate, their frequency and percentage cover were noted.

RESULTS AND DISCUSSION

Distribution and population status:

In the MHNP, barking deer are confined to the central and northern part of the Park. They are associated with the dense scrub forest in the lower scales between 500 and 1200 m elevations. It has also been

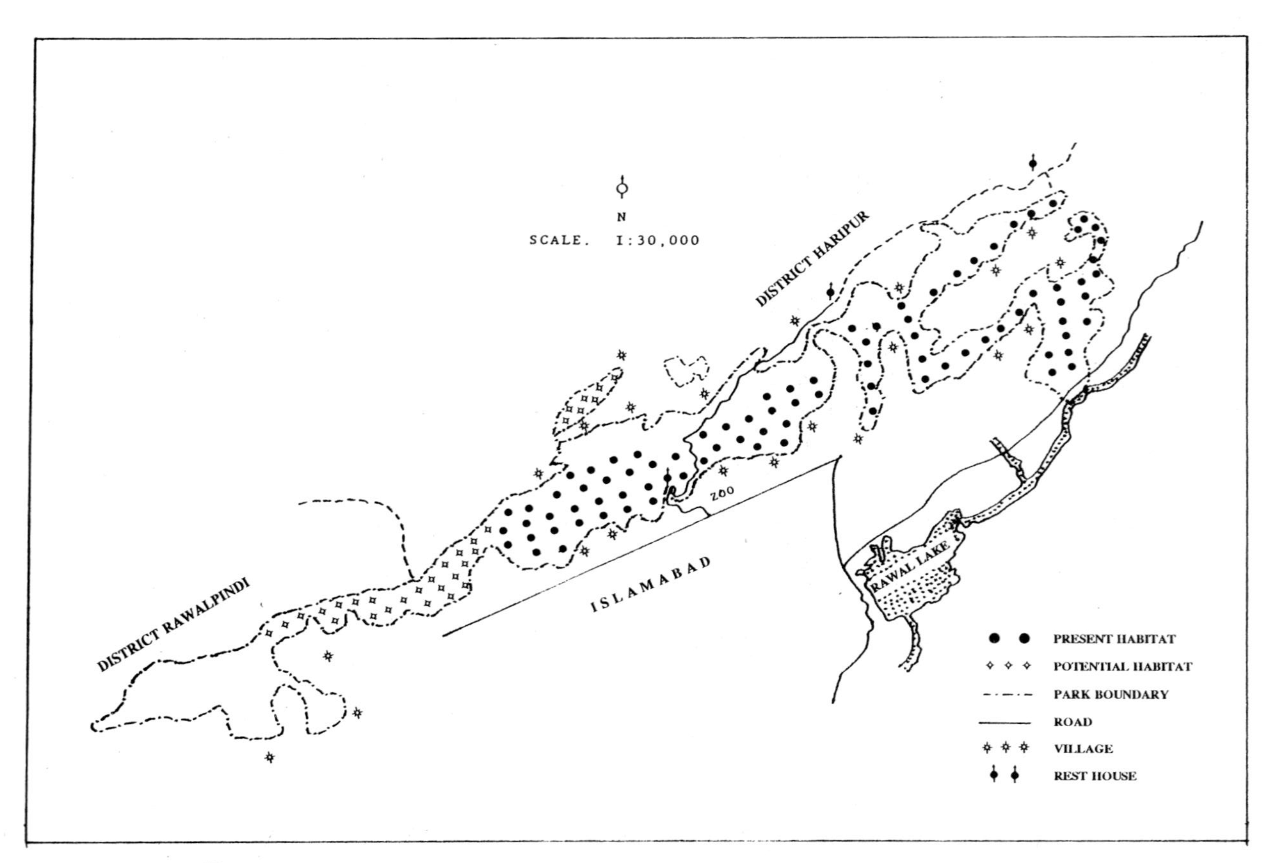

Figure 2. Present and potential Barking deer habitat in the Margalla Hills National Park.

reported by Roberts (1977) that this animal is restricted in range and confined to Himalayan foot hill zone where there are remnants of tropical dry deciduous forests and does not ascend above 1200m elevation and has no association with chir pine forest. They are mostly found in the small valleys between the ridges close to the water points.

Presently, more than 7000 ha (55 % of the total Park area) in the Margalla Hills National Park is occupied by barking deer (Figure 2). Another 3000 ha (23% of total Park area) has potential as barking deer habitat. Therefore, 70% of the total Park area can support this animal. The remaining 22 % of Park area is not suitable for barking deer habitat as it falls out of the natural barking deer range in the chir pine zone. It is occupied by the villages in the Park or used for agriculture crops etc. by the local residents.

Barking deer is a very secretive animal and rarely comes out of the vegetation cover. In a study in Nepal, barking deer showed preference for dense forest cover, rarely coming out in the open grass areas and were limited in habitat selection by high drinking water requirements (Seidensticker, 1976; Dinerstein, 1980). It can only be observed early in the morning at places where there is very little or no disturbance caused by visitors or domestic livestock.

The presence of barkings deer in an area was mainly confirmed by observing their signs such as foot prints and droppings. The foot prints can only be observed near water points on wet soil. Generally, it is very difficult to observe barking deer but it was possible to see the animal on few occasions. Hence, a non-variance-supported population estimate was calculated from the number of locations and number of animals at each location. Within the determined habitat of barking deer in the Park, nine sites (water points) were randomly selected and deer tracks were counted for three consecutive days. Tracks of two animals were counted at four sites and three animals at five sites. In the total barking deer habitat area in the Park, 44 such sites were identified.

Proportionately, it was supposed that at least 2 animals are present at 20 sites and 3 animals at 24 sites. At these 44 sites in the barking deer population was estimated between 100 and 120 individuals (Table I). They usually stick to their territories throughout the year especially during the breeding season and seldom move, so there are very few chances of overlapping of the observation records.

Table I: Barking deer population estimates in the Margalla Hills National Park during 1992-93

Area	Number of locations	Number of animals
2 animals/site	20	40
3 animals/site	24	72
Total	**44**	**112**

Habitat analysis:

In order to determine habitat status of barking deer, vegetation survey of its currently occupied habitat was conducted. Thirty quadrates of 5x5 m each were laid within the barking deer habitat area in the Park on different locations at various aspects and elevations. The data was analyzed to calculate relative abundance, relative cover and relative frequency of each plant species (Table II). *Carissa opaca, Dodonaea viscosa, Adhatoda vasica, Maytenus royleana, Acacia modesta, Olea ferruginea, Zizyphus nummularia, Woodfordia fruticosa, Myrsine africana* made up more than 60% of the total vegetation cover in barking deer habitat.

Table II: Relative abundance, cover and frequency of major plant species in barking deer habitat in the Margalla Hills National Park

Name of Species	RD	RC	RF
Carissa oppaca	21.77	26.29	8.01
Dodonaea viscosa	13.82	11.82	6.23
Adhatoda vasica	07.77	03.71	0.75
Maytenus royleana	04.90	02.52	5.93
Acacia modesta	01.63	07.83	3.86
Mallotus philippensis	02.96	03.58	4.15
Myrsine africana	03.31	02.98	1.78
Olea ferruginea	00.39	04.01	1.48
Grewia optiva	00.97	00.87	2.97
Woodfordia fruticosa	01.06	01.76	2.37
Bauhinia variegata	01.06	00.95	2.08
Zizyphus nummularia	00.35	01.14	1.48
Cassia fistula	00.39	00.62	1.48
Themeda anathera	15.84	12.49	7.42
Heteropogon contortus	03.93	01.95	3.26
Apluda aristata	01.28	00.54	0.89

RD= Relative Density, RC= Relative Cover, RF= Relative Frequency

Carissa opaca, Dodonea viscosa, Adhatoda vasica, Maytenus royleana, Acacia modesta and Myrsine africana provide cover for barking deer and their offspring for hiding against predators and harsh weather conditions. Dense cover is critical component of barking deer habitat due to their secretive habits and it has to be maintained for the protection and conservation of this species. They are found in the areas where more than 70% vegetation cover is available. Water points are found throughout the habitat area, some of which are permanent while others get dried during the summer months.

Management impications:

Presently, barking deer is found in the areas with more than 70 % vegetation cover. They are absent from the areas having same plant species but less hiding cover than required by the species. Vegetation in same areas has been affected by fuelwood cutting and livestock grazing. Dense vegetation cover is a critical element of barking deer habitat which is used for hiding against predators and harsh weather conditions. This cover has to be maintained/improved to provide a good habitat for better conservation of this endangered species. Livestock and human activities have to be controlled in the Park either by reducing their population or restricting them outside the critical barking deer habitat. With protection, 3000 hectares of potential habitat should develope into a dense vegetation cover which can be used by barking deer. Its population will definitely increase with the increase in its distribution range.

Livestock do not appear to compete with barking deer for food as plenty of forage species are available in the Park. But, they certainly compete for space which disturbs barking deer in the Park and hence, may reduce its habitat area and productivity.

Wildlife has been protected since the establishment of Margalla Hills as a National Park. Generally, barking deer and its habitat is quite secure in the Margalla Hills National Park. Occasional incidents of illegal hunting were noted outside but close to the National Park boundary which definitely affect the population of barking deer inside the Park. Close watch is required for the protection of existing barking deer population.

Due to low literacy rate, most villagers are unaware of the importance of wildlife in the ecosystem and in our daily life. A strong awareness campaign is needed to educate the people in and around the Park about the importance of natural resources and involve them in the management of these resources.

REFERENCES

AMIN, A., G. AKBAR, ANWAR, M. AND AHMED, M. 1982. Preliminary survey of the vegetation of the Margalla Hills with relation to wildlife species. *Pakistan J. Sci. Stud.* **2**:1-4.

BARRETTE, C. 1987. The comparative behavior and ecology of chevrotains, musk deer and morphologically onservative deer. Biology and Management of the Cervidae. in C.M. Wemmer, ed., pp. 200-213, Smithsonian Institution Press, New York.

DINERSTEIN, E. 1979. An ecological survey of Royal Kernali-bardi Wildlife Reserve, Nepal: II Habitat and animal interactions. *Biol. Conserv.* **16**: 5-38.

DINERSTEIN, E. 1980. An ecological survey of Royal Kernali-bardi Wildlife Reserve, Nepal: IIIUngulate populations. *Biol. Conserv.* **18:** 5-38.

LINDZEY, F.G., THOMPSON, S.K. AND HODGES, J.I. 1977. Scent station index of black bear abundance. *J. Wildl. Manage.* **40**: 151-153.

MCCAFFERY, K.R. 1976. Deer trail counts as an index to population and habitat use. *J. Wildl. Manage.* **40**: 308-316.

PAKISTAN AGRICULTURAL RESEARCH COUNCIL. 1984. Study on status of habitat and distribution of wildlife in Islamabad District (Margalla Hills, Bannigala, Rawal Lake and surrounding areas). *Annual Report.* 62 pp.

RIDLEY, M.W., AND ISLAM, K.. 1982. *Report on the cheer pheasant re-introduction project, World Pheasant Association, Pakistan.*

ROBERTS, T.J. 1977. *The Mammals of Pakistan.* Ernest Benn Limited, London and Tonbridge, England.

SARRAZIN, J.P.R. AND BIDER, J.R. 1973. Activity, a neglected parameter in population estimates: the development of a new technique. *J. Mammal.* **54**: 369-382.

SEIDENSTICKER, J. 1976. Ungulate populations in Chitwan Valley, *Nepal. Biol. Conserv.* **10**: 183 - 210.

THE STATUS OF THE WOOLLY FLYING SQUIRREL (*EUPETAURUS CINEREUS*) IN NORTHERN PAKISTAN

PETER ZAHLER* AND CHARLES A. WOODS**

**W.C.S. Research Fellow, 217 Cornell St. Ithaca, NY 14850*
*** Florida Museum of Natural History, Gainesville, U.S.A.*
(In association with Naeem Dar, Akhtar Karim, Waseem Ahmad Khan, C.W. Kilpatrick & Mohammad Rafique)

Abstract: The "giant" woolly flying squirrel (*Eupetaurus cinereus*) is an extremely rare and unusual mammal. Most information concerning this species comes from a few study skins collected over a century ago. Virtually nothing is known of its food, reproduction, range, habitat preferences, or behavior, and previous to 1994 many authorities considered it to be extinct. In 1994, however, a live specimen was captured in northern Pakistan. Remains of other specimens were found in the area, and information was gathered from local people concerning the squirrel's natural history and biology. Research was conducted in 1995 and 1996 to determine the distribution and range of the woolly flying squirrel in the Northern Areas of Pakistan, and to gather more information about the squirrel's habits and natural history. Its distribution appears to be centered on side valleys in the central Indus River Valley in the eastern most region of the Hindu Kush Range and around Nanga Parbat, the most westerly main massif in the Himalayan Range. The woolly flying squirrel has many unique adaptations that distinguish it from all other flying squirrels and also appear to allow it to coexist in close association with other flying squirrels in Pakistan. In some areas of northern Pakistan, three species of flying squirrel appear to be sympatric in isolated mountain valleys. This paper discusses the present status and possible past distribution of *Eupetaurus cinereus* , and analyzes how and why this species evolved and what is its closest living relative. It is our opinion that this species is not the last of an ancient radiation that is relictual in its distribution in northern Pakistan, but rather an example of another level of evolution of flying squirrels, and potentially (if protected from extinction) the basis of a new adaptive radiation of flying squirrels in the mountains of Central Asia.

INTRODUCTION

The "giant" woolly flying squirrel (*Eupetaurus cinereus*) is one of the most unusual and least well-known mammals in the world. Only 11 specimens are known, and previous to this work no animals had been collected or reported as having been seen alive since 1924. The pattern of its dentition is unlike that of any other known squirrel, either fossil or recent. In fact, the structure of its cheek-teeth is so divergent from other flying squirrels that two major treatises on mammals proposed placing it in its own rodent family, the Eupetauridae(Grassé & Dekeyser, 1955; Schaub, 1958). However, while dentally it stands far removed from any other flying squirrel, in all other features it is definitely a sciurid. It is probably closely related to the Giant Himalayan Flying Squirrels (*Petaurista*) which also occur in Pakistan.

The English mammalogist J. Oldfield Thomas of the British Museum first described and named *Eupetaurus cinereus* in 1888. However, confusion about the status of the new squirrel began at this point, since the exact location where the type specimen came from was vague. In addition, the drawing in Jentink (1890) of a right lower jaw apparently is incorrect and is in fact a left jaw (McKenna, 1962). This small slip is what, in part, led to the confusion in the literature as to whether the species should be set apart from other rodents in its own distinct family. Finally, Thomas observed that the claws of the squirrel were very blunt and therefore *Eupetaurus cinereus* was probably much less adapted to being arboreal than are other flying squirrels. This observation is not supported by the 4 museum and 6 live specimens examined by the authors, all of which had very sharp claws. But the observation did serve as a "red herring" in that it led investigators initially to search for the woolly flying squirrel in higher and dryer valleys without forest cover.

Little information about this animal has been gathered since Thomas's original description, and most of what is published has been inferred from analyzing a few skins, most collected around the turn of the century in or near what is now northern Pakistan (Blanford, 1891; McKenna, 1962; Ellerman, 1963; Chakraborty and Agrawal, 1977; Roberts, 1977). Thomas (1888) states that one skin was "probably" from Tibet, and Agrawal and Chakraborty (1970) note a skin at the Zoological Survey of India from northern Sikkim. There are two specimens of *Eupetaurus cinereus* at Kunming Institute of Zoology in China from an upper gorge of the Nu River at 3000 meters elevation.The only information gathered concerning a live individual of this squirrel previous to this field work is Lorimer's (1924) brief and anecdotal description. From that time until 1994 there were no records of the woolly flying squirrel by scientists working in the area (C. Woods, T. Roberts, D. Blumstein, G. Schaller, pers. com.). Many people considered the squirrel to be extinct, and as a result of the lack of data it was not listed in references on endangered or threatened species (Thornback and Jenkins, 1982; Inskipp and Barzo, 1987; Lidicker, 1989; U.S. Fish and Wildlife Service, 1992; Greenbridge, 1993).

What little is known about the anatomy and natural history of the woolly flying squirrel is unusual enough to warrant placement in its own genus. It is a very large squirrel, with a head and body length of 450-600 mm, and it has extremely dense, thick fur. The phylogeny of *Eupetaurus* was studied by McKenna (1962), who confirmed that indeed this species has a very distinct dental pattern from other squirrels. The most important dental feature that sets *Eupetaurus cinereus* apart from other squirrels is that it has high-crowned (hypsodont) cheek teeth (pre-molar and molar teeth). The teeth of the lower jaw, rather than meeting and meshing with the upper teeth as in all other squirrels, slide back and forth against the upper toothrow. Therefore, they are flat crowned. High-crowned teeth with flat surfaces are found in a number of rodents that feed on abrasive substances, such as capromyid rodents from the West Indies, spiny rats (echimyids) from South America, porcupines (from both the New and Old World), and cane rats (*Thryonomys*) from Africa, in addition to beavers (Castoridae).

In all of these forms with flat-crowned hypsodont teeth, including *Eupetaurus cinereus*, there are other modifications to the jaw and skull that are associated with grinding food rather than crushing substances. These changes include:

1) The toothrows converge anteriorly;
2) The glenoid fossa is elongated in an antero-posterior axis;
3) The pterygoid fossa is open and posterior in position.

In all of these features *Eupetaurus cinereus* is more like a capromyid rodent than a flying squirrel, so its feeding strategy and food habits should be more like a capromyid rodent than a typical flying squirrel. It appears that *Eupetaurus cinereus* is convergent with capromyids as well as cane rats from Africa and New World spiny rats, rather than part of the same taxonomic group as was once suggested (see McKenna, 1962 for an excellent analysis of the reasons *Eupetaurus cinereus* is a squirrel). However, if it is a flying squirrel as first proposed by Thomas (1888) and confirmed by McKenna (1962), then why is it so divergent from other flying squirrels? What are its unique food habits? And of equal importance, if it is a flying squirrel, to what flying squirrel is it most closely related?

Known Specimens of *Eupetaurus cinereus*

Original locations for skins collected around the turn of the century were mostly from the general region of Gilgit in the area of the confluence of the Himalayan, Karakoram, and Hindu Kush mountain ranges in northern Pakistan. The area is at approximately 36 degrees latitude, and can be characterized as high, cold desert dominated by *Artemisia* and *Juniperus* above 2,000 m, with many valleys having scattered forests of *Pinus* and *Picea* at higher altitudes.

Known Specimens = 11 (+ 2 possible ones from China)

Bombay Natural History Society (4 specimens)

1) BNHS No. 7107 collected May 20, 1916 by Maj. A. MacPherson at Gilgit. This specimen is a skin only, and is in good shape. The pelage color is gray. There are no measurements.

2) BNHS No. 7108 collected May 29, 1924 by Lt. Col. D. Lorimer. The collection site is not associated with the specimen, however, Lorimer (1924) reported that the specimen came from the "Sai Nalah" on the west bank of the Indus River opposite the town of Bunji. Lorimer was stationed in Gilgit. The field number of this specimen is M.1856, which is the same number as the skull photographed in McKenna (1962). No skull is currently associated with this specimen at the BNHS, and there is no record at the BNHS of a skull, but we assume that the skull did exist at one time and is currently misplaced. The skin is in very poor shape, and is little more than a bunch of skin in a plastic bag, but appears to have been gray in color. There are no measurements. A photograph of this individual is in Lorimer (1924), and is reproduced in Nowak (1991).

3) BNHS No. 7109 (no date) collected by H. T. Fulton in the Chitral area. The specimen is dark chocolate brown in color (melanistic), and has a white V-shaped marking on the chest and throat. The tail is broken and quite short. The recorded measurements are: 2051; 31; 12; 07. There is no skull.

4) BNHS No. 7110 collected January 22, 1924 by Maj. L. MacKenzie at the Sai Valley off Right Bank of Indus. It is gray in color. There is no skull.

British Museum of Natural History (4 specimens)

5) BMNH No. 88.9.29.1 Co-type with a skin and fragmentary snout, purchased by R. Lydekker in 1879 from Srinagar and said to be from the Astor District.

6-8) "Of the four specimens of *Eupetaurus cinereus* in the collection, two were collected at Chitral, India, one from Gilgit and the fourth was purchased from Gyantse Bazaar, Tibet. There is very little additional data on these specimens, and no collection dates." (Jenkins, personal communication, 1991)

Indian Museum in Calcutta (2 specimens)

9) IM No. 9492 Skin & Skull (Lectotype) described and figured by Thomas. The specimen was brought to G. M. Giles in Gilgit in 1887. Listed in Agrawal & Chakraborty (1979) as "Syntype" and from "Gilgit, Kashmir". The current tag on the specimen lists the locality as "Gilgit Valley 5000 feet". The specimen is grey.

10) IM No. 19103 Skin. from Sikkim at " 9000 feet elevation" , collected by J. S. Gill. The dorsal hairs are slaty grey.

Leiden Museum in the Netherlands (1 specimen)

11) Specimen described by Anderson (1878) and illustrated in Jentink (1890). Skin and Skull, with location "Tibet". Since the skin and skull are associated, it is likely that this was a wild captured specimen rather than just a skin bought in a local bazar.

Kunming Institute in China (2 specimens)

Two specimens (skins only) from Yunnan, China.

Rediscovery of *Eupetaurus cinereus*

In 1992 and 1994 Peter Zahler and his survey team investigated the original locations for the skins mentioned in the literature, including Shishpar, Naltar, Kargah, Sai, and Astor. Extensive efforts were made to gather information about the animal from local inhabitants at each site visited and, when warranted, efforts were made to capture individuals using Tomahawk 207 live traps. Between 1991-1995 a team of mammalogists from the Pakistan Museum of Natural history and the Zoological Survey Department of Pakistan searched throughout the NWFP and Northern Areas of Pakistan for this species as part of a survey of the small mammals of Pakistan.

On 8 July 1994, a female woolly flying squirrel was captured by two local men in the Sai Valley between Hurkus and Gashu Gah on the south side of the Sai River (Zahler, 1996). In addition, a number of woolly flying squirrel body parts were discovered in the Sai Valley. The parts were all found within 100 m of an eagle owl (*Bubo bubo*) roost at 2,500 m elevation and within 5 km of the woolly flying squirrel capture site. The parts were fresh and from at least two individuals, suggesting that the owl was a regular predator on woolly flying squirrels. While this proved the continued existence of the woolly flying squirrel in northern Pakistan, there was still no information about the range and distribution of the squirrel. Research in 1995 was, therefore, focused on determining where outside of the Sai Valley the woolly flying squirrel could be found.

In 1995 we focused our attention on an area surrounding the 1994 capture site in the Sai Valley. This included the valleys of Kargah, Yasin, and Ishkoman to the northwest; Naltar, Karimabad, and Nagar to the northeast; Babusar, Tangir, and Khanbari to the southwest; and Astor and Gorabad to the southeast. This would effectively enclose the Sai Valley capture site, with the exception of due east toward Skardu and due west toward the Swat Valley. Peter Zahler and fellow research associates Mr. Naeem Dar and Mr. Akhtar Karim visited each of the valleys listed above during July and August of 1995. During each trip we attempted to contact and interview any and all salajit collectors (see below for a discussion of salajit) and professional hunters living in the valley. Salajit collectors were considered especially important due to their practice of visiting steep cliff caves to collect salajit - the same types of caves in which the woolly flying squirrel is apparently found. We also posted a reward for the capture of a live woolly flying squirrel.

Extreme care was taken during conversations with local people to avoid leading questions. Individuals were often asked to describe the medium-sized mammals found in the area, or were asked to describe any animals they had seen in caves. If an animal matched the description of the woolly flying squirrel, further questions were asked to ascertain whether accurate identification could be made. A picture was shown, and information was then gathered from the person being interviewed. A great deal of time and effort was spent explaining to local people about how the squirrel was both rare and unique. Efforts were made to state clearly that the squirrel should never be killed, and that villagers should consider it a source of pride as a remarkable endemic in the valleys where it was found.

A male woolly flying squirrel was captured for us upon our return visit to Gorabad on August 21, 1995. The squirrel is assumed to be a sub-adult due to its size when compared to the female caught in 1994 and from descriptions by local people of having seen and captured much larger males. Measurements, tissue samples for DNA analysis, and observational data were taken and the squirrel was released the following day.

During October and November of 1996 Zahler and Karim visited Swat, Chitral and Balti Gali, a small valley adjacent to Gorabad. No evidence of woolly flying squirrel was found in Swat. There was indication that the squirrel might be found in upper Chitral, but inclement weather did not allow for further investigation. In Balti Gali four woolly flying squirrels were captured with the help of local people. The squirrels consisted of three males and one female.

Measurements for the six *Eupetaurus cinereus* captured are as follows

	1994	1995	1996			
			1	2	3	4
Sex	Female	Male	Male	Male	Male	Female
Head and Body (cm):*	*	*	46	45	42	51
Tail Length (cm):**	54.5	48.3	45	47	43	48
Ear Length (cm):	-	3.5	4.3	4.9	4.4	4.6
Hind Foot Length (cm):	9.0	8.25	9.0	8.3	9.0	9.3
Weight (kg):	2.5	1.75	1.99	1.42	1.65	2.10

* We were not able to measure head and body length precisely in 1994 & 1995 because the live animals were not anesthetized, and were moving about in awkward positions as we measured them. We estimate that the 1994 female had a head and body length approximately equal to tail length, and the 1995 male had a head and body length less than tail length, but at least 40.6 cm.

** In addition a tail collected from the eagle owl roost in 1994 measured 54 cm.

The female was captured on 8 July 1994 between Hurkus and Gashu Gah in the Sai Valley at approximately 35^0 45'N and 74^0 27'E. The capture site was a cave at 3,200 m approximately 800 m above the valley floor on the south side of the Sai River. The cave was about 10 m up an isolated, northwest facing, vertical cliff wall. The cave entrance was 2 by 2.5 m, with a depth of about 3.5 m. It was a known site for collecting salajit. The slope varied between 30 and 60 degrees, and was well vegetated with grass and scattered blue pines (*Pinus wallichiana*) and juniper (*Juniperus macropoda*).

The male was captured on 21 August 1995 about a 12-hour hike northeast of Gorabad village at approximately 35°38'N, 74°33'E. The

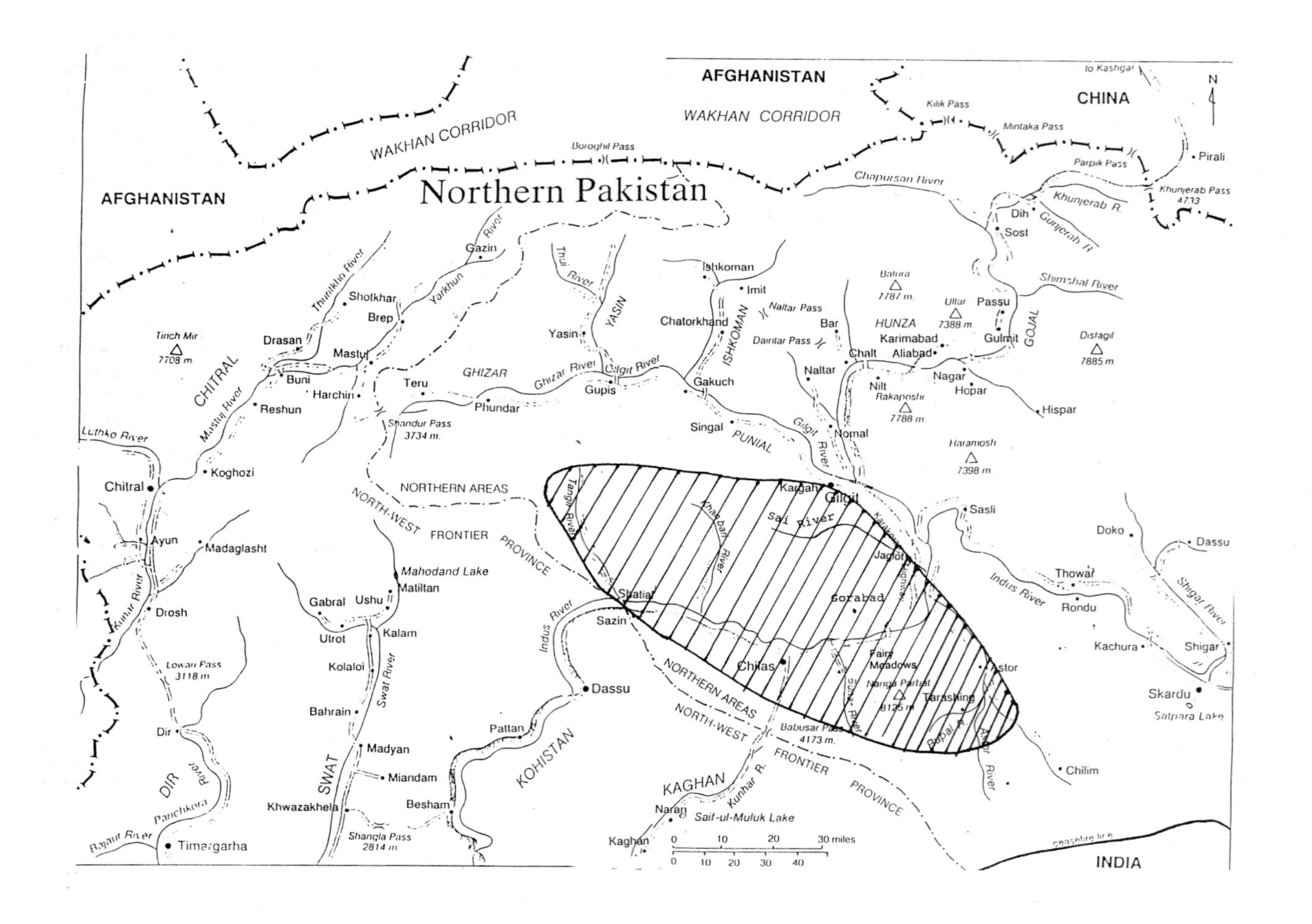
Northern Pakistan
AFGHANISTAN
WAKHAN CORRIDOR
CHINA
INDIA
N
to Kashgar
Kilik Pass
Mintaka Pass
Parpik Pass
Khunjerab Pass
4733
Pirali
Boroghil Pass
Chapursan River
Khunjerab R.
Gunjerab R
Dih
Sost
Shimshal River
Batura
7787 m.
Ultar
7388 m
Passu
Gulmit
GOJAL
HUNZA
Karimabad
Aliabad
Distagil
7885 m
Nagar
Hopar
Hispar
Nilt
Rakaposhi
7788 m
Haramosh
7398 m
Ishkoman
Imit
Naltar Pass
Daintar Pass
Bar
Chalt
Naltar
ISHKOMAN
Chatorkhand
Gakuch
Singal
PUNIAL
Gilgit River
Nomal
Kargah
Gilgit
Sai River
Jaglot
Gorabad
Karakoram Highway
Sasli
Doko
Dassu
Thowar
Rondu
Indus River
Shigar River
Kachura
Shigar
Skardu
Satpara Lake
Astor
Fairy Meadows
Nanga Parbat
8125 m
Tarashing
Astor River
Chilim
Chilas
Babusar Pass
4173 m.
NORTHERN AREAS
NORTH-WEST FRONTIER PROVINCE
Khanbari River
Tangir River
Shatial
Sazin
Indus River
Dassu
KOHISTAN
Pattan
Besham
KAGHAN
Kunhar R.
Naran
Saif-ul-Muluk Lake
Kaghan
ceasefire line
0 10 20 30 miles
0 10 20 30 40
Thui River
YASIN
Yasin
Gupis
Ghizar River
GHIZAR
Teru
Phundar
Shandur Pass
3734 m.
Gazin
Yarkhun River
Thuinkho River
Shotkhar
Brep
Mastuj
Drasan
Buni
Harchin
Reshun
Tirich Mir
7708 m
CHITRAL
Mastuj River
Lutkho River
Koghozi
Chitral
Ayun
Madaglasht
Kunar River
Drosh
Lowari Pass
3118 m
Dir
DIR
Panchkora River
Rajaul River
Timargarha
Mahodand Lake
Matiltan
Gabral
Ushu
Kalam
Utrot
Kolaloi
Swat River
Bahrain
Madyan
SWAT
Miandam
Khwazakhela
Shangla Pass
2814 m

squirrel was captured in a cave on a steep rock wall around 50 m high, at an elevation of about 3,000 meters. The cave was about 25 m, or halfway, up the cliff. The cave was approximately 1 by 1 m wide at the opening, and 15 m deep. A mixed forest of pine (*Pinus gerardiana* and *P. wallichiana*), spruce (*Picea smithiana*), juniper, and oak (*Quercus baloot*) occurred in the area near the cliff.

Four squirrels were captured in 1996, on 2, 8, and 9 October and 5 November. All four were found in or around Balti Gali, at approximately 35°33'N, 74°35'S. Three of the squirrels were found in caves during the day, while the fourth was captured when it was observed in the morning attempting to flee from a human walking nearby. The squirrels were found in an area that has steep wooded cliffs rising about 2,400 to 3,400 m. The woody vegetation consisted of *Juniperus macropoda*, *Pinus gerardiana*, *P. wallichiana*, *Picea smithiana* and *Quercus baloot.*

Current Status of *Eupetaurus cinereus*

Current Distribution: According to the research conducted on the woolly flying squirrel in 1995, the distribution of this rare squirrel appears limited to a surprisingly small area (see map), one that may be as small as approximately 100 miles east-to-west by 50 miles north-to-south. The woolly flying squirrel appears to be extremely rare or absent from most areas north of the Gilgit and Hunza Rivers, including Yasin, Ishkoman, Naltar, Karimabad and Shishpar Valleys. It also appears to be extremely rare or absent in the upper Hunza region south of the Hunza River (i.e., the Nagar Valley, including Hopar and Hamdar). However, we have information that suggests the squirrel may still exist in the upper Hunza Valley near Sost, perhaps in one or more small and isolated populations. The woolly flying squirrel appears to be absent from areas south of the Babusar Pass and south of Besham. This region superficially appears to be acceptable habitat for the woolly flying squirrel, with abundant high-elevation conifer forest. However, local people who are familiar with the squirrel state that appropriate cliffs or caves are rare or absent in that area.

The squirrel does appear to exist in the Chilas region south of Gilgit and east of the Indus River. Accurate descriptions of the squirrel were recorded in upper Kargah near Jut, in the Sai Valley, around Gorabad, and in the Tangir and Khanbari Valleys. West of the Indus River reliable reports of the squirrel came from near Bunji across from the Sai Valley, and in Haramosh region 50 km east of Gilgit. East of Nanga Parbat there are reliable reports from the Astor Valley around Harchu Goh, Gorikot, and in some of the lower valleys. West of Nanga Parbat there are reliable reports from Gunar Farm and the

Bunar Valley (Ges Bala, Ges Pain) near the village of Tato on the route from Raikot Bridge to Fairy Meadows, around Fairy Meadows itself, and from Babusar Village above Chilas.

The distribution of the woolly flying squirrel can be described as an area whose boundaries appear to be the Gilgit River in the north and the Babusar Pass in the south. It seems to occur on both sides of the Indus River through the Chilas region from at least Khanbari in the west to at least Astor in the east. The eastern and western limits of the squirrel's range are still not fully determined. To the east of the Indus it occurs in the Bunar Das Valley south of Nanga Parbat. On the west side of the Indus it occurs in the upper Tangir Valley as well as the upper Khanbari Valley. Although upper Swat is effectively part of the mountain ranges that continue to the confluence of the Gilgit and Hunza Rivers, it appears that the squirrel does not occur in the Swat Valley. It is possible that it occurs in the Kandia Valley.

Chitral is also listed as part of the historical range of this squirrel, and the drainage area on the south side of the Gilgit River east of Kargah to the Shandur Pass and beyond needs to be studied as well (however, see below for a discussion of Chitral). The lower Skardu Valley and the region between that valley and Rakaposhi are also areas that need to be investigated in the future to determine the woolly flying squirrel's eastern limits.

While northern and western Chitral (the Kafiristan-Chitral border) is regularly mentioned in the literature as a location for one of the museum skins, a check of the original paper by Thomas (1888) reveals the following quote: "At last in 1887, Mr. G. M. Giles, . . . when on the Kafiristan-Chitral Mission . . . had brought to him *at Gilgit* a living example of the present form . . ." (italics ours). From this description it appears that while Mr. Giles was on a mission to northern Chitral, he received the specimen in Gilgit. A live specimen of woolly flying squirrel is unlikely to have survived a long trek through the hot valleys of this area. We have traveled this route in a jeep and can testify how arduous this trip is. Thus, the specimen in question was probably collected somewhere around Gilgit, and not near the Kafiristan-Chitral border, as is frequently mentioned. (If the specimen came from "near the Kafiristan-Chitral border" it would be from the area now known as the Kalash valleys 35°40'N, 71°40'E approximately 30 km southwest of the town of Chitral).

A skin of the woolly flying squirrel in the Bombay Natural History Museum (BNHS No. 7109) attributed to H. T. Fulton (with no date), is also listed as coming from Chitral. This specimen is dark uniformly chocolate brown in coloration (melanistic), which is quite distinct from the usual gray coloration of most other known specimens. We also have evidence suggesting that the squirrel may still exist in upper Chitral, although this needs further confirmation. At this point the presence of *Eupetaurus cinereus* in side valleys in Chitral, especially in the Hindu Raj area, cannot be ruled out as part of the current range of the woolly flying squirrel.

Closest living relatives of Eupetaurus cinereus: According to McKenna (1962), the dental pattern of *Eupetaurus cinereus* is most similar to *Petaurista xanthotis* of China. The latter is placed in the same genus as *Petaurista petaurista*, which occurs in the mountains of Pakistan. *Petaurista xanthotis* was considered a subspecies of *Petaurista leucogenys* by Ellerman (1948). However, the cheek teeth of *P. xanthotis* are much more complex than any other species of *Petaurista*. It is so distinct that McKenna (1962) concluded that it could be the type of a new genus. While its cheek teeth are low crowned, and the occlusal surfaces rounded (semi-lophodont), there is a suggestion of the pattern found in *Eupetaurus cinereus*. Indeed, if the cheek teeth of *Petaurista xanthotis* were to become high crowned as a result of a change of food habits, the resulting pattern would look much like the pattern found in *Eupetaurus*. This is the only squirrel with a dental pattern that has any similarity to that of *Eupetaurus*.

Biogeographical and evolutionary implications: The common ancestor of both *Eupetaurus* and *Petaurista* is likely to have originated in the eastern Himalayas and Indo-China, since the center of diversity of *Petaurista* is in this area, and the known range of *Petaurista xanthotis* is western China from Yunnan and eastern Tibet to Gansu. The pattern of mountain-building that has characterized the Himalayas during the middle to late Cenozoic would have led to the isolation of some populations of the common ancestor of *Eupetaurus cinereus* and *Petaurista xanthotis* in patches of mesic forest in the dry ranges of northwestern Pakistan (valleys of Kohistan, the Hindu Kush, the Hindu Raj and the Western Himalayas in the Astor Valley). Once isolated in these very disjunct patches of suitable habitat, the chewing apparatus of the proto-*Eupetaurus* may have become modified to include high-crowned teeth and a suite of cranial characters associated with feeding on abrasive materials. Since *Eupetaurus* is so morphologically distinct, it is likely that it has been isolated for a long period of time.

While it is possible that *Eupetaurus* developed its unique morphological and dental features due to isolation in marginal habitat, it is also possible that the strong selective pressure that led to the unique morphological and dental features found in *Eupetaurus cinereus* was associated with competition between *Eupetaurus cinereus* and *Petaurista petaurista* (and possibly *Eoglaucomys*

[=*Hylopetes*] *fimbriatus*). We have observed that both *Eupetaurus* and *Petaurista* may occur together in several valleys in north-central Pakistan, such as the Khanbari Valley. *Eoglaucomys fimbriatus* also occurs in the Khanbari Valley. The selective pressure caused by the competition of three species of flying squirrel living in the same isolated habitats, surrounded by kilometers of unsuitable habitat associated with snowy mountain peaks, xeric and hot river valleys, and large active glaciers could have led to important functional shifts in the feeding mechanism of the proto-*Eupetaurus* as it evolved into *Eupetaurus cinereus*.

At the moment *Eupetaurus cinereus* is clearly part of the same petauristine radiation that includes *Petaurista*. However, it has reached a new level of morphological distinction in its masticatory apparatus, and is now distinct from all other flying squirrels. Given its isolated distribution, the fragmented pattern of its range, and the many unique features of its masticatory apparatus, *Eupetaurus cinereus* is poised to form the base of an adaptive radiation leading to a new group of squirrels. Rather than being a relict species in the process of becoming extinct, it is just as possible to view *Eupetaurus cinereus* as the root of a new radiation of flying squirrels in the mountains of Central Asia.

Reasons for the unique dentition of Eupetaurus cinereus: The chewing apparatus of *Eupetaurus cinereus* is very distinct from that of other flying squirrels living in Pakistan. This suggests that its feeding strategy and food sources should also be quite distinct. At this time it is impossible with the data available to answer questions concerning the diet. We do not know what *Eupetaurus* feeds on, or even where it forages for food when outside rock crevices and caves. Thomas (1888) suggested that the animal feeds mainly on lichens, mosses and other plants associated with rocky areas. The only known first-hand account of food habits is the observation made by Zahler that the specimen captured in the Gorabad Valley in August 1995 fed overnight in its cage on spruce buds (*Picea smithiana*). Local people believe that *Eupetaurus cinereus* feeds on the seeds, needles and bark of conifers. Based on comparison with other animals with similar modifications to the masticatory apparatus (Woods and Howland, 1979), such as West Indian capromyid rodents and the nutria from Argentina (family Myocastoridae), we suggest that *Eupetaurus cinereus* must feed on abrasive materials. A good "ecomorph" is *Plagiodontia aedium* from Hispaniola. This animal also lives in crevices and caves in rocky areas at high elevations. The food habits of *Plagiodontia* include chewing on bark (including the pine, *Pinus occidentalis*), as well as feeding on buds, leaves and root tubers.

Association of Eupetaurus cinereus with Salajit: As pointed out by Zahler (1996), the woolly flying squirrel found in the Sai Valley in 1994 was located by a salajit collector. Four of the five squirrels found near Gorabad in 1995 and 1996 also were located by salajit collectors. Salajit is a substance found in caves and rock crevices. It can accumulate in large blocks in the caves frequented by *Eupetaurus cinereus*. These caves are usually on nearly inaccessible vertical cliffs. Ancient Persian medicinal writings (M. Tooti, pers. com.) speculated that the substance was indurated urine produced by animals (these squirrels) living in the caves where the substance was collected. Since the substance is still prized for its medicinal value as a cure for pain of the joints and as an aphrodisiac (Khan, 1981), there is enough of a market for salajit to support a modest local industry, and salajit collectors regularly search for caves and rock crevices with the substance.The salajit collectors are the individuals who most frequently encounter the squirrel.

We interviewed a number of salajit collectors and shop keepers who bought and sold salajit in Chilas and Gilgit. The results indicate that salajit collectors regularly see woolly flying squirrels in caves. Three separate dealers reported that squirrels were brought to them in the past five years and that they attempted to keep them in captivity without success. We also examined blocks of salajit in the hands of salajit dealers. Of ten blocks examined, nine had fecal pellets and hairs of *Eupetaurus cinereus* lodged in crevices and hollows. However, none of the blocks showed signs of having been chewed.

The assumption that woolly flying squirrels actually produce salajit seems very unlikely. Salajit dealers in Chilas and Gilgit report that blocks of salajit weighing more than 40 kg are regularly brought into their shops by salajit collectors, and some store rooms of dealers had over 100 kg of salajit. The substance is of such high value that it rarely remains in the stockrooms very long, so these large deposits are not just accumulations over long periods of time. We are currently analyzing samples of salajit. Data available at this time indicate that it is not organic in origin. It appears to come from crevices in cliffs, especially in the area of north-central Pakistan near Chilas, and to be limited to this zone of hot summer temperatures where the rock faces of the mountains heat up to very high temperatures.

In conclusion, it does not appear to us at this time that there is any evidence that salajit is actually produced by the woolly flying squirrel. However, squirrels are found in many of the same caves and crevices where salajit is collected, and natural blocks of salajit almost always contain hairs and feces of *Eupetaurus*.The caves and crevices where salajit is collected by salajit collectors are most often on steep rock faces and nearly inaccessible except with ropes or by experienced climbers. These inaccessible locations might be the last refuges of *Eupetaurus cinereus*, which once occupied adjacent forest areas in large numbers, or they might be the historical preferred habitat as suggested by Oldfield Thomas in his original paper describing the woolly flying squirrel. This question cannot be answered until further field studies are completed. However, after four years of searching for *Eupetaurus cinereus* by two teams of mammalogists in a variety of habitats in northern Pakistan, the only reliable data and almost all of the confirmed living specimens of the woolly flying squirrel were associated with salajit collectors. There appears to be a very close association between salajit and the woolly flying squirrel.

Note- It is difficult to know how to spell salajit, since there is variation in pronunciation of the substance in the region, and we have never seen the word written. All of the participants associated with the project feel that the closest phonetic spelling is "salajit." In Zahler (1996) it was spelled as salagit. It could also be spelled salajeet.

Estimate of Current Abundance

It is difficult to estimate the number of squirrels living in the known area of distribution of *Eupetaurus cinereus*. However, we feel that it is important for us to give some indication of what we believe to be their current abundance in order to use this information as the basis for a conservation program for the species, and as a first step in developing a species recovery plan. Salajit collectors reported from 2 to 20 squirrels per 50 caves visited, with the most common estimate being 10 squirrels. We estimate that there are about 200 professional salajit collectors in the region. Using these figures we would expect there to be about 2,000 woolly flying squirrels in the area of known distribution.

Most experienced local people in the area we surveyed stated that they thought there were 20-50 squirrels in their "area" of the valley. Using this information, and extrapolating to all of the potentially suitable habitat in a triangle using Tangir, Astor and Gilgit as the corners, one would estimate that there were between 3,000-5,000 squirrels present. However, the squirrel apparently has a spotty distribution within its known range, and is not found in all suitable habitats. For example, in the Sai Valley, local residents insist that the squirrel is only found in a 25 km stretch of this ca. 70 km valley. We also know that the squirrel is very dependent on caves that are patchy in distribution both in a vertical and horizontal gradient. Therefore, we believe that 3,000-5,000 is too large an estimate for the number of woolly flying squirrels living in the area within the Tangir-Astor-Gilgit triangle.

We believe that a minimum estimate of the number of woolly flying squirrels living in the area is 1,000-2,000, but that the maximum number is not greater than 3,000. If the squirrel is confirmed to be present beyond this known range, and occurs in areas of the upper Hunza and Chitral valleys, then the above estimates could be doubled. So, the good news is that *Eupetaurus cinereus* survives; it is present in low to moderate numbers within a restricted area of north-central Pakistan. The bad news is that the squirrel is very limited in its distribution within this area; it appears to be dependent on a patchy habitat that contains conifers and isolated caves and rock crevices. Many populations may also be isolated from one another by hostile intervening terrain. As a result, the squirrel is vulnerable to local extirpation from human disturbance or habitat loss.

Natural History of *Eupetaurus cinereus*

Information gathered from local people in 1995 and 1996 agrees with what was discovered about the woolly flying squirrel in 1994 (Zahler, 1996). All locations and areas described as having populations of the woolly flying squirrel were between approximately 2,400 and 3,600 meters in elevation, the same elevation that *Pinus girardiana*, *Pinus wallichiana*, *Picea smithiana*, and other conifers are found. The squirrel is almost always found during the day in caves on steep cliff walls, not among jumbled boulders as is sometimes mentioned in the literature. One person noted seeing two squirrels that fled from a hole in the top of the trunk of a dead conifer (but see discussion of *Petaurista* below). Like other flying squirrels, the woolly flying squirrel is nocturnal. All captive squirrels became active in the late evening and became inactive at around 4:00 am. This agrees with virtually all information gathered from local people, with the exception of one man who accurately described a woolly flying squirrel that he saw moving on a cliff at approximately 10:00 am. Most local people stated that the woolly flying squirrel climbs trees to feed on seeds and needles of the various conifers in the area. All captives refused all fruits, nuts and other foods offered them but the male in 1995 did feed during the night on the buds of spruce (*Picea smithiana*) from branches that were left in the cage.The male drank by lapping water from a stream for over one minute previous to its release the following day. Like the female specimen captured in 1994, the male captured in 1995 was slow and deliberate in its movements. Upon release the following day the squirrel moved away from its captors at a speed equivalent to a human's fast walk. While caged, the male in 1995 regularly made a hissing growl as a warning cry when the cage was approached. This differed from the female caught in 1994, who was silent except for emitting a few grunting noises and a single chirring growl when she was first deposited in the cage. Like the female, the male performed a slashing motion with both front paws when the cage was touched. Local people interviewed in 1995 and 1996 stated that the woolly flying squirrel usually has 2-3 young, which are seen in spring. One person stated that very young squirrels were seen in late summer. If this is true, it suggests that there may be two litters a year, but this statement needs to be verified.

The discovery of the giant red flying squirrel (*Petaurista petaurista*) in Khanbari is interesting from a number of perspectives.The discovery makes the Khanbari area one of the few places on earth that has three sympatric genera of flying squirrels (*Petaurista*, *Eupetaurus*, and the Kashmir flying squirrel *Eoglaucomys* [*Hylopetes*]). However, it presents a problem from a research perspective as *Petaurista* is almost the same size as *Eupetaurus*, and it becomes difficult to determine if local people are able to make the distinction between the two species when and if they occur together. (This is a problem even for the Kashmir flying squirrel, which is only half the size of the woolly flying squirrel.) For example, a local's description of two large squirrels seen escaping from a hole in the trunk of a dead conifer may well have alluded to *Petaurista*, a tree hole nester, rather than *Eupetaurus*.

Another interesting fact uncovered in 1995 that also presents research difficulties is the discovery that the Kashmir flying squirrel may occasionally utilize caves in the Northern Areas. An adult female Kashmir flying squirrel was captured in a cave at Dormushk in Kargah during our 1995 research. Many people in Ishkoman and Yasin stated that the Kashmir flying squirrel uses caves and rock crevices to rest in during the day when raiding walnut and fruit crops in fall and winter. As there are many caves and no wild trees for some distance from the fruit trees, it is likely that this species is using caves for refuges when necessary. This is the first suggestion that the Kashmir flying squirrel may occasionally utilize caves, and it presents a problem for *Eupetaurus* research, as it proves that two flying squirrels may be found living in caves.

Conservation Recommendations

Future research should focus on determining the woolly flying squirrel's range in the southwest from Tangir to the Swat Valley, in the northwest from Kargah to Chitral, and in the east from the Indus into the Skardu Valley and up to Rakaposhi. Behavioral work needs to be continued to gather more precise information about individual home range, food habits, and habitat use.

We feel that it is necessary to implement a conservation plan in the immediate future. The woolly flying squirrel appears to have an extraordinarily small range. The area where the species is known to occur is undergoing rapid deforestation, although the Fairy Meadows area has apparently just been identified as a new national park in Pakistan (CNPPA, 1995). Many valleys in the area are under extreme stress from overgrazing and heavy logging; for example a huge logging operation is underway in the Sai Valley. In Tangir the village elders, despite a reputation for ignoring or rejecting outside help, asked us for assistance in controlling outside forces that are rapidly deforesting the valley. The deforestation is apparently going on with the tacit or direct involvement of local government officials. If this activity and similar activities in other valleys are not curtailed, these valleys will soon lose their forests and subsequently their wildlife, including the woolly flying squirrel.

At this point we can make four recommendations to ensure the survival of *Eupetaurus*:

1). Increasing local people's awareness of the importance of wildlife conservation, perhaps through a local community-based environmental education program and the use of conservation posters featuring the giant woolly flying squirrel;

2). Getting local people in these valleys to agree to sustainable activities such as replanting native conifers and rotational grazing to allow plant regeneration;

3). Putting an end to illegal or excessive deforestation through the active policing of the valleys, and the prosecution of individuals involved in illegal deforestation;

4). Creating a protected area in the Sai-Gorabad region.

Most people in the Northern Areas are surprisingly knowledgeable and proud of their wildlife, but they seem unaware that the recent increase in human population and weapons in their area have driven many species to the edge of extinction. A common belief is that when a species disappears from one valley the animals have simply moved to the next valley. Sadly, this attitude can be found in valley after valley, with the local people not comprehending that the species has actually been wiped out over a large section of its range. Educating local people about environmental concerns, for example through a traveling winter slide show/lecture presentation, would go a long way toward stopping some of the loss of habitat and killing of wildlife at its source.

The woolly flying squirrel apparently depends upon a habitat that exists as a patchy, thin band at middle elevations on mountain slopes. With hostile terrain above and below, the local populations of flying squirrels may be isolated from one another. If a local population of squirrel is extirpated, it would be difficult for other populations to immigrate and re-colonize the area. While the woolly flying squirrel is not actively hunted, it is occasionally killed by salajit collectors or others who encounter it. A greater danger is the squirrel's possible dependence upon the rapidly disappearing pine and spruce trees upon which it feeds. As these valleys may be the only region in the world containing the woolly flying squirrel, the loss of the forest may well mean the local, and then the complete extinction for the largest and most unusual flying squirrel in the world.

ACKNOWLEDGMENTS

We thank WWF-Pakistan for supporting this research in 1994, 1995 and 1996 and we thank the Wildlife Conservation Society and WWF-U.S. for their support in 1996. Field work by Akhtar Karim, Naeem Dar, Mohammad Rafique and Waseem Khan was instrumental in our success. We also wish to thank Chantal Dietemann, Brigadier Mukhtar Ahmed, Ali Hasan Habib, Tehmina Ali, and Masood Arshad for their logistic support, assistance, and encouragement. We thank the U.S. Fish and Wildlife Service for support of the Small Mammals of Pakistan Project, and the Pakistan Museum of Natural History. Drs. Shehzad Mufti and S. Azhar Hasan of the PMNH provided essential support and assistance, which we acknowledge with pleasure, along with thanks to Dr. M. Farooq Ahmad, Director of the Zoological Survey Department of Pakistan. We thank S. Chakraborty and his colleagues at the Zoological Survey of India (Calcutta) for their assistance and permission to examine specimens. We also thank A. Rahmani, N. Chaturvedi, M. Muni and V. Hedge for their assistance and permission to study specimens at the Bombay Natural History Society. At Kunming Institute of Zoology we thank Weizhi Ji, Wang YingXing, Yu Fahong and Peng Yan Zhang for assistance and permission to examine specimens.

REFERENCES

ANDERSON, J., 1878. Anatomical and zoological researches: comprising an account of the zoological results of two expeditions to western Yunnan in 1868 and 1878; and a monograph of the two cetacean genera, *Platanista* and *Orcella*. 2 vols. pp. 25 + 980 + 41, 81 color plates, B. Quaritch, London.

AGRAWAL, V. C. & CHAKRABORTY, S., 1970. Occurrence of the woolly flying squirrel *Eupetaurus cinereus* Thomas (Mammalia: Rodentia: Sciuridae) in North Sikkim. *J. Bombay Nat. Hist. Soc.,* **66**: 615-616.

AGRAWAL, V. C. & CHAKRABORTY, S., 1979. Catalogue of mammals in the Zoological Survey of India. Rodentia. Part I. Sciuridae. *Records of the Zoological Survey of India,* **74**(4):330-481.

BLANFORD, W. T., 1891. *The Fauna of British India, Mammalia.* Taylor and Francis, London, 617 pp.

CHAKRABORTY, S. & AGRAWAL, V. C. 1977. A melanistic example of woolly flying squirrel *Eupetaurus cinereus* Thomas (Rodentia: Sciuridae). *J. Bombay Nat. Hist. Soc.,* **74**:346-347.

CORBET, G. B. & HILL, J.E., 1992. *The Mammals of the Indomalayan Region; A Systematic Review.* Oxford University Press, 488 pp.

CNPPA., 1995. New parks for Pakistan. Commission on National Parks and Protected Areas. *Newsletter (Gland, Switzerland),* **66**:5.

ELLERMAN, J. R., 1948. Key to the rodents of South-West Asia. *Proc. Zoological Society London,* **118**:765-816.

ELLERMAN, J. R. 1963. *The fauna of India (including Pakistan, Burma, and Ceylon). Mammalia 3 (Rodentia).* Government of India, Delhi, 482 pp.

FISH AND WILDLIFE SERVICE. 1992. *Endangered and threatened wildlife and.plants.* 50 CFR 17.11 & 17.12. U.S. Fish and Wildlife Service, Washington, D.C., 38 pp.

GRASSÉ, P. P. & DEKEYSER., P.L., 1955. Ordre des Rongeurs. pp. 1331-1525 in P. P. Grassé, (ed.) *Traité de Zoologie.* Paris, Masson, **17**(2).

GREENBRIDGE, B., 1993. *1994 IUCN Red List of Threatened Animals. International Union for the Conservation of Nature and Natural Resources,* Cambridge, U.K., 286 pp.

INSKIPP, T. & BARZO, J., 1987. *World checklist of threatened mammals.* Nature Conservancy Council, Cambridge, 125 pp.

JENTINK, F.A., 1890. Observations relating *Eupetaurus cinereus*, Oldfield Thomas. *Notes of the Leiden Museum (Netherlands),* **12** (20):143-144, pl 7, figs. 1-6.

KHAN, S. M., 1981. *Salajit: It's compositions and biological actions.* Unpublished MS thesis, University of Peshawar. 101 pp.

LIDICKER , JR. W. Z., 1989. *Rodents. A world survey of species of conservation concern.* IUCN/Species Survival Commission Rodent Specialist Group, Gland, Switzerland, 60 pp.

LORIMER, D. L., 1924. Woolly flying squirrel. *J. Bombay Nat. Hist. Soc.,* **30**:219.

MCKENNA, M. C., 1962. *Eupetaurus* and the living Petauristine Sciurids. *American Museum of Natural History Novitates* **2104**:1-38.

NOWAK, R. M., (ed.) 1991. *Walker's Mammals of the World.* vol. **1** 5th edition. Johns Hopkins University Press, Maryland, 642 pp.

SHAUB, S., 1958. Simplicidentata (Rodentia). *J. Piveteau, Traité de Paléontologie.* Paris, Masson, **6** (2): 659-818.

ROBERTS, T. J., . 1977. *The Mammals of Pakistan.* Ernest Benn Ltd., London, 361 pp.

THOMAS, O., 1888. *Eupetaurus*-a new form of flying squirrel from Kashmir. *J. Asiat. Soc. Bengal,* **57**:256-260.

THORNBACK, J. & JENKINS, M., 1994. The IUCN mammal red data book. Part 1. IUCN Conservation and Monitoring Centre, Gland, Switzerland, 516 pp.

WANG, Y-X. & YANG, G-H., 1986. Data Collection 1951-1986 for Control and Research Institute of Epidemic Diseases of Yunnan Province, China, 134-151 pp.

WOODS, C. A. & HOWLAND, E.B.,. 1979. Adaptive radiation of capromyid rodents: Anatomy of the masticatory apparatus. *Journal of Mammalogy* **60** (1):95-116.

ZAHLER, P., 1996. Rediscovery of the Woolly Flying Squirrel (*Eupetaurus cinereus*). *Journal of Mammalogy* **77**(1):54-57.

BIODIVERSITY IN MUROIDS (RODENTIA, MAMMALIA) DURING MIOCENE, POTWAR PLATEAU, PAKISTAN

I. U. CHEEMA, A. R. RAJPAR AND S. M. RAZA*

Pakistan Museum of Natural History, Garden Avenue, Shakarparian, Islamabad

** Oil and Gas Development Corporation, Islamabad*

Abstract: The Siwalik Group rocks in the Potwar Plateau contain the best Neogene rodent fossil record for southern and southeast Asia. Rodents provide a long, detailed record of biological changes which, in turn can be related with various geological and climatic events. This aspect makes the entire mammalian fauna of the Siwaliks of special interest for understanding the course of evolution and biodiversification in the region.

In Potwar, the interval between 18 and 7 Ma has been most intensively studied and changes in biodiversity and relative abundance of muroid rodents are documented with temporal resolution of 200,000 years. Within this interval, diversity varies considerably in muroids at two stages i.e. at 13 and at 10 Ma. Megacricetodontine decrease at 12 Ma, smaller declines among myocricetodontine and copemyine rodents after 16 Ma and an abrupt increase in murids at 12 Ma are recorded. An increase of dendromurine rodents at 15.5 Ma is also observed. Appearence of murid *Antemus* at 14 Ma and more derived murid *Progonomys* probably evolved from *Antemus chinjiensis* at ca 12 Ma. A new species *Progonomys hussaini* described from ca 11 Ma old, Jalalpur area in southeastern Potwar is a distinct species with mosaic of some primitive murid characters.

The issues of diversity and relative abundance among muroid rodents during the Miocene have also been investigated. It is concluded that a) a rapid increase in muroid diversity was largely due to immigration, whereas in situ speciation had only a secondary role, b) during intervals of increasing diversity, resident lineages did not have higher than average rates of in situ speciation. c) during those intervals, it is evident that greater extinction did not accompany increasing diversity and finally, d) changes in biodiversity are not necessarily linked to changes in other aspects of ecological structure.

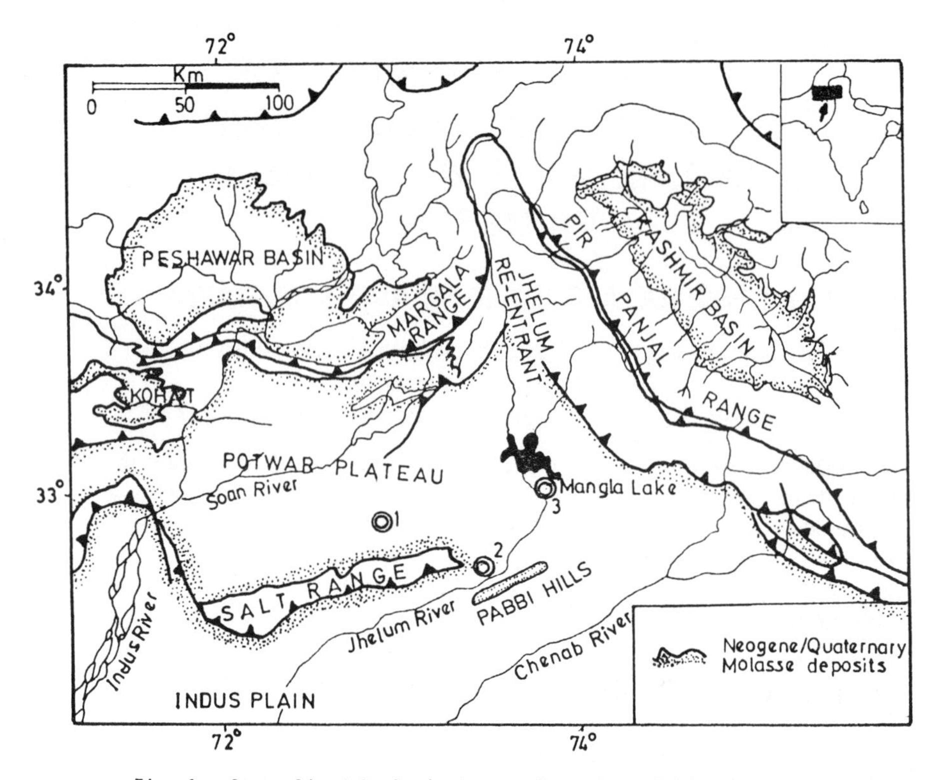

Fig. 1. Generalised Geological map of Northern Pakistan showing the extent of Neogene-Quatenary mollase deposits. The areas studied in this report are: 1. Kallar Kahar - Dhok Thalian, 2. Jalalpur, 3. Mirpur, Azad Kashmir.

Modified after Johnson et al (1986) and Burbank et al (1986).

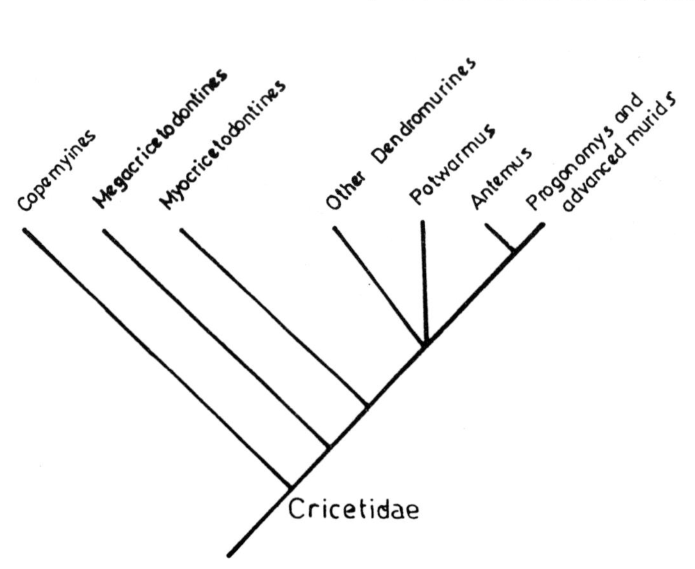

Fig. 2. Postulated phylogenetic relationship of Muroid rodents. Dendromurines include Potwarmus. Murids include Antemus.

INTRODUCTION

The Siwalik Group rocks are thick sequence of continental molassic sediments which comprise the foot- hills of the Himalaya and associated mountain ranges in the northern India and Pakistan. They range in age from Miocene through Pleistocene.

Siwalik sediments, over 7000 meters in thickness, are predominantly fluvial in origin. They were shed from the rising Himalaya mountains, the result of the collision of the Indo- Pakistan subcontinent with Eurasian Plate (Powel and Conaghan, 1973). Continued northward movement of the Indian Plate has deformed the Siwalik Group rocks.

The most complete representation of Siwalik sediments is found in the Potwar Plateau, Pakistan. The Potwar Plateau is bounded to the north by the Kala Chitta and Margala hills, to the south by the Salt Range, to the west by the Indus river and to the east by the Jehlum river (Fig. 1).

A major strength of the Siwalik fossil record is the superb temporal control provided by magnetic polarity stratigraphy. The integration of systematics palaeontology with well constrained temporal control provide a unique insight into the evolutionary and dispersal pattern of some of the more common vertebrate groups found in the Siwalik rocks. Rodents are one of those groups which have been extensively studied in the last two decades. The extensive studies of Jacobs, 1978; Flynn, 1982 and Lindsay, 1988 have established broad evolutionary and migration patterns of murids, rhizomyids and cricetids in the south Asia during the Neogene times.

EVOLUTION PATTERN IN THE MUROIDS

The base of terrestrial Miocene sequence in Potwar Plateau, Pakistan is older than 18.3 Ma. and the top is less than 0.6 Ma.(Early Miocene through Pleistocene). In this nearly 19 Million years interval muroid rodents are very common fossils. Muroids comprise five groups (1) copemyines (or democricetodontines) (2) megacricetodontines (3) myocricetodontines (4) *Potwarmus* and other dendromurines and (5) murids. However the first four groups have been collectively referred to as "Cricetids" (Fig. 2).

DIVERSITY IN CRICETIDS

The Miocene is certainly the acme of Cricetids in Asia. Lindsay (1994) divided the 19 Million years long Miocene interval into five divisions, to better show diversity, sequence of appearances and faunal turnover (Table I). Chronologic resolution for these divisions is based primarily on the Siwalik sequence, of the Potwar Plateau, where the sites are correlated by magnetostratigraphy.

Table I: Siwalik cricetid record

Early Miocene (24 - 22 Mya)
Upper Chitar Wata Formation
Eucricetodon sp.
Primus sp.
Spanocricetodon
Murree Formation
Spanocricetodon khani
Spanocricetodon lii
Primus microps
Prokanisamys arifi
Early Middle Miocene (22 - 18 Mya)
Lower Vihowa Formation
Spanocricetodon sp.
Democricetodon sp.
Myocricetodon sp.
Manchar Formation
Spanocricetodon lii
Democricetodon cf. franconicus
Myocricetodont gen. 1
Myocricetodont gen. 2
Prokanisamys arifi
Middle Miocene (18 - 14 Mya)
Kamlial Formation, Potwar Plateau
Democricetodon sp. A
Democricetodon sp. B-C
Democricetodon kohatensis
Democricetodon sp. E
Megacricetodon aguliari
Megacricetodon mythikos
Punjabemys downsi
Punjabemys mikros
Myocricetodon sivalensis
Myocricetodon sp.
Dakkamyoides lavacoti
Potwarmus primitivus
Manchar Formation
Democricetodon cf. franconicus
Democricetodon aff. kohatensis
Myocricetodon cf. parvus
Myocricetodon sp.

? Potwarmus sp.
Antemus chinjiensis
Prokanisamys arifi
Late Middle Miocene (14 - 10 Mya)
Chinji Formation, Potwar Plateau
Democricetodon spp. B-C
Democricetodon kohatensis
Democricetodon sp. E
Democricetodon sp. F
Democricetodon sp. G
Democricetodon sp. H
Megacricetodon aguilari
Megacricetodon daamsi
Megacricetodon mythikos
Punjabemys downsi
Punjabemys mikros
Myocricetodon sivalensis
Dakkamyoides lavacoti
Dakkamyoides perplexus
Dakkamys baaryi
Dakkamys asiaticus
Paradakkamys chinjiensis
Antemus chinjiensis
Manchar Formation
Democricetodon kohatensis
Myocricetodont gen. 1
Myocricetodont gen. 2
Potwarmus primitivus
Dakkamys ? sp.
Antemus chinjiensis
Daud Khel
Democricetodon sp.
Myocricetodon sp.
Antemus cf. chinjiensis
Nagri and Dhok Pathan Formation
Democricetodon spp. B-C
Democricetodon kohatensis
Democricetodon sp. E
Democricetodon sp. F
Democricetodon spp. H
Dakkamys asiaticus
Paradakkamys chinjiensis

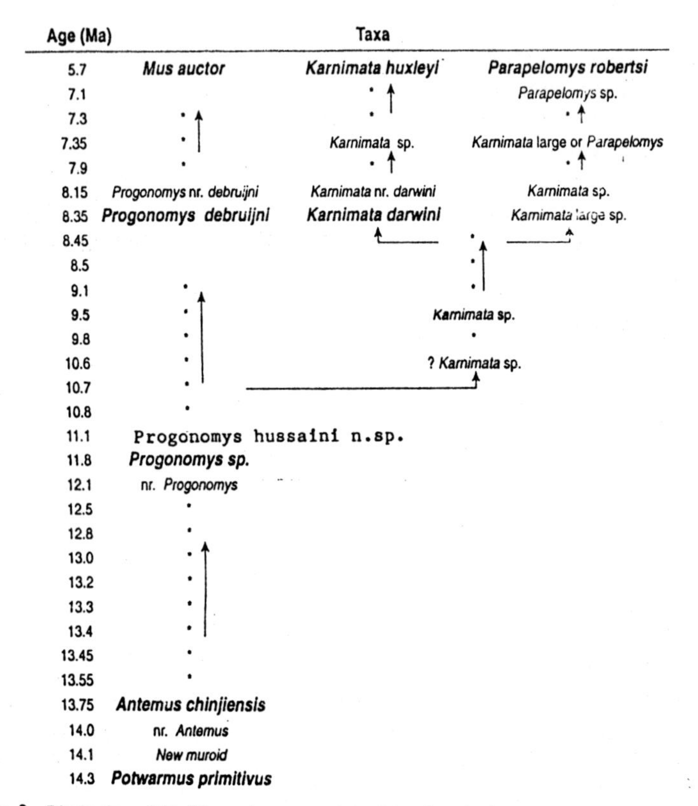

Fig. 3. Distribution of Siwalik murine taxa and localities through time. Arrows show hypothesized evolutionary lineages. Dots signify a taxon present at the locality that is similar to the preceding sample.

Early Miocene of Southern Asia:

(22-24 Ma) records the appearance of *Primus, Spanocricetodon, Prokanisamys* along with *Eucricetodon* from Chitarwata and Murree Formations. Both *Primus* and *Eucricetodon* are primitive.

Early middle Miocene:

(18-22 Ma) *Megacricetodon, Democricetodon, Myocricetodon* appear in southern Asia from lower Vihowa Formation and Gaj River (Manchar Formation). This interval also includes the last appearance of *Primus* and *Spanocricetodon.*

The Siwalik record begins at 18.3 Ma which also approximates the middle Miocene interval (14-18 Ma). The Siwalik Kamlial record includes four species of *Democricetodon*, two species of *Megacricetodon*, two species of *Punjabemys*, three species of Dendromurine,*Potwarmus primitivus* that gave rise to the murids in the late middle Miocene, the 10-14 Ma. interval. *Punjabemys, Dakkamyoides* and *Antemus* appear in southern Asia during the same interval.

Late middle Miocene:

In (10-14 Ma) interval, six species of *Democricetodon*, four species of *Megacricetodon*, two species of *Punjabemys*, two genera of *Myocricetodon* and three species of *Dakkamyoides* are widely distributed in Siwaliks . In this time *Dakkamys, Paradakkamys* and *Protatera* appear in southern Asia.

Cricetids are poorly represented from Siwaliks during the early Pliocene, also probably the result of increasing competition with murid rodents in southern Asia.

DIVERSITY IN MURIDS

The most primitive murine species known is *Antemus chinjiensis*, whose oldest certain record is 13.75 Ma. This is based on a monophyletic definition of murids as having two lingual cusps associated with the anterior two chevrons on the upper first molars (Jacobs, 1977).

Antemus chinjiensis persists for 1.3 million years than a similar anagenetic pattern occurs at the transition from *Antemus* to *Progonomys,* the genus that first exhibits an essentially modern grade of murine dental morphology. The youngest *Antemus* (12.5 Ma) is separated from the oldest *Progonomys* (11.8 Ma). This transformation took no more than 700,000 years.

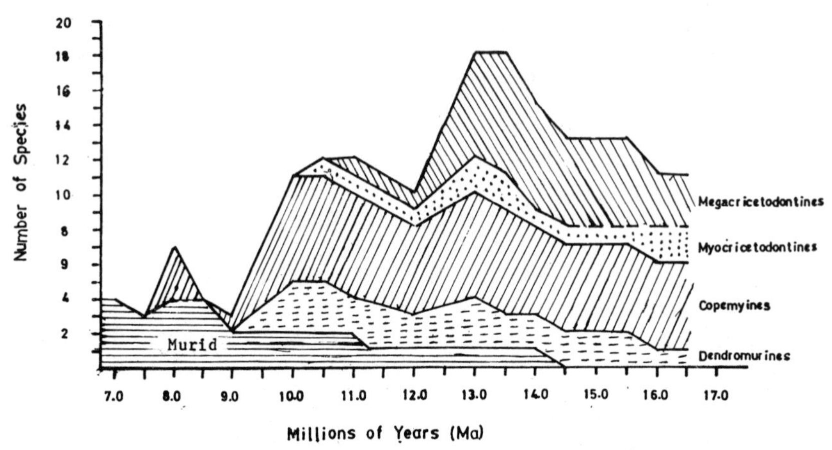

Figure 4. Number of species of muroid rodents found or inferred to be present in each 0.5 my

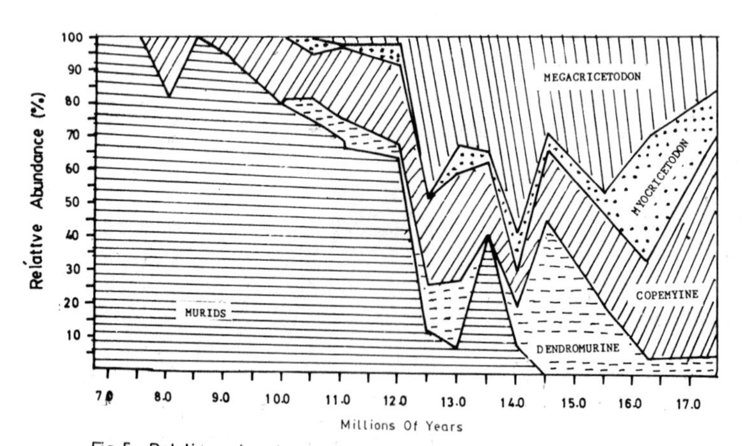

Fig. 5. Relative abundance of muroid rodents.

Progonomys is first clearly present in the Potwar Plateau at locality Y-634 at 11.6 Ma age (Fig: 3). The last record of *Antemus* and first record of *Progonomys* are close temporally. Based on these Siwalik records, *Progonomys* is very likely to have evolved in south Asia, probably at ca 11.8 Ma. However, the first definite *Progonomys* species, the *Progonomys hussaini* comes from about 11 Ma old locality in JAL-101 Jalalpur area. It is primitive in low crown height without strong inclination of cusps, in weak cusp connections and in the broad posterior lobe of lower third molar, which may indicate homology with entoconid.

The last record of Siwalik *Progonomys* is *P. debruijni* at 7 Ma. If the Siwalik record is an accurate reflection, *Mus* evolved from *Progonomys* by 5.7 Ma in an anagenetic transition that took no more than 600,000 years.

Progonomys debruijni and *Karnimata darwini* were first described from a locality dated 8.35 Ma. Two morphologically similar taxa can be recognized in two old sites (9.1 and ca 9.5 Ma). A similar cladogenetic pattern is seen in the origin of *Parapelomys* from *Karnimata* between 8.45 and 8.6 Ma. During the interval in which *Parapelomys* evolved, the size range of Siwalik murine increases.

Faunal turnover can also be investigated using biostratigraphic ranges to determine number of appearance and extinction over specified intervals. In this study, we use published updated information on the number of species of muroid rodents, as well as new data on their relative abundance.

Faunal turnovers in Siwaliks resulted from local extinction, immigration, anagenetic evolution and cladogenetic speciation. The last two are referred as in situ speciation. Immigrants are species or lineages that dispersed from other geographic regions during periods of biotic interchange. By contrast, in situ speciation, whether anagenetic or cladogenetic, takes place within the province. When close related ancestral species are known from older horizons within the province, a species is presumed to have arisen through in situ speciation.

DIVERSITY

The number of species in Table II are shown in Fig. 4, where the most important feature is the contrast between the low diversity of murids and the larger more variable numbers in the cricetid subfamilies. Muroid diversity peaks at 13 Ma with 18 species and subsequently falls into steps as first megacricetodontines and myocricetodontines and then dendromurines and copemyines decline. During the period of higher diversity, the changes are largely due to the megacricetodontines and dendromurines, which rise from four to nine

Table II: Number of Muroid species (Sp) and number of individuals (Ind), based on the total number of molars. There are no sites with rodents in the 9.5-, 11.5- 15-, and 17.5-Ma levels. p, present; T. sp, total species and T.Ind, total individuals.

Interval	Copemyines		Myocricetodontines		Megacricetodontines		Dendromurines		Murids		T. Sp.	T.Ind.
Ma	sp	ind	sp	ind	sp	ind	sp	Ind	sp	ind		
7.0	0	0	0	0	0	0	0	0	4	80	4	80
7.5	0	0	0	0	0	0	0	0	3	295	3	295
8.0	3	43	0	0	0	0	0	0	4	298	4	298
9.0	1	7	0	0	0	0	0	0	2	123	3	123
9.5	-	-	-	-	-	-	-	-	-	-	-	-
10.0	6	98	0	0	0	0	3	9	2	395	11	502
10.5	6	7	1	3	0	0	3	4	2	41	12	55
11.0	6	40	1	2	1	4	3	18	2	223	13	287
11.5	-	-	-	-	-	-	-	-	-	-	-	-
12.0	5	51	1	12	1	5	2	8	1	137	10	213
12.5	?	22	?	1	?	41	?	12	1	11	?	87
13.0	6	263	2	64	6	267	3	170	1	69	18	883
13.5	6	155	2	21	7	219	2	11	1	289	18	695
14.0	5	42	1	40	6	215	2	40	1	33	15	370
14.5	5	97	1	23	5	131	2	210	0	0	13	461
15.0	-	-	-	-	-	-	-	-	-	-	-	-
15.5	5	14	1?	3	5	24	2	11	0	0	13	52
16.0	5	9	2?	6	3	5	1	1	0	0	11	21
16.5	5	23	2	34	3	27	1	4	0	0	11	99
17.0	?	3	?	1	?	p	?	1	0	0	?	5
17.5	-	-	-	-	-	-	-	-	-	-	-	-
18.0	?	18	?	5	?	6	?	1	0	0	?	30

species between 16 and 13 Ma. After 9 Ma, species diversity does not exceed seven species and murids are more dominent than cricetids by this time, both in diversity and abundance. Table I gives the composition of cricetid subfamilies at different time intervals in the Miocene mollasic sequence from different key areas in Pakistan.

RELATIVE ABUNDANCE

The relative abundance of the muroids was computed from the number of molars in each interval. Table II is shown in Fig. 5. The important feature of Fig. 5 is the abrupt jump in the relative abundance of murids at 12 Ma, followed by a steady rise until they comprise 80-100% of the assemblages. Among the cricetids, the copemyines and myocricetodontines are most abundant in the oldest levels but first decline abruptly between 16 and 15.5 Ma and then more gradually until the late Miocene. The myocricetodontines in particular become a very minor group. The dendromurines and megacricetodontines are most abundant between 15.5 and 12.5 Ma and then decline very rapidly. It is also noted that the abundance of murids appear inversely correlated to that of their sister group dendromurines, and both togather are inversely correlated to the more distantly related megacricetodontines.

SPECIES ORIGIN

To understand how diversity and abundance trends are related to biotic interchange, it is first necessary to decide which species were immigrants and which evolved in situ. The status of each species is indicated in Table III.

Table III: Ages of First and Last appearances of muroid rodents in the Siwalik formations of Potwar Plateau, Pakistan. Is, in situ; Im, immigrant; >, appears before.

Family / species	First appear (Ma)	Last appear (Ma)	Duration (m.y.)	Status
Copemyinae				
Democricetodon kohatensis	>16.3	9.8	6.5	-
Democricetodon sp. A	>16.3	13.2	3.1	-
Democricetodon sp. B	*>16.3*	*9.2*	*7.1*	-
Democricetodon sp. C	>16.3	9.8	6.5	-
Democricetodon sp. E	>16.3	9.8	6.5	-
Democricetodon sp. E ?	7.9	7.8	0.1	Im
Democricetodon sp. F	12.0	9.8	2.2	Is
Democricetodon sp. F ?	7.9	7.8	0.1	Im
Democricetodon sp. G	11.1	9.8	1.3	Im
Democricetodon sp. G ?	7.9	7.8	0.1	Im
Democricetodon sp. H	13.7	13.2	0.5	Is

Megacricetodontinae				
Megacricetodon aguilari	>16.3	13.6	2.7	-
Megacricetodon sivalensis	13.7	13.2	0.5	Im
Megacricetodon daamsi	15.3	10.8	4.5	Im
Megacricetodon mythikos	>16.3	13.2	3.1	-
Punjabemys downsi	14.1	13.2	0.9	Is
Punjabemys leptos	15.3	13.2	2.1	Is
Punjabemys mikros	>16.3	13.2	3.1	-
Myocricetodontinae				
Myocricetodon sivalensis	>16.3	?15.3	1.0	-
? Myocricetodon, indet.	16.3	?10.6	5.7	-
Dakkamyoides lavocati	13.7	13.5	0.2	Im?
Dakkamyoides perplexus	13.7	12.8	0.9	Im?
Dendromurinae				
Dakkamys asiaticus	11.1	9.8	1.3	Im
Dakkamys barryi	13.7	12.8	0.9	Im
Dakkamys, small sp.	12.8	9.8	3.0	Im
Paradakkamys chinjiensis	13.7	9.8	3.9	Is?
Potwarmus primitivus	>16.3	14.0	2.3	-
Potwarmus minimus	15.3	14.0	1.3	Im
Muridae				
Antemus chinjiensis	14.1	12.5	2.1	s
Progonomys spp.	12.0	7.1	4.8	Is
Karnimata small spp.	10.6?	8.2	2.4	Is
Karnimata large sp.	8.4	7.1	1.3	Is
Parapodemus sp.	8.4	7.1	1.3	Im
cf. Parapelomys sp.	7.1	-	-	Is

We interpret the appearance and most of the subsequent evolution of the murids as in situ speciation. In all, murids had five or more in situ speciation events and, in the late Miocene, one immigration, that is *Parapodemus.* The status of *Potwarmus primitivus* is uncertain, but the later appearing *Potwarmus minimus* is more primitive and we consider it an immigrant. The remaining dendromurines, *Dakkamys* and *Paradakkamys,* are most similar to species in Turkey. It is assumed that the two lineages were immigrants into southern Asia. Subsequently there may have been one additional immigration and one in situ speciation within *Dakkamys.* The myocricetodontines also have in situ and immigrant species, with two immigrations and one in situ speciation. Early Siwalik megacricetodontines share many similarities with species in Europe and China, suggesting they are mostly immigrant taxa. Four species appear after 16 Ma, two of *Punjabemys* that may have evolved in situ and two *Megacricetodon* that are more likely to be immigrants. Finally, the copemyines have not yet been analysed in detail but appear to be a complex of long ranging species having three in situ off shoots. The reappearance of three species at 8 Ma after their apparent local extinction are immigration events.

MECHANISM / CAUSES OF BIODIVERSITY

The above discussion has clearly indicated the tempo and mode of taxonomic diversity, relative abundance and speciation events of muroids during Miocene in the Potwar Plateau. This also reflects a general pattern prevailing in other parts of south Asia during the Miocene, from where not so complete rodent record is available. In this context these are the following principal concerns: how diversity is related to immigration, speciation and extinction and how diversity relates to changes in the ecological structure of the fossil assemblages, as evidenced by trends in abundance and origin.

Diversity and ecological structure:

Two major revolutions transform Siwalik assemblage during Miocene. a) the middle Miocene (14-18 Ma) radiation of cricetids and murids and b) the late Miocene (10-14 Ma) domination of rodent assemblages by murids. It is therefore, expected that change in relative abundance should accompany changes in diversity. This expectation is met by the megacricetodontines and dendromurines, whose abundance tracks diversity. However, there are exceptions such as at 14 Ma, murids have shown significant increase in abundance without apparent changes in diversity. Copemyines, myocricetodontines exhibit the same pattern to a lesser extent, at times decreasing in abundance but not diversity.

Diversity and Immigration:

The first question is whether the observed diversity increase was principally due to in situ speciation or immigration. For all time intervals, the ratio between in situ speciation and immigration events is 11:12 indicating both processes were of equal importance overall.

The conclusion is that taxa increase in diversity by means of immigration, not in situ radiations. This principle is best illustrated by dendromurines and megacricetodontines. Murids are an exception, although their diversification came long after their initial appearance.

Diversity and speciation:

Between 15.5 and 7 Ma there were 11 in situ speciation events, an average of 1.5 event per 0.5 million years interval. In the five intervals of increasing diversity encompassing 15.5 to 13.5 Ma, there are three in situ speciation events. It is concluded that, the rate of speciation did not increase with rising diversity.

CONCLUSION

We have documented changes in the number of species and relative abundance for muroid rodents in a Siwalik sequence are documented. Between 18 and 7 Ma the diversity of Muroids varied considerably. The most important changes include an abrupt rise in the diversity of megacricetodontines between 15 and 13 Ma, and a decline in cricetid diversity in two steps after 13 and 10 Ma. Among the muroids, an abrupt increase in relative abundance of murids and decrease of megacricetodontines at 12 Ma is very striking. Decreases in the relative abundance of myocricetodontines and copemyines after 16 Ma and an increase for dendromurines at 15.5 Ma are also noteworthy.

It is concluded that 1. the rapid increase in diversity documented in the Siwaliks was largely due to immigration of lineages. In situ speciation had only a secondary role 2. the temporal patterns of speciation events suggest that during intervals of increasing diversity, the rate of in situ speciation within resident lineages was unchanged, 3. it is evident that greater extinction did not accompany increasing diversity. Finally, changes in diversity are not necessarily linked to changes in other aspects of ecological structure.

REFERENCES

BARRY, J.C., AND FLYNN, L.J., 1990. Key biostratigraphic events in the Siwalik sequence.pp 557-571. In (eds) E.H.Lindsay et al. *European Neogene Mammal Chronology*. Plenum Press, New York.

BARRY, J.C., MORGAN, M.E., WINKLER, A.J., FLYNN, L.J., LINDSAY, E.H., JACOBS, L.L., AND PILBEAM, D., 1991. Faunal interchange and Miocene terrestrial vertebrates of southern Asia. *Paleobiology,* **17:** 231- 245.

FLYNN, L.J., JACOBS, L.L., AND LINDSAY, E.H.,1985. Problems in muroid phylogeny relationship to other rodents and origin of major groups. In Luckeet,W.P., and Hartenberger, J.L., (eds) *Evolutionary relationships among rodents*. Plenum press, New York. pp. 589-616.

FLYNN,L.J.,PILBEAM,D., JACOBS,L.L., BARRY,J.C., BEHRENSMEYER, A.K., AND KAPPELMAN, J.W.,1990. The Siwalik of Pakistan:Time and Faunas in a Miocene terrestrial setting. *Journal of Geology,* **98:** 589-604, The University of Chicago.

JACOBS,L.L., FLYNN,L.J., DOWNS,W.R., AND BARRY, J. C. 1990. QUO VADIS? The Siwalik Muroid record (eds) E. H. Lindsay et al., *European Neogene mammal chronology* Plenum Press, New York. pp. 573-586.

JACOBS,L.L., AND DOWNS, W.R., 1994. The evolution of murine rodents in Asia. pp 149-156. (eds) Y.Tomida., C.K. Li., and T. Setguchi., Rodent and lagomorph families of Asian origins and diversification. *National Science Museum Monographs*, No. 8, Tokyo.

LINDSAY, E.H., 1994. The fossil record of Asian cricetidae with emphasison Siwalik cricetids. pp 131-147. (eds) Y.Tomida.,C. K. Li., and T. Setoguchi., Rodent and Lagomorph families of Asian origins and diversification. *National Science Museum Monographs* No: 8, Tokyo.

UNGULATES OF PAKISTAN

MUHAMMAD FAROOQ AHMAD

Zoological Survey Department, Pakistan Secretariat, Karachi.

Abstract: The paper deals with the distribution and present status of ungulates found in Pakistan. These large mammals are now much reduced in number and in some cases threatened by a combination of habitat loss and over hunting. Innovative programmes of sustainable management and even carefully designed hunting may be the only way to ensure the survival of some of the species such as Markhor, Urial and Ibex which also face the problem of genetic swamping by closely related domestic livestock.

INTRODUCTION

The mammalian orders Perissodactyla and Artiodactyla are represented in Pakistan by 1 and 17 species respectively (Ahmad and Ghalib, 1975; Ahmad et al., 1986). They are popularly known as ungulates and included in the category of larger game animals. Most of them are threatened with extinction because of the following two major factors:

i) hunting

ii) habitat loss due to urbanisation and use of land for agricultural purposes

A brief account of the various species of ungulates found in Pakistan is given below:

Equus hemionus (Asiatic Wild Ass)

It occurs in marshy areas of Rann of Kutch near Nagarparkar and Chachro on the border of Pakistan and India. According to an aerial count by K. S. Dharama-kumar Sinhji in 1970, the total Indian population was estimated to be about 400 (Roberts, 1977). In Pakistan, no reliable data are available, but it is believed to be threatened with extinction. As is the case with other wild species, man is the main predator on wild ass. It is considered a pest, damaging agricultural fields in the areas of its distribution.

Sus scrofa (Wild Boar)

Wild boars are associated with the Indus basin riverine tracts throughout Punjab and Sindh down to Indus mouth. They also occur around Peshawar, Mardan, Bannu and Dera Ismail Khan. Wild boar can be found up to 3000 ft elevation in the Margalla hills and the foothills of the Murree range. They are serious pest to agricultural crops adjacent to reverain forests. Wild boars are not traditionally hunted, nor do they have any natural predators in Pakistan. No effective measures for its control, have been adopted. The species, therefore, is well established in the areas of its occurrence.

Moschus moschiferus (Musk Deer)

Musk deer occur from 10,000 to 13,000 ft. elevation within the sub-alpine scrub zone. They once were found in the Murree hills, but were exterminated in the 1930s. Until the early 1960s, Musk Deer were found in southern Chitral, northern Hazara, Indus Kohistan and the Gilgit area (Robert's, 1977). At present, they occur in Indus Kohistan and the Hushe valley in Baltistan. Man is the most serious predator on this little deer and is primarily responsible for its disappearance from most of the areas of its range.

Muntiacus muntjak (Barking Deer)

In Pakistan, the muntjak is confined to Margalla hills from Nurpur Shahan to Chattar. It has not been recorded west to Margalla hills but is believed to be still found at Sagian and Sapalahi in Azad Kashmir. Its occurrence in Indus Kohistan has been confirmed by Naeem Ashraf of Himalayan Jungle Project (Pers. Comm. Nov. 1995). The animal is threatened with extinction because of its persecution by local villagers while the fawns are occasionally taken by jackals. The habitat is otherwise much disturbed due to increasing human activity and grazing by domestic cattle.

Axis porcinus (Hog Deer or Para)

Hog deer are confined to thickets of *Tamarix* and *Saccharum* along Indus river. They also occur in the riverine forest reserves like Rajhari, Kethi Shhu, Khaddi and Nara. A few survive along Chenab river around Suleimanki headworks on Sutlaj and Batapur on Ravi. Hog deer which were once plentiful throughout the riverine tracts of Sindh and Punjab, now face extirpation within Pakistan territory. Special reserves should be created for its preservation in the range of its distribution.

Cervus elaphus (Kashmir Stag)

This deer does not occur permanently in Pakistan territory including Azad Kashmir, however, there are unconfirmed reports of its existence in Neelam valley close to cease-fire line.

Boselaphus tragocamelus (Nilgai or Blue Bull)

In Pakistan no resident population of nilgai exists today, but occasionally it is reported close to Indian border near Kasur, Bahawalnagar and Sialkot.

In the Thar desert a few individuals survive close to the Indian border. It is possible that nilgai could establish a foothold in the Bahawalnagar and Tharparkar regions if total protection is provided.

Antilope cervicapra (Black Buck)

The stronghold of the black buck in Pakistan was northern Cholistan around Bahawalnagar and Fort Abbas where it was sighted until 1968. Small numbers also used to occur on the edge of the Thar desert near Nagarparkar. In addition to these natural populations the black buck was introduced in the desert portion of Khairpur by the former Mir in the early part of this century. After complete disappearance of the natural population in Cholistan, the black buck was reintroduced in Lal Suhanra National Park by obtaining a donation from Taxas ranchers through the courtesy of WWF-Pakistan. In Lal Suhanra they are kept in a fence and have not been released into nature.

Gazella subgutturosa (Goitred Gazelle)

Goitred gazelle is found in the flat plains between mountain ranges from 3500 ft. to 7000 ft. elevation in Chagai, Nushki and Chaman areas. Population ·of gazelles have been greatly reduced due to persecution by man. Presently it is believed to be threatened with extinction due to hunting by Afghan refugees.

Gazella gazella (Chinkara)

The Chinkara survives in Pakistan from sea level to 5000 ft. elevation and is widespread in plains, sand dunes and rocky areas of Sindh, Punjab and Balochistan. Once it was plentiful in Dera Ghazi Khan, Dera Ismail Khan, Mardan and Peshawar, but now it has been exterminated from all these areas. It is also reduced to the point of extinction along the eastern border of Pakistan. However, it is believed to be safe in parts of Kalabagh and Kala Chitta hills where motorized hunting is impossible due to difficult terrain. Chinkara would increase if permanent protection were to be provided in Cholistan and the Tharparkar areas.

Naemorhedus goral (Grey Goral)

Grey goral occur in the Murree foothills and Margalla range from 2700 to 5000 ft. and in Swat up to 6500 ft. elevation where there is dense cover of thorny bushes. They also occur in parts of Neelam valley and Indus Kohistan. In the Margalla hills it is rarer than the barking deer. It is not likely to survive if total protection is not provided, particularly in southern Swat.

Capra aegagrus (Wild Goat or Sindh Ibex)

Wild goats are found in all the mountain ranges of southern Balochistan from Makran coastal range at Pasni to Sindh Kohistan and Kirthar range in the east. In the north, their territory extends up to Koh-i-Maran 72 km south of Quetta. The Kirthar range is the main stronghold where its population is estimated to be more than 4000 individuals.Despite the enforcement of game laws, constant hunting by local villagers is common. Leopards were the main predator but now the wild goat needs protection from man and domestic goats. Jakals, hyaenas and wolves occasionally take the young kids.

Capra ibex siberian (Siberian Ibex or Himalayan Ibex)

Siberian ibex occurs above the tree line between 12000 and 22000 ft. During winter it descends to about 7000 ft. elevation where it is hunted. Its range includes mountains of Baltistan (Korakoram, Haramosh and Deosai ranges). They are also seen in parts of Gilgit, Ishkoman, Yasin, Hunza and Chitral. Baltistan is the main stronghold of Siberian ibex. Due to inaccessibility of its habitat, future survival is not endangered in Pakistan.

Capra falconeri (Markhor)

There is much controversy concerning the actual number of valid subspecies of markhor. The configuration of their horns are the primary forms of variation. I have chosen to recognise five sub-species, although fewer are recognised by Schaller (1977) and others. The species as a whole is threatened with extinction.

a) ***Capra f. falconeri*** (Astor Markhor)

The Astor markhor is confined to Gilgit region. The real stronghold is the Nanga Parbat massif. Kargah Nullah holds about 500 head and is protected as a hunting preserve.

b) ***Capra f. cashmiriensis*** (Pir Panjal or Kashmir Markhor)

It occurs in Chitral, Dir and Swat Kohistan. It has been almost exterminated from the Neelam valley close to the cesefire line by the troops stationed there. The Pir Panjal range on the Indian side is the extreme limit of its occurrence.

c) ***Capra f. megaceros*** (Kabul Markhor)

It occurs in southern Chitral, southwestern Swat, Sakra range and Safed Koh range in upper Kurram valley. This subspecies used to occur on hills around Bannu and Khyber, but has now been exterminated. It is threatened with extinction throughout its range.

d) ***Capra f. jerdoni*** (Straight-horned or Suleiman Markhor)

Among the subspecies of markhors, the Suleiman or straight horned markhor is the most widespread, but is now severely restricted in numbers. It occurs in scattered isolated populations on all the major mountain ranges of Pakistan north and east of Quetta, such as Murdar, Takhatu, Zarghun, Khaliphat, Torghar and Koh Suleiman.

e) ***Capra f. chialtanensis*** (Chiltan Markhor)

The greatest number of this subspecies survive in Chiltan. A few individuals are also found on Koh Murdar to the east of Chiltan and Koh-i-Maran to the south of Chiltan. The population estimates are around 200 to 300 head on Chiltan, while 12 to 15 on each of the Koh Murdar and Koh-i-Maran.

Pseudois nayaur (Bharal or Blue Sheep)

Bharal are found in Hunza and Korakoram range in Baltistan from 15,000 to 21,500 ft. elevation. They once occured in considerable numbers, but are under increased hunting pressure especially by border troops. Bharal and Siberian ibex frequent the same areas in Baltistan, but shapu apparently do not occur in the same areas.

Ovis ammon polii (Marco Polo Sheep)

They are found in the Karakoram range of Pakistan from 15,000 ft. to 20,000 ft. elevation. The Marco Polo Sheep is not a permanent resident in Pakistan but migrates in early winter from China across the Khunjerab Pass into northern Hunza. In early spring, they again migrate northwards to China or Afghan territory and the lambs are born outside of Pakistan. They are believed to have been much reduced in numbers by Chinese army personnel serving in the border areas. Their position is better in further northwest, since they are protected in Russian territory. In the Pamir range, wolves are the natural predators on this sheep.

Ovis orientalis (Urial, Shapu, Gad)

Three subspecies of this wild Sheep are recognized:

a) ***Ovis o. vignei*** (Shapu)

This subspecies occurs in hilly tracts of Khyber and Malakand agencies, Kohat, Gilgit, Baltistan and adjoining areas up to an altitude of 9000 ft. They average larger in size than the Punjab and Balochistan populations. The populations in Gilgit and Chitral valleys have been subjected to severe hunting pressure and this species is liable to become extinct.

b) ***Ovis o. punjabiensis*** (Punjab Urial)

The horns of *Ovis orientalis punjabiensis* are more massive at their base than those of other Urial in Salt Range, Kala Chitta hills and Kala Bagh areas.

c) ***Ovis o. blanfordi*** (Balochistan Urial or Gad)

This subspecies is widely distributed throughout most of the higher hill ranges of Balochistan, Las Bela, Sindh Kohistan, Pab and Kirthar hills. The rams have horns which often develop more than a complete arc. Record length of horns from Zhob (Balochistan) is 45 inches. Their numbers have considerably declined in all the areas of their distribution due to hunting pressure.

Pakistan is fortunate in having such a diversified variety of ungulates. Unfortunately, most of them are now rare. With the exception of the wild boar, all other species are generally on the decline. The reasons for this decline are hunting and poaching on one hand and habitat destruction on the other. There is an urgent need to enforce effective conservation measures by the provincial wildlife departments to protect the existing populations. Due to increase in livestock grazing wild species face competition for food and may face the problem of genetic swamping by closely related domestic livestock. Therefore, extensive research on genetic resources is immediately required. Innovative programmes to save markhor and urial have been initiated in Balochistan and NWFP. In these programmes, hunting of trophy "heads" promises to raise the value of some forms. In this way the local people receive more benefits by maintaining wild goats and sheep in their regions than domestic livestock. Such programmes must be carefully designed keeping in view CITES regulations. It is a controversial proposition, but may prove to be a long term solution to the survival of some of the Pakistan's most majestic ungulates.

REFERENCES

AHMAD, M. F. & GHALIB, S. A., 1975. A Checklist of mammals of Pakistan. *Rec. Zool. Sur. Pak.* **VII** (1&2): 1 - 34.

AHMAD, M. , KHANUM, Z. AND AHMAD, M.F., 1986. Wild Hoofed Mammals of Pakistan (Urdu). *Rec. Zool. Sur. Pak.*

ROBERTS, T. J., 1977. *Mammals of Pakistan.* Ernest Benn Ltd. London and Tombridge, England.

SCHALLER, G. B., 1977. *Mountain monarchs: Wild sheep and goat of the Himalaya.* The Univ. of Chicago Press, Chicago.

RECOMMENDATIONS OF THE SYMPOSIUM

The following recommendations were made by the participants of the Symposium:

- Much more taxonomic work is needed to be carried out so as to have a better understanding of the fauna and flora of Pakistan, alongwith their distribution pattern.
- Modern techniques should be adopted for such taxonomic studies such as DNA studies.
- There is an urgent need to reduce and eventually eliminate all factors which are contributing towards habitat destruction. This is especially relevant to our forests and our fresh water and marine habitats.
- There is a special need to preserve and enhance species of economic importance such as fish and medically important plants.
- Pakistan is fortunate in having some important species of animals and plants which are however, threatened with extinction. There is urgent need to take up all necessary measures to save such species from extinction.
- There is need to carry out extensive research on genetic resources found in this country, so that we can produce plants and animals which are highly resistant to disease and infection and also have high yield characteristics.
- An integrated approach should be adopted to save various habitats of Pakistan, especially its forests which happen to be mostly located in the northern areas.
- Work should be done on palaeoenvironment of Pakistan so that a better understanding of its present environment and management can be achieved.
- There is a need for the training of manpower along with improvement in biodiversity research facilities.
- There is an urgent need to educate general public and students about natural resources and their conservation.